Third Edition

STATISTICAL THINKING

Elementary Statistics

[**Leonard C. Onyiah**]

D1451255

Kendall Hunt

publishing company

Cover image © Shutterstock, Inc.

Kendall Hunt
publishing company

www.kendallhunt.com
Send all inquiries to:
4050 Westmark Drive
Dubuque, IA 52004-1840

Text and Maple TA ISBN 978-1-5249-1594-0
Text Alone ISBN 978-1-5249-1595-7

Published in the United States of America

Contents

CHAPTER 4   Elements of Probability and Related Topics   97

CHAPTER 5 Discrete Probability Distributions   133

CHAPTER 6 Continuous Distributions   159

CHAPTER 9 **Tests of Hypotheses** **259**

CHAPTER 10 **Simple Linear Regression Analysis** **299**

CHAPTER **11** Correlation

CHAPTER **12** Two Sample Confidence Intervals

Preface

There is no question that statistics as a subject is difficult for many people; some who have to take it even dread it. It is more so if one is taking a statistics class for the first time, and especially so if the individual does not have a good mathematics background. There is no need to dread doing statistics. Although the field is replete with formulas, methods, and special names, a student who applies himself/herself to the subject can and does easily become familiar with the norms and practices in the field, its notations, and methodologies. Such students eventually become comfortable and start enjoying it. The aim of this book is to introduce statistics to students in as gentle a manner as possible, without compromising rigor or learning. One does not need to possess knowledge beyond high school algebra to benefit from the material in this book. The following five p's could be very useful pieces of advice to those contemplating, or actually taking, a statistics class.

I studied statistics under a teacher who emphasized to the class that in order for us to understand, statistical concepts and do well in the field, we needed to be disciplined to follow the five p's. The p's are:

Punctuality: Always be punctual to your statistics class.
Practice: Regularly practice your statistical concepts and methodologies.
Patience: Be patient, make sure you give yourself enough time to understand the statistical concepts and methodology; Rome was not built in a day.
Persistence: Be persistent in pursuing the goal of doing well in your statistics courses.
Present: Always be present when the course material is taught.

This book has been written to be example driven. Quite unlike some texts, for each chapter of this book there are several examples. Most of the time, in order to help with the understanding of a concept or methodology, more than one example is offered. At the end of the analyses (which may involve calculation or computation) explanations are offered for the results. Traditional manual analysis is provided in order to lay basic foundations for understanding of the concepts. At the end of each chapter the use of technology is discussed, and quite a few examples of analyses using technology are presented. In fact, for some examples in which analyses were carried out manually, a technology approach using MINITAB is presented. The steps required to leverage technology for the analyses are presented as templates that can be used for similar problems the reader would encounter. In short, not just the output of computations, but also the code, is presented. There are also several exercise problems at the end of each chapter. Full-blown solutions showing steps that lead to the final results, as opposed to only answers, are presented for at least half of the exercises. In addition, online exercise materials are presented. Most assignments and tests would be carried out online. Each purchaser of the text receives an access code that will enable them to access the online material.

To take away the fears some have of statistics as a subject, the approach is methodic. It emphasizes manual analysis that utilizes scientific calculators without stating calculator-syntaxes. It was deliberately decided not to present methodology by using a particular calculator in order to avoid making a prescription that all those who use the text should purchase the particular calculator. The fact is that a cheap scientific calculator (as distinct from adding machines which can mostly add, subtract, and divide) will be sufficient for all the calculations that need to be done. The calculator must have features that have scientific keys such as e^x, ln, $\sqrt{x}$, x^{-1}, y^x, $n\,C_r$ and memory that would do STAT calculations such as $\bar{x}$, s, and others.

These days, there is the tendency of some books at this level to hide data. They use samples of sizes so small as to make it sound like statisticians only take samples of less than ten, or in that ball park. The argument is that if a person can calculate standard deviation for a sample of size seven, the person can do it for larger samples. While there may be something to be said about this approach, it also leaves the impression that real life practice of statistics does not encounter large data sets. To avoid that impression, all data have been included in this text—small and large samples. The use of calculators for this course should make computations easier. Teachers using this text should teach students the benefits of using spreadsheets such as EXCEL for computing, which virtually removes tedium from the calculation as 'copy and paste' and 'dragging' of formulas entered in one cell and transference across several cells, reduce computation drastically. Large samples do not therefore carry any extra burden than entering them into the computational software. Our discussion of MINITAB as a tool with explicit codes again makes it easier to deal with samples large or small.

Statisticians deal with data collection, data analyses, and reaching conclusions about data. It is important that data they use are quality data collected scientifically. This book starts by dealing with scientific methods for collecting data, and treating probability sampling methods. It progresses through exploratory methods for presenting collected data in a frequency table and sundry pictorial presentations, such as histograms, polygons, and ogives, along with pie and bar graphs. Next, it treats measures of the center along with the five number summary and other descriptive statistics that have to do with measuring variation. Set theoretical notions are revised and linked with the concept of events and its application in probability theory. Probability, its definitions and axioms, all the way to Bayes' Theorem, are treated. The book also treats discrete and continuous distributions, emphasizing the binomial, Poisson, uniform, and the Normal. Sampling distributions and central limit theorem are treated, leading to parameter estimation, and discussion of point and confidence interval estimates, especially for means and proportions. First, estimation for a single population mean and proportion are treated. Hypotheses testing for single population mean and proportion are treated next. Confidence interval estimation for difference between two population means and two population proportions are treated. Hypotheses testing about differences between two means for pooled variance when variances are believed to be equal and unknown, or for unknown variances that are not assumed equal, up to matched pair comparison are treated. All these are for data that are normally distributed. What to do when normality could not be assumed and sample sizes are large is discussed. Linear least squares regression and Pearson product moment correlation are treated in order to explore relationships between two variables that may be linearly related. Attendant hypotheses about slope and correlation are also discussed.

Material treated in this book is more than sufficient for a Statistical thinking course. In fact, for a Statistical thinking course one could cover up to chapter eleven. One could go all the way to chapters twelve and thirteen which cover two sample confidence intervals and tests of hypotheses. Some may elect to exclude both chapters. Others may, in addition, also exclude material in chapter three related to grouped data. This book was primarily written, without loss of generality, to address the contents of the STAT 193 course at SCSU, a Statistical thinking course. The author places material at the levels he has found students can cope with through experience. For STAT 193 at SCSU, we aim to thoroughly cover chapters one to eleven. These chapters cover everything up to the tests of hypotheses about a single population mean or proportion. They also cover basic regression and correlation theory that explores linear relationships between two variables. The topics covered provide enough material that introduces students to the world of statistics, without involving calculus. Chapters twelve and thirteen can be covered, depending on the amount of time available and the ability of the students in the class.

The software used dominantly for computing in this text is the MINITAB. There are uses of EXCEL for calculating p-values when testing hypotheses that require test statistics that are t-distributed. It is recommended that the teacher help students who are deficient in spreadsheet computation to learn the basics that would ease the calculations they make. Analyses carried out manually, which really implies the using of calculators, are used in every chapter without providing any calculator syntax or referencing of any particular calculator. MINITAB analyses for material in each chapter are presented either within or at the end of each chapter. Chapter twelve analyses are treated at the end of chapter thirteen alongside MINITAB treatment of material in chapter thirteen. In most cases, detailed MINITAB commands needed to carry out analyses are presented alongside outputs. MINITAB codes supplied can easily be adapted to questions similar to the ones addressed. Discussions of outputs and conclusions of MINITAB analyses are presented with a view to aiding the students understanding of the material.

Relevant tables have been provided. These are tables of random digits, standard normal distribution, and the t-distribution, which appear in Appendices I, II, and III respectively. These tables have also been inserted within the text, closest to the part of the text where they are most needed.

Leonard C. Onyiah

Acknowledgement

The writing and revising of this book involved contributions from many people. Some gave encouragement, others a comment or two, still others made some explicit suggestions about how to make the text work. We are indebted to all who contributed, in one way or another, to making the writing of this text possible. We acknowledge the students of STAT 193 classes at St. Cloud State University, who have provided inspiration for writing the text based on the author's sixteen years of teaching this course to different streams of these students by both direct on-campus classroom contact and online arrangements.

Special thanks are due to Uche Nwoke and Sui Xinyue who helped proof-read the material.

Special thanks are due to MINITAB Inc for use of their copyrighted material. Codes and outputs used in this text are printed with the permission of MINITAB® Statistical Software Inc. MINITAB® and all other trademarks and logos for the Company's products and services are the exclusive property of Minitab Inc.

I want to thank the staff of the Kendall/Hunt Publishing, particularly Leah Schneider and Ryan Schrodt, whose hard work has led to the production of this text.

I am grateful to all members of my family, particularly my wife, Constance Onyiah, who made sacrifices for me to find time to revise or rewrite the text. Above all, I thank the Lord Jesus Christ by whose grace I have found the health, strength, and understanding to write this book.

Leonard C. Onyiah

CHAPTER 1

Elements of Sampling

Statistics is a field of study and practice which deals with scientific methods for collecting, organizing, analyzing, and reaching conclusions about data. In a study, data are usually collected on an attribute, or some attributes, of the large group of individuals or objects which is being studied. In Statistics the large group being studied is called a population. The attributes whose values are measured or counted in the studies are called variables; this is because their values change from one individual or object to another. Part of the analysis of the data so collected has to do with making conclusions about the population. This sub-field of statistics which seeks to reach some conclusions or make decisions about a population is called statistical inference.

Reaching conclusions about a population value (called a parameter) for a variable is done by taking a small subset of the population (called a sample). Counts, or measurements, of the values of the variable are made on the individual members of the sample. Analysis of the collected sample data is carried out in order to reach a broader conclusion or make a decision about the population from which the sample was drawn. The analysis often identifies a sample value (called a statistic) which is to be used to make a decision about the parameter. In this approach some assumptions are made about the underlying probabilistic law or distribution which governs the values that the statistic could take. This approach is known as parametric inference. Another type of inference is made without assuming an underlying distribution for attributes of the sample. These distribution-free methods have given rise to another field of statistical inference called the nonparametric inference.

Since the sample is very central to decisions or conclusions reached by statisticians, care is usually taken to draw good samples. There is a whole body of methods which ensures that good samples are selected and used for inference. It is referred to as probability or scientific sampling methods. Sometimes studies involving samples are called surveys.

Sometimes, but very rarely, conclusions about the population are made by obtaining information from every member of the population. This approach is called a census. The census is a process that elicits information from every member of the population. In the United States, a population census is conducted every ten years. In between these ten years, several surveys (samplings) are conducted in order to elicit information useful for making estimates and decisions about the entire population.

1.1   Data Collection by Sampling

It is clear from what has been said in the previous section that a large chunk of the work that statisticians do involves dealing with numbers. In particular, they often use numbers to make decisions which may have far-reaching consequences. For instance, in drug development, clinical trials data are collected and analyzed on the efficacy, safety, and other characteristics of a chemical compound before it can be declared safe enough to be prescribed by physicians as a medication for treating a disease. There are dire consequences for using unreliable data for such a process.

In politics, regular surveys of the population are used to redraw the boundaries of congressional districts, or to allocate funds to states or counties. Surveys are used to determine changes in prices of goods and commodities, and to measure inflation—increases in prices which result in decreases in the purchasing power of money. Surveys are used for collecting data, on the basis of which governments prioritize projects, plan for adequate provisions for the populace, and budget for future development. The practice of statistics has pervasive influence on our daily lives.

Still in politics, we are familiar with the use of opinion polls to try to predict performances of parties and individuals in various races for elected offices. Most of these polls, when the samples are selected scientifically, do give good predictions within their margins of error. There are some out there wittingly or unwittingly conducting biased opinion polls not based on the scientific methods that statisticians advocate. These are examples of the abuse of the statistical methodology, and underlines why we study and advocate the proper practice of the statistics profession. Scientific opinion polls can be used as invaluable tools for social change and political mobilization.

The process of manufacturing items requires that certain standards be met both for safety of the product, and efficient performance of the manufactured article. In the production line in manufacturing, the statistician is usually appointed as the quality control officer who periodically takes samples from the production line and does some analysis to ensure that the final product is within specification. If not, the process is stopped and adjustments are made, or the products are not released for consumption. For instance, when a patient obtains a prescription for 180 mg of a medication to be taken three times daily, and gets it filled at a pharmacy, it is important that each caplet is within the limits of, or is exactly 180 mg. This ensures that the patient does not take an overdose with severe consequences. The statistician as the quality control officer provides this service in the manufacturing process by setting up control limits, and the monitoring process.

An issue that has engaged our attention in recent years is racial profiling in police stops. For good policing, statistics can be used. Following outcries on the possibility of such profiling, some police authorities wanted to make sure that in their counties or cities, such is not the case. Others wanted to make changes and improvements where necessary. Recently, the St. Cloud City Police engaged some professors of criminal justice, statistics, and sociology in St. Cloud State University,

St. Cloud, MN, to study racial profiling in police stops in the city. In the study, ordinary people (nonpolice personnel) were used to make determinations of which motorists to stop, staying in the kinds of positions that police officers would normally be staying. The racial profiles of stopped motorists were compared with records of police stops previously carried out by police officers from the same positions. The results were used by the city police to improve their work in this area.

In business, government, and industry, accuracy and reliability are required of the data collected, in order to reach valid conclusions or decisions on issues that impact our lives. The proper practice of statistics ensures reliable data for these purposes.

In order to make valid conclusions about an issue, it is required that data which have been collected in a scientific way be used. To collect such data we need to use scientific methods. The scientific methods we need are named **probability sampling methods.** Probability sampling is the process of choosing a good **sample** from a **population.** It usually involves a method which allows us to determine a-priori the chance that any member of the population would be chosen as a member of the very sample that is used in actual study of the population. Sometimes, the chance of being chosen as a member of the sample is the same for all members of the population. However, the chances do not always have to be equal.

Data elicited from a sample drawn by using different probability sampling methods are usually analyzed. The analysis may be carried out in order to make some deductions about the entire population from which it was drawn. This process is called **statistical inference.** The sample data are used to obtain a **statistic** which is used to estimate the **parameter.** Thus, in the general scheme of studying attributes of populations, sampling and probability sampling methods are invaluable tools.

1.1.1 Some Definitions

Before going further, let us define terms that have so far been alluded to or that could be used in this text.

Data: The raw material of statistics is *data.* We define data as numbers which arise (a) from measurements, or (b) from counting. When a doctor takes the blood pressure of a patient, he makes a measurement. When the midwife in a maternity keeps records of the number of babies born per day, she is counting. Information arising from such measuring and counting are the data that statisticians use.

Variable: If we find that a characteristic takes different values for different individuals, places, or things, then we say that the characteristic is a *variable.*

Some variables are measurable and are said to be *quantitative,* such as *weights, heights, age,* etc.; they give information about the amount.

Other variables cannot be measured but can be categorized and are said to be *qualitative,* such as *color of hair, color of skin, type of disease, type of medication,* etc.

Random Variable: Suppose we determine the age, or weight, or height of an individual; we would refer to it as the *value.* If the values of a variable obtained arise due to chance factors so that they cannot be exactly predicted in advance, the variable is said to be a *random variable.* For instance, the weight of an adult depends on so many things (genetics, eating habits, exercise, etc.) that we cannot predict it in advance. Weight of an adult is therefore a random variable, as is height. We sometimes refer to measured values as *observations.*

Discrete Random Variables: The values of discrete random variables take integer or point values such as $-8, -7, 9, 0, 1, 3$, etc. For instance, the number of people in a household, number of cars passing through a junction per unit time, number of children born in a hospital per day, number of telephone calls through an exchange per unit time, etc. *We cannot have 2.33 babies born in an hour at a hospital. It is either 2 or 3 babies. Thus, a discrete variable is characterized by not having gaps in the values it can assume.* The listing of the values of a discrete random variable can be finite or infinite.

Continuous Variables: *A continuous variable is characterized by gaps in the values it can assume. It assumes any value within the specified interval of values that it can assume.* The height of an individual is a continuous variable because no matter how close two heights are, we can find another height between them.

Unit: A single individual or object from which we can elicit some information; this information is obtained either by making a measurement on the unit or by getting the unit to respond to some questions or stimuli.

Population: Entire set of units under study from which we would like to obtain some information.

Sample: The subset of the population which is used in the actual study; these are the units that we actually measure or allow to make responses to survey questions which we then use in our study.

Parameter: A population value obtained by making measurements on all members of the population.

Statistic: A value obtained by using measurements made on a sample. Usually, this value is used as an estimate of the population value, the parameter.

Statistical Inference: In general, the process of reaching conclusions or making decisions concerning a population on the basis of random samples is called statistical inference. In particular, the process of using a statistic to reach conclusions about a parameter is called parametric inference—this relies heavily on the assumed distribution of the statistic. There is another type of distribution-free inference referred to as nonparametric inference.

Bias: Statisticians measure bias of an estimator as the difference between its mathematical expectation and the true population value it estimates. When the bias is zero, the estimator is said to be unbiased.

1.2   Some Unrecommended Sampling Methods

There are two broad types of sampling methods that can be adopted. They are Probability and Nonprobability sampling methods. The major difference between these sampling methods has to do with the assumptions designed into the sampling process. In probability sampling, each member of the population being sampled has an equal chance (probability) of being chosen to belong to the sample. In nonprobability sampling it is believed that any sample chosen from the population will be representative of the population, no matter how the sample is chosen. To further illustrate, in nonprobability sampling there is an assumption that all members of the population are somewhat evenly spread on a continuum, and that choosing the sample from the beginning, middle, later part, or indeed any other part of the continuum, would lead to similar samples. In nonprobability sampling the probability of choosing each individual in the population is therefore not the same, because where the researcher starts depends on their whim. In this illustration of nonprobability sampling, randomization is assumed to be the nature of the population resulting in an even distribution of the members on the continuum, so to say. In probability sampling, randomization is not assumed to be the nature of the population, and the process of choosing the sample has an in-built randomization. Each sample chosen under the probability sampling methods can thus be said to be a random sample. That is not the case for samples chosen under nonprobability sampling methods.

Some of the studies done in popular media and elsewhere do not meet the stringent requirements of the probability sampling and are not recommended for scientific enquiry. They fall into the category of nonprobability sampling. In such studies, choosing the sample is arbitrary, and it is hard to measure the probability of selecting any particular member of the population to be part of the sample. For all we know, because of this arbitrariness, some members of the population may have zero probability of being chosen while others may have certainty of being chosen. Also since there is no assurance that each member has a chance of being included, it is not possible to estimate sampling variability or determine whether there is a possible **bias.** Also, the reliability of a sample in nonprobability sampling cannot be determined; the only way to measure this may be to compare information gathered from such a sample with information others have previously gathered about the population. Some of these nonprobability sampling methods are treated here before discussing probability sampling methods.

Biased Samples

One of the issues that need to be guarded against in taking samples is bias. Bias in a sample can arise in subtle ways. If a researcher takes a sample made up of women in order to draw conclusions on a research idea where gender differences may be an issue, then the sample may be biased in gender. If in St. Cloud State University, a study which intends to elicit the opinions of all students across the campus collects opinions by sampling students who come through the Engineering and Computer Center (ECC), the study would be biased as it is most likely to obtain information from students who are studying Mathematics, Statistics, Computer Science, or Engineering to the exclusion of a vast majority of students in other colleges and disciplines elsewhere on the campus. The opinions would therefore be biased towards the disciplines that frequent the ECC.

The big problem with biased samples is that they lead to unreliable conclusions and results. Take for instance the issue of political ideology regarding taxation. It is known that in the USA most Republicans are ideologically opposed to taxing people according to their wealth in order to provide services. They favor letting people "in general" keep as much of their money as possible. Similarly, the Democrats are ideologically opposed to what the Republicans stand for. They believe in a fair share tax, and in citizens paying as much taxes as is necessary to provide social amenities and for the common good. Suppose that an opinion poll is to be commissioned to determine what the people of the USA think about taxes. If the sample used is made up of mostly Republicans, the outcome will support ideas that Republicans favor. On the other hand, loading the sample with Democrats will result in an outcome favoring ideas that Democrats support. In both cases, we illustrated classical cases of biased samples leading to biased outcomes, which in the end are unreliable, or really uninformative.

Biased samples arise because the constituent subsections of the population are not adequately or equitably represented in the sample. One would think that in important debates about evolution versus creationism, the scientists on both sides are unbiased. That is certainly not true. Both sides believe passionately in their causes, and so both sides work meticulously to prove their point and assiduously seek to collect information that appears to support their view. They are mostly uncharitable to information to the contrary.

Some biased samples arise out of prejudice. Wittingly or not, prejudiced people look out and are alert to events and issues that support their prejudice. Asked to study in the area of that prejudice, they will mostly collect information that favors it. Other biased samples arise out of laziness, as people seek to take samples that are convenient or easy to obtain.

Clearly bias is relative to the purpose of the sample. In sampling it is important to guard against hidden prejudices towards the purpose of the sample that could lead to biasing, or bending, the sample.

Convenience Sampling or Opportunity Sampling

Convenience samples involve the selection of those individuals or objects that are readily available. For instance, one could stand at the entrance to an underground train station in London on Tuesday morning and try to get the opinions on the train services of those going in or out of the station. Convenience samples are in general biased samples, but we treat them here because they are often used and there are circumstances in which information from such a survey might be useful. In the above example information obtained may be useful as long as it is restricted to the time, the place, and those who used the train during that time and in that place. It fails to be useful if the idea was to get the general opinion on train services from the general population.

In another form of this sampling method, called opportunity sampling, the researcher grabs anyone available, and could very easily use friends, family, classmates, and coworkers. It suffers from the same bias, since the sample may not adequately represent the targeted population. This leaves outcomes of such studies as not very useful.

Volunteer Sampling

Members of this sample select themselves to participate in the survey. The participants become part of the research because they chose to make themselves available, because someone asked them or they saw an advertisement, sometimes money was offered to participants. Although this type of approach can quickly assemble a sample for study and may even boast a sample made up of a variety of individuals, participants in such a sample are most likely to be unrepresentative of the population targeted for the study. The result of the study therefore fails to be generalizable. Volunteer samples are often used in clinical trials to determine the efficacy of a medication or therapy for a disease. To avoid some of the flaws of volunteer sampling in clinical trials, a sample of volunteers from a targeted population with the desired characteristics is taken.

Consider one type of volunteer sampling that we often encounter. A radio or television news outfit discusses a political issue and invites the audience to telephone or text the studio to express their opinion on the issue. This highlights some of the flaws of volunteer sampling. In such sampling, only those who feel strongly about the issue will respond. The population participating will also be limited to mostly those who heard the discussion of the issue. Often in this kind of sampling, no limit is placed on the number of times an individual may register his/her opinion. The result of these flaws is that the outcome of the sampling (the opinions held) is skewed away from what it would have been with a more representative sample.

Quota Sampling

In quota sampling, selection is made until targeted numbers of individuals are chosen from each of the subpopulations. It is one of the non-probability sampling methods that is often used. It is mostly used because of a desire to make the overall sample representative of all of the component parts of the population. To this end, targeted numbers from each subpopulation may be based on the proportions of the subpopulations within the general population.

Quota sampling should not be confused with stratified sampling. There is a similarity in that each stratum in a population on which stratified sampling is applied is a subpopulation in which similar units are grouped together, which is somewhat true in quota sampling. The difference is that in stratified sampling, probability sampling is involved. Individuals within each stratum are randomly selected with simple random sampling or systematic sampling, while in quota sampling the choice of individuals within the subpopulation is arbitrary. The arbitrariness of the selection under quota sampling leads to bias. Making inferences with data collected by quota sampling requires an assumption that the individuals selected in the sample are similar to those not selected. This is very hard to justify, and makes this sampling method less useful for inferential purposes.

1.3   Probability Sampling Methods

Sampling is the science of choosing a subset of a population. The object is to use the subset in the actual study of the population. This study will eventually lead to estimation and inference about population values. In order to obtain a good sample, it is recommended that probability sampling methods be used. Other sampling methods suffer from deficiencies which make the findings that are based on them unreliable. Probability sampling methods enable us to use a small sample to obtain reliable information which is accurate in its reflection of the actual population information. For instance, it is possible and does happen that with probability sampling we can use a small sample of a couple of thousands voters to accurately estimate the voting intentions of several millions of voters.

In probability sampling, each member of the sample is selected from a population at random. Randomization is deliberately used in probability sampling as part of the process. In short, each member of the population gets to be chosen to join the sample by chance, which is quantifiable. When compared to other sampling regimens, probability sampling takes more time and costs more to accomplish. However, it is free of the biases that plague other types of sampling. Since the probability of choosing every member of the population can be obtained, the estimates based on the sample are reliable. Sampling error can be obtained so that conclusions or decisions can be made about the population, based on the sample.

Probability sampling is made up of the following sampling methods; which one can be used is dependent on how appropriate they are for each study:

- Simple random sampling
- Systematic sampling
- Stratified sampling

- Cluster sampling
- Multi-stage sampling

In the next few sections we discuss these sampling methods.

1.3.1 Simple Random Sampling

The simplest of the probability sampling methods is the simple random sample. In a simple random sample which selects n individuals from a population of size N, each individual has to be equally likely to be selected in the sample with a probability of n/N. This also means that each sample of size n from this population has equal chance of being the selected sample.

To obtain a simple random sample we need a table of random numbers, or table of random digits. The table of random digits can be found in Appendix I (page 417).

Table of Random Numbers: The table of random digits has the quality that the numbers are so randomly arranged that if a finger is put anywhere within the table, with eyes closed, then any of the numbers:

- 0 1 2 3 4 5 6 7 8 9 (one-digit numbers) have equal probability of being on that spot
- 00 01 02 03 . . . 99 (two-digit numbers) have equal probability of being on that spot
- 000 001 003 004 . . . 999 (three-digit numbers) have equal probability of being on that spot
- 0000 0001 0002 . . . 9999 (four-digit numbers) have equal probability of being on that spot
- And so on

Due to this quality of the table of random digits we can label all members of a population as one-digit numbers (if the population size is not more than ten) and each of them would then have equal probability of being chosen in our sample no matter where we start in the table of random digits, as long as we count one digit at a time. Similarly, we can label all members of a population as two-digit numbers (if the population size is not more than one hundred) and each of them would then have equal probability of being chosen in our sample no matter where we start in the table of random digits, as long as we count one digit at a time, and so on. It is important to note that if our population is of size $N = 10$, we are right in using labels 0, 1, 2, . . . , 9. If $N = 100$, we are right to use 00, 01, 02, . . . , 99, and so on.

If $N = 10$, and we write labels 1, 2, 3, . . . , 10, we are mixing the two-digit number 10, with one-digit numbers 1, 2, 3, . . . , 9. The probability of a two-digit number is not the same as the probability of a one-digit number in the table of random digits. If we do that, we violate one of the basic rules of probability sampling method and of simple random sampling, which requires that the members of the population all should have an equal probability of being chosen to be in the sample. Further, if we write labels 1, 2, . . . , 10, we have omitted 0, which should have the same probability as 1, 2, 3, . . . , 9.

Note: When using the table of random digits, moves in the table when counting can be made either row-wise or column-wise, but switching between both is not acceptable. For consistency, and without discounting column-wise movements, the counting

moves in this text will be only row-wise in all the sampling discussed. Next, we illustrate how the simple random sample can be taken from a population.

■ **EXAMPLE 1.1** **Simple Random Sampling**

Consider the following class of Fashion Marketing students in a Western European university with twenty-eight students, as listed in Table 1.1, and use a simple random sampling method to draw a sample size of nine from this population.

■ **TABLE 1.1**

Name	Height (ft)	Span (cm)	Shoe Size	Hair Color	Eye Color	Age
Jan	5.5	18	5.5	Auburn	Blue	20
John	5.33	21.1	5.5	Brown	Brown	20
Allan	5.25	19.7	3.5	Brown	Brown	22
Pamela	5.5	21.3	5.5	Brown	Hazel	20
Ian	5.17	20.5	3.5	Brown	Green	21
Grace	5.38	20.4	5.5	Brown	Green	20
Brian	5.3	18.5	8	Brown	Brown	20
Steve	5.41	20.3	5	Brown	Blue	20
Liz	5.41	18.2	5	Brown	Green	20
Gordon	5	18.9	3	Brown	Blue	19
Phil	5.64	20.1	7	Brown	Brown	19
Graeme	5.5	20.6	4	Brown	Brown	21
Jean	5.58	20.8	6	Brown	Blue	20
Anne	5.5	21.1	6	Brown	Green	20
Alice	5.58	20.9	6.5	Brown	Grey	20
Edna	5	16.5	4	Brown	Green	20
Sharon	5.33	20.6	5	Blonde	Blue	19
Angela	5.37	20	4.5	Brown	Blue	19
Daniel	5.45	20.2	8	Brown	Hazel	20
Scot	5.61	19.1	9	Brown	Blue	23
Jamie	5.5	19	6	Brown	Hazel	21
Alison	5.17	19.1	4.5	Brown	Blue	20
Joy	5.92	20.9	7	Brown	Brown	19
Lenny	5.84	21.1	9.5	Brown	Blue	19
Stuart	5.5	20.4	7	Brown	Blue	20
Gail	5.48	20.3	4.5	Brown	Blue	20
Olga	5.33	20.2	4.5	Blonde	Blue	20
Matt	5.66	20.7	7.5	Black	Brown	22

To draw a simple random sample of size nine from this class, labels are attached to all the members of this class. Thereafter, the sample is drawn. To label the population, we note that $N = 28$ (a two-digit number). Therefore, labels of two digits need to be assigned to all the members, so that the labels go from 00, 01, 02, . . . , 27. The labeled population is presented in Table 1.2:

■ **TABLE 1.2**

Labels	First Name
00	Jan
01	John
02	Allan
03	Pamela
04	Ian
05	Grace
06	Brian
07	Steve
08	Liz
09	Gordon
10	Phil
11	Graeme
12	Jean
13	Anne
14	Alice
15	Edna
16	Sharon
17	Angela
18	Daniel
19	Scot
20	Jamie
21	Alison
22	Joy
23	Lenny
24	Stuart
25	Gail
26	Olga
27	Matt

Using the table of random digits, sampling can be started at an arbitrary position in the table. One way to do so is by closing your eyes, putting a finger in the table somewhere, and then opening them to find out where the finger is. Let us suppose that it was at column 24, row 13. Then the pivot number is 0. Then, counting off two digits at a time, we begin from this position, going right on that row and continuing on to the next row if that row is exhausted, we carry over any fragments from each row to the next row. If a two-digit number is found between 00 and 27, it is chosen as a member of the sample. This process is continued until the full sample is chosen, in this case nine individuals. If during the process a particular two-digit number within the range of labels occurs more than once, it is recorded only once, and ignored every other time it occurs. Other two-digit numbers would continue to be chosen until the sample is made up.

■ **TABLE 1.3**

Rows	___	___	___	___	___	Columns	___	___	___	___	___	___
	1–5	6–10	11–15	16–20	21–25	26–30	31–35	36–40	41–45	46–50	50–55	56–60
13	03813	55452	12072	65137	68304	68835	96418	38610	10270	28783	91173	93877
14	36635	26966	00037	47517	11038	92559	13220	62684	25891	07377	84848	06536
15	10433	21728	44543	76598	55639	67408	07560	53371	80692	92056	56721	64087
16	53496	62915	73129	53761	90505	45837	85376	49571	96246	58693	41725	90269
17	65705	73230	23348	93900	98303	58430	86614	72905	54098	44340	32081	50058
18	77435	12048	82677	30529	60360	29855	43089	37825	28210	59090	14250	06752
19	20070	92043	24037	83583	25355	55480	11193	54184	18361	51270	98499	61398
20	55342	49934	11082	54818	79576	36977	37721	41914	39538	30179	49244	36056

The sample is 04, 18, 01, 02, 11, 26, 00, 10, and 20. It should also be noted that when a row is exhausted, sampling continues on the next row while carrying the fragments left from the row that is exhausted. In the sampling above, row 13 was exhausted and the sampling continued on to row 14 while carrying on the fragment 7 in order to read the next two-digit number as 73, combining the fragment 7 and the value in column 1 of row 14, which is 3. Sampling stopped on realizing 20, the ninth member of the sample. Thus, the chosen students were those to whom were assigned the labels 04, 18, 01, 02, 11, 26, 00, 10, and 20. In practice, after these individuals were chosen in the sample, they would be measured to elicit the information in the table below relating to height, span, shoe size, dress size, and hair color. The information was supplied for teaching purposes, but in real life we have to go and collect the information by making the measurements for those chosen to be in the sample.

TABLE 1.4

Labels	First Name	Height (ft)	Span (cm)	Shoe Size	Hair Color	Eye Color	Age
00	Jan	5.5	18	5.5	Auburn	Blue	20
01	John	5.33	21.1	5.5	Brown	Brown	20
02	Allan	5.25	19.7	3.5	Brown	Brown	22
04	Ian	5.17	20.5	3.5	Brown	Green	21
10	Phil	5.64	20.1	7	Brown	Brown	19
11	Graeme	5.5	20.6	4	Brown	Brown	21
18	Daniel	5.45	20.2	8	Brown	Hazel	20
20	Jamie	5.5	19	6	Brown	Hazel	21
26	Olga	5.33	20.2	4.5	Blonde	Blue	20

The individuals in the simple random sample were Jan, John, Allan, Ian, Phil, Graeme, Daniel, Jamie, and Olga. The actual sample information that could be used is their heights, spans, shoe sizes, hair colors, eye colors, and age. These are the data that will be further analyzed.

EXAMPLE 1.2 **Simple Random Sampling**
Out of the twenty-eight students, take a simple random sample of size 10 starting from row 37, column 15 of the table of random digits (Appendix I, p. 417).

Solution

TABLE 1.5

	Columns											
Rows	1–5	6–10	11–15	16–20	21–25	26–30	31–35	36–40	41–45	46–50	51–55	56–60
37	74241	20930	55156	32173	74289	18739	89341	18749	95080	92976	77994	47274
38	98135	25411	49569	54907	69233	02176	23906	84356	56102	25131	65558	05969
39	79131	93858	23598	82938	85525	44416	73168	49378	02654	66410	84477	49229
40	14834	33594	82125	23681	02892	63118	96853	35632	33014	44085	66664	29900
41	94539	90867	63392	56092	72658	83677	31880	01922	52559	13759	49878	70038
42	94082	41975	02433	55758	41188	86952	35027	40867	82724	55949	03205	55231

First, we arrange the population of twenty-eight students with labels as in Table 1.2. Table 1.5 is also an extract from the table of random digits (Appendix I, p. 417). We see that the pivot number is 6, but we are counting two digits at a time, so going row-wise to the right, the first two-digit number is 63, which is not between 00 and 27, so it is discarded. Our first choice is the next two digits, 21 (which is within 00 . . . 27), so we choose it as the first member of the sample. Continuing in the same manner, our chosen sample is made up of class members who were labeled

21, 11, 08, 09, 13, 07, 23, 10, 22, and 19. In the process of choosing this sample, the numbers 11, 21, 23, and 13 occurred at least twice. They were chosen only once, and subsequent occurrences were ignored. Table 1.6 shows the sample and the associated values for the variables of interest.

■ **TABLE 1.6**

	Name	Height (ft)	Span (cm)	Shoe Size	Hair Color	Eye Color	Age
07	Steve	5.41	20.3	5	Brown	Blue	20
08	Liz	5.41	18.2	5	Brown	Green	20
09	Gordon	5	18.9	3	Brown	Blue	19
10	Phil	5.64	20.1	7	Brown	Brown	19
11	Graeme	5.5	20.6	4	Brown	Brown	21
13	Anne	5.5	21.1	6	Brown	Green	20
19	Scot	5.61	19.1	9	Brown	Blue	23
21	Alison	5.17	19.1	4.5	Brown	Blue	20
22	Joy	5.92	20.9	7	Brown	Brown	19
23	Lenny	5.84	21.1	9.5	Brown	Blue	19

1.3.2 Systematic Sampling

This type of probability sampling is often called the 1 in k systematic sampling. Basically the sampling process determines that from the population, 1 out of every k individuals would have to be chosen. For instance, in surveying students from a university with 14,000 students, the researcher may be interested in using about 1,000 students in the study. Roughly, he/she needs to use 1 in every 14 students in the survey. Therefore for this study, $k = 14$. The number k plays a very crucial role in the choice of a systematic sample.

The process for carrying out a 1 in k systematic sampling requires the following steps:

1. Find the entire group of individuals (members of the target population) to be sampled, and arrange them in some order such as alphabetical, serial, or some other convenient way. It is important to make sure that while arranging data in order in this way that the units are random with respect to the attributes being studied.
2. Determine what k is for our systematic sample.
3. Write labels for k; which may be 0, 1, 2, . . . , k if k is single-digit or 00, 01, . . . , k if k is double-digit, etc.
4. Go to the table of random digits (Appendix I, p. 417), start at an arbitrary point (close your eyes and put your finger in the table somewhere; for examples and exercises, we will specify the row and column where we will start), and choose the first one-digit number within the range of 0, 1, to k if k is one-digit, or choose the first two-digit number if it is within the range of 00, 01, . . . , k if k is two-digit, etc.

5. The first number chosen would be the first member of the 1 in k sample to be used in the study. Call this number c. We therefore go to the table with all the population arranged in order and choose the individual in that position. Suppose the number is $c = 2$, then the individual in position 2 is the first member of the sample.

6. Next, we choose individuals in the ordered table in positions $c + k$, $c + 2k$, $c + 3k$, $c + 4k$, etc., until the population is exhausted, or the number left is less than k. Thus, if $c = 2$, $k = 6$, and population size is 50, we would find the sample is made up of individuals in the ordered population in the positions: 2, 8, 14, 20, 26, 32, 38, 44, 50.

■ **EXAMPLE 1.3** Suppose that we had the population below made up of 81 individuals, and we want to take a sample of about ten out of the eighty-one. We are interested in finding estimates for the means for heights and weights for the population. It makes sense to consider using a 1 in 8 sample from the population. So we do. Therefore $k = 8$ in this case.

Serial Number	First Name	Serial Number	First Name	Serial Number	First Name
1	Nicole	28	Jan	55	Holly
2	Eric	29	John	56	Greg
3	Peter	30	Allan	57	Nick
4	Brian	31	Pamela	58	Shawn
5	Brooke	32	Ian	59	Paula
6	Hannah	33	Brian	60	Cassie
7	Abby	34	Steve	61	Blaine
8	Robert	35	Liz	62	Stan
9	Cathy	36	Gordon	63	Jolie
10	Joe	37	Phil	64	Keith
11	Jennifer	38	Graeme	65	Charles
12	Megan	39	Jean	66	Luke
13	Brianna	40	Anne	67	Chad
14	Tyler	41	Alice	68	Blake
15	Jessica	42	Edna	69	Andy
16	Brenda	43	Sharon	70	Kelly
17	Zach	44	Angela	71	Kate
18	Amanda	45	Daniel	72	Rachel
19	Krystal	46	Scot	73	Matt
20	Theresa	47	Jamie	74	Austin
21	Leslie	48	Alison	75	Pete
22	Kim	49	Joy	76	Ricky

Continued.

Serial Number	First Name	Serial Number	First Name	Serial Number	First Name
23	Kristie	50	Lenn	77	Cory
24	Tom	51	Stuart	78	Jessie
25	Pat	52	Gail	79	Juan
26	Ben	53	Olga	80	Eduardo
27	Stefan	54	Matt	81	Lance

Solution   The data is already arranged in serial order, so we proceed to write the labels 0, 1, 2, 3, 4, 5, 6, 7 to correspond to the numbers 1, 2, 3, 4, 5, 6, 7, 8 that exist between 1 and 8. We close our eyes and place a finger on the table of random digits (Appendix I, p. 417) and find that it is placed at row 22, and column 32. Thus, the pivot number is 9 which do not lie between 0 and 7, so we move to the next one-digit number which is 3 as highlighted below.

Rows	\multicolumn{12}{c}{Columns}

Rows	1–5	6–10	11–15	16–20	21–25	26–30	31–35	36–40	41–45	46–50	50–55	56–60
21	52383	44746	77059	51975	09785	37041	01019	77380	90769	23518	02198	95592
22	92043	37438	04560	81999	38557	16839	29378	07988	91050	95668	38101	52314
23	67098	47316	11626	69053	75493	86847	72258	24375	18816	60796	64402	30683
24	39672	59529	29220	79082	07038	78190	10638	14937	77841	01929	38865	45337

Now label 3 is equivalent to serial number 4 because the numbers between 1 and 8, that is 1, 2, 3, 4, 5, 6, 7, and 8, are considered to correspond to 0, 1, 2, 3, 4, 5, 6, and 7. Therefore our sample starts with the person in position 4, namely, Brian. The full sample is made up of those in positions: 4, 4 + k, 4 + 2k, 4 + 3k, etc. The people chosen should therefore be those in positions: 4, 12, 20, 28, 36, 44, 52, 60, 68, and 76. The actual individuals are Brian, Megan, Theresa, Jan, Gordon, Angela, Gail, Cassie, Blake, and Ricky. Following the identification of the sample, we measure the members of the sample and obtain the following weights and heights:

Serial Number	Name	Height (ins.)	Weight (lbs.)
4	Brian	63	120
12	Megan	65	145
20	Theresa	70	132
28	Jan	66	137
36	Gordon	58	158
44	Angela	69	164
52	Gail	67	156
60	Cassie	64	158
68	Blake	65	144
76	Ricky	68	151

With the data collected we can calculate the means (averages) for the heights and weights. The means are 65.5 inches and 146.5 pounds, respectively. They are used as estimates for the population means for height and weight.

The **systematic sample uses simple random sampling to select the first member of the sample.** Also each member of the population in a systematic sample has equal probability of being chosen as a member of the sample. **However, systematic sampling is not the same as simple random sampling.** Additionally, we note that the sample size of a systematic sample is not fixed. It depends on the starting position chosen by using the table of random digits. Suppose that we had chosen 0 as the first unit in the above example, this would have corresponded to the number 1, so that the sample would have been those in positions:

1, 9, 17, 25, 33, 41, 49, 57, 65, 73, 81

The sample size in this case is eleven, which is different from the sample size of ten we found above.

1.3.3 Cluster Sampling

Cluster sampling is appropriate for populations which appear to be naturally divided into clusters (groups). In this sampling method, the clusters which make up the population are labeled, and a simple random sample of the clusters is taken. Thereafter, each of the members of each chosen cluster is then considered a member of the sample to be used in the actual study. The information needed is elicited from each member of each chosen cluster. Thus, if a cluster is chosen all individuals within it is a chosen member of the sample. It is important to note that the clusters are usually not of equal size, and therefore the sample size is not known until the clusters are selected. Usually, the numbers in the samples are added up to determine the sample size.

The following are the steps needed in order to carry out cluster sampling from a population:

1. Identify the target population for the study and identify all the clusters that make up the population.
2. Decide how many of the clusters should be chosen to make up the sample.
3. Label the clusters using methods of labeling described earlier for labeling individuals in the simple random sampling, select the clusters that are in the sample.
4. Every member of all the chosen clusters is a member of the sample to be used for the study. Elicit the required information or measure the attribute of interest from every member of every chosen cluster.

■ **EXAMPLE 1.4**   A polling organization wished to survey a church denomination in a local area about their opinion regarding gay issues (should gays be allowed to marry, should gays be ordained as ministers, etc.). There are thirty-two churches affiliated to the denomination in the area, with memberships as described below:

Serial Number	Label	Church Name (Membership Size)	Serial Number	Label	Church Name (Membership Size)
1	00	St. Joseph (239)	17	16	St. John Parish (221)
2	01	All Saints (125)	18	17	Resurrection (245)
3	02	St. Peter (110)	19	18	Victory Parish (265)
4	03	St. Bath (226)	20	19	St. Mary (441)
5	04	St. Nicholas (324)	21	20	St. Paul (289)
6	05	All Hallows (561)	22	21	St. Francis (376)
7	06	St. Luke (378)	23	22	St. Mark (446)
8	07	St. Philip (421)	24	23	City Assembly (504)
9	08	Ascension (571)	25	24	Stanley Memorial (622)
10	09	Blessed Hope (332)	26	25	St. Appolos (223)
11	10	Transfiguration (354)	27	26	St. Charles (429)
12	11	Annunciation (397)	28	27	Redeemed Assembly (337)
13	12	Biggard Memorial (338)	29	28	St. Christopher (551)
14	13	Bigham Cathedral (1227)	30	29	Patterson Memorial (388)
15	14	Peterbrough (357)	31	30	Blaine Assembly (537)
16	15	Udi Parish (448)	32	31	God's Love (377)

They decided to use cluster sampling, and to choose a sample of five clusters, considering each church as a cluster.

Solution Since there are thirty-two churches, that is thirty-two clusters in the population. In order to choose a sample of five clusters, we label all the churches with two-digit labels so that their labels would be 00, 01, 02, 03, . . . , 31; with 00 as the label of St. Joseph, 01 the label of All Saints, . . . , and 31 the label of God's Love. Suppose they use the table of random digits (Appendix I), and start from row 35, column 13, then the chosen clusters would be those highlighted in the table below, that is, 00, 28, 24, 02, 03.

	Columns											
Rows	1–5	6–10	11–15	16–20	21–25	26–30	31–35	36–40	41–45	46–50	50–55	56–60
35	73805	58967	53850	02887	24026	99077	88445	80076	03387	18983	01881	85633
36	82184	00144	36534	27061	20621	36213	89024	20944	83539	18079	04923	62871
37	74241	20930	55156	32173	74289	18739	89341	18749	95080	92976	77994	47274
38	98135	25411	49569	54907	69233	02176	23906	84356	56102	25131	65558	05969
39	79131	93858	23598	82938	85525	44416	73168	49378	02654	66410	84477	49229
40	14834	33594	82125	23681	02892	63118	96853	35632	33014	44085	66664	29900
41	94539	90867	63392	56092	72658	83677	31880	01922	52559	13759	49878	70038
42	94082	41975	02433	55758	41188	86952	35027	40867	82724	55949	03205	55231

The corresponding churches and their memberships are: St. Joseph (239), St. Christopher (551), Stanley Memorial (622), St. Peter(110), and St. Bath (226). These five churches make up the required sample. Information required would have to be obtained from each of the members of each of these churches, and used to make conclusions about the views of the entire denomination. Thus, the sample size is made up of all the members of the five churches; this means that the sample size is $n = 239 + 551 + 622 + 110 + 226 = 1748$. Thus, the number of people polled on the issues was 1748.

■ **EXAMPLE 1.5** Recently, the Department of Statistics of St. Cloud State University has been teaching STAT 193 using the help of LAs (learning assistants). To survey the students to see how helpful the program is, it was decided to use cluster sampling. There are roughly twelve classes of STAT 193 in an academic year, with each class being made up of roughly sixty students. However, towards the end of the semester some classes do not have exactly sixty students; mostly they have less. In order to survey roughly 180 students, we can decide to consider each class as a cluster and select three clusters.

The classes are identified in their names in the table below according to whether they are taught in first semester (s1), second semester (s2), or during the summer (sum):

Class	S193s1-01	S193s1-02	S193s1-03	S193s1-04	S193s1-05	S193s2-01	S193s2-02
Size	58	56	60	60	57	55	60
Class	S193s2-03	S193s2-04	S193s2-05	S193sum1-01	S193sum1-02		
Size	57	60	60	56	60		

Solution The classes are considered to be clusters. To select a sample of three clusters that will make up the sample from the population of STAT 193 students, the first step is to assign labels to the classes as follows:

Class	S193s1-01	S193s1-02	S193s1-03	S193s1-04	S193s1-05	S193s2-01	S193s2-02
Class-Size	58	56	60	60	57	55	60
Labels	00	01	02	03	04	05	06
Class	S193s2-03	S193s2-04	S193s2-05	S193sum1-01	S193sum1-02		
Class-Size	57	60	60	56	60		
Labels	07	08	09	10	11		

Using row 3, column 13 of table random digits,

					Columns							
Rows	1–5	6–10	11–15	16–20	21–25	26–30	31–35	36–40	41–45	46–50	50–55	56–60
1	30120	13850	81903	56587	39129	94727	78226	95207	76354	38718	01669	33878
2	69696	81799	27328	33287	35476	56650	57330	02079	82972	82287	75634	55692
3	17784	00005	25584	51364	02330	44697	39343	75351	34890	87080	64632	49892

Continued.

	Columns											
Rows	**1–5**	**6–10**	**11–15**	**16–20**	**21–25**	**26–30**	**31–35**	**36–40**	**41–45**	**46–50**	**50–55**	**56–60**
4	35821	49630	87686	53852	56806	57379	26797	94186	19280	64342	78912	96759
5	75763	40570	04655	30679	00398	95493	91902	11365	83316	07188	27891	31001
6	87138	59718	98465	91803	82212	98364	92735	35410	78370	12608	21966	61052

We see that the selected three classes have the labels: 02, 04, and 08. Thus the chosen classes are:

Class	S193s1-02	S193s1-04	S193s2-03
Class-Size	60	57	60
Labels	**02**	**04**	**08**

The classes (clusters) chosen have 60, 57, and 60 students, respectively. Each student in every cluster is a member of the sample. All are interviewed to elicit all the relevant responses from each one of them. The total number of students in the sample is 177.

1.3.4 Stratified Sampling

The word stratum connotes layer. Sometimes, it is possible to see that some of the populations to be studied appear to be made up of strata (layers). If a population is made up of layers, each layer should appear to be more homogeneous than the aggregate population. When this is true, then the population is stratified, and better results will be obtained by applying stratified sampling to the population than would have been obtained by employing simple random sampling. The purpose of using the stratified sampling method is to ensure that the chosen sample reflects the diverse characteristics that may be inherent in the different strata.

The process involves identifying all the strata in the population and taking samples from each of the strata to make up the sample. Suppose that a population to be studied has k strata, and it is desired to take a sample of size n from the population. Equal samples or an equal number of observations can be taken from each stratum. If this is done, the number of observations taken from each stratum would be roughly n/k. Often however, proportionate allocation is used. This method is found useful if the strata are of different sizes. With this strategy, more observations are taken from larger strata, and less taken from the smaller strata, providing the needed balance in the sample. In such cases different numbers, n_1 is chosen from stratum 1, n_2 chosen from stratum 2, $\ldots$, n_k is chosen from stratum k, where $n = n_1 + n_2 + \ldots + n_k$.

To choose a sample as described above, at each of the k strata the simple random sampling method is used to select the number of objects required. The systematic sampling method can also be used in each stratum to select the samples. Thereafter, the information desired is elicited from each member of the sample chosen from the k strata. Thus, if the entire number of individuals or objects in stratum 1 is N_1 then the simple random sampling method is used to select n_1 from N_1 in stratum 1, n_2 from N_2 in stratum 2, . . . , n_k from N_k in stratum k. The following picture illustrates how the sample is chosen when the population is perceived to be made up of four strata:

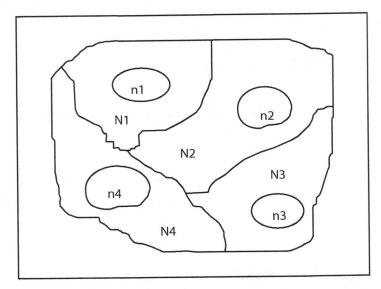

In the picture we illustrate a population made up of four strata, with subpopulations of N_1, N_2, N_3, and N_4 individuals, so that the total population under study contains $N = N_1 + N_2 + N_3 + N_4$. From each of these strata, samples of sizes n_1, n_2, n_3, and n_4 are respectively drawn, leading to a total sample size of $n = n_1 + n_2 + n_3 + n_4$. All the n individuals in the sample are used to obtain the required information which would in turn be used to make conclusions about the entire population. If equal samples were chosen from each stratum, then $n_1 = n_2 = n_3 = n_4$. Adoption of proportionate allocation would lead to unequal sample sizes, n_i ($i = 1, 2, 3, 4$) corresponding to the four strata. Next, the stratified sampling method is treated by an example.

Formulae for weighted mean, and weighted proportion
Weighted sample mean, $\overline{X}_w$:

$$\overline{X}_w = \frac{N_1}{N}(\overline{X}_1) + \frac{N_2}{N}(\overline{X}_2) + \frac{N_3}{N}(\overline{X}_3) + \frac{N_1}{N}(\overline{X}_4)$$

Weighted Proportion, $\hat{p}_w$:

$$\hat{p}_w = \frac{N_1}{N}(\hat{p}_1) + \frac{N_2}{N}(\hat{p}_2) + \frac{N_3}{N}(\hat{p}_3) + \frac{N_1}{N}(\hat{p}_4)$$

■ **EXAMPLE 1.6**   The following data represent (in thousands of dollars) the investments made by people in the biomedical professions, aged between thirty-five and forty-four, in 2001. The professionals were divided into two groups based on age; so that ages 35–39 form one group, while the other is made up of those aged 40–44. Each group contains twenty-one professionals, whose genders are indicated. Consider the two groups as strata (layers) of the population, and take a stratified sample from the data. Take a sample of size seven from stratum I (those aged 40–44); using row 13, column 10 of the table of random digits on page 417. Similarly, take a sample of size seven from stratum II (those aged 35–39) starting from row 15 of column 3 of the same table of random digits. Using the data obtain the weighted mean investment of a professional. From sample data we obtain the weighted estimate of the proportion of females in the population.

Biomedics Aged 40–44 (Stratum I)	Investment	Gender	Biomedics Aged 35–39 (Stratum II)	Investment	Gender
1	13.9	M	1	13.8	M
2	10.4	M	2	9.5	M
3	5.2	M	3	22.7	M
4	6	F	4	10	F
5	14	F	5	16	F
6	15	F	6	28	F
7	10	M	7	16	M
8	11	M	8	15	M
9	8	M	9	20	M
10	9.4	M	10	13.5	M
11	21.6	M	11	16.8	M
12	11.7	M	12	12.8	M
13	14.7	M	13	12	M
14	4.7	M	14	8	M
15	11.3	F	15	11.5	F
16	14.9	F	16	13.8	F
17	14	F	17	8	F
18	9.1	F	18	15.4	F
19	8.1	F	19	14.6	F
20	9.7	F	20	11.3	F
21	14.8	F	21	9.8	F

Solution We can start anywhere in the table of random digits to select the samples from the different strata, but for exercises we state the columns and row at which we start, for teaching purposes. In order to choose the sample from stratum I we use row 13, column 10. We label all the members of that stratum 00, 01, 02, . . . , 20.

	Columns											
Rows	1–5	6–10	11–15	16–20	21–25	26–30	31–35	36–40	41–45	46–50	50–55	56–60
13	03813	55452	12072	65137	68304	68835	96418	38610	10270	28783	91173	93877
14	36635	26966	00037	47517	11038	92559	13220	62684	25891	07377	84848	06536
15	10433	21728	44543	76598	55639	67408	07560	53371	80692	92056	56721	64087
16	53496	62915	73129	53761	90505	45837	85376	49571	96246	58693	41725	90269
17	65705	73230	23348	93900	98303	58430	86614	72905	54098	44340	32081	50058

Continued.

Rows	Columns											
	1–5	6–10	11–15	16–20	21–25	26–30	31–35	36–40	41–45	46–50	50–55	56–60
18	77435	12048	82677	30529	60360	29855	43089	37825	28210	59090	14250	06752
19	20070	92043	24037	83583	25355	55480	11193	54184	18361	51270	98499	61398
20	55342	49934	11082	54818	79576	36977	37721	41914	39538	30179	49244	36056

The selected sample from stratum I has the labels: 20, 13, 04, 18, 01, 02, and 11. Similarly, by labeling all members of stratum II, 00, 01, 02, 03, . . . , 20 and using column 3, row 15 to select the sample from the stratum, the labels of the selected sample from this stratum are: 17 08 07 05 20 16 15.

The final sample from both strata is:

Biomedics Aged 40–44 (Stratum I)	Investment	Gender	Biomedics Aged 35–39 (Stratum II)	Investment	Gender
01	10.4	M	05	28	F
02	5.2	M	07	15	M
04	14	F	12	12	M
11	11.7	M	15	13.8	F
13	4.7	M	16	8	F
18	8.1	F	17	15.4	F
20	14.8	F	20	9.8	F

With this sample, the weighted estimates of the mean and proportion of females can be obtained as follows:

$$\text{weighted mean} = \overline{X} = \frac{21}{42}\left(\frac{68.9}{7}\right) + \frac{21}{42}\left(\frac{102}{7}\right) = 12.20714$$

$$\text{weighted proportion of females} = \hat{p}_{\text{weighted}} = \frac{21}{42}\left(\frac{3}{7}\right) + \frac{21}{42}\left(\frac{5}{7}\right) = 0.5714$$

■ **EXAMPLE 1.7** Suppose that instead of using simple random sampling to select members of the sample in each stratum, systematic sampling had been adopted as suggested above. Determine the sample that would have been chosen using the same starting point in the table of random digits that were used in Example 1.6 for the different strata. Obtain the weighted estimates.

Solution In this case to choose seven from each stratum containing twenty-one individuals, it would be tantamount to choosing one in every three individuals in each stratum. Here the 1 in k systematic sampling is 1 in 3, making the value of $k = 3$.

For stratum I: Labels 0, 1, 2 are used (this is equivalent to writing labels 1 to 3). Starting from row 13, column 10, the first of the labels chosen is 2 (which is equivalent to position 3 in the ordered data), thus the sample chosen from stratum I are in positions 3, 6, 9, 12, 15, 18, and 21.

For stratum II: Labels 0, 1, 2 are used. Starting from row 15, column 3 again, the first of the labels chosen is 2, thus the sample chosen from stratum II are in positions 3, 6, 9, 12, 15, 18, and 21. The resulting samples and attendant information are given below:

Biomedics Aged 40–44 (Stratum I)	Investment	Gender	Biomedics Aged 35–39 (Stratum II)	Investment	Gender
3	5.2	M	3	22.7	M
6	15	F	6	28	F
9	8	M	9	20	M
12	11.7	M	12	12.8	M
15	11.3	F	15	11.5	F
18	9.1	F	18	15.4	F
21	14.8	F	21	9.8	F

$$\text{weighted mean} = \overline{X} = \frac{21}{42}\left(\frac{75.1}{7}\right) + \frac{21}{42}\left(\frac{120.2}{7}\right) = 13.95$$

$$\text{weighted proportion of females} = \hat{p}_{\text{weighted}} = \frac{21}{42}\left(\frac{4}{7}\right) + \frac{21}{42}\left(\frac{4}{7}\right) = 0.5714$$

1.3.5 Multi-stage Sampling

The only other probability sampling method left is the multi-stage sampling. This method uses one of the previous four (simple random, systematic, cluster, and stratified) at the different stages of the sampling process to arrive at the sample. For instance, sampling the USA by multi-stage sampling, one could use simple random sampling at stage I to select five states, then at stage II use systematic sampling in each of the five states to select counties, and at stage III use stratified sampling to divide each county into cities and select individuals from those.

Exercises

s/n	Grade 9	Expense	Gender	Grade 8	Expense	Gender	Grade 7	Expense	Gender
1	Jan	21	F	Geun	18	F	Sarah	6	F
2	Eric	29	M	Kortney	21	F	Ben	17	M
3	Sven	26	M	Hoe	24	F	Jordan	25	M
4	Scot	30	M	Frank	15	M	Daniel	13	M
5	Miles	25	M	Wei	17	F	Cory	6	M
6	Grace	25	F	Tom	17	M	Matthew	8	M
7	Phil	25	M	Breann	21	F	Joseph	7	M
8	Ramon	18	M	Megan	10	F	Derek	17	M
9	Crystal	20	F	Courtney	21	F	Judith	24	F
10	Lona	27	F	Brian	11	M	Lacy	19	F
11	Dan	16	M	Kelly	12	F	Kayla	17	F
12	Paul	21	M	Rachel	25	F	Travis	23	M
13	Amber	22	F	Carrie	6	F	Greg	13	M
14	Matt	30	M	Robert	17	M	Nicholas	6	M
15	Alonzo	16	M	John	24	M	Angela	21	F
16	Dennis	11	M	Amanda	10	F	Lauren	10	F
17	James	11	M	Sarah	19	F	Graham	20	M
18	Lance	23	M	Loretta	14	F	Amy	9	F
19	Pete	28	M	Heather	6	F	Merrisa	19	F
20	Dana	23	F	Brent	21	M	Ashley	21	F
21	Anne	19	F	David	18	M	Mark	23	M
22	Brianna	22	F	Tyler	7	M	Sandra	5	F
23	Stacy	18	F	David	10	M	Matthew	18	M
24	Ashley	27	F	Leslie	15	F	Joshua	23	M
25	Henry	11	M	Collin	20	M	Fatah	22	F
26	Paula	17	F	Devan	7	F	Richard	20	M
27	Lu	29	F	Adam	25	M	Tiffinee	16	F
28	Aisha	20	F	Patrick	13	M	Lucas	23	M
29	Bertha	16	F	Matthew	24	M	Daniel	22	M
30	Margaret	18	F	Mark	15	M	Robin	17	F

Continued.

s/n	Grade 9	Expense	Gender	Grade 8	Expense	Gender	Grade 7	Expense	Gender
31	Penny	20	F	Amy	10	f	Hillary	14	F
32	Olga	13	F	Ewa	14	f	Matthew	11	M
33	Mary	21	F	Cory	9	m	Bobbi	12	F
34	Harriet	17	F	Karen	12	f	Scott	18	M
35	Henrieta	13	F	Lydia	14	f	Katie	7	F
36	Liana	29	F	Freaderico	14	m	Joshua	11	M
37	Cassie	13	F	Nicholas	12	m	Allison	21	F
38	Allison	13	F	Alberto	24	m	Lea	22	F
39	Brian	13	M	Jesse	16	f	Alexander	16	M
40	Elizabeth	18	F	Curtis	17	m	Jason	16	M
41	Erin	17	F	Adil	18	m	Brice	23	M
42	Briahn	18	M	Hadiza	17	f	Pratima	7	F
43	Heng	20	F	Ethan	9	m	Paul	17	M
44	Tina	24	F	Jason	12	m	Joel	14	M
45	Graham	12	M	Kelly	17	f	Vern	20	M
46	Kirstin	14	F	Joseph	12	m	Brian	18	M
47	Anna	19	F	Ahmed	10	m	Amanda	23	F
48	Kara	24	F	Chris	8	m	Kimberly	20	F
49	Madelyn	19	F	Humphrey	15	m	Steven	15	M
50	Mary	13	F	Holly	6	f	Jeremy	20	M

In the exercises below, identify every member of the sample selected and carry out the requested analysis.

1.1 Using row 7, column 12 of the table of random digits (Appendix I, page 417), obtain a simple random sample of size ten from Grade 9 students. Obtain (i) the mean expenses, and (ii) proportion of females in the sample.

1.2 Using row 5, column 9 of the table of random digits (Appendix I, page 417), obtain a simple random sample of size ten from Grade 8 students from the data of Exercise 1.1. Obtain (i) the mean expenses, and (ii) proportion of males in the sample.

1.3 Using row 15, column 19 of the table of random digits (Appendix 8, page 417), obtain a simple random sample of size ten from Grade 7 students from the data of Exercise 1.1. Obtain (i) the mean expenses, and (ii) proportion of females in the sample.

1.4 By considering the grades as strata, obtain a stratified sample from the data of Exercise 1.1 by using row 1, column 12 to select a random sample of size nine from Grade 9; using row 6, column 10 to select a sample of size nine from Grade 8; and using row 4, column 8 to select a sample of size nine from Grade 7. (a) Obtain a weighted mean expense for the sample, and (b) obtain a weighted proportion of females in the sample.

1.5 By considering the grades as strata, obtain a stratified sample from the data of Exercise 1.1 by using row 12, column 10 to select a random sample of size nine from Grade 9; using row 16, column 9 to select a sample of size eight from Grade 8; and using row 14, column 5 to select a sample of size seven from Grade 7. (a) Obtain a weighted mean expense for the sample, and (b) obtain a weighted proportion of females in the sample.

1.6 By considering the grades as strata, obtain a stratified sample from the data of Exercise 1.1 by using row 9, column 12 to select a random sample of size six from Grade 9; using row 9, column 12 to select a sample of size six from Grade 8; and using row 13, column 15 to select a sample of size six from Grade 7. (a) Obtain a weighted mean expense for the sample, and (b) obtain a weighted proportion of females in the sample.

Reconsider the Weekly snacks expenses of grades 7–9 students as a single data set.

s/n	Grade 9	Expense	Gender	s/n	Grade 8	Expense	Gender	s/n	Grade 7	Expense	Gender
1	Jan	21	F	51	Geun	18	F	101	Sarah	6	F
2	Eric	29	M	52	Kortney	21	F	102	Ben	17	M
3	Sven	26	M	53	Hoe	24	F	103	Jordan	25	M
4	Scot	30	M	54	Frank	15	M	104	Daniel	13	M
5	Miles	25	M	55	Wei	17	F	105	Cory	6	M
6	Grace	25	F	56	Tom	17	M	106	Matthew	8	M
7	Phil	25	M	57	Breann	21	F	107	Joseph	7	M
8	Ramon	18	M	58	Megan	10	F	108	Derek	17	M
9	Crystal	20	F	59	Courtney	21	F	109	Judith	24	F
10	Lona	27	F	60	Brian	11	M	110	Lacy	19	F
11	Dan	16	M	61	Kelly	12	F	111	Kayla	17	F
12	Paul	21	M	62	Rachel	25	F	112	Travis	23	M
13	Amber	22	F	63	Carrie	6	F	113	Greg	13	M
14	Matt	30	M	64	Robert	17	M	114	Nicholas	6	M
15	Alonzo	16	M	65	John	24	M	115	Angela	21	F
16	Dennis	11	M	66	Amanda	10	F	116	Lauren	10	F
17	James	11	M	67	Sarah	19	F	117	Graham	20	M
18	Lance	23	M	68	Loretta	14	F	118	Amy	9	F
19	Pete	28	M	69	Heather	6	F	119	Merrisa	19	F
20	Dana	23	F	70	Brent	21	M	120	Ashley	21	F
21	Anne	19	F	71	David	18	M	121	Mark	23	M
22	Brianna	22	F	72	Tyler	7	M	122	Sandra	5	F
23	Stacy	18	F	73	David	10	M	123	Matthew	18	M
24	Ashley	27	F	74	Leslie	15	F	124	Joshua	23	M

Continued.

s/n	Grade 9	Expense	Gender	s/n	Grade 8	Expense	Gender	s/n	Grade 7	Expense	Gender
25	Henry	11	M	75	Collin	20	M	125	Fatah	22	F
26	Paula	17	F	76	Devan	7	F	126	Richard	20	M
27	Lu	29	F	77	Adam	25	M	127	Tiffinee	16	F
28	Aisha	20	F	78	Patrick	13	M	128	Lucas	23	M
29	Bertha	16	F	79	Matthew	24	M	129	Daniel	22	M
30	Margaret	18	F	80	Mark	15	M	130	Robin	17	F
31	Penny	20	F	81	Amy	10	F	131	Hillary	14	F
32	Olga	13	F	82	Ewa	14	F	132	Matthew	11	M
33	Mary	21	F	83	Cory	9	M	133	Bobbi	12	F
34	Harriet	17	F	84	Karen	12	F	134	Scott	18	M
35	Henrieta	13	F	85	Lydia	14	F	135	Katie	7	F
36	Liana	29	F	86	Freaderico	14	M	136	Joshua	11	M
37	Cassie	13	F	87	Nicholas	12	M	137	Allison	21	F
38	Allison	13	F	88	Alberto	24	M	138	Lea	22	F
39	Brian	13	M	89	Jesse	16	F	139	Alexander	16	M
40	Elizabeth	18	F	90	Curtis	17	M	140	Jason	16	M
41	Erin	17	F	91	Adil	18	M	141	Brice	23	M
42	Briahn	18	M	92	Hadiza	17	F	142	Pratima	7	F
43	Heng	20	F	93	Ethan	9	M	143	Paul	17	M
44	Tina	24	F	94	Jason	12	M	144	Joel	14	M
45	Graham	12	M	95	Kelly	17	F	145	Vern	20	M
46	Kirstin	14	F	96	Joseph	12	M	146	Brian	18	M
47	Anna	19	F	97	Ahmed	10	M	147	Amanda	23	F
48	Kara	24	F	98	Chris	8	M	148	Kimberly	20	F
49	Madelyn	19	F	99	Humphrey	15	M	149	Steven	15	M
50	Mary	13	F	100	Holly	6	F	150	Jeremy	20	M

1.7 Using row 17, column 7, obtain a simple random sample of size ten from the table. Obtain (i) the mean expenses, and (ii) proportion of females in the sample.

1.8 Using row 13, column 5, obtain a simple random sample of size twelve from the table. Obtain (i) the mean expenses, and (ii) proportion of males in the sample.

1.9 Using row 7, column 15, obtain a simple random sample of size eleven from the table. Obtain (i) the mean expenses, and (ii) proportion of females in the sample.

1.10 Using row 8, column 6, obtain a 1 in 9 systematic sample from the table. Obtain (i) the mean expenses, and (ii) proportion of females in the sample.

1.11 Using row 3, column 8, obtain a 1 in 15 systematic sample from the table. Obtain (i) the mean expenses, and (ii) proportion of males in the sample.

1.12 Using row 8, column 13, obtain a 1 in 10 systematic sample from the table. Obtain (i) the mean expenses, and (ii) proportion of females in the sample.

1.13 Consider the data of Example 1.6. Take a sample of size six from stratum I (those aged 40–44), use row 10, column 8 of the table of random digits, Appendix I on page 415). Similarly, take a sample of size six from stratum II (those aged 35–39) starting from row 13 of column 4 of the same table of random digits. Using the data, obtain the weighted mean investment of a professional.

1.14 Consider the data of Example 1.6. Take a sample of size six from stratum I (those aged 40–44), use row 12, column 7 of the table of random digits, Appendix I on page 415. Similarly, take a sample of size six from stratum II (those aged 35–39) starting from row 11 of column 9 of the same table of random digits. Using the data, obtain the weighted mean investment of a professional.

1.15 Consider the data of Example 1.5; using row 14, column 16 select a sample of four clusters. How do you obtain the required information following sampling?

1.16 Consider the data of Example 1.4 and using row 7, column 10, select a sample of five clusters.

1.17 Consider the data of Example 1.4 and using row 16, column 17, select a sample of five clusters.

1.18 Weights of 120 boys:

30	57	40	46	78	59
34	46	40	49	52	54
35	42	61	53	62	44
64	58	55	49	61	66
61	41	61	71	54	46
64	37	77	51	62	34
55	37	46	51	45	50
50	38	44	36	47	36
66	45	44	52	34	50
49	56	34	39	43	44
56	43	32	52	71	67
57	46	47	33	31	47
44	76	61	52	44	68
56	43	42	72	41	42
52	43	42	44	32	69
35	51	32	75	51	48
57	34	37	56	53	69
34	38	51	72	42	46
43	39	50	33	59	71
45	65	80	76	83	81

Using the data as presented and proceeding row-wise in the data, obtain a 1 in 11 systematic sample of the data and obtain its mean using row 6, column 11 of the table of random digits. Find the mean.

1.19   Using the data as presented in Exercise 1.18, and proceeding row-wise in the data, label the data and obtain a simple random sample of size 21 using row 14, column 19 of the table of random digits. Find the mean.

1.20   Consider the data in the following table:

Serial Number	First Name	Weight	Height	Serial Number	First Name	Weight	Height	Serial Number	First Name	Weight	Height
1	Nicole	126	64	28	Jan	130	69	55	Holly	145	67
2	Eric	143	68	29	John	147	71	56	Greg	173	71
3	Peter	171	70	30	Allan	175	74	57	Nick	170	73
4	Brian	168	73	31	Pamela	172	71	58	Shirley	172	67
5	Brooke	200	70	32	Ian	204	65	59	Paula	158	65
6	Hannah	156	64	33	Brenda	160	67	60	Cassie	172	67
7	Abby	170	66	34	Steve	174	65	61	Blaine	180	69
8	Robert	178	64	35	Liz	182	69	62	Stan	178	67
9	Cathy	176	68	36	Gordon	180	73	63	Jolly	214	71
10	Joe	212	72	37	Phil	216	57	64	Keith	190	65
11	Jennifer	188	56	38	Graeme	192	71	65	Charles	188	72
12	Megan	186	70	39	Jean	190	70	66	Luke	178	71
13	Brianna	176	69	40	Anne	180	72	67	Chad	214	63
14	Tyler	212	71	41	Alice	216	66	68	Blake	157	67
15	Jessica	155	65	42	Edna	159	61	69	Andy	184	62
16	Brenda	182	60	43	Sharon	186	73	70	Kelly	172	64
17	Zach	170	72	44	Angela	174	68	71	Kate	169	69
18	Amanda	167	67	45	Daniel	171	62	72	Rachel	147	63
19	Krystal	145	61	46	Scot	149	65	73	Martha	167	66
20	Theresa	165	64	47	Jamie	169	71	74	Austin	180	72
21	Leslie	178	70	48	Alison	182	71	75	Pete	170	72
22	Kim	168	70	49	Joy	172	66	76	Ricky	171	67
23	Kristie	169	65	50	Lenny	173	72	77	Cory	194	60
24	Tom	192	71	51	Stuart	196	64	78	Jessie	190	61
25	Pat	188	63	52	Gail	192	70	79	Juan	191	62
26	Ben	189	69	53	Olga	193	68	80	Eduardo	192	63
27	Stefan	190	67	54	Matt	194	65	81	Lance	170	64

Using row 7, column 20 of the supplied the table of random digits draw a simple random sample of size twenty. Present their names, heights, and weights. Obtain the mean weights and heights for your sample.

1.21   Using the data of Exercise 1.20, draw a 1 in 6 systematic sample using row 12 and column 14 of the table of random digits. Present their names, heights, and weights. Obtain the mean weights and heights for your sample.

■ TABLE OF RANDOM DIGITS

Rows	1–5	6–10	11–15	16–20	21–25	26–30	31–35	36–40	41–45	46–50	50–55	56–60
1	30120	13850	81903	56587	39129	94727	78226	95207	76354	38718	01669	33878
2	69696	81799	27328	33287	35476	56650	57330	02079	82972	82287	75634	55692
3	17784	00005	25584	51364	02330	44697	39343	75351	34890	87080	64632	49892
4	35821	49630	87686	53852	56806	57379	26797	94186	19280	64342	78912	96759
5	75763	40570	04655	30679	00398	95493	91902	11365	83316	07188	27891	31001
6	87138	59718	98465	91803	82212	98364	92735	35410	78370	12608	21966	61052
7	11654	48648	36173	91029	30989	50408	12641	21246	16638	36266	85720	71967
8	41754	43010	16749	73270	70472	54849	40547	08786	29933	99319	62931	49428
9	20754	31665	87389	81339	79106	92621	89831	06127	45080	82142	94014	68695
10	11198	19216	82884	95655	63248	20721	32433	75861	77861	55700	83148	89194
11	19976	38351	52917	64368	80502	23253	88104	10826	06580	92391	49517	42460
12	30418	63656	60308	20088	86266	04800	71798	70611	16329	84639	31713	15717
13	03813	55452	12072	65137	68304	68835	96418	38610	10270	28783	91173	93877
14	36635	26966	00037	47517	11038	92559	13220	62684	25891	07377	84848	06536
15	10433	21728	44543	76598	55639	67408	07560	53371	80692	92056	56721	64087
16	53496	62915	73129	53761	90505	45837	85376	49571	96246	58693	41725	90269
17	65705	73230	23348	93900	98303	58430	86614	72905	54098	44340	32081	50058
18	77435	12048	82677	30529	60360	29855	43089	37825	28210	59090	14250	06752
19	20070	92043	24037	83583	25355	55480	11193	54184	18361	51270	98499	61398
20	55342	49934	11082	54818	79576	36977	37721	41914	39538	30179	49244	36056
21	07567	03612	81928	05111	43492	75714	97754	64447	25100	72756	02198	95592
22	68140	15564	89299	88433	55768	65473	32746	61493	19649	80919	38101	52314
23	90081	94223	88193	70464	63503	40140	97620	01805	92242	40242	64402	30683
24	63887	28878	95742	07029	52663	93720	89502	55551	41135	99199	38865	45337
25	06718	36267	04006	90154	86114	47543	82086	45230	58218	39799	06650	80884
26	02050	84631	38553	67622	11489	26119	62581	96146	93108	44378	32050	14470
27	19352	65919	06096	82444	20818	64342	87826	37332	77672	90399	32535	61002
28	91882	47249	86196	69169	22571	71419	58379	91656	44157	28443	16095	09186
29	21890	49206	28120	56209	21158	24639	17042	37034	86757	47673	02127	66913
30	71266	90921	67705	63770	41962	83737	63284	14900	54646	50966	47298	59165

Continued.

■ TABLE OF RANDOM DIGITS—cont'd

Rows	1–5	6–10	11–15	16–20	21–25	26–30	31–35	36–40	41–45	46–50	50–55	56–60
						Columns						
31	06910	70975	81644	87240	93841	47206	50645	82994	72155	43492	03554	02371
32	67192	42127	36387	54886	50590	75099	43463	87116	04279	70709	94437	33201
33	13710	33875	80634	36230	55774	66277	52248	05794	93379	68338	07635	97764
34	92866	39641	43754	21920	10555	87509	68595	86524	54432	17106	85324	76684
35	73805	58967	53850	02887	24026	99077	88445	80076	03387	18983	01881	85633
36	82184	00144	36534	27061	20621	36213	89024	20944	83539	18079	04923	62871
37	74241	20930	55156	32173	74289	18739	89341	18749	95080	92976	77994	47274
38	98135	25411	49569	54907	69233	02176	23906	84356	56102	25131	65558	05969
39	79131	93858	23598	82938	85525	44416	73168	49378	02654	66410	84477	49229
40	14834	33594	82125	23681	02892	63118	96853	35632	33014	44085	66664	29900
41	94539	90867	63392	56092	72658	83677	31880	01922	52559	13759	49878	70038
42	94082	41975	02433	55758	41188	86952	35027	40867	82724	55949	03205	55231
43	63571	75636	64594	24715	74438	29793	60304	40566	79428	31355	49690	50023
44	23735	26942	09467	86699	46783	68004	49807	28067	58183	27562	04510	38564
45	23058	41489	09596	48957	78581	37003	02835	70991	00637	56205	55988	86716
46	04611	22913	53999	22825	29524	35725	32667	75573	63286	62191	19604	67222
47	27280	44306	61397	06112	84833	35907	84332	56427	18291	16725	05287	13393
48	43281	83120	04399	67542	04021	71163	61568	36141	08417	61576	47224	31488
49	81145	49102	26158	22946	56398	20575	58013	80801	52669	28443	02232	47913
50	02472	93734	25928	54042	23864	16400	12609	66281	97155	58773	79130	14579
51	21841	00021	49649	55396	49026	05092	86167	79487	97287	70064	58844	31775
52	61578	24025	05690	19185	65233	31338	68038	10503	09360	27241	83379	13417
53	80794	08643	61406	70009	31906	54962	02415	05169	24386	57836	03376	21700
54	80104	02474	11481	54034	73147	01883	05586	88572	81821	52034	82960	01114
55	80191	95373	08098	83779	39557	38779	79934	50648	51626	45459	55074	48268
56	45850	94026	49830	11281	83396	76894	63339	60038	44412	39201	10852	50561
57	62608	26249	74921	75064	81342	32050	81234	96110	96774	05872	71230	98739
58	78611	82103	22524	87069	57691	71013	11912	79660	14745	69357	40463	58575
59	42695	67012	53281	85270	03758	20525	56774	77059	51607	67027	73280	91777
60	91505	75468	25148	60872	09303	29766	46042	19720	82766	24178	96874	48006

CHAPTER 2

Basic Data Presentation and Analysis

2.0 Introduction

In the previous chapter probability sampling methods were discussed. The methods were practically demonstrated by using them for sampling from populations. Once samples were chosen, information (data) was elicited from the individual members of the sample. The object of sampling was to obtain data which would provide sample information which would be used to estimate the same information about the population from which it was drawn. Once sample data are collected, there is a need to make sense of the data if the general public or targeted audience is to benefit from the inherent information in the sample. This information will not be clear until the data is analyzed. Analysis can vary from the rudimentary to the sophisticated; that is, from data presentation to inference about population parameters. In this chapter the rudimentary, data presentation, is explored.

In order to give a clearer understanding of the nature of the information contained in a population which is being researched, data presentation uses tables, graphs, and pictures to elucidate sample information. These include stem-and-leaf plots, frequency tables, bar and pie charts, histograms, ogives, and more. This is part of the general area of statistical analyses referred to as exploratory data analysis. Statisticians essentially study and use these presentation methods. The job of a **statistician** is to look at a set of numbers after data collection, and try to make sense of them. One of the elementary aspects of the job is the presentation of data in a form which makes sense to all, especially the non-statistician.

2.1 Presentation by Charts (Bar Charts and Pie Charts)

One of the elementary ways of presenting data is by the use of diagrams, often called pictorial presentation. Data in its most basic form can easily lend itself to presentation in the form of a picture. If the basic data relates to categories of the population, then its inherent information may sometimes be better conveyed to people if it is presented in a pictorial form. The different types of pictorial presentations which may be suitable for data which are categorized are graphs, **bar** charts, and **pie** charts. Other forms of pictorial presentation, such as **histogram, polygons, ogives,** etc., are more adequate for data which are mostly of the continuous type but which could be grouped in classes.

The following data was collected from twenty-eight students on the Fashion Marketing Degree Course who attended a statistics lecture. We can immediately see that some of the data lend themselves to presentation in the form of one chart or another, since the students could easily be grouped into categories.

Height (ft)	Span (cm)	Shoe Size	Dress Size	Hair Color	Eye Color	Age
5.5	18	5.5	12	Auburn	Blue	20
5.33	21.1	5.5	10	Brown	Brown	20
5.25	19.7	3.5	8	Brown	Brown	22
5.5	21	5.5	12	Brown	Hazel	20
5.17	20.5	3.5	10	Brown	Green	21
5.38	20.4	5.5	12	Brown	Green	20
5.2	18	3	8	Brown	Brown	20
5.41	20.3	5	10	Brown	Blue	20
5.41	18.2	5	12	Brown	Green	20
5	18.9	3	10	Brown	Blue	19
5.54	21.1	6	12	Brown	Brown	19
5.5	20.6	4	12	Brown	Brown	21
5.58	20.8	6	10	Brown	Blue	20
5.5	21.1	6	12	Brown	Green	20
5.58	20.9	6.5	12	Brown	Grey	20
5	16.5	4	10	Brown	Green	20
5.33	20.6	5	10	Blonde	Blue	19
5.17	20	4.5	10	Brown	Blue	19
5.5	20	5.5	14	Brown	Hazel	20
5.41	19.1	4	10	Brown	Blue	23
5.5	19	6	14	Brown	Hazel	21
5.17	19.1	4.5	10	Brown	Blue	20
5.92	20.9	7	14	Brown	Brown	19
5.84	21.1	9.5	16	Brown	Blue	19
5.5	20	5	10	Brown	Blue	20
5.58	20.3	6	10	Brown	Blue	20
5.33	20.2	4.5	8	Blonde	Blue	20
5.66	20.7	7.5	16	Black	Brown	22

2.1.1 Pie Charts

Pie charts are used to represent data which are in categories. We can illustrate a pie chart by using data on hair and eye colors to construct pie charts. From the above table, we can see that of the twenty-eight students, one student had an auburn colored hair, twenty-four had brown hair, two had blonde hair and one had black hair.

■ **EXAMPLE 2.1**   Construct a pie chart to depict the distribution of the hair color of the students of the Fashion Marketing class, according to its categories.

Solution   A pie chart is a circle in the form of a pie divided in proportion to the counts in each category of hair color. Classically, in order to manually construct the pie chart, the 360° in the circle is divided by the total number of counts for all the categories, and multiplied by the number in each category to determine the number of degrees to be assigned to each category, as shown in Table 2.1 below. Thereafter a protractor and dividers is used to construct the pie chart. Take the hair color data from the table:

■ **TABLE 2.1**   Categorization of Hair Color for Fashion Marketing Students

Hair Color	Count	Degrees
Auburn	1	12.86
Blonde	2	25.71
Brown	24	308.57
Black	1	12.86
Total	**28**	**360**

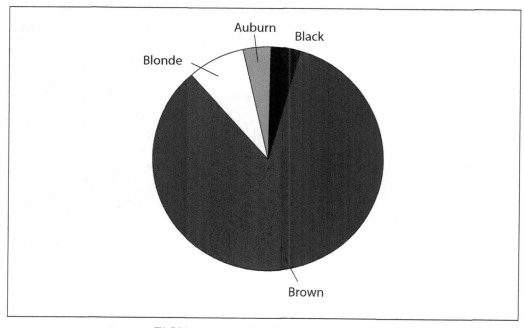

■ **FIGURE 2.1   Students' Hair Colors**

A number of software programs exist, such as MINITAB, EXCEL, SAS, and JMP, which can be used to construct the pie charts. Some graphing calculators have the capability to construct pie charts. Each of them can be used to construct a pie chart; however, there is some value in learning how to do it manually. In fact, even with the use of software, the information in the above table, or part of it, may need to be entered into the computer before software can construct the pie chart.

To construct a similar pie chart for eye color, Table 2.2 shows the division of 360° in the circle among eye color categories:

■ **TABLE 2.2   Categorization of Eye Color for Fashion Marketing Students**

Eye Color	Count	Degrees
Blue	13	154.29
Brown	7	90
Green	5	64.29
Grey	1	12.86
Hazel	3	38.57
Total	**28**	**360**

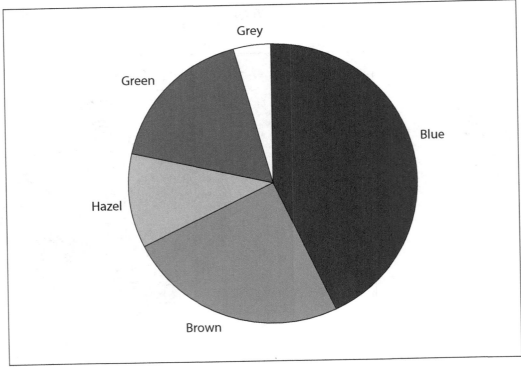

■ **FIGURE 2.2   Eye Colors of Students**

2.1.2 Bar Charts (or Graphs)

Bar charts can be used to represent data which are in categories, just as the pie charts can. However, there are different varieties of bar charts. Some bar charts have single components in a bar, while others have multiple components on each bar. The basic idea behind each bar in a chart is that its size should be proportional to the number of observations in the category it represents. Even when a number of components are represented on a single bar, the size of the bar devoted to each component should reflect the number of observations belonging to that component.

The data on students' eye colors is used here to illustrate the basic bar chart.

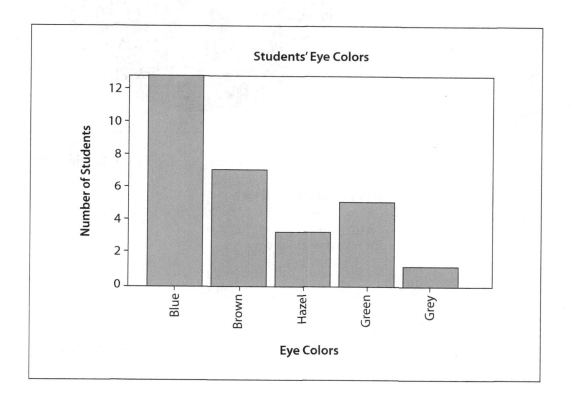

Use of Minitab to Construct Pie and Bar Charts

In order to obtain a pie chart for hair color for the Fashion Marketing students, enter the data for different categories of hair color in column 1 of Table 2.1, into column 1 of MINITAB. Enter their corresponding counts into column 2 in MINITAB. Enter the following commands at the MINITAB prompt:

Minitab Code for Pie Chart

```
MTB>   Name c1 'hair_color'
MTB>   PieChart (c2)*c1;
SUBC>  Combine 0.02;
SUBC>  Panel.
```

The outcome will be the following pie chart:

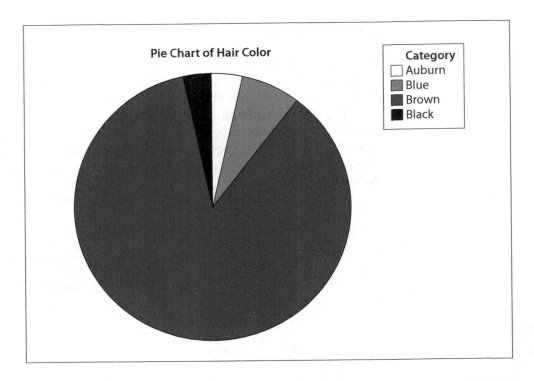

To obtain a bar chart for hair color, enter the following commands into MINITAB:

Minitab Code for Bar Chart

```
MTB> name c1 'hair_color' c2 'student_counts'
MTB> chart (c2)*c1;
MTB> summarized;
MTB> bar.
```

The following output will follow:

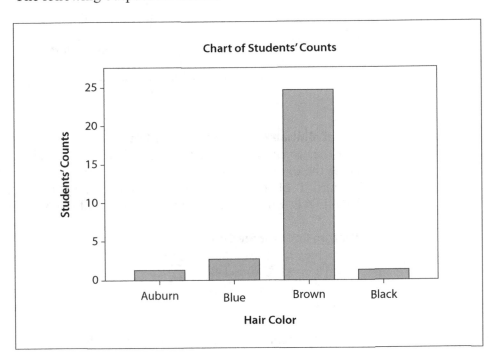

Construction of Multiple Component Bar Graphs

The multiple component bar chart is illustrated by using data on the production of GCH Farm, Inc. in the years 1988–1993. The data is presented in the table, followed by a multiple-component bar chart of the data:

■ **TABLE 2.3** Production of GCH Farms, Inc., 1998–1993

Year	Wheat	Corn	Rice
1988	210	85	110
1989	195	95	120
1990	235	110	110
1991	260	95	122
1992	275	110	132
1993	300	120	145

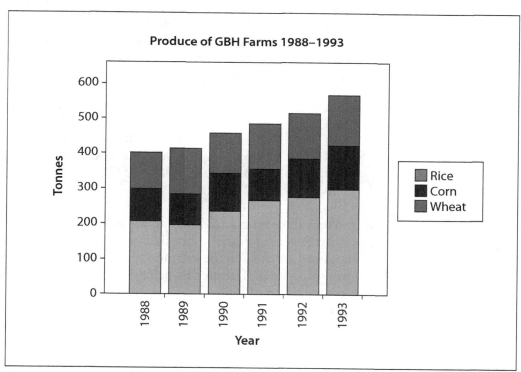

■ **FIGURE 2.3** Multiple-Component Bar Chart of Production of GCH Farms, Inc., 1998–1993

There are other types of charts, such as horizontal charts, 3D charts, and etc., which will not be formally treated in this text save to say that they are logical extensions from the basic charts which have been discussed and presented.

2.2 Summarizing a Raw Numeric Data Set Pictorially

As indicated earlier, data sets do not make much sense until they are organized. Most data sets initially come to us in the form of **raw data.** A number of tools have been developed for the purpose of dealing with such raw data and presenting them in more lucid forms. These include graphs, charts, frequency tables, and other computational ways of summarizing data. In this section, grouping data in frequency tables is treated.

Raw data refers to collected data which have been organized into figures. An example of raw data is the birth weights of a sample of 192 premature babies born in a group of hospitals from a large area, which is presented below:

```
0.60 3.00 3.00 3.00 3.00 3.00 3.10 3.10 3.10 3.10 2.20
3.10 3.10 1.20 1.20 3.20 3.20 3.20 3.20 1.60 2.00 2.00
2.00 1.60 1.60 1.60 1.70 1.80 1.80 1.90 1.90 3.30 3.30
3.30 2.30 2.30 2.30 2.30 2.30 2.40 2.40 2.40 2.40 2.40
2.40 2.50 2.50 2.50 2.50 2.50 2.50 2.60 2.60 2.60 2.60
2.70 2.70 2.70 0.70 3.70 3.70 3.70 3.70 3.70 3.70 3.70
2.70 2.70 2.80 2.80 2.80 2.80 2.80 2.80 2.90 2.90 2.90
2.90 2.90 2.90 2.90 2.90 2.90 2.90 2.90 2.90 2.90 2.90
2.90 2.90 3.00 3.00 3.00 3.00 3.00 3.00 3.00 3.00 3.00
3.00 3.00 3.10 3.10 3.20 2.70 2.70 2.70 2.80 2.80 2.80
2.80 3.20 3.20 3.20 3.20 3.20 3.20 3.20 3.20 4.10 4.20
4.30 3.20 3.30 3.30 3.30 3.30 3.30 3.30 3.30 3.30 3.30
3.30 3.40 3.40 3.40 3.40 3.60 3.60 3.60 3.00 3.40 3.40
3.40 3.40 3.40 3.50 3.50 3.50 3.50 3.50 3.50 3.50 3.50
3.50 3.60 3.60 3.60 3.60 3.60 3.60 3.60 3.60 3.80 3.80
3.80 3.80 3.80 3.80 3.90 3.90 3.90 4.00 4.00 4.00 4.10
3.80 3.80 3.90 3.90 3.90 1.00 1.20 2.00 2.00 3.90 3.90
3.90 4.00 2.10 2.10 0.60
```

Raw data can be rearranged into an **array.** This is usually the first step taken in order to make some sense of a raw data set—the data is presented in order according to size, most often from the least to the largest. It is the first step taken towards organizing a data set. The process helps a great deal in interpreting the data and further organization of it.

An **array** is an arrangement of raw data into ascending or descending order of magnitude. The following will be the result if the data for the birth weights of the babies is presented in the form of an array:

```
0.60 0.60 0.70 1.00 1.20 1.20 1.20 1.60 1.60 1.60 1.60
1.70 1.80 1.80 1.90 1.90 2.00 2.00 2.00 2.00 2.00 2.10
2.10 2.20 2.30 2.30 2.30 2.30 2.30 2.40 2.40 2.40 2.40
2.40 2.40 2.50 2.50 2.50 2.50 2.50 2.50 2.60 2.60 2.60
2.60 2.70 2.70 2.70 2.70 2.70 2.70 2.70 2.70 2.80 2.80
2.80 2.80 2.80 2.80 2.80 2.80 2.80 2.80 2.90 2.90 2.90
2.90 2.90 2.90 2.90 2.90 2.90 2.90 2.90 2.90 2.90 2.90
```

```
2.90  2.90  3.00  3.00  3.00  3.00  3.00  3.00  3.00  3.00  3.00
3.00  3.00  3.00  3.00  3.00  3.00  3.00  3.00  3.10  3.10  3.10
3.10  3.10  3.10  3.10  3.10  3.20  3.20  3.20  3.20  3.20  3.20
3.20  3.20  3.20  3.20  3.20  3.20  3.20  3.20  3.30  3.30  3.30
3.30  3.30  3.30  3.30  3.30  3.30  3.30  3.30  3.30  3.30  3.40
3.40  3.40  3.40  3.40  3.40  3.40  3.40  3.40  3.50  3.50  3.50
3.50  3.50  3.50  3.50  3.50  3.50  3.60  3.60  3.60  3.60  3.60
3.60  3.60  3.60  3.60  3.60  3.60  3.70  3.70  3.70  3.70  3.70
3.70  3.70  3.80  3.80  3.80  3.80  3.80  3.80  3.80  3.80  3.90
3.90  3.90  3.90  3.90  3.90  3.90  3.90  3.90  4.00  4.00  4.00
4.00  4.10  4.10  4.20  4.30
```

The array does provide some information by telling the least and highest weights of premature babies in the sample, 0.6 and 4.3. The range of weights can also be deduced from it as: 3.7 = 4.3 − 0.6. Beyond that, not much more can be learned from an array at this stage. With the arrangement of data in the form of an array, it can be seen that other ways of summarizing the data may be needed in order to convey the information in the data more clearly.

2.2.1 Using Minitab to Obtain an Array

To use Minitab to obtain an array, we put all the raw data in column 1 (C1) in Minitab, then:

1. In the MENUS, first click Editor
2. Then click the Enable command, the prompt "MTB>" appears; then type at the prompt:
3. MTB> sort c1 c2
4. MTB>Print c2

   The Minitab output is the same as the ARRAY above.
   The array above is an example of the **presentation** of the data in an **ungrouped frequency table.** Later we shall present the same data in a grouped frequency table.

Stem-and-Leaf Plots

There are several functional forms which describe the distribution of data sets. The functions are used to model different types of data. When the functions are plotted, they lead to different graphical shapes. Now that the data is put in the form of an array, it is useful to get an idea of the approximate shape of the distribution of the data. This can be achieved by using a frequency table, and after that the graphical form of the frequency table which is called a histogram. The shape can also be explored by using an exploratory graph called the stem-and-leaf plot.

Consider numbers that are made up of two digits. To obtain a stem-and-leaf plot of a data set made up of two-digit numbers, the first thing to do is to list the first digit of the numbers (that is the digit on the left the numbers) in a column. Thereafter, a vertical line is drawn to the right of the column of first digits. Next, the second digit of each number is listed to the right of the vertical line, associating each one to its left digit. This is carried out until all the numbers in the data set have been listed.

■ **EXAMPLE 2.2**   The following data show the scores of ninety-one children in a cognitive test, and obtain an array and a stem-and-leaf plot of the data:

```
69 49 67 96 67 94 67 91 66 88 64 67 63 49 62
43 62 41 62 40 62 39 61 35 61 34 61 29 60 28
60 20 59 59 59 29 58 96 58 94 57 91 56 88 56
85 55 78 55 76 55 75 53 57 53 56 51 38 51 37
51 31 49 39 48 38 43 54 42 46 36 84 80 80 82
83 81 80 70 72 73 73 73 72 46 46 46 48 43 75
76
```

The data is presented as an array:

```
20 28 29 29 31 34 35 36 37 38 38 39 39 40 41
42 43 43 43 46 46 46 46 48 48 49 49 49 51 51
51 53 53 54 55 55 55 56 56 56 57 57 58 58 59
59 59 60 60 61 61 61 62 62 62 62 63 64 66 67
67 67 67 69 70 72 72 73 73 73 75 75 76 76 78
80 80 80 81 82 83 84 85 88 88 91 91 94 94 96
96
```

Stem-and-leaf plot

```
2 | 0899
3 | 145678899
4 | 012333666688999
5 | 1113345556667788999
6 | 00111222234677779
7 | 02233355668
8 | 0001234588
9 | 114466
```

The numbers to the left of the vertical line of the stem and leaf plot are called stems, while those to the right are called leaves.

Sometimes in order to get a better picture of the shape of the data, the stem and leaf is constructed in such a way that the stems are repeated on the left of the vertical line. For the first of the repeated stems, only second digits of the number which are between 0 and 4 are associated with it. Second digits of the numbers between 5 and 9 are associated with the second of the repeated stems. The stem and leaf plot is again reconstructed for the data on cognitive tests, adopting the second approach:

Stem-and-leaf with repeated stems:

```
2 | 0
2 | 899
3 | 14
3 | 5678899
4 | 012333
4 | 666688999
```

```
5 | 111334
5 | 5556667788999
6 | 00111222234
6 | 677779
7 | 022333
7 | 55668
8 | 0001234
8 | 588
9 | 1144
9 | 66
```

What happens if the data is three digits? It depends on whether the three digits are close enough that they lend themselves to the construction of a stem-and-leaf plot.

■ **EXAMPLE 2.3**   Consider the following data set:

290 295 302 300 304 307 309 294 293 296 299 293 293 307 300 302
317 313 316 310 310 304 305 309 306 301 301 304 301 312 311 291 295
299 295 300 305 311 310 314 307 308 300 302 314 318 318 313 300 300 305

Obtain an array and a stem-and-leaf plot of the data.

Solution

Array:

290 291 293 293 293 294 295 295 295 296 299 299 300
300 300 300 300 300 301 301 301 302 302 302 304 304
304 305 305 305 306 307 307 307 308 309 309 310 310
310 311 311 312 313 313 314 314 316 317 318 318

The values are between 290 and 318. A stem-and-leaf plot is constructed using the first two digits as stems and the third digit as leaves, as follows:

```
29 | 013334
29 | 555699
30 | 000000111222444
30 | 5556777899
31 | 0001123344
31 | 6788
```

■ **EXAMPLE 2.4**   Obtain a stem-and-leaf plot for the birth weights of the 192 premature babies discussed previously.

Solution   The method described previously is slightly modified because the data involves decimals. On the leading digits called *stems,* the integer part of the number is plotted while the decimal part of the number is used as the *leaf.* In this case we split the number of decimal units on each stem so that 0, 1, 2, 3, 4 are attached to the first use of the stem in the plot, while units such as 5, 6, 7, 8, 9 are attached to the second use of the stem. This is equivalent to using two lines for each stem as described previously.

We apply the method to the data on weights of premature babies and obtain the following stem-and-leaf plot:

```
0 | 6 6 7
1 | 0 2 2 2
1 | 6 6 6 6 7 8 8 9 9
2 | 0 0 0 0 1 1 2 3 3 3 3 3 4 4 4 4 4 4
2 | 5 5 5 5 5 5 6 6 6 6 7 7 7 7 7 7 7 7 8 8 8 8 8 8 8 8 8 9 9 9 9 9 9 9 9 9 9 9 9 9 9 9
3 | 0 0 0 0 0 0 0 0 0 0 0 0 0 0 0 0 1 1 1 1 1 1 1 1 2 2 2 2 2 2 2 2 2 2 2 2 2 2 3 3 3 3 3 3 3 3 3 3 3 3 4 4 4 4 4 4 4 4 4
3 | 5 5 5 5 5 5 5 5 5 6 6 6 6 6 6 6 6 6 6 7 7 7 7 7 7 7 8 8 8 8 8 8 8 8 9 9 9 9 9 9 9 9 9
4 | 0 0 0 0 1 1 2 2
```

2.2.2 Use of Minitab to Construct Stem-and-Leaf Plots

We can use Minitab to construct the stem-and-leaf plot for the data.

Applying this to the data on weights of premature babies, we illustrate the use of the following command in Minitab to obtain the plot:

```
MTB > stem-and-leaf c1
```

This is the output from Minitab:

Stem-and-Leaf Display: premature baby-weights

Stem-and-leaf of premature baby-weight $N = 192$
Leaf Unit = 0.10

```
   3   0  667
   7   1  0222
  16   1  666678899
  35   2  0000011233333444444
  79   2  5555556666777777778888888889999999999999999
 (61)  3  0000000000000000011111111122222222222222223333333333333334444444444
  52   3  555555555566666666666777777788888888999999999
   8   4  00001123
```

The first column records the counts. It is the cumulative count up to a row below the row that contains the median, The line in which the median is has a the cumulative count for that line enclosed in parenthesis. The second column is the stem. The lines are the leaf part of the plot.

The number in the first column enclosed in parenthesis represents the count for the row which contains the median value. The counts for rows above and below the median are cumulative. The count for a row above the median represents the total count for that row and all the rows above it. The number indicated for a row below the median represents the total count for that row and all the rows below it. That means that total number in sample is: 79 + 61 + 52 = 192.

Next, we discuss how to put data in a grouped frequency.

2.3 | Grouping Data

The array shown previously was an example of the **presentation** of data in an **ungrouped frequency table.** Frequency simply refers to the number of occurrences of a value. The data can also be presented in a **grouped frequency table.** This is usually achieved by dividing the entire data into **classes** or **groups.** To divide a set of data into classes, the **range** of the observations need to be known. This is the difference between the magnitude of the largest and the smallest observations, which the array provides. This range is divided into **sub-intervals** which enable the classes or groups to be defined.

For a large data set it is advisable to have at least seven classes. It is easier in practice to have classes of equal width. This simplifies any arithmetic that needs to be done with the figures. However, in some circumstances, it may be more beneficial to have classes which are of larger width than the rest. It is better to merge two classes into one to obtain a class with a larger width than the rest, instead of having a class with zero frequency and another class with positive frequency. In the example below, an attempt has been made to obtain a frequency table of equal width. As a result, 0.5 has been included, which is 0.1 lower than the smallest observation of 0.6, and 4.4 which is 0.1 above the largest observation which is 4.3.

■ **TABLE 2.4   A Basic Frequency Table for Birth Weights of Premature Babies**

Classes or Groups	Class-Boundary	Frequency (f_i)
0.50–0.90	0.45 and under 0.95	3
1.00–1.40	0.95 and under 1.45	4
1.50–1.90	1.45 and under 1.95	9
2.00–2.40	1.95 and under 2.45	19
2.50–2.90	2.45 and under 2.95	44
3.00–3.40	2.95 and under 3.45	61
3.50–3.90	3.45 and under 3.95	44
4.00–4.40	3.95 and under 4.45	8
		192

Often the numbers denoting the intervals for each group in Column 1 of Table 2.4 are called **class limits** while the ends of the intervals for each group shown in Column 2 are called **class boundaries.** It is clear that the lower class boundaries were obtained by subtracting 0.05 from the lower limits of the classes, and the upper class boundaries were obtained by adding 0.05 to the upper limits of the classes. If the numbers in the class intervals were integers, the usual thing would be to subtract and add 0.5 as described above in order to obtain the equivalent class boundaries.

We can also explore other ways of summarizing data in a **frequency table** by including information such as **cumulative frequencies,** which refer to the sum of the frequency of a class and the frequencies of other classes below that class. Such tables may also include **relative frequencies** which refer to the number of occurrences in that class relative to the total number of observations in the set. Extending the above table to provide this extra information, we obtain the table below. It is also common to find a column containing the **class marks** or **mid-points.** For purposes of statistical analysis it is often assumed that the frequency of each class is concentrated at the midpoint of that class. If for a typical class its upper class boundary is denoted by U_B and its lower class boundary by L_B then the class **mark or midpoint (often denoted by X)** of each class is obtained as:

$$\text{mark } (X) = \text{midpoint} = \frac{U_B + L_B}{2}$$

$$\text{class} - \text{width} = U_B - L_B$$

Class Limits	Class Boundaries	Class Midpoint (mark)	Count or Frequency	Cumulative Count or Frequency	Relative Frequency	Cumulative Relative Frequency
0.5–0.9	0.45–0.95	0.70	3	3	3/192	3/192
1.0–1.4	0.95–1.45	1.20	4	7	4/192	7/192
1.5–1.9	1.45–1.95	1.70	9	16	9/192	16/192
2.0–2.4	1.95–2.45	2.20	19	35	19/192	35/192
2.5–2.9	2.45–2.95	2.70	44	79	44/192	79/192
3.0–3.4	2.95–3.45	3.20	61	140	61/192	140/192
3.5–3.9	3.45–3.95	3.70	44	184	44/192	184/192
4.0–4.4	3.95–4.45	4.20	8	192	8/192	192/192

At this stage the data set certainly makes more sense than it did in its raw form. It is easier to interpret the data to anyone who is interested than it would have been in its raw state.

2.3.1 Use of Minitab as an Aid for Preparing Frequency Tables

We need this data divided into seven classes as suggested. We put all the data for the babies in say column one (c1) in Minitab, and then carry out the following instructions in Minitab to tally the data:

1. In the MENUS, first click Editor
2. Then click Enable command, the prompt "MTB>" appears; then type at the prompt:
3. MTB> code (0.5:0.9)0 (1.0:1.4)1 (1.5:1.9)2 (2.0:2.4)3 (2.5:2.9)4 (3.0:3.4)5 (3.5:3.9)6 (4.0:4.4)7 c1 c2
4. MTB> Tally c2;
5. SUBC> counts;
6. SUBC> cumcounts.

We obtain an output like this:

```
MTB > code (0.5:0.9)0 (1.0:1.4)1 (1.5:1.9)2 (2.0:2.4)3 (2.5:2.9)4
(3.0:3.4)5 (3.5:3.9)6 (4.0:4.4)7 c1 c2
MTB > Tally c2;
SUBC> counts;
SUBC> cumcounts.
```

Tally for Discrete Variables: c2

C2	Count	CumCnt
0	3	3
1	4	7
2	9	16
3	19	35
4	44	79
5	61	140
6	44	184
7	8	192

$N = 192$

With the above output we can prepare a full frequency table as presented above (listing all the classes, their limits, boundaries, midpoints, frequencies, cumulative frequencies, relative frequencies, and cumulative relative frequencies).

2.3.2 Histograms

In the previous sections, some pictorial forms of data presentation were discussed. At this stage, with the data arranged in frequency tables, it is found that certain other pictorial presentations of data are useful. Chief among these are the histograms and the ogives.

The histogram is a pictorial representation of the **distribution** of a data set according to classes which have been created in the **frequency table.** A histogram is a graphical plot of the frequencies of the classes in the frequency table created by constructing rectangles which touch each other. The rectangles for adjacent classes touch at the class boundary which is common to both of them. The heights of the rectangles are equal to the frequencies of the classes. The rectangles all have bases of equal size (the width of the class) if all the classes are of equal widths. The bases are of unequal size if some classes have different widths. If classes are of equal width, the class boundaries or the midpoints can be used as the tick marks on the axis which represents the values of the variable (classes). The base of each rectangle in the histogram starts on the lower class boundary of each class and ends at the upper class boundary of the same class, ensuring that each class dove-tails into the next, and that each rectangle is balanced at the mark or midpoint for each class. It is pretty straight forward to use the method described to construct the histogram for the above data on premature baby weights. All that is needed in order to construct the histogram is a graph sheet and an appropriate choice of scales (see the hand-drawn histogram and frequency polygon below):

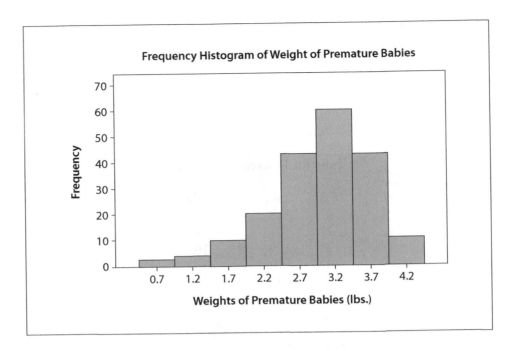

In the histogram of the data we can see that all the rectangles representing the frequencies for each class is centered at the midpoint of each class and that the height of each rectangle is proportional to the number of frequencies in each class.

2.3.3 The Frequency Polygon

This is an exploratory graph which can be used to assess the shape of the distribution of the data. It is constructed via the histogram by joining the midpoints of the tops of the rectangles that make up the histogram, and hypothetically extended by dotted lines to touch the baselines for unobserved classes as shown in the hand-drawn example for the birth weights of premature babies shown below:

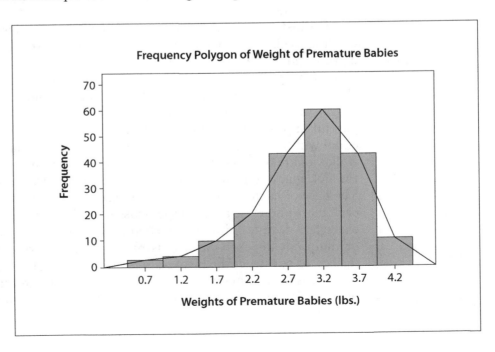

2.3.4 Use of Minitab to Construct Histograms

Many statistics software, including MINITAB, can construct histograms. The histogram can be plotted by using the midpoints or class boundaries as the ticks on the data axis. The use of MINITAB to construct histograms is illustrated here by using the data on the birth weights of premature babies. After the data has been put in c1 in MINITAB, in order to use the midpoints as cutpoints on the bases of the rectangles of the histogram the commands to give on MINITAB are as follows:

```
MTB > name c1 'premature baby-weights';
MTB > histogram c1;
SUBC> midpoints 0.7:4.2/0.5.
```

First the data in c1 was named, and then it was requested that the midpoints be used as tick marks in a SUBCOMMAND of the HISTOGRAM command. It was indicated that the midpoints would go from 0.7 to 4.2 incremented by 0.5. The result is the following histogram:

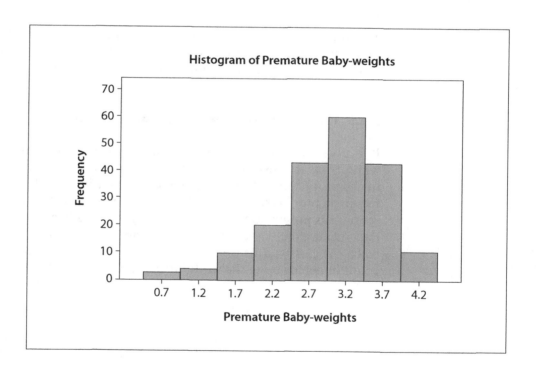

If it is desired to use the class boundaries as the ticks on the histogram, then the following command is given in Minitab:

```
MTB > histogram c1;
SUBC> cutpoints 0.45 0.95 1.45 1.95 2.45 2.95 3.45 3.95 4.45.
```

The result is the next histogram. Sometimes all the ticks may not appear; just double click on the ticks, choose "Position of ticks" in the dialog box, and then enter all the cutpoints.

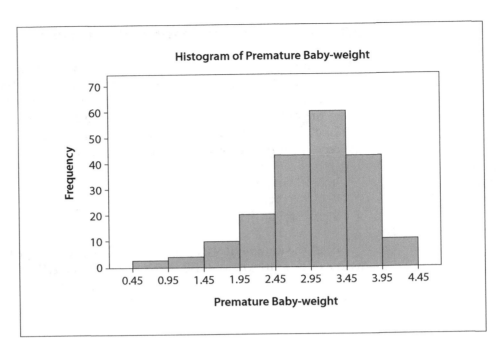

This form of the histogram explains the usefulness of the class boundaries in the frequency table. It shows how the classes dove-tail into one another, allowing us to convert data which is essentially discrete into a continuous distribution.

2.3.5 Histogram with Rectangles of Unequal Bases (Class Widths)

It is possible to construct histograms when one or more classes are of different width from the rest of the classes. In this case, the height of the rectangle still remains proportional to the frequency, but the width is taken into account so that the rectangle produced is also made proportional to the width. Suppose we had a data set such as presented in the next table. We can easily construct the histogram for the data, using exactly the same principles applied in the construction of the histogram when the class widths were equal, but now taking the width of the classes into account in determining the width of the rectangle which represents each class.

Classes	Class Marks or Midpoints (X)	Frequencies (f_i)
0.50–0.90	0.7	2
1.00–1.40	1.2	2
1.50–1.90	1.7	6
2.00–2.40	2.2	20
2.50–2.90	2.7	44
3.00–3.40	3.2	60
3.50–3.90	3.7	45
4.00–4.40	4.2	8
4.50–5.40	4.95	8
		195

One of the classes in the above table is of double the size of the rest of the classes. In accordance with our discussion, we construct the frequency histogram reflecting both the widths and the frequency of each class:

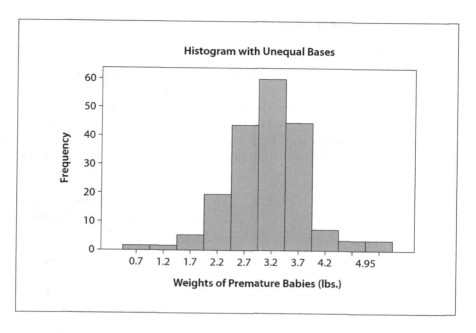

Shapes of Distributions and Construction of a Frequency Polygon

Once we have produced the frequency histogram, it is easy to use the result to construct a frequency polygon for the data. In order to get an idea of the possible shape of the underlying distribution of the data, the frequency polygon is produced by joining the midpoints of the tops of the rectangles which make up the histogram. At both ends they are joined to the assumed midpoints of a class below the lowest class and a class above the highest class. The frequency polygon constructed in this way is supposed to give an idea of the type of statistical distribution from which the data might have arisen. Here again we use the same data for birth weights of babies and extend the histogram we have constructed to produce the frequency polygon:

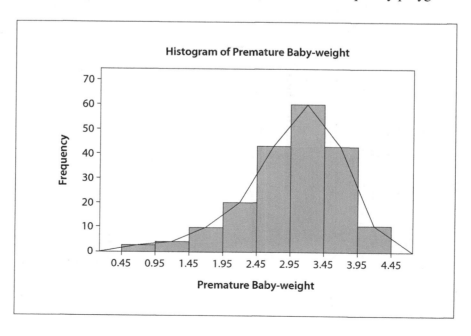

The joining of the midpoints is achieved in MINITAB by adjusting the graph, using graphical tools.

2.3.6 Ogives

Ogives are cumulative frequency curves which have been smoothed. Thus, we can have two different types of ogives: the **"or more"** and the **"less than"** ogives. These two ogives arise from the fact that in constructing cumulative frequency tables we can arrange the data in two different forms which, when graphed, produce one of the two types of ogives we have mentioned. For the data on birth weights of babies, we present both the "or more" and the "less than" frequency tables, and subsequently we shall produce the "or more" and "less than" ogives.

Birth-Weights	"Or more" Cumulative Frequency
0.5 or more	190
1.0 or more	188
1.5 or more	186
2.0 or more	177
2.5 or more	157
3.0 or more	113
3.5 or more	53
4.0 or more	8
4.5 or more	0

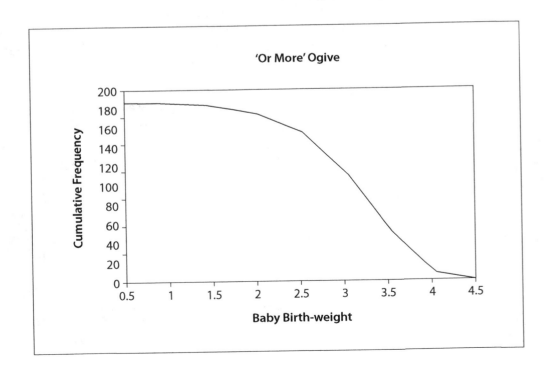

Birth-Weights	"Less Than" Cumulative Frequency
0.5 or less	0
1.0 or less	2
1.5 or less	4
2.0 or less	13
2.5 or less	33
3.0 or less	77
3.5 or less	137
4.0 or less	182
4.5 or less	190

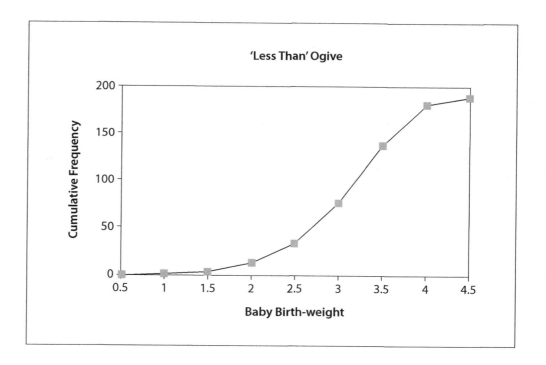

If we had put the data in percentages, we would have been able to construct percentage polygons and ogives as well. This is achieved by converting the frequencies into percentages and proceeding as described earlier. The construction of percentage polygons and ogives is left as an exercise for the interested reader.

Exercises

The following data show the heights of 132 patients treated at Central Hospital over a period:

```
60 59 64 64 59 49 45 48 67 62 66 64 72 59 59
61 44 58 45 55 61 57 70 54 61 53 59 45 62 44
47 58 52 71 76 79 58 90 88 91 62 56 62 62 74
69 71 75 73 81 78 71 61 56 58 77 58 69 64 67
57 71 82 50 69 64 64 51 49 59 43 45 35 55 50
52 46 45 64 44 55 61 48 37 58 55 52 46 43 55
59 56 45 30 54 47 55 40 47 43 61 61 57 43 55
31 49 57 58 64 57 56 62 58 33 45 66 49 51 53
63 59 46 59 59 35 31 40 66 62 49 63
```

2.1 Obtain a frequency table of the data in seven classes of equal width of 10, showing classes, class boundaries, class midpoints, class frequencies, relative frequencies, cumulative frequencies, and relative cumulative frequencies.

2.2 Using a graph sheet, plot a histogram of the data using the seven classes shown in the table of 2.1, with midpoints as the markers on the base of the histogram.

2.3 Plot a frequency polygon for the data.

The following data represent the scores of ninety students in a graduated placement test:

```
120 130 70 80 70 91 89 90 95 67 90 100 30
60 85 57 68 58 52 62 140 150 100 110 20 52
120 130 90 81 80 85 89 92 101 51 44 52 50
57 60 65 78 82 50 91 89 90 95 30 125 110
95 90 110 90 94 65 78 85 95 96 100 110 123
80 85 89 92 104 75 70 76 80 68 60 65 78
82 110 95 96 100 110 105 70 75 81 85 30
```

2.4 Obtain a frequency table of the data in seven classes of equal width 20, showing classes, class boundaries, class midpoints, class frequencies, relative frequencies, cumulative frequencies, and relative cumulative frequencies.

2.5 Using a graph sheet, plot a histogram of the data using the seven classes shown in the table of 2.4, with class boundaries as markers on the base of the histogram.

2.6 Plot a frequency polygon for the data.
The following are the weights of the people in Exercise 1.20:

126 143 171 168 200 156 170 178 176 212 188 186 176
212 155 182 170 167 145 165 178 168 169 192 188 189
190 130 147 175 172 204 160 174 182 180 216 192 190
180 216 159 186 174 171 149 169 182 172 173 196 192
193 194 145 173 170 172 158 172 180 178 214 190 188
178 214 157 184 172 169 147 167 180 170 171 194 190
191 192 170

2.7 Obtain a frequency table of the data in ten classes of equal width 10, showing classes, class boundaries, class midpoints, class frequencies, relative frequencies, cumulative frequencies, and relative cumulative frequencies.

2.8 Using a graph sheet, plot a histogram of the data using the ten classes shown in the table of 2.7, with class boundaries as markers on the base of the histogram.

2.9 Plot a frequency polygon for the data.
The following are the heights of the people in Exercise 1.20

64 68 70 73 70 64 66 64 68 72 56 70 69 71 65
60 72 67 61 64 70 70 65 71 63 69 67 69 71 74
71 65 67 65 69 73 57 71 70 72 66 61 73 68 62
65 71 71 66 72 64 70 68 65 67 71 73 67 65 67
69 67 71 65 72 71 63 67 62 64 69 63 66 72 72
67 60 61 62 63 64

2.10 Obtain a frequency table of the data in seven classes of equal width 3, showing classes, class boundaries, class midpoints, class frequencies, relative frequencies, cumulative frequencies, and relative cumulative frequencies.

2.11 Using a graph sheet, plot a histogram of the data using the ten classes shown in the table of 2.10, with class boundaries as markers on the base of the histogram.

2.12 Plot a frequency polygon for the data.
The following data show the expenses of students on snacks of Exercise 1.1:

21 29 26 30 25 25 25 18 20 27 16 21 22 30 16
11 11 23 28 23 19 22 18 27 11 17 29 20 16 18
20 13 21 17 13 29 13 13 13 18 17 18 20 24 12
14 19 24 19 13 18 21 24 15 17 17 21 10 21 11
12 25 6 17 24 10 19 14 6 21 18 7 10 15 20
 7 25 13 24 15 10 14 9 12 14 14 12 24 16 17
18 17 9 12 17 12 10 8 15 6 6 17 25 13 6
 8 7 17 24 19 17 23 13 6 21 10 20 9 19 21
23 5 18 23 22 20 16 23 22 17 14 11 12 18 7
11 21 22 16 16 23 7 17 14 20 18 23 20 15 20

2.13 By adding a class which includes the maximum, obtain a frequency table of the data in six classes of equal width of 5, showing classes, class boundaries, class midpoints, class frequencies, relative frequencies, cumulative frequencies, and relative cumulative frequencies.

2.14 Using a graph sheet, plot a histogram of the data using the ten classes shown in the table of 2.13, with class boundaries as markers on the base of the histogram.

2.15 Plot a frequency polygon for the data.

The following are the heights of the 120 boys of Exercise 1.18:

```
30 57 40 46 78 59 34 46 40 49 52 54 35 42 61
53 62 44 64 58 55 49 61 66 61 41 61 71 54 46
64 37 77 51 62 34 55 37 46 51 45 50 50 38 44
36 47 36 66 45 44 52 34 50 49 56 34 39 43 44
56 43 32 52 71 67 57 46 47 33 31 47 44 76 61
52 44 68 56 43 42 72 41 42 52 43 42 44 32 69
35 51 32 75 51 48 57 34 37 56 53 69 34 38 51
72 42 46 43 39 50 33 59 71 45 65 80 76 83 81
```

2.16 Obtain a frequency table of the data in eleven classes of equal width of 5, showing classes, class boundaries, class midpoints, class frequencies, relative frequencies, cumulative frequencies, and relative cumulative frequencies.

2.17 Using a graph sheet, plot a histogram of the data using the ten classes shown in the table of 2.16, with class boundaries as markers on the base of the histogram.

2.18 Plot a frequency polygon for the data.

2.19 Obtain a stem-and-leaf plot of the data in Exercises of 2.1.

2.20 Obtain a stem-and-leaf plot of the data in Exercises of 2.4.

2.21 Obtain a stem-and-leaf plot of the data in Exercises of 2.7.

2.22 Obtain a stem-and-leaf plot of the data in Exercises of 2.10.

2.23 Obtain a stem-and-leaf plot of the data in Exercises of 2.13.

2.24 Obtain a stem-and-leaf plot of the data in Exercises of 2.16.

CHAPTER 3

Descriptive Measures for a Data Set

3.0   Introduction

We often need to numerically summarize a data set. This is done by finding the center of the data set. A measure of the center needs to be qualified by providing some other ancillary information, which alongside the measure of the center would make the features of the data set more lucid. To describe a data set by one of the measures of the center called the mean, we also need to know the amount of variation in the data set. For instance, a five number summary incorporates one of the measures of the center called median. The five number summary explains the median. Descriptive statistics measure the center of a data set along with related information. They also measure, among other things, the amount of variation in the data set. In the next sections, we study the measures of the center, variation, and other descriptive measures for a data set. In this section we discuss the descriptive measures for samples and populations.

3.1   The Measures of the Center (Mean, Mode, Median) for a Sample

We have seen that one of the major efforts that statisticians make is to present data in a clearer and more informative way than that suggested by the raw data. One of the ways of achieving this aim is by summarizing data as have been attempted in the frequency tables and in the pictorial presentations discussed earlier. A lot of information can be conveyed about a data set by finding one of the **measures of the center of the data.** There are three measures of the center of a data set. They are the **mean, median,** and **mode.** Learning how to find them is the purpose of this section. By far the most popular of the measures of the center of a data set is the **mean.** The mean or, more precisely, the arithmetic mean is the quantity referred to in common man's parlance as the **average.**

3.1.1 The Sample Mean

As its common name (average) suggests, the mean is the average of a set of observations. This implies that it is the sum of a set of observations divided by the total number of observations in that set. The sample mean is the sum of all values in a sample divided by the number of observations in the sample. The mean is very easily affected by extreme values (small or large), and therefore is not useful for measuring the center of some data sets.

Using the shorthand notation of statisticians to aid computation, we write a set of observations containing n units as: $X_1, X_2, X_3, \ldots, X_n$ so we can write the mean as:

$$\overline{X} = \frac{X_1 + X_2 + X_3 + \ldots + X_n}{n}.$$

The same quantity can be written in a more concise manner as:

$$\overline{X} = \frac{\sum_{i=1}^{n} X_i}{n}.$$

To further elucidate, consider Example 3.1.

■ **EXAMPLE 3.1** Take the sample of heights of students in inches shown below. Let us call the variable (height) X. Height is called a variable because it takes different values for different individuals:

66 56 70 56 57 60 67 67 68 72 71 55 69 54 70

There are $n = 15$ observations in the sample of heights above. The mean is denoted by $\overline{x}$ (pronounced X-bar) and is obtained (for the above data) as:

$$\overline{x} = \frac{66 + 56 + 70 + 56 + 57 + 60 + 67 + 67 + 68 + 72 + 71 + 55 + 69 + 54 + 70}{15}$$

$$= \frac{958}{15} = 63.867 \text{ inches.}$$

In mathematics, the symbol Σ is used for denoting summation. Suppose it was decided to write the data so that:

$$x_1 = 66, x_2 = 56, x_3 = 70, x_4 = 56, x_5 = 57, x_6 = 60, x_7 = 67, x_8 = 67,$$
$$x_9 = 68, x_{10} = 72, x_{11} = 71, x_{12} = 55, x_{13} = 69, x_{14} = 54, x_{15} = 70.$$

Then the mean could have been represented by:

$$\overline{x} = \frac{x_1 + x_2 + x_3 + x_4 + x_5 + x_6 + x_7 + x_8 + x_9 + x_{10} + x_{11} + x_{12} + x_{13} + x_{14} + x_{15}}{15}$$

$$= \frac{958}{15} = 63.867.$$

With this arrangement, the mean can be written as $\overline{x} = \dfrac{\sum\limits_{i=1}^{15} x_i}{15} = \dfrac{958}{15} = 63.867.$

This means that $\sum\limits_{i=1}^{15} x_i = x_1 + x_2 + x_3 + x_4 + x_5 + x_6 + x_7 + x_8 + x_9 + x_{10} + x_{11} + x_{12} + x_{13} + x_{14} + x_{15}.$

Extending the above principle, in general, if there are n observations in a sample then the sample mean is written as:

$$\overline{x} = \frac{x_1 + x_2 + \ldots + x_n}{n} = \frac{\sum\limits_{i=1}^{n} x_i}{n},$$ that is the sum of all the n observations in the sample divided by n.

■ **EXAMPLE 3.2**   Using the above formula, find the mean for the birth weights of the 192 premature babies born in the hospital group from Section 2.2.

Solution   The mean is:

$$\overline{X} = \frac{(0.6 + 0.6 + 1.2 + \ldots + 4.3)}{190} = \frac{575}{192} = 2.99479$$

The mean is so important as a measure of central tendency that parametric inferences about populations are often based on it. Some examples of these inferences will be seen later in the text. The mean is one of the most frequently misused and misinterpreted quantities. It should be noted that the mean, because it is derived from the sum of all observations in a sample, is liable to be influenced by extreme observations since one large observation can increase its size, while a small value can drastically reduce its magnitude. **The mean can be calculated for a sample or for the entire population.** The best way to use the mean is to quote its value along with the **standard deviation** of the data from which it was obtained—the **standard deviation is a measure of the spread of the data, which we shall discuss shortly.**

3.1.2 The Sample Median

The next measure of the center of a data set is the median. The median, as its name implies, is the observation which lies in the middle of the data set. Just as the mean, the median can be calculated for a sample, or for an entire population. Fifty percent of the observations in the data set lie below the median and 50% lie above the median. For discrete observations, it is easy to obtain the median. To obtain the median there is a presupposition that the data is first arranged as an **array** (from the smallest to the largest observation). Starting from the smallest, the observation in the middle of the array is called the median.

Written mathematically, let $X_{(1)}, X_{(2)}, X_{(3)}, \ldots, X_{(n)}$ represent the arrangement of n $(n = 2k + 1$, if n is odd) observations in ascending order of magnitude, then the median is $X_{(k+1)}$. If the set of observations contains n $(n = 2k$, if n is even), then the median is obtained as:

$$\textbf{Median} = \frac{X_{(k)} + X_{(k+1)}}{2}$$

The median of a data set is robust (unaffected) to extreme values, and is therefore more useful for measuring the center than the mean in some situations (it is better to talk about median income than mean income).

■ EXAMPLE 3.3   The following represent the scores of nineteen students in a test whose maximum mark is 100:

55 51 44 48 67 65 66 54 58 77 75 81 64 73 87 80 91 88 64

Obtain the median.

Solution   Rearranging the data in ascending order of magnitude, we obtain:

44 48 51 54 55 58 64 64 65 66 67 73 75 77 80 81 87 88 91

which are the equivalents of $X_{(1)} = 44$, $X_{(2)} = 48$, $X_{(3)} = 51$, ..., $X_{(19)} = 91$ so that the median in this case is $X_{(10)} = 66$. As a measure of central tendency this compares with the mean, which for the 19 scores is 67.79

Suppose that instead of nineteen students we had twenty of them whose scores were:

55 51 44 48 67 65 66 54 58 77 75 81 70 64 73 87 80 91 88 64

in this case, the number of observations is even, so that rearranging we obtain:

44 48 51 54 55 58 64 64 65 66 67 70 73 75 77 80 81 87 88 91
$X_{(1)} = 44$, $X_{(2)} = 48$, $X_{(3)} = 51$, ..., $X_{(20)} = 91$

We can now obtain the median as $M = \left[\dfrac{X_{(10)} + X_{(11)}}{2} \right] = \dfrac{66 + 67}{2} = 66.50$

■ EXAMPLE 3.4   Consider the students' heights from the earlier example, with $n = 15$ observations. In this case **sample size n is odd.**

66 56 70 56 57 60 67 67 68 72 71 55 69 54 70

Solution   We rearrange the data as an array as follows:

54 55 56 56 57 60 66 67 67 68 69 70 70 71 72

$n = 15$, the median is in the middle, $\left(\dfrac{n+1}{2} \right)$th $= \left(\dfrac{15+1}{2} \right)$th $= $ 8th position, which is 67. Median is therefore 67 inches.

■ EXAMPLE 3.5   Recall the sample data we obtained when treating **systematic sampling.** Their heights and weights in inches and pounds are shown in the table below. We obtain the medians for the heights and weights.

Serial Number	Name	Height (ins.)	Weight (lbs.)
4	Brian	63	120
12	Megan	65	145
20	Theresa	70	132
28	Jan	66	137
36	Gordon	58	158
44	Angela	69	164
52	Gail	67	156
60	Cassie	64	158
68	Blake	65	144
76	Ricky	68	151

Solution   There are $n = 10$ observations, so that **sample size is even.** We rearrange the weights and heights in arrays as follows:

Heights: 58 63 64 65 **65 66** 67 68 69 70
Weights: 120 132 137 144 **145 151** 156 158 158 164

The median height is half of the sum of $\left(\frac{n}{2}\right)$th and $\left(\frac{n+2}{2}\right)$th, that is half of the sum of the 5th and the 6th numbers, in this case; median height $= \frac{65 + 66}{2} = 65.5$ inches.

Similarly, for the same sample, the median height $= \frac{145 + 151}{2} = 148$ lbs.

EXAMPLE 3.6   Obtain the median for the birth weights of premature babies discussed previously.

Solution   First we arrange the data in an array:

0.60 0.60 0.70 1.00 1.20 1.20 1.20 1.60 1.60 1.60 1.60
1.70 1.80 1.80 1.90 1.90 2.00 2.00 2.00 2.00 2.00 2.10
2.10 2.20 2.30 2.30 2.30 2.30 2.30 2.40 2.40 2.40 2.40
2.40 2.40 2.50 2.50 2.50 2.50 2.50 2.50 2.60 2.60 2.60
2.60 2.70 2.70 2.70 2.70 2.70 2.70 2.70 2.70 2.80 2.80
2.80 2.80 2.80 2.80 2.80 2.80 2.80 2.80 2.90 2.90 2.90
2.90 2.90 2.90 2.90 2.90 2.90 2.90 2.90 2.90 2.90 2.90
2.90 2.90 3.00 3.00 3.00 3.00 3.00 3.00 3.00 3.00 3.00
3.00 3.00 3.00 3.00 3.00 3.00 3.00 3.00 3.10 3.10 3.10
3.10 3.10 3.10 3.10 3.10 3.20 3.20 3.20 3.20 3.20 3.20
3.20 3.20 3.20 3.20 3.20 3.20 3.20 3.20 3.30 3.30 3.30
3.30 3.30 3.30 3.30 3.30 3.30 3.30 3.30 3.30 3.30 3.40
3.40 3.40 3.40 3.40 3.40 3.40 3.40 3.40 3.50 3.50 3.50
3.50 3.50 3.50 3.50 3.50 3.50 3.60 3.60 3.60 3.60 3.60
3.60 3.60 3.60 3.60 3.60 3.60 3.70 3.70 3.70 3.70 3.70
3.70 3.70 3.80 3.80 3.80 3.80 3.80 3.80 3.80 3.80 3.90
3.90 3.90 3.90 3.90 3.90 3.90 3.90 4.00 4.00 4.00
4.00 4.10 4.10 4.20 4.30

There are 192 babies. The sample size is even, so that $192 = 2(96) \Rightarrow k = 96$. The median will be obtained as the average of the 96th and the 97th observations in the array counting from the least. That is median $= \dfrac{3.00 + 3.10}{2} = 3.05$.

3.1.3 The Sample Mode for Ungrouped Observations

For a discrete data set, the mode is the most frequently occurring observation. It is therefore possible to have one or more modes for a set of data. Thus we say that a data set is bimodal if there are two modes, trimodal if there are three, and so on.

■ EXAMPLE 3.7   Find the mode for the scores of nineteen students.

Solution   44 48 51 54 55 58 64 64 65 66 67 73 75 77 80 81 87 88 91

The number 64 appears twice and no other number has that frequency. So the mode is unique and is 64.

■ EXAMPLE 3.8   For the sample of fifteen student heights (Example 3.4) find the mode.

Solution   66 56 70 56 57 60 67 67 68 72 71 55 69 54 70

Array: 54 55 56 56 57 60 66 67 67 68 69 70 70 71 72

There are three modes: 56, 67, and 70. This distribution is trimodal.

■ EXAMPLE 3.9   Find the mode for the birth weights of premature babies referred to in Example 3.5.

Solution   Putting the data in array helps:

0.60 0.60 0.70 1.00 1.20 1.20 1.20 1.60 1.60 1.60 1.60
1.70 1.80 1.80 1.90 1.90 2.00 2.00 2.00 2.00 2.00 2.10
2.10 2.20 2.30 2.30 2.30 2.30 2.30 2.40 2.40 2.40 2.40
2.40 2.40 2.50 2.50 2.50 2.50 2.50 2.50 2.60 2.60 2.60
2.60 2.70 2.70 2.70 2.70 2.70 2.70 2.70 2.70 2.80 2.80
2.80 2.80 2.80 2.80 2.80 2.80 2.80 2.80 2.90 2.90 2.90
2.90 2.90 2.90 2.90 2.90 2.90 2.90 2.90 2.90 2.90 2.90
2.90 2.90 3.00 3.00 3.00 3.00 3.00 3.00 3.00 3.00 3.00
3.00 3.00 3.00 3.00 3.00 3.00 3.00 3.00 3.10 3.10 3.10
3.10 3.10 3.10 3.10 3.10 3.20 3.20 3.20 3.20 3.20 3.20
3.20 3.20 3.20 3.20 3.20 3.20 3.20 3.20 3.30 3.30 3.30
3.30 3.30 3.30 3.30 3.30 3.30 3.30 3.30 3.30 3.30 3.40
3.40 3.40 3.40 3.40 3.40 3.40 3.40 3.40 3.50 3.50 3.50
3.50 3.50 3.50 3.50 3.50 3.50 3.60 3.60 3.60 3.60 3.60
3.60 3.60 3.60 3.60 3.60 3.60 3.70 3.70 3.70 3.70 3.70
3.70 3.70 3.80 3.80 3.80 3.80 3.80 3.80 3.80 3.80 3.90
3.90 3.90 3.90 3.90 3.90 3.90 3.90 3.90 4.00 4.00 4.00
4.00 4.10 4.10 4.20 4.30

The most frequently occurring value is 3.0, and is the mode.

3.2 The Five Number Summary and Interquartile Range

The five number summary is obtained as the minimum, the first quartile, the median, the third quartile, and the maximum for a data set. Thus to obtain the five number summary, the data is first arranged as an array, and then the minimum, the maximum, the median, and the quartiles are determined. They represent the five number summary.

The Interquartile range is obtained by first finding the quartiles for a data set. It is the difference between the third and first quartile of a data set.

Next, we learn how to obtain the quartiles, and thus the five number summary, and the interquartile range. The median is the second quartile; that is, the observation in the middle of the data set. It is the value in the array of the data set that is in the 50th percentile position. Thus, 50% of the data set is comprised of all values from the minimum observation up to the median. When data are arranged in an array, the first quartile, Q_1, is the 25th percentile of the data set; or 25% of the data set is comprised of all values from the minimum observation up to the Q_1. Similarly, in an array, the third quartile, Q_3, is in the 75th percentile; or 75% of the data set is made up of values from minimum to Q_3.

Between the minimum and Q_1 lie 25% of the data; between Q_1 and the median lie 25% of the data; between the median and Q_3 lie another 25% of the data. The remaining 25% of the data lie between Q_3 and the maximum. Thus, if we consider the array of the data set from minimum to median only, then its median is Q_1. Similarly, if we consider the array of the data set from median to maximum alone, its median is Q_3. The calculation of the quartiles is carried out as if we are calculating the separate medians for the lower and upper halves of the data set. Just like the method used for calculation of the median was different for cases when n, the sample size, was even or odd, we observe the same rules for calculating quartiles when the size of each half of the data set is even or odd.

Next, we show two examples to illustrate the calculation of the quartiles, and hence the interquartile range (IQR).

■ EXAMPLE 3.10 Again, we consider the students' heights from earlier with $n = 15$ observations. In this case **sample size n is odd.**

66 56 70 56 57 60 67 67 68 72 71 55 69 54 70

Find the interquartile range.

Solution We rearrange the data as an array as follows:

54 55 56 56 57 60 66 **67** 67 68 69 70 70 71 72

We already know that the median is 67. To obtain the first quartile, Q_1, we consider only the following data (minimum to median):

54 55 56 56 57 60 66 **67**

We find the median of this half of the data set in order to find the first quartile, or $Q_1 = \dfrac{56 + 57}{2} = 56.5$ for all the sample data.

Similarly, considering only the following portion of the data set (median to maximum):

67 67 68 69 70 70 71 72

We find its median which is also the third quartile $Q_3 = \dfrac{69 + 70}{2} = 69.5$ for all the sample data.

The five number summary is:

■ FIVE NUMBER SUMMARY

Minimum	Q_1	Median	Q_3	Maximum
54	56.5	67	69.5	72

Thus we can obtain the **interquartile range** as IQR $= Q_3 - Q_1 = 69.5 - 56.5 = 13$.

■ EXAMPLE 3.11   **Again,** recall the sample data we obtained when treating **systematic sampling.** Their heights and weights in inches and pounds are shown in the table below. We obtain the medians for the quartiles and the interquartile range for the heights and weights of the sample.

Serial Number	Name	Height (ins.)	Weight (lbs.)
4	Brian	63	120
12	Megan	65	145
20	Theresa	70	132
28	Jan	66	137
36	Gordon	58	158
44	Angela	69	164
52	Gail	67	156
60	Cassie	64	158
68	Blake	65	144
76	Ricky	68	151

Solution   There are $n = 10$ observations, so **sample size is even.** We rearrange the weights and heights in arrays as follows:

Heights: 58 63 64 65 **65 66** 67 68 69 70
Weights: 120 132 137 144 **145 151** 156 158 158 164

The sample size is even, $n = 10$. The medians for height and weight were 65.5 inches, and 148 pounds, respectively. However, since they are not actual observations, they are not considered in calculating the Q_1 and Q_3. Thus we find first quartile for heights by considering only:

58 63 64 65 **65**

whose median is $64 = Q_1$ for all the heights. Similarly, by considering only:

66 67 68 69 70

We find that the median is $68 = Q_3$.
The five number summary is:

▪ FIVE NUMBER SUMMARY

Minimum	Q_1	Median	Q_3	Maximum
58	64	65.5	68	70

Thus, the interquartile range for the entire sample is:

$$\text{IQR} = Q_3 - Q_1 = 68 - 64 = 14.$$

Similarly, we find first quartile for weights by considering only:

120 132 137 144 **145**

whose median is $137 = Q_1$ for all the heights. Similarly, by considering only **151** 156 158 158 164, we find that the median is $158 = Q_3$.
The five number summary is:

▪ FIVE NUMBER SUMMARY

Minimum	Q_1	Median	Q_3	Maximum
120	137	148	158	164

Thus, the interquartile range for the entire sample is:

$$\text{IQR} = Q_3 - Q_1 = 158 - 137 = 21.$$

3.2.1 Boxplots and Outliers

A **boxplot** is a graphical representation of the five number summary. Two or more boxplots, on the same scales for comparison of the data sets or populations from which they arose, can be plotted on the same sheet.

The boxplot can be used to determine whether a data set contains one or more **outliers.** Outliers are observations which are either too large or too small compared to the rest of the data. The potential outlier is determined by calculating the upper and lower limits (sometimes called upper and lower fence). If any observation is found to lie outside the limits, or fence, then it is a potential outlier. The limits are obtained as:

$$\text{Lower limit} = Q_1 - 1.5 \times \text{IQR}$$
$$\text{Upper limit} = Q_3 + 1.5 \times \text{IQR}$$

■ **EXAMPLE 3.12** Again, we consider the students' heights above with n=15 observations.

66 56 70 56 57 60 67 67 68 72 71 55 69 54 70

The five number summary is:

■ **FIVE NUMBER SUMMARY**

Minimum	Q_1	Median	Q_3	Maximum
54	56.5	67	69.5	72

The IQR $= 13$, so that the limits are:

$$\text{Lower limit} = Q_1 - 1.5 \times \text{IQR} = 56.5 - 1.5 \times 13 = 37$$
$$\text{Upper limit} = Q_3 + 1.5 \times \text{IQR} = 69.5 + 1.5 \times 13 = 89$$

We plot the boxplot for the above data set.

■ **EXAMPLE 3.13** The following is a sample of the heights of seedlings of a plant after six weeks:

38 76 51 52 44 44 39 66 49 37 43 55 46 34 37
33 41 34 52 45 53

Obtain the five number summary for the heights, and plot a boxplot for the data.

Solution **ARRAY:**

33 34 34 37 37 38 39 41 43 44 44 45 46 49 51
52 52 53 55 66 76

Minimum $= 33$, $Q_1 = 38$, Median $= 44$, $Q_3 = 52$, Maximum $= 76$.
That is the five number summary.
The boxplot for this sample can be found on the following page.

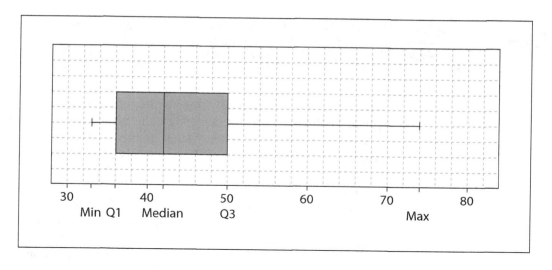

$$\text{IQR} = 52 - 38 = 14$$
$$1.5 \times \text{IQR} = 21$$

The fences are $F1 = Q_1 - 1.5\text{IQR} = 38 - 1.5(14) = 17$ and $F2 = Q3 + 1.5(14) = 73$. From this, the value of 76 will be flagged as a possible outlier.

We can redraw the same boxplot with fences to show that there is an outlier by marking the fences in the boxplot. This exercise is left for the reader.

Multiple Boxplots

We can present multiple boxplots on the same graph sheet, using the same scales, in order to compare them. We demonstrate this in the next example:

■ **EXAMPLE 3.14**   The degrading times (in days) were sampled for two types of biomasses (A and B).

A: 55 47 45 77 70 52 64 63 61 74 46 79 60 66 48 51 56 63 67
B: 63 68 64 57 60 64 69 56 65 59 66 61 63 69 65 72 70 58 55

Solution   **Array A:** 45 46 47 48 51 52 55 56 60 61 63 63 64 66 67 70 74 77 79
Min = 45 Q_1 = 51.5   Median = 61 Q_3 = 66.5   Max = 79
Array B: 55 56 57 58 59 60 61 63 63 64 64 65 65 66 68 69 69 70 72
Min = 55 Q_1 = 59.5   Median = 64 Q_3 = 67   Max = 72

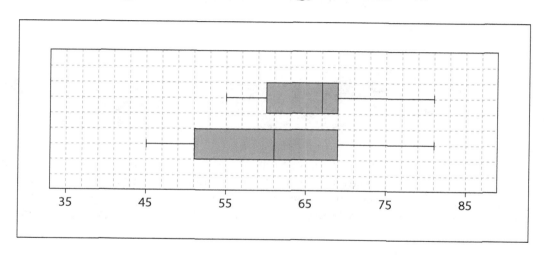

3.3   Measure of Dispersion or Spread

When analyzing a data set, one factor which should be of interest to us is the spread of the data. In summarizing data, we are often interested in speaking about a measure of central tendency such as mean. We often say things like "the mean height of the students in XYZ University is 1.6m," "the mean annual income of the males in Anytown, aged thirty and above, is $19,500.00," and so on. We obviously obtained this information using a larger data set. The summary is, however, incomplete until we provide a measure of spread. The measure of spread helps us have an idea of what percentages of students in XYZ University would have heights in ranges below or above the mean of 1.6m. Similarly, if information on spread is made available, we can also estimate percentages of incomes of males in Anytown with incomes in the ranges below or above the mean income of $19,500.00. A number of measures are used for describing the spread of a data set. Among these are the **range, semi-interquartile range, mean absolute deviation,** and **the standard deviation.** The range is simply the difference between the largest and the smallest observation in a sample or population. By far the most popular of the measures of spread is the **standard deviation.** This is mainly because it lends itself to further statistical manipulation and constitutes part of the tools used in influential areas of statistics such as **statistical inference.** We shall encounter the use of this tool later on in the text.

3.3.1 The Semi-Interquartile Range

The semi-interquartile range is often used as a measure of dispersion. It is defined as:

$$Q = \frac{Q_3 - Q_1}{2},$$

where Q_1 and Q_3 are the first and third quartiles. Sometimes, the interquartile range $(Q_1 - Q_3)$ is used as a measure of dispersion. However, the semi-interquartile range is more popularly used.

3.4   The Standard Deviation

Just as we have pointed out for the mean and median, we can obtain the standard deviation of a complete population as well as for a sample. To differentiate between these, it is conventional to represent the population standard deviation by the Greek letter, σ, while the sample standard deviation is represented by s. The standard deviation is a measure of the spread in a set of data and is often expressed mathematically as the square root of the variance of that data set. If we denote a random variable by X with population mean, μ, and its population size as N, we can obtain the variance of the N observations in the population and the standard deviation as:

$$\sigma^2 = \frac{1}{N} \sum_{i=1}^{N} (X_i - \mu)^2; \sigma = \sqrt{\left[\frac{1}{N} \sum_{i=1}^{N} (X_i - \mu)^2 \right]};$$

while the variance and standard deviation from a sample of size n drawn from the population are obtained as:

$$s^2 = \frac{1}{n-1} \sum_{i=1}^{n} (x_i - \overline{x})^2 \text{ and } s = \sqrt{\left(\frac{1}{n-1} \sum_{i=1}^{n} (x_i - \overline{x})^2 \right)};$$

where $\overline{x}$ is the sample mean. The factor $n - 1$, is used in the sample variance in order to obtain an unbiased estimator for σ^2, the population variance. Sometimes, we find that standard deviation is defined as:

$$s^2 = \frac{1}{n} \sum_{i=1}^{n} (x_i - \overline{x})^2.$$

This sample standard deviation is biased and is not recommended for use. However, we recognize that as n becomes larger, the difference between the reciprocals of n and $n - 1$ becomes very small, and that both formulae are equivalent.

The above formulae are used for discrete data.

3.4.1 Calculation of Standard Deviation.

In the next examples, we demonstrate the application of this formula for calculating the variances of some data sets.

■ **EXAMPLE 3.15**   Recall the Scores:

55 51 44 48 67 65 66 54 58 77 75 81 64 73 87 80 91 88 64

They are the scores of nineteen students in a test used previously in calculating the median. We shall now calculate the standard deviation of their scores.

Since we already know that:

$\Sigma x = 1288$, then $\overline{x} = \dfrac{1288}{19} = 67.78947$, we may adopt, for ease of calculation,

the procedure outlined on the following page.

i	x	x − x̄	(x − x̄)²
1	55	−12.7895	163.5706
2	51	−16.7895	281.8864
3	44	−23.7895	565.9391
4	48	−19.7895	391.6233
5	67	−0.7895	0.623269
6	65	−2.7895	7.781163
7	66	−1.7895	3.202216
8	54	−13.7895	190.1496
9	58	−9.7895	95.8338
10	77	9.2105	84.8338
11	75	7.2105	51.99169
12	81	13.2105	174.518
13	64	−3.7895	14.36011
14	73	5.2105	27.14958
15	87	19.2105	369.0443
16	80	12.2105	149.097
17	91	23.2105	538.7285
18	88	20.2105	408.4654
19	64	−3.7895	14.36011

From the above table we see that $\sum_{i=1}^{19} (x - \bar{x})^2 = 3533.158$ and we can calculate the variance as:

$$s^2 = \frac{1}{19 - 1} \sum_{i=1}^{19} (x - \bar{x})^2 = \frac{1}{18}(3533.158) = 196.2865;$$

and the standard deviation is:

$$s = \sqrt{\frac{1}{19 - 1} \sum_{i=1}^{19} (x - \bar{x})^2} = \sqrt{\frac{1}{18}(3533.158)} = \sqrt{196.2865} = 14.01023.$$

The method used above is one way of calculating the standard deviation and some people find it simpler to use. However, others find it tedious and more time consuming than other forms of the formula.

For instance, a little algebraic manipulation of the formula for the variance of the sample of ungrouped observations shows that it is possible to write that formula as:

$$s^2 = \frac{1}{n-1}\left[\sum_{i=1}^{n} x_i^2 - n\bar{x}^2\right] = \frac{\sum_{i=1}^{n} x_i^2 - n\bar{x}^2}{n-1}.$$

We can calculate the standard deviation of a data set of observations using this form of the formula. If we use this form of the formula, all we need to know is the number of observations n, the square of each observation, x_i^2, and the sample mean, $\bar{x}$, in order to calculate the standard deviation or variance.

We illustrate the use of this form of the formula for the same set of observations; namely, the scores of the nineteen students in the test.

All we need is a table containing the observations and their squares:

x	x^2
55	3025
51	2601
44	1936
48	2304
67	4489
65	4225
66	4356
54	2916
58	3364
77	5929
75	5625
81	6561
64	4096
73	5329
87	7569
80	6400
91	8281
88	7744
64	4096
1288	**90846**

Since $n = 19$, and from the table we know that $\bar{x} = \dfrac{1288}{19} = 67.78947$, and $\displaystyle\sum_{i=1}^{19} x_i^2 = 90846$, we can calculate the variance s^2 and hence, the standard deviation, s, as follows:

$$s^2 = \frac{1}{19 - 1}[90846 - 19(67.78947)^2] = 196.2871$$

and hence,

$$s = \sqrt{196.2871} = 14.0102.$$

3.4.2 Mean Absolute Deviation

Mean absolute deviation measures the average of the absolute deviations of all observations from the mean of all the observations in the sample. Mathematically, the formula for this measure of dispersion of n discrete observations in a sample is:

$$\text{Mean Absolute Deviation} = \frac{\displaystyle\sum_{i=1}^{n} |x_i - \bar{x}|}{n}.$$

The x_i represents the individual observations, while $|x_i - \bar{x}|$ is the absolute value of the difference between each observation and the sample mean.

In the next example, we illustrate the calculation of mean absolute deviation:

■ **EXAMPLE 3.16**   The following data represent the intervals between the arrivals of twenty-two customers at a service point (measured to the nearest minute):

1, 7, 2, 9, 5, 1, 6, 4, 3, 8, 8, 8, 4, 9, 5, 3, 5, 6, 9, 10, 10, 7, 5

Find the mean absolute deviation for the intervals.

Solution   This is the calculation of the Mean Absolute Deviation (discrete case)
We note that:

$$\bar{x} = \frac{135}{23} = 5.8696$$

i	x_i	$|x_i - \bar{x}|$
1	1	4.8696
2	7	1.1304
3	2	3.8696
4	9	3.1304
5	5	0.8696

Continued.

i	x_i	$\lvert x_i - \bar{x} \rvert$
6	1	4.8696
7	6	0.1304
8	4	1.8696
9	3	2.8696
10	8	2.1304
11	8	2.1304
12	8	2.1304
13	4	1.8696
14	9	3.1304
15	5	0.8696
16	3	2.8696
17	5	0.8696
18	6	0.1304
19	9	3.1304
20	10	4.1304
21	10	4.1304
22	7	1.1304
23	5	0.8696
		53.1304

From the above table, we see that: $\sum_{i=1}^{23} \lvert x_i - \bar{x} \rvert = 53.1304$, hence,

$$\text{M.A.D.} = \frac{\sum_{i=1}^{23} \lvert x_i - \bar{x} \rvert}{23} = \frac{53.1304}{23} = 2.31.$$

Illustration of Appropriate Measure of Central Tendency: Application to Students Data

The data on the Fashion Marketing students which we used earlier in studying pictorial presentation of data can also be analyzed by calculating some measures of central tendency for the data. For four sets of data it is inappropriate for us to use the mean as a measure of central tendency. These are the data on eye color, shoe sizes, dress sizes, and hair color. In these cases, it is better to speak of the mode as the most appropriate measure of central tendency. The mean of shoe sizes would not make sense, neither would the mean of dress sizes if they turned out to be sizes that are not manufactured, such as size 4.7 for shoes and 11.75 for dresses. Although the average shoe size and the average dress size in these cases turned out to be 6.5 and 14, it is still safer to use the mode as the measure of central tendency as it provides information on the sizes that are most likely to be acquired by these students, or demanded in the shops by real customers. This piece of information is of interest to a fashion designer or seller. Similarly, information on mode of hair color might be of interest to manufacturers and marketers of hair products, and hair-dressing salons. As for the sets of data on height, span, and age, we can freely use the mean. The mean height for the students happened to be 5.58 feet with a standard deviation of 0.113. The mean and standard deviations for span and age are (19.35, 1.909) and (20.35, 2.121) respectively. The mean absolute deviation (or average deviation) for height, span, and age are 0.08, 1.35, and 1.5 respectively. The verification of these calculations is left as an exercise to the reader.

As a project, verify the solutions presented above.

3.5 | Relationships Between Measures of Dispersion

It has been found empirically that for moderately skewed distributions:

$$\text{Mean absolute deviation} = \frac{4}{5} \text{ (standard deviation), and}$$

$$\text{Semi-interquartile range} = \frac{2}{3} \text{ (standard deviation)}$$

These relationships were derived from the empirical knowledge that for the normal distribution, the semi-interquartile range is equal to 0.7979 (standard deviation) and mean absolute deviation is 0.6745 (standard deviation).

3.6 | Descriptive Measures for a Population

A descriptive measure obtained from a sample is called a **statistic.**

A descriptive measure obtained from a population is called a **parameter.**

The number of subjects under observation in a population is usually denoted by N. Out of N, we usually draw a sample of size n. That is why the mean for the sample is written as:

$$\bar{x} = \frac{\sum\limits_{i=1}^{n} x_i}{n}.$$

The mean of the population is obtained by using all the N members of the population and the formula is written as:

$$\mu = \frac{\sum\limits_{i=1}^{N} x_i}{N}.$$

Similarly, the population variance and standard deviation are obtained as:

$$\sigma^2 = \frac{\sum_{i=1}^{N}(x_i - \mu)^2}{N} = \frac{\sum_{i=1}^{N}x_i^2}{N} - \mu^2 \text{ and } \sigma = \sqrt{\frac{\sum_{i=1}^{N}(x_i - \mu)^2}{N}} = \sqrt{\frac{\sum_{i=1}^{N}x_i^2}{N} - \mu^2}.$$

Thus, $\bar{x}$ is a statistic, while μ is a parameter.

■ **EXAMPLE 3.17**   The white blood corpuscle measurements for all the fifty patients in a hospital are:

8.90 5.25 5.95 10.05 6.50 9.45 5.45 5.30 7.65
6.40 5.15 16.60 5.75 5.55 11.60 6.85 6.65 5.90
9.30 6.30 8.55 6.40 10.80 4.85 4.90 7.85 8.75
7.70 5.30 6.50 6.90 4.55 7.10 8.00 4.70 4.40
9.75 4.90 10.75 11.00 4.05 9.05 5.05 6.40 4.05
7.60 4.95 9.60 3.00 9.10

Obtain the mean, variance, and standard deviation for this population.

$$\sum_{i=1}^{50} x_i = 357.05; \mu = \frac{357.05}{50} = 7.141$$

$$\sum_{i=1}^{50} x_i^2 = 2857.91; \sigma^2 = \frac{2857.91}{50} - 7.141^2 = 6.164319$$

$$\sigma = \sqrt{\frac{2857.91}{50} - 7.141^2} = 2.482805$$

3.7	**Analyzing and Finding Descriptive Statistics for Grouped Data**

The rules for analyzing grouped data are different from the methods described earlier for ungrouped data. Here we treat grouped data, and the methods used. Sometimes data comes to us already grouped without the benefit of us being able to refer to the raw data. We can use the techniques described herein to analyze the data and gain some further insight into the data set that was grouped.

3.7.1 Sample Mean for Grouped Data

If data are supplied to us in the form of a grouped frequency table, so that we have no access to the raw observations from which the group frequency table was formed, then we can find the sample mean as we discuss in this section. For such grouped data, the class midpoints are very useful for calculating the mean for the data. Obtaining the mean in this situation adopts the method usually employed for calculating weighted mean, which is:

$$\overline{X} = \frac{\sum_{i=1}^{k} w_i X_i}{\sum_{i=1}^{k} w_i} \quad \text{where } w_i \text{ is the weight associated with } X_i.$$

The formula for the calculation of the mean is similar, and is of the form:

$$\overline{X} = \frac{\sum_{i=1}^{k} f_i X_i}{\sum_{i=1}^{k} f_i}.$$

Here, it is assumed that there are k different groups or classes in the frequency table, f_i is the frequency of the ith class while X_i is the class mark or midpoint for the ith class. The f_i acts as the weight for each class.

■ **EXAMPLE 3.18** We can illustrate the use of this formula by calculating the mean of the grouped form of the birth weights of the 192 babies.

Class-Limits	f_i	Class Marks or Midpoints (X_i)	$f_i X_i$
0.50–0.90	3	0.7	2.1
1.00–1.40	4	1.2	4.8
1.50–1.90	9	1.7	15.3
2.00–2.40	19	2.2	41.8
2.50–2.90	44	2.7	118.8
3.00–3.40	61	3.2	195.2
3.50–3.90	44	3.7	162.8
4.00–4.40	8	4.2	33.6
	192		574.4

Solution It can be seen that $\sum_{i=1}^{8} f_i X_i = 574.4$; $\sum_{i=1}^{8} f_i = 192$ and $\overline{X} = \frac{574.4}{192} = 2.991667$.

In this case, the mean calculated from the grouped data happens to be fairly close to the value obtained using discrete observations. It is not always so.

3.7.2 Sample Median for Grouped Data

When data is arranged in a grouped frequency table the median can still be obtained, but it requires a different formula. The median obtained this way would be an approximate value. As much as possible, obtain the median using ungrouped frequency data. If the only information available is in the form of a grouped frequency table, then use the method described below. The formula required is of the form:

$$\text{Median} = L_{\text{med}} + \frac{\left[\left(\dfrac{\sum_{i=1}^{k} f_i}{2}\right) - (\Sigma f_i)_{\text{bmc}}\right] \times c}{f_{\text{med}}},$$

where:

c = class width of the median class; that is, the difference between the upper and lower class boundaries of the median class

f_{med} = frequency of the median class

$(\Sigma f_i)_{\text{bmc}}$ = sum of frequencies of all classes below the median class

$\dfrac{\sum_{i=1}^{k} f_i}{2}$ = half the sum of the frequencies for all the classes

L_{med} = lower class boundary of the median class

The **median class** is the class whose cumulative frequency exceeds half of the sum of all frequencies for the first time, counting from the lowest class.

■ **EXAMPLE 3.19**   Use the grouped frequency data for the premature babies (Table 3.1) to obtain the median.

■ **TABLE 3.1**

Classes or Groups	Class Boundaries	Frequency (f_i)	Cumulative Frequency
0.50–0.90	0.45 and under 0.95	3	3
1.00–1.40	0.95 and under 1.45	4	7
1.50–1.90	1.45 and under 1.95	9	16
2.00–2.40	1.95 and under 2.45	19	35
2.50–2.90	2.45 and under 2.95	44	79
3.00–3.40	2.95 and under 3.45	61	140
3.50–3.90	3.45 and under 3.95	44	184
4.00–4.40	3.95 and under 4.45	8	192

Solution   The median class is the sixth class starting from the lowest class. This is because the cumulative frequency exceeds half the sum of all frequencies (which is 96) for the first time in this class. From this class we obtain all we need to calculate the median:

$$L_{\text{med}} = 2.95; f_{\text{med}} = 61; c = 3.45 - 2.95 = 0.50; \frac{1}{2}\sum_{i=1}^{8} f_i = 192/2 = 96$$

$$(\Sigma f_i)_{\text{bmc}} = 3 + 4 + 9 + 19 + 44 = 79$$

Hence

$$\text{Median} = 2.95 + \left[\frac{96 - 79}{61}\right] \times 0.50 = 3.089344$$

The median, unlike the mean, does not feature often in further statistical analysis. It finds most application only in some aspects of non-parametric inference and quality control. Interestingly, the median of this data set obtained by grouping the data (3.089344) compares very well with that obtained by using the ungrouped frequency (3.05).

3.7.3 The Sample Mode for Grouped Data

For grouped data it is easy to identify the modal class; that is, the class with the highest frequency. To obtain the mode itself, we need to construct the histogram of the grouped frequency table.

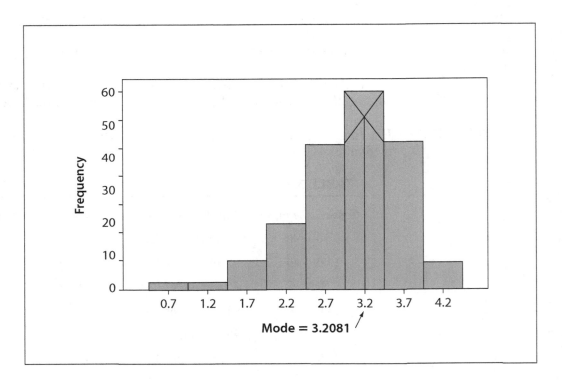

Calculation of Mode by Using Co-Ordinate Geometry
The construction of the histogram shown, as well as the subsequent pictorial identification of the mode, can be avoided. This is achieved by using coordinate geometry to solve for the mode. In this case, the data in the frequency table is used to construct the equations of the two lines which join the top vertices of the rectangle representing the frequency of the modal class in the histogram, and the equations are solved simultaneously to obtain the x co-ordinate of the point at which they intersect, which would be the mode. The ordinates of the vertices of the two straight lines are used. In the above example, we can easily obtain the vertices. These are obtained by reference to the frequency table.

■ **EXAMPLE 3.20**   Find the mode for the grouped data on premature babies.

Solution   The line joining the top right vertex of the rectangle has a positive slope and passes through the points (61, 3.45) and (44, 2.95) (44 is the frequency of the class immediately below the modal class). Hence the equation of this line is:

$$\frac{y - 61}{x - 3.45} = \frac{61 - 44}{3.45 - 2.95}$$
$$\Rightarrow y = 34x - 56.3.$$

Similarly, the other line, which joins the left vertex of the rectangle representing the frequency of the modal class, passes through the points (60, 2.95) and (44, 3.45) so that its equation is:

$$\frac{y - 45}{x - 3.45} = \frac{61 - 45}{3.45 - 2.95}$$
$$\Rightarrow y = -34x + 161.3.$$

The mode occurs at the point where both lines cross. The value of x at that point is the mode.

By subtracting the second equation from the first, we obtain:

$$68x = 217.6 \Rightarrow x = 3.2.$$

The mode is 3.2. It turns out to be the class midpoint in this case.

3.7.4 Percentile for Grouped Data

Consider the calculation of the **first quartile.** We amend the formula for the median to produce a new formula for the first quartile as follows:

$$Q_1 = L_{Q_1} + \frac{\left[\left(\dfrac{\sum_{i=1}^{k} f_i}{4} \right) - (\Sigma f_i)_{bQ_1 c} \right] \times c}{f_{Q_1}}.$$

L_{Q_1} is the lower class boundary of the first quartile class; f_{Q_1} is the frequency of the first quartile class; $(\Sigma f_i)_{bQ_1 c}$ is the sum of the frequencies of all classes below the first quartile class while c is the width of the first quartile class.

Similarly, the formula for the third quartile, Q_3, is obtained from modification of the formula for the median as:

$$Q_3 = L_{Q_3} + \frac{\left[\left(\dfrac{3 \sum_{i=1}^{k} f_i}{4} \right) - (\Sigma f_i)_{bQ_3 c} \right] \times c}{f_{Q_3}}.$$

L_{Q_3} is the lower class boundary of the third quartile class; f_{Q_3} is the frequency of the third quartile class; $(\Sigma f_i)_{bQ_3c}$ is the sum of the frequencies of all classes below the third quartile class while c is the width of the first quartile class.

■ **EXAMPLE 3.21**   Obtain the first and third quartiles of the grouped frequency table of the data on premature babies.

Solution   We can calculate the first and third quartiles for the grouped frequency form of the birth weights of the 192 babies as follows:

Classes or Groups	Class Boundaries	Frequency (f_i)	Cumulative Frequency
0.50–0.90	0.45 and under 0.95	3	3
1.00–1.40	0.95 and under 1.45	4	7
1.50–1.90	1.45 and under 1.95	9	16
2.00–2.40	1.95 and under 2.45	19	35
2.50–2.90	2.45 and under 2.95	44	79
3.00–3.40	2.95 and under 3.45	61	140
3.50–3.90	3.45 and under 3.95	44	184
4.00–4.40	3.95 and under 4.45	8	192

For Q_1

$L_{Q_1} = 2.45$; $(\Sigma f_i)_{bQ_1c} = 35$; $f_{Q_1} = 44$; $c = 0.5$;

$$Q_1 = 2.45 + \frac{\left(\dfrac{192}{4} - 35\right) \times 0.5}{44} = 2.5977$$

For Q_3

$L_{Q_3} = 3.45$; $(\Sigma f_i)_{bQ_3c} = 140$; $f_{Q3} = 44$; $c = 0.5$;

$$Q_3 = 3.45 + \frac{\left(\dfrac{3(192)}{4} - 140\right) \times 0.5}{44} = 3.4955$$

Calculation of Other Percentiles for Grouped Data

It is possible to calculate any percentile of the grouped data by simply modifying the formula for the median using information related to the percentile, such as percentile class, frequency of percentile class, sum of frequencies below percentile class, and class width of percentile class.

3.7.5 Calculating the Variance and Standard Deviation for Grouped Data

Often, we have the data in grouped frequency tables so that different forms of the formulae are required for calculating the variance and standard deviation. In such situations the versions of the formulae used for population variance and standard deviation is:

$$\sigma^2 = \frac{1}{\sum\limits_{i=1}^{k} f_i} \sum_{i=1}^{k} f_i(X_i - \mu)^2; \quad \sigma = \sqrt{\left[\frac{1}{\sum\limits_{i=1}^{k} f_i} \sum_{i=1}^{k} f_i(X_i - \mu)^2\right]}.$$

In the above formulae, it is assumed that there are k classes whose midpoints are denoted by X_i and their frequencies denoted by f_i ($i = 1, 2, \ldots, k$). When a sample is taken from the population, and divided into k classes, with class-marks X_i and class frequencies, f_i, the formulae to use are:

$$s^2 = \frac{1}{\left(\sum\limits_{i=1}^{k} f_i - 1\right)} \sum_{i=1}^{k} f_i(x_i - \overline{x})^2; \quad s = \sqrt{\left[\frac{1}{\left(\sum\limits_{i=1}^{k} f_i - 1\right)} \sum_{i=1}^{k} f_i(x_i - \overline{x})^2\right]}.$$

In this case, $\overline{X} = \dfrac{\sum\limits_{i=1}^{k} f_i X_i}{\sum\limits_{i=1}^{k} f_i}$, as defined earlier in the text. This is usually obtained from the sample.

The standard deviation for grouped data is sometimes not as accurate as it should be, due to grouping. It is therefore necessary in some situations to apply some correction to the variance to compensate for this error. A quantity, $\dfrac{c^2}{12}$, called the **Sheppard's correction** is subtracted from the variance obtained for grouped data. Here c is the class interval size. Statisticians are not totally agreed as to whether the Sheppard's correction is a good idea because it often over-corrects and introduces another error in the process of correcting one. It is therefore advisable that the data and the resulting variance be examined properly before deciding to apply the correction. In situations where it is suitable to apply the correction, we have the formulae as follows:

$$s^2 = \frac{1}{\left(\sum\limits_{i=1}^{k} f_i - 1\right)} \sum_{i=1}^{k} f_i(x_i - \overline{x})^2 - \frac{c^2}{12}; \quad s = \sqrt{\left[\frac{1}{\left(\sum\limits_{i=1}^{k} f_i - 1\right)} \sum_{i=1}^{k} f_i(x_i - \overline{x})^2 - \frac{c^2}{12}\right]}$$

EXAMPLE 3.22 Using the grouped frequency table for the grouped data on birth weights of babies, calculate the variance and standard deviation of the weights of the babies.

Solution To calculate the standard deviation for the grouped data on birth weights of babies, we may set up the table as follows:

f_i	x_i	$x_i - \bar{x}$	$(x_i - \bar{x})^2$	$f_i(x_i - \bar{x})^2$
3	0.7	−2.2917	5.25189	15.7557
4	1.2	−1.7917	3.21019	12.8408
9	1.7	−1.2917	1.66849	15.0164
19	2.2	−0.7917	0.62679	11.909
44	2.7	−0.2917	0.08509	3.74391
61	3.2	0.2083	0.04339	2.64672
44	3.7	0.7083	0.50169	22.0743
8	4.2	1.2083	1.45999	11.6799
				95.6667

In this table we used the value of $\bar{x} = 2.9917$ (See Example 3.18) which was calculated previously. We can see from the table that $\sum_{i=1}^{8} f_i(x_i - \bar{x})^2 = 95.6667$, so that $s^2 = \dfrac{1}{192 - 1}(95.6667) = 0.500873$, and consequently, $s = \sqrt{0.500873} = 0.70773$.

Just as in the case when data are ungrouped, a little algebra will show that the variance s^2 can be written as:

$$s^2 = \frac{1}{\left(\sum_{i=1}^{k} f_i - 1\right)}\left[\sum_{i=1}^{k} f_i x_i^2 - \left(\sum_{i=1}^{k} f_i\right)\bar{x}^2\right].$$

To use this form of the formula for calculating the variance for the grouped data on birth weights of babies, we need a simpler table which is presented below:

f_i	x_i	x_i^2	$f_i x_i^2$
3	0.7	0.49	1.47
4	1.2	1.44	5.76
9	1.7	2.89	26.01
19	2.2	4.84	91.96
44	2.7	7.29	320.76
61	3.2	10.24	624.64
44	3.7	13.69	602.36
8	4.2	17.64	141.12
			1814.08

Since we know that $\bar{x} = 2.9917$ from our previous calculation, and from the above table that:

$$\sum_{i=1}^{8} f_i x_i^2 = 1814.08, \text{ then:}$$

$$s^2 = \frac{1}{(192 - 1)}[1814.08 - 192(2.9917)^2] = 0.50067$$

and:

$$s = \sqrt{0.50067} = 0.70758.$$

It is clear that the two methods are equivalent for calculating standard deviation whether data is grouped or ungrouped. However, from the point of view of ease of computation, one may find one of these methods is simpler and therefore should use it.

If we decided to apply the **Sheppard's correction,** then we have:

$$s^2 = \frac{1}{(192 - 1)}[1814.08 - 192(2.9917)^2] - \frac{(0.5)^2}{12} = 0.500563$$

which, in this case, is no better an approximation of the true variance than that calculated by using the actual data.

If the data set for which the variance or standard deviation is being calculated is normally distributed, we can easily establish a pattern of spread of data once we know the mean and standard deviation. Even in cases where it is known that observations are not normally distributed, the distribution of the data may approach normality if the sample size is large. We shall discuss the implications of the standard deviation when data is normally distributed when we treat the normal distribution.

3.7.6 Mean Absolute Deviation for Grouped Data

When data is presented in grouped frequency form, the formula is changed accordingly and the resulting form is:

$$\text{Mean absolute deviation (M.A.D.)} = \frac{\sum_{i=1}^{k} f_i |x_i - \bar{x}|}{\sum_{i=1}^{k} f_i}$$

In the above formula, $|x_i - \bar{x}|$ means the absolute value of the difference between the class marks, x_i, and the sample mean, $\bar{x}$.

■ **EXAMPLE 3.23** The weights gained by 520 babies (in hundreds of grams) who were fed for a specified period with a formula manufactured by XYZ Company are shown in the frequency table below. Calculate the mean absolute deviation of the weight gains.

Class Boundaries (Weights gained)	Frequency
6.0–6.5	41
6.5–7.0	46
7.0–7.5	50
7.5–8.0	52
8.0–8.5	60
8.5–9.0	64
9.0–9.5	65
9.5–10.0	70
10.0–10.5	72

Solution

Class Boundaries	Frequency	Class Mark or Class Midpoint (x_i)	$f_i x_i$	$x_i - \bar{x}$	$f|x_i - \bar{x}|$
6.0–6.5	41	6.25	256.25	−2.2288	91.3808
6.5–7.0	46	6.75	310.5	−1.7288	79.5248
7.0–7.5	50	7.25	362.5	−1.2288	61.44
7.5–8.0	52	7.75	403	−0.7288	37.8976
8.0–8.5	60	8.25	495	−0.2288	13.728
8.5–9.0	64	8.75	560	0.2712	17.3568
9.0–9.5	65	9.25	601.25	0.7712	50.128
9.5–10.0	70	9.75	682.5	1.2712	88.984
10.0–10.5	72	10.25	738	1.7712	127.5264
			4409		**567.9664**

From the above table we see that $\bar{x} = \dfrac{\sum\limits_{i=1}^{9} f_i x_i}{\sum\limits_{i=1}^{9} f_i} = \dfrac{4409}{520} = 8.4788$ and consequently,

$$\text{M.A.D.} = \dfrac{\sum\limits_{i=1}^{9} f_i |x_i - \bar{x}|}{\sum\limits_{i=1}^{9} f_i} = \dfrac{567.9664}{520} = 1.0922$$

3.8   Minitab Solutions

We can use MINITAB to obtain all the descriptive statistics that we discussed in the above sections and carry out a boxplot for each data set. However, MINITAB uses a different approach from the methods described above to calculate the quantities Q_1, Median, and Q_3. So it should not be a surprise if values obtained by MINITAB vary from what we get by calculation using our methods. Their boxplots, based on the three quantities, would most likely look different. There is just a principled disagreement between a vast majority of professors of statistics and MINITAB on the method they use. Here is the use of MINITAB to find the descriptive statistics for the data for the heights of fifteen students. We can use this method or use the pull-down menus in MINITAB to achieve the same result.

```
MTB > set c1
DATA> 66 56 70 56 57 60 67 67 68 72 71 55 69 54 70
DATA> end
MTB > name c1 'heights'
MTB > describe c1
```

3.8.1 Descriptive Statistics: Heights

```
Variable N N* Mean SE Mean StDev Minimum  Q1  Median  Q3  Maximum
 heights 15 0 63.87  1.72   6.66   54.00  56.00  67.00 70.00  72.00
```

To obtain the boxplot we simply use the following command:

```
MTB > boxplot c1
```

3.8.2 Boxplot of Heights

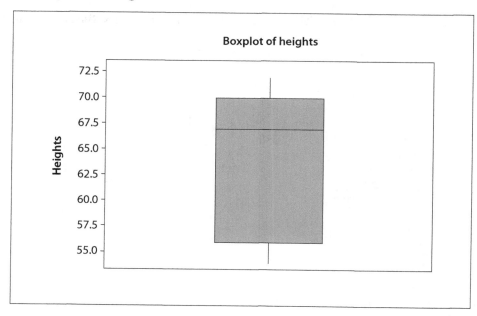

As anticipated from the calculations of the limits above, there is no outlier indicated in the boxplot, which would normally be depicted with an asterisk (*) in the plot because no observation is outside the calculated limits.

```
MTB> name c3 'seed-heights'
MTB > Boxplot 'seed-heights';
SUBC> IQRBox;
SUBC> Outlier.
```

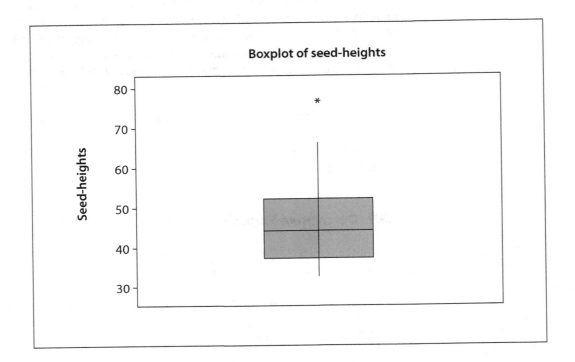

To use MINITAB to do multiple boxplots, the commands to enter in MINITAB are:

```
MTB> name c1 'A' c2 'B'
MTB > Boxplot c1 c2;
SUBC> Overlay;
SUBC> Scale 2;
SUBC> LDisplay 1 1 1 1;
SUBC> IQRBox;
SUBC> Outlier.
```

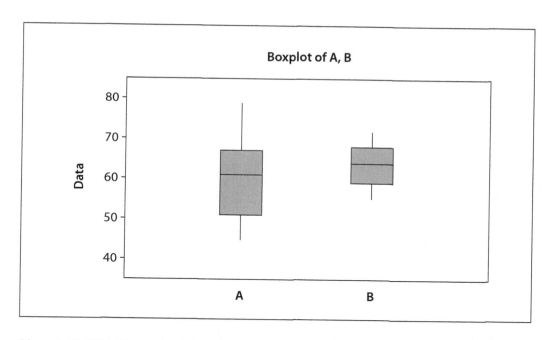

Since MINITAB's method for calculating Q_1, Median, and Q_3, some members of the five number summary, is different from methods used by many professors, it is better to find out what method your professor prefers and use it.

■ **EXAMPLE 3.24**   Use MINITAB to obtain the descriptive statistics for data of Example 3.16.

Solution

```
MTB > set c1
DATA> 8.90 5.25  5.95 10.05  6.50  9.45 5.45 5.30 7.65
DATA> 6.40 5.15 16.60  5.75  5.55 11.60 6.85 6.65 5.90
DATA> 9.30 6.30  8.55  6.40 10.80  4.85 4.90 7.85 8.75
DATA> 7.70 5.30  6.50  6.90  4.55  7.10 8.00 4.70 4.40
DATA> 9.75 4.90 10.75 11.00  4.05  9.05 5.05 6.40 4.05
DATA> 7.60 4.95  9.60  3.00  9.10
DATA> end
MTB > describe c1
```

3.8.3 Descriptive Statistics: c1

Variable	N	N*	Mean	SE Mean	StDev	Minimum	Q1	Median	Q3	Maximum
c1	50	0	7.141	0.355	2.508	3.000	5.225	6.500	8.938	16.600

In the output, we have sample size N, number of missing values, N*, the mean, standard error of the mean $(s/\sqrt{n})$, standard deviation, minimum, Q_1, median, Q_3, and maximum.

Exercises

For questions 1–9, refer to the following table:

Sex	Age	Red BCC	Hemoglobin	Platelets
F	30	4.39	13.5	224
M	29	5.18	15.9	264.5
M	21	4.88	13.9	360
M	33	5.94	15.85	384.5
F	24	4.48	13.55	364.5
F	28	3.58	10.95	468
M	57	4.4	12.5	171
M	28	4.88	15.6	328.5
F	48	4.22	13.05	323.5
F	60	4.36	12.2	306.5
F	20	4.59	14.25	264.5
F	19	3.63	11.5	233
F	53	4.54	14	254.5
M	34	4.92	14.05	267
F	26	3.92	11.4	463
M	18	5.44	14.65	238
M	26	5.07	15.4	251
F	25	4.24	12.8	282.5
F	48	4.21	14.3	307.5
M	24	4.8	14.7	321.5
F	48	5.06	14.9	360.5
M	28	5.14	15.2	282.5
F	52	4.71	14.9	315
F	32	4.23	11.5	284
F	46	4.68	14.45	259.5
M	20	5.99	14.05	291.5
F	23	3.83	12.2	259.5
M	26	4.82	15.35	164
M	27	4.57	14.15	199.5

Continued.

Sex	Age	Red BCC	Hemoglobin	Platelets
M	46	4.59	14.5	220
F	37	3.95	11.1	369
M	27	4.77	14.85	245
M	25	5.12	14.6	266
M	55	5.37	15.6	369
M	27	5.17	15.45	210.5
M	53	4.97	14.6	234
F	50	3.87	12.75	471
M	18	4.8	15.2	244.5
M	36	5.05	15.65	365.5
M	52	5.11	15.3	265
F	29	4.2	12.75	198
F	31	4.07	13.1	390
F	17	3.74	11.4	269.5
F	24	4.72	13.3	344.5
F	37	4.63	13.7	386.5
F	48	4.63	15.1	256
F	23	4.3	13.55	226
M	45	5.13	15.95	225
F	21	4.07	11.9	259
F	22	4.6	14.2	271.5

3.1   Obtain the five number summary for age in the above table.

3.2   Obtain the five number summary for Red BCC in the above table.

3.3   Obtain the five number summary for Hemoglobin in the above table.

3.4   Obtain the five number summary for Platelets in the above table.

3.5   Find the means of (a) age (b) Red BCC (c) Hemoglobin (d) Platelets.

3.6   Find the variance and standard deviation of the first fifteen ages in the table above.

3.7   Find variance and standard deviation of the first twelve Red BCC from the above table.

3.8   Find the variance and standard deviation for the first thirteen of Hemoglobin.

3.9   Find the variance and standard deviation for the first fourteen Platelets.

3.10 Obtain the mean, variance, and standard deviation of (a) data set A and (b) data set B of Example 3.13.

3.11   The following are the scores on an aptitude test of candidates that interviewed for a job: 42 57 33 47 32 35 33 39 47 49 42 41 38 43 48 45 61 65 64 56 52 58 53 54 65 74 80 73 85 82 78 89
Obtain the mean, variance, and standard deviation of the data.

3.12   For the values: 42 61 65 64 39 52 58 33 54 49 57 80 61 85 47 78 74
Obtain the variance and standard deviation.

3.13   Select a 1 in 11 systematic sample from the data in the table presented in Exercise 1.20 (page 30) starting from row 7, column 10 of table of random numbers (Appendix I, page 417). Find the mean, variance, and standard deviation for (a) the weight, and (b) the height of the sample.

3.14   For the solution (sample obtained) for Exercise 1.18, find the five number summary.

3.15   For the solution (sample obtained) for Exercise 1.18, find the variance and standard deviation.

3.16   For the solution (sample obtained) for Exercise 1.19, find the five number summary.

3.17   For the solution (sample obtained) for Exercise 1.19, find the variance and standard deviation.

3.18   For the solution (sample obtained) for Exercise 1.20, find the five number summary.

3.19   For the solution (sample obtained) for Exercise 1.20, find the variance and standard deviation.

3.20   For the solution (sample obtained) for Exercise 1.21, find the five number summary.

3.21   For the solution (sample obtained) for Exercise 1.21, find the variance and standard deviation.

Exercises on Grouped Data

Classes	Frequency
0–4	2
5–9	5
10–14	7
15–19	9
20–24	12
25–29	14
29–34	10
35–39	8
40–44	5
45–49	3

3.22   Obtain the median for the data set.

3.23   Obtain the first quartile for the data set.

3.24 Obtain the third quartile for the data set.

3.25 Obtain the mean, variance, and standard deviation for the data set.

Classes	Frequency
26–30	4
31–35	6
36–40	10
41–45	12
46–50	20
51–55	24
56–60	15
61–65	10
66–70	7

3.26 Obtain the median for the data set.

3.27 Obtain the first quartile for the data set.

3.28 Obtain the third quartile for the data set.

3.29 Obtain the mean, variance, and standard deviation for the data set.

Classes	Frequency
20–29	3
30–39	5
40–49	8
50–59	15
60–69	13
70–79	10
80–89	8
90–99	4

3.30 Obtain the median for the data set.

3.31 Obtain the first quartile for the data set.

3.32 Obtain the third quartile for the data set.

3.33 Obtain the mean, variance, and standard deviation for the data set.

3.34 Obtain the mean absolute deviation for the data set.

CHAPTER 4

Elements of Probability and Related Topics

4.0 Introduction

Set theoretical notions are similar to the mathematical concepts that play a very important part in understanding some aspects of modern Probability theory, especially those related to events, event space, and relationships between events. In summary, modern probability theory regards possible outcomes (results or responses) of a statistical experiment as points within a space S. The space S may be n-dimensional, where $n \geq 1$ (n, integer). Each of the outcomes of an experiment is associated with a non-negative number which is the chance, or probability, of that outcome; and the sum of the probabilities for all outcomes in the space S is unity. To prepare a solid basis for our study of probability theory, we think it is wise at this stage to outline some of the basic set theoretical ideas which could help our understanding of probability before we proceed to formally study this topic.

4.1 Sets

A **set** is a collection of well-defined and identifiable **objects.** A set is composed of members or elements. The elements which belong to a set may either be listed or defined by a rule. It is conventional to enclose elements of a set in "curly" brackets. A set may contain other sets or it may be possible to define other sets which are entirely contained in a set. Such sets are then **subsets** of the original sets from which they are defined.

■ **EXAMPLE 4.1** A: the set of all integers from 1 to 5; this is conventionally written as
$A = \{1, 2, 3, 4, 5\}$.
B: all professors, which can be written as $B = \{$all university professors$\}$.

Subsets can be obtained from A and B above as follows:

B_1 = {all professors of mathematics}
B_2 = {all male professors}
B_3 = {all female professors of mathematics}
A_1 = {2, 3, 4}
A_2 = {1, 3, 5}
A_3 = {4}

A_1, A_2, and A_3 are proper subsets of the set A. They are called proper subsets of A because all the elements in each of them are contained within A. Likewise, B_1, B_2, and B_3 are proper subsets of B.

The set of all possible elements which could be considered is called the **universal set.** For instance, A, A_1, A_2, and A_3 are all subsets of the universal set S = {all real numbers}. Also, B, B_1, B_2, and B_3 are subsets of the universal set {all professors}. We denote the universal set here as S instead of U which is conventional, so as to use S for sample space when we discuss probability without causing any confusion.

The set which contains nothing is called the **empty set,** and denoted by ϕ = { }. This empty set is a subset of all sets.

We say that two sets are equal if every member of one set is a member of the other set.

It is conventional to denote the number of members in a given set, D, by $n(D)$. We can see that $n(A) = 5, n(A_1) = 3, n(A_2) = 3, n(A_3) = 1$.

4.1.1 Use of Venn Diagrams

It is possible to use **Venn or Euler Diagrams** to show the relationships between sets. We shall use this method whenever we wish to illustrate the relationships between sets. The example below illustrates the relationship between three sets: A, B, and C, and the universal set, S.

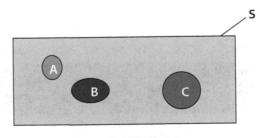

■ **FIGURE 4.1**

The above Venn diagram illustrates the universal set with three subsets: A, B, and C.

■ **EXAMPLE 4.2**   Suppose that in CTK School we choose a sample of 120 students who are studying mathematics and discover that thirty of them scored an A in Mathematics.

Here S = set of all students in CTK School
  M = sample of 120 students studying mathematics, so that $n(M) = 120$

A = set of students who scored an A in mathematics; $n(A)$ = 30. Clearly, M is a subset of S, and A is a subset of M, as well as of S. We can illustrate this by using some Venn diagrams.

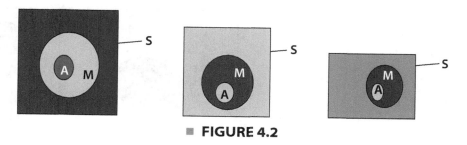

■ **FIGURE 4.2**

4.1.2 The Complement of a Set

The complement of a set B, usually denoted as either B' or $\overline{B}$ is the set comprising all the elements of the universal set which are not elements of B. Hence, we can show the complement of B by Venn diagram as follows:

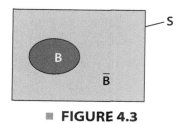

■ **FIGURE 4.3**

The grey area shows the complement of B in S.

■ **EXAMPLE 4.3**　If in a container of 130 eggs, thirty-five are damaged during transportation, then:

$n(S)$ = number of eggs = 130; $n(B)$ = number of eggs damaged during transportation. We can obtain $n(\overline{B}) = n(S) - n(B) = 130 - 35 = 95$.

The two pictures in Figure 4.4 illustrate the complements of a set when two subsets within a universal set are involved.

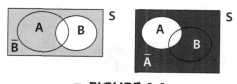

■ **FIGURE 4.4**

All the shaded area in the first picture shows the complement of the set B. Similarly, the shaded area in the second picture shows the complement of the set A.

4.1.3 The Intersection of Sets

The two sets A and B are said to intersect if some elements which belong to the set A are also members of the set B. The common elements are referred to as the intersection; the set of the common elements is denoted as $A \cap B$. The following Venn diagrams illustrate the intersection of two sets, A and B, and the intersection of three sets, A, B, and C.

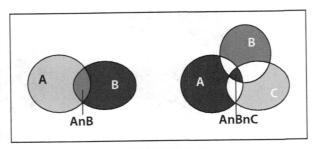

■ **FIGURE 4.5**

In this case, there are some elements which are common to A and B so that $n(A \cap B) > 0$. Similarly, there are elements which are common to A, B, and C in the illustration which leads to the conclusion that $n(A \cap B \cap C) > 0$.

If two sets do not intersect, they are said to be disjoint. If two sets are disjoint, such as A and B in Figure 4.6, then their intersection is the empty set. That is written in mathematical notation as:

$$A \cap B = \phi \Rightarrow n(A \cap B) = 0.$$

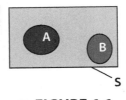

■ **FIGURE 4.6**

Similarly, we know that B and $\overline{B}$ are disjoint (as illustrated in Figure 4.3, see p. 99), as illustrated in the second picture below, then:

$$B \cap \overline{B} = \phi \text{ and } n(B \cap \overline{B}) = 0.$$

We can illustrate intersections of sets by using intersections arising between two sets and their complements with the aid of the three pictures in Figure 4.7:

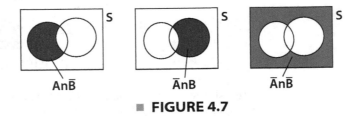

■ **FIGURE 4.7**

4.1.4 Union of Sets

The union of two sets A and B is the set which comprises all elements which belong to the sets A and B. It is a listing of all elements which belong to both sets without repeating the elements which are common to both. From this definition it should be clear that A and B are subsets of the union of A and B. Conventionally, the union of two sets A and B is written as $A \cup B$.

EXAMPLE 4.4   Consider the two sets, A and B, whose members are listed below:

$A = \{5, 7, 10, 16, 26, 31, 44, 49\}; B = \{6, 8, 10, 16, 33, 44, 48\ 50\};$

Find the union of the two sets.

Solution   We simply list all the elements of A and B without repeating any, so that:

$A \cup B = \{5, 6, 7, 8, 10, 16, 26, 31, 33, 44, 48, 49, 50\}$
$n(A) = 8, n(B) = 8, n(A \cup B) = 13$

We can see that although 10, 16, and 44 appear in both A and B, they are listed only once in their union. It is therefore not right to say that $n(A) + n(B) = n(A \cup B)$. This is only true if $A \cap B = \phi$. We also note that the union of any set and its complement will give us the universal set. For instance, $(B \cup \overline{B}) = S$. Similarly, $(A \cup \overline{A}) = S$. Using Venn diagrams, we can illustrate the union of two sets A and B as shown below in the picture. Here the shaded part represents the union of A and B.

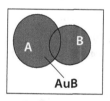

FIGURE 4.8

It is clear from the above discussion that:

$n(A \cup B) = n(A) + n(B) - n(A \cap B)$.

Using our earlier example, we see that $n(A) = 8, n(B) = 8$ and that $n(A \cap B) = 3$ so that $n(A \cup B) = 8 + 8 - 3 = 13$. Again we further illustrate other forms of unions between two sets and their complements using the Venn diagrams below. The shaded sections of each Venn diagram represents the union of sets being illustrated:

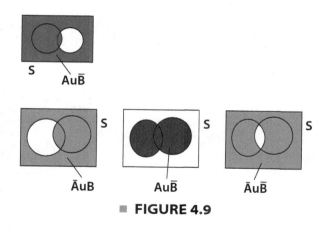

FIGURE 4.9

■ **EXAMPLE 4.5**   A box of spare parts made by CBH Co. is known to contain forty items, of which ten are defective. Fourteen of the items are wrongly labeled, of which six are also defective. Find how many items are in the sets containing (i) parts which are defective, wrongly labeled, or both (ii) parts which are not defective, or are wrongly labeled, or are both not defective or wrongly labeled (iii) parts which are defective or are not wrongly labeled or are both defective and not wrongly labeled (iv) the parts which are both not defective or not wrongly labeled or are both not defective and not wrongly labeled.

Solution   On inspection, we can see that the Venn diagrams in Figure 4.9 do depict the sets required in the above questions. Let A stand for defectives and B for wrongly labeled. Then the following sets represent the four sets required in (i)–(iv) of Example 4.5:

(i) $A \cup B$ (ii) $\overline{A} \cup B$ (iii) $A \cup \overline{B}$ and (iv) $\overline{A} \cup \overline{B}$

We note that $n(S) = 40$, $n(A) = 10$, $n(B) = 14$, $n(\overline{A}) = 30$, $n(\overline{B}) = 26$, $n(A \cap B) = 6$. We find the number of parts in the sets as follows:

(i) $n(A \cup B) = n(A) + n(B) - n(A \cap B) = 10 + 14 - 6 = 18$
(ii) $n(\overline{A} \cup B) = n(\overline{A}) + n(B) - n(\overline{A} \cap B) = 30 + 14 - 8 = 36$
   [note that $n(\overline{A} \cap B) = 8$]
(iii) $n(A \cup \overline{B}) = n(A) + n(\overline{B}) - n(A \cap \overline{B}) = 10 + 26 - 4 = 32$
   [note that $n(A \cap \overline{B}) = 4$]
(iv) $n(\overline{A} \cup \overline{B}) = n(\overline{A}) + n(\overline{B}) - n(\overline{A} \cap \overline{B}) = 30 + 26 - 22 = 34$
   [note that $n(\overline{A} \cap \overline{B}) = 22$]

4.1.5 Exhaustive Sets

Sometimes, it is possible to divide the universal set, S into k ($k > 1$) sets, A_1, $A_2, \ldots, A_k$ so that each element in S belongs to one and only one of the sets. The k sets are then said to be **exhaustive.**

■ **EXAMPLE 4.6**   A bag contains seventy-five balls of which twenty-four are blue, eighteen are brown, sixteen are red, and seventeen are green. We can constitute the contents of this bag into sets based on the colors of the balls, and therefore show that they are exhaustive.
   Let A_1 = set of brown balls, A_2 = set of red balls, A_3 = set of blue balls, and A_4 = set of green balls. We can present the sets in form of a Venn diagram as follows:

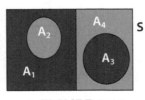

■ **FIGURE 4.10**

Here, $k = 4$; and the sets A_i account for all the elements of S, and therefore form an exhaustive set. We can also see that any two complementary sets form an exhaustive set.

4.1.6 Laws of Set Operations

Based on what we have studied so far about sets, namely the operations of union, intersections, and complement, it is possible to identify a number of set theoretical laws which govern the manipulations of sets. We will simply state these laws here in their different forms. For the laws, we assume that A, B, and C are three sets and that S is the universal set while ϕ is the empty set and A' is the complement of A.

Associative Laws
    (i) $(A \cup B) \cup C = A \cup (B \cup C)$ (ii) $(A \cap B) \cap C = A \cap (B \cap C)$

Commutative Laws
    (i) $A \cap B = B \cap A$ (ii) $A \cup B = B \cup A$

Distributive Laws
    (i) $A \cup (B \cap C) = (A \cup B) \cap (A \cup C)$
    (ii) $A \cap (B \cup C) = (A \cap B) \cup (A \cap C)$

De Morgan's Laws
    (i) $(A \cup B)' = A' \cap B'$ (ii) $(A \cap B)' = A' \cup B'$

Idempotent Laws
    (i) $A \cap A = A$ (ii) $A \cup A = A$

Identity Laws
    (i) $A \cup S = S$ (ii) $A \cup \phi = A$ (iii) $A \cap \phi = \phi$ (iv) $A \cap S = A$

Complement Laws
    (i) $A \cap A' = \phi$ (ii) $A \cup A' = S$ (iii) $(A')' = A$ (iv) $S' = \phi$
    (v) $\phi' = S$

We shall use some of these laws when using set theoretical ideas in our study of probability.

4.2 Basic Concepts of Probability

In everyday language we often use the words chance, odds, or likelihood in our speech, without necessarily thinking that we are talking about probability. There is a good chance that he will get the job. The odds are in favor of his horse winning the race. The likelihood that it will rain today is high. There is a good chance that you will have an accident if you are texting while driving. All of these are definitely references to probability. Somehow, we know how "good," "high," or "low" the probability is, without necessarily knowing how to quantify it.

It is very hard to provide a single definition of probability which fully encapsulates all aspects of the concept. **Mathematical probability** can be defined as the measure or estimate of the degree of confidence one has that an event will occur, measured on a scale from zero to one. Probability is the mathematical likelihood that an event will occur. Probability is a number expressing the likelihood of the occurrence of a given event. It is a fraction which expresses the proportion of times an event will occur in a given number of trials or experiments which have other possible outcomes. It is zero if it is impossible for the event to occur or one if it is certain that the event will occur.

A statistical **experiment** generally leads to one or more outcomes; usually the outcomes of an experiment are all known and can be specified in advance. Also, in an experiment the occurrence of each outcome depends on chance. An experiment can be made up of one or a series of trials. Typical experiments performed by statisticians include flipping a coin or rolling a die, among others.

Flipping a coin is referred to as a **trial**. An experiment can be made up of a single flip of a coin or several flips of a coin. In flipping a coin the list of all possible outcomes is a set S made up of two simple events so that $S = \{head, tail\}$. The occurrence of either head or tail is uncertain and depends on chance. Similarly, in rolling a die, the list of possible outcomes is a set $S = \{1, 2, 3, 4, 5, 6\}$.

In real life the statistician will regard a pregnancy, for instance, as a trial whose outcome is either a girl or a boy. If a study is about the number of children in a family, the choice of each family is a trial whose outcome is an integer value representing the number of children in the family.

To establish the probability that an event will occur, we can approach it from the **classical, empirical (relative frequency),** or **subjective** points of view. In each case, probability is established in relation to an **event.**

We shall present the concepts of probability based on these three approaches. Before then, let us connect **set theory** with probability by defining an event.

4.2.1 What is an Event?

Any phenomenon that can occur, or take place, is **an event** denoted by a capital letter. An event is a set, or collection of points, within the universal set, S — the **space** of all possible outcomes of an experiment. An event is therefore a subset of the entire event space, S. It is clear that an event is therefore subject to the theories and set operations which we have discussed in the previous section. If we have two sets E_1 and E_2, we can have a union of events, $E_1 \cup E_2$; the intersection of two or more events such as $E_1 \cap E_2$. ϕ is now used to represent the **impossible event (whereas it was the empty set in set theory),** so that if we say that $E_1 \cap E_2 \cap E_3 = \phi$, then the probability of joint occurrence of the events E_1, E_2, and E_3 is impossible, which is usually written as $P\{E_1 \cap E_2 \cap E_3\} = 0$. Similarly, if we say that $E_1 \cup E_2 \cup E_3 = S$, then we have defined the **sure event** because, $P\{E_1 \cup E_2 \cup E_3\} = 1.0$.

For instance, in the experiment of flipping a coin, the space of all possible outcomes $S = \{head, tail\}$. $A = \{head\}$ can be defined as an event under this experiment. If we roll a die with six faces, we can define the events $A = \{1, 2, 4, 6\}$ and $B = \{2, 3, 6\}$ with the event space of $S = \{1, 2, 3, 4, 5, 6\}$ for this experiment. Events are tied up with experiments or observations.

A simple event: A simple event is an event which cannot be broken down into two or more events. The total number of outcomes represented in an event space, S, is the number of simple events in that event space.

4.2.2 Random Variable

This definition of an event enables us to define a random variable. Elementarily, we can conceptualize a random variable as the numeric result of performing a haphazard or nondeterministic experiment. In higher treatment of probability a random variable is defined as a function. **A random variable is a real-valued function whose domain is the sample (or event) space S.** For instance, if we throw a die, there are six faces, 1, 2, 3, 4, 5, or 6, which can turn up. We can now define a random variable X based on that experiment. We define a random variable X, which is the value of the face of the die which turns up. With that we can assign some values to the event that $X \leq 3$ and can calculate $P\{X \leq 3\}$ on the basis of our knowledge of the probability of each face turning up.

4.2.3 Classical Definition of Probability

Classical probability approach assigns probability to an event by using the number of outcomes favorable to the event, divided by the number of all possible outcomes of the experiment. One does this by assuming all outcomes are equally likely. This method fails if all the outcomes are not equally likely.

Definition of Classical probability: if a random experiment (a process with an uncertain outcome) can lead to n mutually exclusive and equally likely outcomes, also if x of these outcomes have an attribute E, then the probability of E is the fraction x/n.

For instance, if we assume that in flipping a coin a head is equally likely as a tail, suppose we then flip three of these coins. The equally likely outcomes of this experiment are {HHH, HHT, HTH, HTT, THH, THT, TTH, TTT}. Then by classical probability definition each of these outcomes has probability of 1/8. This method of calculating probability breaks down if we know that there is a way that flipping the coin is being influenced in favor of a head or tail outcome.

Apart from difficulties associated with ensuring "equally-likeliness" of outcomes, the classical approach relies heavily on the assumption that the number of mutually exclusive and equally likely outcomes, n, is finite. It has no theory for when n is infinite.

If an event E can occur in k ways out of n possible equally likely outcomes of an experiment, then we say that the probability of occurrence of E is:

$$p = P(E) = \frac{k}{n}$$

We also say that the probability of non-occurrence of E or failure of E, is actually the probability of the complement of E, and is obtained as:

$$q = P(E') = \frac{n - k}{n} = 1 - \frac{k}{n}$$

Probability of an event is a non-negative number which must lie between zero and unity, and so also is the probability of its complement. As you would have observed in our definition of events and the explanations that followed, S and ϕ are extremes of an event space and respectively have the maximum and minimum probabilities assignable, 1 and 0.

Probability Expressed as Odds

Sometimes, the word **odds** is used to represent probability. In this case, it must be made clear what is being said. If we say that the odds in favor of the E occurring is $p{:}q$ (p to q), and the odds against its occurrence is $q{:}p$ (q to p) it means that the probability of occurrence of E is $\dfrac{p}{p+q}$, while the probability of its non-occurrence is $\dfrac{q}{p+q}$, which is p and q respectively because $p + q = 1$. We can also say that odds in favor of an event E is 3:2 while the odds against it is 2:3. This simply means that the probability of occurrence of such an event is 3/5 and the probability of its failure is 2/5.

4.2.4 Relative Frequency Definition of Probability

The major argument against the classical definition of probability is that it uses the words "equally likely" which makes the definition of probability circular, since it gives the impression that we are depending on the concept of probability to define probability. For this reason, it is proposed that statistically, we should define probability from the relative frequency point of view. The **empirical probability** of an event is regarded as the relative frequency of the occurrence of that event when the number of trials or observations is large. The actual probability is then taken as the limit of this empirical probability as the number of trials or observations approaches infinity. While this definition is useful in practice, mathematically the actual limiting number which is the probability of an event may not exist. For this reason, probability is often defined axiomatically.

Empirical probability: Determination of empirical probability is based on direct observations or experimental results. The empirical probability of an event is therefore an estimate of the probability or chance that the event will happen, based on how often the event occurs after collecting data. The estimate gets better if the experiment involves a large number of trials, or if the number of observations from which the probability is deduced is large.

Practically, the **Empirical probability** approach looks at the past, counts the number of outcomes favorable to the event, and divides it by the total number of trials, in order to determine the probability that the event will occur in the future. It is defined as the proportion of outcomes observed in a sample of past trials which favors an event.

For instance, if we observe one hundred pregnancies, and find that the number of girls born is fifty-one, then we determine that the probability of giving birth to a girl is $\dfrac{51}{100}$. This is an empirical probability of the event that a girl is born.

Suppose we look into the medical records and find that out of 10,000 patients who presented with a certain adverse medical condition, 879 recovered without medical intervention. Based on the Empirical probability approach, we will conclude that the probability of future recovery from the condition without treatment would be $\dfrac{879}{10000}$.

Clearly empirical probability uses the relative frequency idea that we have encountered while grouping data into classes in chapter two.

Subjective Definition of Probability
Subjective probability is assigned on the basis of any information available.

An event on the other hand is tied up with **experiments.** Probability is tied up with the concepts of events and experiments. We will discuss the concept of experiments.

We now look at some examples which are applications of the definitions or concepts we have treated up to this point.

■ EXAMPLE 4.7   In a typical experiment, a coin is flipped. Obtain the event space and simple events for the experiment.

Event space $S = \{H, T\}$. There are two simple events: (i) H, and (ii) T

■ EXAMPLE 4.8   If a die is rolled, obtain the event space and all the simple events.

The event space for this experiment is $S = \{1, 2, 3, 4, 5, 6\}$. We have six simple events made up of the values of the six faces: $\{1\}, \{2\}, \{3\}, \{4\}, \{5\}, \text{and } \{6\}$.

■ EXAMPLE 4.9   A coin is flipped four times repeatedly. Use a tree diagram to obtain the event space S for this experiment.

Solution

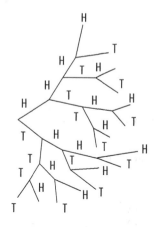

$$S = \{\text{HHHH, HHHT, HHTH, HHTT, HTHH, HTHT, HTTH, HTTT,}$$
$$\text{THHH, THHT, THTH, THTT, TTHH, TTHT, TTTH, TTTT}\}$$

Out of a typical event space we can define other events that are not necessarily simple. Out of the experiment on flipping the coin, we can define events $E_1 = H$ and $E_2 = T$. For instance, in Example 4.8 above on the rolling of a die, we can define an event E_1 that faces 1, 2, or 5 occur. Then $E_1 = \{1, 2, 5\}$. Another event is E_2 so that sum of any two faces is 9 or more. In this case, $E_2 = \{4, 5, 6\}$.

■ EXAMPLE 4.10   Out of event space S for Example 4.9, obtain (i) event A, that no more than two heads occur; (ii) event B, that exactly three heads occur; and (iii) event C, that two or three heads occur.

Solution $A = \{\text{HHTT, HTHT, HTTH, HTTT, THHT, THTH, THTT, TTHH, TTHT,}$
$\text{TTTH, TTTT}\}$

$$B = \{HHHT, HHTH, HTHH, THHH\}$$
$$C = \{HHHT, HHTH, HHTT, HTHH, HTHT, HTTH, THHH, THHT,$$
$$THTH, TTHH\}$$

When outcomes of an experiment are equally likely, then the probabilities of events are obtained as relative frequencies; that is, the number of elements in an event, divided by the total number of elements in the event space. We show how it works:

Consider the experiment of **Example 4.7.** If all outcomes are equally likely, $P(H) = P(E_1) = 1/2$; $P(T) = P(E_2) = 1/2$; $P(S) = P(H) + P(T) = 1/2 + 1/2 = 1.0$. Similarly for the experiment of **Example 4.8;**

If all faces are equally likely, then $p(1) = p(2) = \ldots = p(6) = 1/6$
$$\Rightarrow P(S) = p(1) + p(2) + p(3) + p(4) + p(5) + p(6) = 1.0$$

$$P(E_1) = p(1) + p(2) + p(3) = \frac{1}{6} + \frac{1}{6} + \frac{1}{6} = \frac{1}{2}$$

$$P(E_2) = p(4) + p(5) + p(6) = \frac{1}{2}$$

■ **EXAMPLE 4.11**   Consider the following set of workers:

Linda	Marcus	Phil
Ray	Carol	Sue

Out of the above set of workers, we want to randomly choose two, without replacement, to attend a conference.

(a) What is the probability that Sue and Marcus are chosen?
(b) What is the probability that Sue is in the two chosen?

Solution   **This is an experiment in which all the possible pairs that can be chosen are:**

$S = \{$(Linda, Marcus), (Linda, Phil), (Linda, Ray), (Linda, Carol),
(Linda, Sue), (Marcus, Phil), (Marcus, Ray), (Marcus, Carol),
(Marcus, Sue), (Phil, Ray), (Phil, Carol), (Phil, Sue), (Ray, Carol),
(Ray, Sue), (Carol, Sue)$\}$

Each pair represents a simple event in the event space.

(a) There are only fifteen possible pairs of workers that can be chosen. Each pair is equally likely, so (Sue, Marcus) or (Marcus, Sue) and similar pairs are the same and so appear only once. Therefore, the probability of choosing a pair is 1/15.
(b) There are five possible pairs which contain Sue. Therefore, the probability that Sue would be included is 5/15 or 1/3.

■ **EXAMPLE 4.12**   Similarly, consider the frequency table for the weights of 192 babies discussed previously when we treated frequency tables. We obtained the relative frequencies for each class as:

$$\text{relative frequency} = \frac{\text{frequency}}{n}$$

Where n = number in the sample. In this case $n = 192$, and the relative frequency for each class is the probability of the event that the weight of the baby would lie in between the lower and upper limits of that class.

Class Limits	Class Boundaries	Count or Frequency	Relative Frequency
0.5–0.9	0.45–0.95	3	3/192
1.0–1.4	0.95–1.45	4	4/192
1.5–1.9	1.45–1.95	9	9/192
2.0–2.4	1.95–2.45	19	19/192
2.5–2.9	2.45–2.95	44	44/192
3.0–3.4	2.95–3.45	61	61/192
3.5–3.9	3.45–3.95	44	44/192
4.0–4.4	3.95–4.45	8	8/192

From the above table based on relative frequencies, among other possibilities, we see that:

$$P(0.45 \leq X < 0.95) = 3/192$$
$$P(0.95 \leq X < 1.45) = 4/192$$
$$P(1.45 \leq X < 1.95) = 9/192$$

4.3 Defining Probability Axiomatically

The above examples illustrate the **relative frequentist interpretation of probability.** It will not do for a small number of trials because the value will keep changing from one sampling to another. The definition of probability of an event E from this point of view states that if an event E occurs $n(E)$ times out of n trials, then the probability of E is given as $P(E) = \dfrac{n(E)}{n}$. This definition would not do as a definition of probability because **it would only be true if n is infinite,** which cannot be achieved practically.

Owing to the problems associated with the classical and relative frequency definitions, we need other ways of defining probability or showing what probability is. The modern approach adopted by statisticians is to define probability axiomatically. Axiomatic approach is an attempt to provide definition for probability without the drawbacks that attend the classical and relative frequency approaches. The classical and relative frequency theory of probability simply leaves the concept of probability undefined. Axiomatic approach gives us a mathematical definition which enables us to identify which functions can be called probability, without telling us the actual probability $P\{.\}$ to be assigned to a particular event E. The word axiom means rules; the axiomatic approach to probability definition provides us with a set of practical rules by which we can say that something is or is not a probability.

Statisticians decided to define probability by using a set of **axioms** (rules). In the next few sections we state the axioms. Axiomatically, therefore, let S be the sample space, a **probability function, $P\{.\}$** is a set function with domain, E (an algebra of events: this can be extended to a sigma-algebra of events; See Mood et al., 1974, pp. 22–23 and counter-domain [0, 1] which satisfies the following axioms:

Axiom 1: The probability of an event E is always a number between 0 and 1. We state that $0 \leq P(E) \leq 1$.

Axiom 2: The probability of an event E which cannot occur is 0. E is termed **an impossible event. The impossible event is denoted as ϕ and $P(\phi) = 0$.**

Axiom 3: The probability of an event E which must occur is 1.0 and E is termed **a certain event. An event which must occur can be denoted as S, so that $P(S) = 1.0$.**

From the above three axioms, we can see that Probability has three basic properties:

1. The probability of an event E, denoted by $P(E)$ is always between zero and one. Therefore the probability of an event is non-negative. It is either zero, a positive fraction, or exactly one.
2. The probability of an event that cannot occur is zero. Such an event is called an impossible event.
3. The probability of an event that must occur or which occurs with certainty is exactly one. This type of event is called a certain event.

We illustrate the applications of these rules by examples:

For practical purposes, it is of interest to us to keep applying the useful classical definition in evaluating probability. With this, we can use the axioms in the evaluation of probabilities of events, and other combinations of events which arise, as a result of set operations on a set of events.

■ **EXAMPLE 4.13**   If two dice are rolled, assuming all outcomes are equally likely:

  (i) Obtain the event space S for the experiment and its probability.
  (ii) Obtain the event E_1 that the sum of two faces is equal to 8 or more, and its probability.
  (iii) Obtain the event E_2 that difference between face of die 1 and face of die 2 is greater than zero, and its probability.

Solution   (i) If we roll two dice, all the outcomes of the experiments which make up the event space S can be stated in the 36 tuples, each of which has probability 1/36.

$$S = \{(1, 1), (1, 2), (1, 3), (1, 4), (1, 5), (1, 6)$$
$$\mathbf{(2, 1), (2, 2), (2, 3), (2, 4), (2, 5),} (2, 6)$$
$$(3, 1), (3, 2), (3, 3), (3, 4), (3, 5), (4, 6)$$
$$(4, 1), (4, 2), (4, 3), (4, 4), (4, 5), (4, 6)$$
$$(5, 1), (5, 2), (5, 3), (5, 4), (5, 5), (5, 6)$$
$$(6, 1), (6, 2), (6, 3), (6, 4), (6, 5), (6, 6)\}$$
$$P(S) = 1.0;$$

(ii) $E_1 = \{(2, 6)$

       $(3, 5), (3, 6)$

       $(4, 4), (4, 5), (4, 6)$

       $(5, 3), (5, 4), (5, 5), (5, 6)$

       $(6, 2), (6, 3), (6, 4), (6, 5), (6, 6)\}$

   $P(E_1) = 15/36$

(iii) $E_2 = \{(2, 1),$

       $(3, 1), (3, 2),$

       $(4, 1), (4, 2), (4, 3),$

       $(5, 1), (5, 2), (5, 3), (5, 4),$

       $(6, 1), (6, 2), (6, 3), (6, 4), (6, 5)\}$

   $P(E_2) = 15/36$

We have so far considered Axioms 1–3 of probability. Next, in order to understand other axioms of probability, we illustrate the relationship between events. Thereafter, we will discuss other axioms of probability.

4.3.1 Relationships between Events and Use of Venn Diagrams

Here, we pictorially describe the relationships between events which will feature in the next axioms of probability which we describe. In each picture, the shaded part is the one being described. S is the event space.

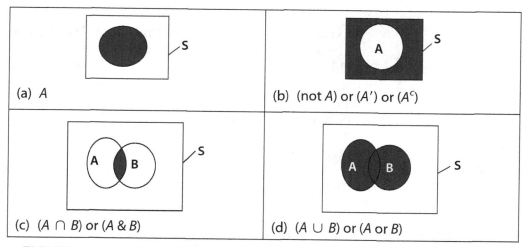

(a) A

(b) (not A) or (A') or (A^c)

(c) ($A \cap B$) or (A & B)

(d) ($A \cup B$) or (A or B)

■ **FIGURE 4.11 (a–h)** *A,* This represents the event A in the event space S. *B,* This represents the complement of the event A in S. It is made up of all the simple events in S which are not in A. We write it as (A') or (A^c) or (not A). *C,* This represents the intersection between two events, A and B. It is made up of all the simple events in S which are common to the two events A and B. We write the intersection as ($A \cap B$) or (A & B). *D,* This represents the union between two events, A and B. It is made up of all the simple events in S which are either in A or B without repetition. We write the intersection as ($A \cup B$) or (A or B).

Continued.

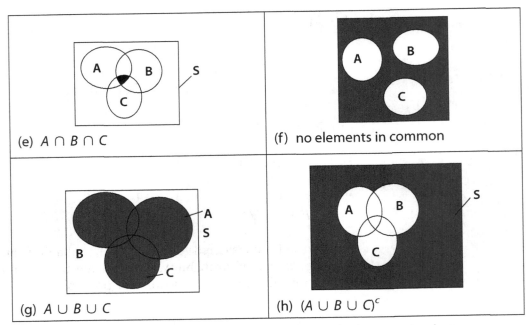

(e) $A \cap B \cap C$

(f) no elements in common

(g) $A \cup B \cup C$

(h) $(A \cup B \cup C)^c$

■ **FIGURE 4.11 (a–h)—cont'd** ***E,*** This represents the intersection between three events *A*, *B*, and *C*. It is written as (*A* & *B* & *C*) or (*A* ∩ *B* ∩ *C*). ***F,*** This represents the three events *A*, *B*, and *C* which are mutually exclusive. The events have no elements in common. For this case, the following are true: $A \cap B = \phi$; $A \cap C = \phi$; $B \cap C = \phi$; $A \cap B \cap C = \phi$ ***G,*** This represents the union of three events *A*, *B*, and *C* in the event space *S*. It is written as (*A* ∪ *B* ∪ *C*) or (*A* or *B* or *C*). ***H,*** This represents the complement of the union of the three events event *A*, *B*, and *C* in *S*. It is made up of all the simple events in *S* which are not in either *A* or *B* or *C*. We write it as $(A \cup B \cup C)^c$.

Axiom 4: The Addition Rule of Probability The addition rule enables us to evaluate the probability of the union of a set of events.

This rule states that if two events, *A* and *B*, are in *S*, then the probability of the union of *A* and *B* (Figures 4.11 (a) and (b) above) is the sum of the probability of *A* and probability of *B* minus the probability of their intersection (picture (*c*)). This can be written in mathematical symbols for two events as:

$$P(A \cup B) = P(A) + P(B) - P(A \cap B).$$

If three events were involved, then it is a little more complicated, but the formula is:

$$P(A \cup B \cup C) = P(A) + P(B) + P(C) - P(A \cap B) - P(A \cap C) - P(B \cap C) + P(A \cap B \cap C).$$

Axiom 5: The Complementation Rule of Probability This rule allows us to obtain the probability of an event in *S* which is not in the particular event being considered.

If *A* is in *S*, then A^c = (not *A*) is the complement of *A* in *S*, and the following rule applies:

$$P(A) = 1 - P(A^c) \text{ and } P(A^c) = 1 - P(A)$$

Next, we consider examples to illustrate Axioms 4 and 5 and the relationships between events.

■ **EXAMPLE 4.14** Consider the data of Example 4.13 and the definitions of E_1 and E_2. Using the tuples in S, find the following probabilities by first finding the event and considering every outcome (tuples) as equally likely:

(i) $P(E_1 \cap E_2)$
(ii) $P(E_1 \cup E_2)$
(iii) Using the probabilities of (i) and previous results in Example 4.11, show that the answer for two can be obtained by the addition rule stated above.

Solution

(i) $(E_1 \cap E_2) = \{(5, 3), (5, 4), (6, 2), (6, 3), (6, 4), (6, 5)\} \Rightarrow P((E_1 \cap E_2)$
       $= 6/36 = 1/6.$

(ii) $(E_1 \cup E_2) = \begin{Bmatrix} (2, 1), (2, 6), (3, 1), (3, 2), (3, 5), (3, 6), (4, 1), (4, 2), \\ (4, 3), (4, 4), (4, 5), (4, 6), (5, 1), (5, 2), (5, 3), (5, 4), \\ (5, 5), (5, 6), (6, 1), (6, 2), (6, 3), (6, 4), (6, 5), (6, 6) \end{Bmatrix}$

       $P(E_1 \cup E_2) = 24/36 = 2/3$

(iii) $P(E_1) = 15/36; P(E_2) = 15/36; P(E_1 \cap E_2) = 6/36$

        $P(E_1 \cup E_2) = P(E_1) + P(E_2) - P(E_1 \cap E_2) = 15/36 + 15/36 - 6/36$
        $= 24/36 = 2/3$

■ **EXAMPLE 4.15** The following data show the percentage of schools in a state which failed to meet the standards for "no child left behind law" (NCLBL), and the number of states:

A = event that the state has a failure percentage of NCLBL of at least 1% but no more than 4%

B = event that the state has a failure percentage of NCLBL of greater than 2% but less than 7%

C = event that the state has a failure percentage of NCLBL of greater than or equal to 5%

D = event that the state has a failure percentage of NCLBL of less than or equal to 8%

Failure (%) for NCLBL	Frequency
0	2
1–2	5
3–4	9
5–6	14
7–8	15
9–10	5
Total	50

(Note: in the table 3–4 means 3 and 4 inclusive). Find the probabilities of:

(a) A (b) B (c) C (d) D (e) $A \cap B$ (f) $A \cup B$ (g) $B \cap C$ (h) $B \cup C$
(i) $C \cap D$ (j) $C \cup D$ (k) $(B \cap D)$ (l) $(B \cup D)$ (m) $(A \cap D)$ (n) $(A \cup D)$
(o) $(A \cap B \cap D)$ (p) $(A \cup B \cup D)$ (q) $(A \cup B \cup D)^c$

Solution   We obtain the relative frequency table:

Failure (%) for NCLBL	Frequency	Relative Frequency
0	2	2/50
1–2	5	5/50
3–4	9	9/50
5–6	14	14/50
7–8	15	15/50
9–10	5	5/50
Total	50	1.0

The relative frequencies represent the probabilities for the ranges of values:

(a) $P(A) = \dfrac{5 + 9}{50} = \dfrac{14}{50}$;

(b) $P(B) = \dfrac{9 + 14}{50} = \dfrac{23}{50}$;

(c) $P(C) = \dfrac{14 + 15 + 5}{50} = \dfrac{34}{50}$;

(d) $P(D) = \dfrac{45}{50}$;

(e) $P(A \cap B) = P \text{ (greater than 2\% but less than or equal 4\%)} = \dfrac{9}{50}$

(f) $P(A \cup B) = P(A) + P(B) - P(AB) = \dfrac{14}{50} + \dfrac{23}{50} - \dfrac{9}{50} = \dfrac{28}{50}$

(g) $P(B \cap C) = P \text{ (of 5 to 6\%)} = \dfrac{14}{50}$

(h) $P(B \cup C) = P(B) + P(C) - P(B \cap C) = \dfrac{23}{50} + \dfrac{34}{50} - \dfrac{14}{50} = \dfrac{43}{50}$

(i) $P(C \cap D) = P \text{ (greater than 5\% or less than or equal to 8\%)} = \dfrac{14 + 15}{50} = \dfrac{29}{50}$

(j) $P(C \cup D) = P(C) + P(D) - P(C \cap D) = \dfrac{34}{50} + \dfrac{45}{50} - \dfrac{29}{50} = 1.0$

(k) $P(B \cap D) = P(B) = \dfrac{23}{50}$

(l) $P(B \cup D) = P(B) + P(D) - P(B \cap D) = \dfrac{23}{50} + \dfrac{45}{50} - \dfrac{23}{50} = \dfrac{45}{50}$

(m) $P(A \cap D) = P(A) = \dfrac{14}{50}$

(n) $P(A \cup D) = P(A) + P(D) - P(A \cap D) = \dfrac{14}{50} + \dfrac{45}{50} - \dfrac{14}{50} = \dfrac{45}{50}$

(o) $P(A \cap B \cap D) = P$ (greater than 2% and no more than 4%) $= \dfrac{9}{50}$

(p) $P(A \cup B \cup D) = P(A) + P(B) + P(D) - P(AB) - P(AD) - P(BD)$
$+ P(A \cap B \cap D) = \dfrac{14}{50} + \dfrac{23}{50} + \dfrac{45}{50} - \dfrac{23}{50} - \dfrac{14}{50} - \dfrac{23}{50} + \dfrac{9}{50} = \dfrac{31}{50}$

(q) Here, we use the complementation rule, so that $P(A \cup B \cup D)^c = P(\text{not}$
$A \cup B \cup D) = 1 - P(A \cup B \cup D) = 1 - \dfrac{31}{50} = \dfrac{19}{50}$

■ **EXAMPLE 4.16** A class is made up of 120 students, sixty-five of whom are studying physics, fifty-five are studying mathematics, while fifty-eight are studying chemistry as their major subjects. Of the physics students, forty take mathematics courses, of which ten take chemistry courses as well. Of those who major in mathematics, twenty take chemistry courses. Twenty-eight of the chemistry majors take physics courses. Find the probabilities that a student chosen from the group is taking:

(a) either physics or chemistry courses.
(b) either mathematics or chemistry courses.
(c) either physics or mathematics courses.
(d) at least one of the subjects.

Find also the probability that a student is:

(e) not taking physics, chemistry, or mathematics.
(f) either not taking chemistry or not taking physics or not taking mathematics.

Solution Let A = set of physics major students; B = set of mathematics major students; and C = set of chemistry major students. Clearly, $n(A) = 65, n(B) = 55, n(C) = 58, n(S) = 120, n(AB) = 40, n(AC) = 28, n(BC) = 20, n(ABC) = 10$.

(a) $P(A \text{ or } C) = P(A \cup C) = \dfrac{n(A \cup C)}{n(S)} = \dfrac{n(A) + n(C) - n(A \cap C)}{n(S)}$
$= \dfrac{65 + 58 - 28}{120} = \dfrac{95}{120} = \dfrac{19}{24}$

(b) $P(B \text{ or } C) = P(B \cup C) = \dfrac{n(B \cup C)}{n(S)} = \dfrac{n(B) + n(C) - n(B \cap C)}{n(S)}$
$= \dfrac{55 + 58 - 20}{120} = \dfrac{93}{120}$

(c) $P(A \text{ or } B) = P(A \cup B) = \dfrac{n(A \cup B)}{n(S)} = \dfrac{n(A) + n(B) - n(A \cap B)}{n(S)}$
$= \dfrac{65 + 55 - 40}{120} = \dfrac{80}{120} = \dfrac{2}{3}$

(d) $P(A \text{ or } B \text{ or } C) = P(A \cup B \cup C) = \dfrac{n(A \cup B \cup C)}{n(S)}$

$$= \dfrac{n(A) + n(B) + n(C) - n(A \cap B) - n(A \cap C) - n(B \cap C) + n(A \cap B \cap C)}{n(S)}$$

$$= \dfrac{65 + 55 + 58 - 40 - 20 - 28 + 10}{120} = \dfrac{100}{120} = \dfrac{5}{6}$$

(e) We need $P(A' \text{ and } B' \text{ and } C') = P(A'B'C')$. We can apply De Morgan's Law for three sets to obtain the solution. We note that under this law, $(A \cup B \cup C)' = A' \cap B' \cap C'$. Since we know from the solution of (d) that $n(A \cup B \cup C) = 100 \Rightarrow n(A \cup B \cup C)' = n(S) - n(A \cup B \cup C) = 120 - 100 = 20$, then $P(A' \cap B' \cap C') = 20/120 = 1/6$.

(f) We need $P(A' \text{ or } B' \text{ or } C') = P(A' \cup B' \cup C')$. We can also obtain the solution to this problem using De Morgan's Law. We note that $A' \cup B' \cup C' = (A \cap B \cap C)'$ by applying the DeMorgan's Law. Since we know from the solution of (d) that $n(A \cap B \cap C) = 10$, then $n(A \cap B \cap C)' = n(S) - n(A \cap B \cap C) = 120 - 10 = 110 \Rightarrow n(A' \cup B' \cup C') = 110$ and $P(A' \cup B' \cup C') = 110/120 = 11/12$.

If we look back at the solutions of Example 4.14 for (e), we can show pictorially, that $A' \cap B' \cap C'$ is the shaded part of the first Venn diagram in Figure 4.12, from which it is clear that $A' \cap B' \cap C'$ is obtained as the complement of $A \cup B \cup C$. Similarly, for (f) $A' \cup B' \cup C'$ is depicted in the second Venn diagram, as the shaded part of diagram, and shows that it is obtained as the complement of $A \cap B \cap C$ which is the unshaded part of the diagram.

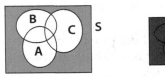

■ **FIGURE 4.12   Illustrations of DeMorgan's Law**

Axiom 6: Conditional Probability Sometimes, we need to find the probability that an event E_2 will occur, given that E_1 has occurred. This is known as conditional probability. We are interested in the probability of occurrence of E_2 conditioned on the fact that E_1 has occurred. This probability is written conventionally as $P[E_2 \mid E_1]$, often read as "probability of E_2 given E_1" for short.

If event E_1 has already occurred, and we are interested in the occurrence of the event E_2, then there is only one possible way that E_2 could occur when E_1 has already occurred, that is, as part of the intersection of the two, namely, $E_1 \cap E_2$. This means that the sample space is now restricted to E_1 and the number of elements therein, $n(E_1)$. The occurrence of E_2 is restricted to the number in the intersection, $n(E_1 \cap E_2)$. Hence we can write $P[E_2 \mid E_1]$ as:

$$P(E_1 \mid E_2) = \dfrac{n(E_1 \cap E_2)}{n(E_1)} = \dfrac{n(E_1 \cap E_2)}{n(S)} \times \dfrac{n(S)}{n(E_1)} = \dfrac{n(E_1 \cap E_2)}{n(S)} \Big/ \dfrac{n(E_1)}{n(S)}$$

$$= \dfrac{P(E_1 \cap E_2)}{P(E_1)}$$

We have shown that

$$P(E_1 \mid E_2) = \frac{P(E_1 \cap E_2)}{P(E_1)}$$

provided that $P(E_1) \neq 0$. This means that the conditional probability of E_2 given that E_1 has occurred is the ratio of the probability of the joint occurrence of E_1 and E_2 and the marginal (total, as obtained in the margin) probability of E_1. Multiplying the right hand and left hand by $P(E_1)$, then

$$P(E_1 \cap E_2) = P(E_1) \cdot P(E_2 \mid E_1)$$

Written this way, we find the **probability of joint occurrence of the events E_1 and E_2** as the product of the conditional probability of E_2 given E_1 and the marginal probability of E_1. We can similarly obtain the probability of joint occurrence of the events E_1 and E_2 as:

$$P(E_1 \mid E_2) = \frac{P(E_1 \cap E_2)}{P(E_2)} \Rightarrow P(E_1 \cap E_2) = P(E_2) \cdot P(E_1 \mid E_2)$$

■ **EXAMPLE 4.17** A box contains 130 items of which fifty are defective. If two items are chosen at random from the box without replacement, find (a) the probability that the first is defective and the second is not, (b) the first is not defective but the second is, and (c) one of the two items is defective.

Solution Let D = defective $\Rightarrow D_1$ means that the first item is defective; D'_1 = first item is not defective; D'_2 means that the second is defective, means the second item is not defective.

(a) We need to find $P(D_1 \cap D'_2) = P(D_1)P(D'_2 \mid D_1)$, but $P(D_1) = \dfrac{50}{130}$;
$P(D'_2 \mid D_1) = \dfrac{80}{129}; \Rightarrow P(D_1 \cap D'_2) = \dfrac{50}{130} \times \dfrac{80}{129} = \dfrac{400}{1677} = 0.2385$

(b) We need to find $P(D'_1 \cap D_2) = P(D'_1)P(D_2 \mid D'_1)$, but $P(D_1) = \dfrac{80}{130}$;
$P(D_2 \mid D'_2) = \dfrac{50}{129}; \Rightarrow P(D'_1 \cap D_2) = \dfrac{80}{130} \times \dfrac{50}{129} = \dfrac{400}{1677} = 0.2385$

(c) Here either $D_1 \cap D'_2$ or $D'_1 \cap D_2$ will satisfy the requirement that one of the two items be defective. The required probability is: $P(D_1 \cap D'_2) +$
$P(D'_1 \cap D_2) = \dfrac{50}{130} \times \dfrac{80}{129} + \dfrac{80}{130} \times \dfrac{50}{129} = 0.4770$

Note that whenever it is either/or, we use both (that is union of both). We can illustrate the above experiment in Example 4.17 using a Tree Diagram. Each branch of the tree represents a step in the experiment. In this case, the first branch = 1st draw, and the second branch = 2nd draw. The tree diagram which applies is given below:

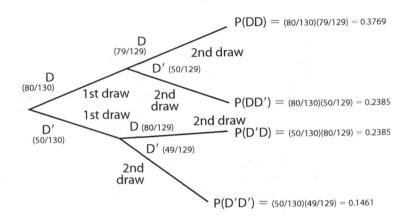

From the above we obtain a general rule which applies to any two sets E_1 and E_2:

$$P(E_1 \mid E_2) = \frac{P(E_1 \cap E_2)}{P(E_1)}$$

Axiom 7: Independence of k Events ($k \geq 2$) The events E_1 and E_2 in S are said to be independent events; otherwise, they are dependent events. If two events, E_1 and E_2 are independent, then:

$$P(E_1 \cap E_2) = P(E_1)P(E_2).$$

We can extend this definition to three events, E_1, E_2, and E_3. If they are dependent events then:

$$P(E_1 \cap E_2 \cap E_3) = P(E_1) \cdot P(E_2 \mid E_1) \cdot P(E_3 \mid E_1 \cap E_2).$$

However, when the three events are independent events, then:

$$P(E_2 \mid E_1) = P(E_2)$$
$$P(E_3 \mid E_1 \cap E_2) = P(E_3)$$
$$\Rightarrow P(E_1 \cap E_2 \cap E_3) = P(E_1) \cdot P(E_2) \cdot P(E_3).$$

In other words the probability of the joint occurrence of the three independent events, E_1, E_2, and E_3, is equal to the product of the three marginal probabilities.

■ **EXAMPLE 4.18**   In casting a fair die, the following events were defined: E_1 = a "six" turns up in the first cast, E_2 = a "four" turns up in the sixth cast, and E_3 = a "three" turns up in the tenth cast. Obtain the probability that either E_1 or E_2 or E_3 occur.

Solution $P(E_1 \text{ or } E_2 \text{ or } E_3) = P(E_1 \cup E_2 \cup E_3) = P(E_1) + P(E_2) + P(E_3) - P(E_1E_2) - P(E_2E_3) - P(E_1E_3) + P(E_1E_2E_3);$

But E_1, E_2, E_3 are independent events so that each of them has probability of 1/6.

$$P(E_1 \cap E_2) = P(E_1) \cdot P(E_2) = \frac{1}{6} \cdot \frac{1}{6}$$

$$P(E_1 \cap E_3) = P(E_1) \cdot P(E_3) = \frac{1}{6} \cdot \frac{1}{6}$$

$$P(E_2 \cap E_3) = P(E_2) \cdot P(E_3) = \frac{1}{6} \cdot \frac{1}{6}$$

$$P(E_1 \cap E_2 \cap E_3) = P(E_1) \cdot P(E_2) = \frac{1}{6} \cdot \frac{1}{6} \cdot \frac{1}{6}$$

$$P(E_1 \cup E_2 \cup E_3)$$
$$= P(E_1) + P(E_2) + P(E_3) - P(E_1 \cap E_2) - P(E_1 \cap E_3) - P(E_2 \cap E_3) + P(E_1 \cap E_2 \cap E_3)$$
$$= \frac{1}{6} + \frac{1}{6} + \frac{1}{6} - \frac{1}{6} \cdot \frac{1}{6} - \frac{1}{6} \cdot \frac{1}{6} - \frac{1}{6} \cdot \frac{1}{6} + \frac{1}{6} \cdot \frac{1}{6} \cdot \frac{1}{6} = \frac{91}{216}$$

Axiom 8: Mutually Exclusive Events Two or more events are said to be mutually exclusive if they are mutually disjoint. In set theoretical language, the events have no elements which are common to all of them so that the occurrence of one of them excludes the occurrence of the other events. Mathematically, the implication of this definition for two mutually exclusive events, E_1 and E_2, is that:

$$P(E_1 \cap E_2) = 0.$$

Hence the probability of either E_1 or E_2 occurring when they are mutually exclusive becomes:

$$P(E_1 \cup E_2) = P(E_1) + P(E_2).$$

If there are three mutually exclusive events, then:

$$P(E_1 \cup E_2 \cup E_3) = P(E_1) + P(E_2) + P(E_3).$$

If two or more mutually exclusive events are exhaustive of the event space S, then the events are partitions. Figure 4.13a shows two mutually exclusive and exhaustive events A and B, while Figure 4.13b shows three mutually and exhaustive events A, B, and C.

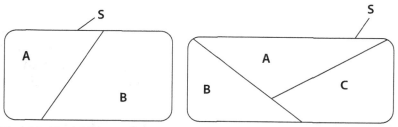

■ **FIGURE 4.13 (A and B)   Illustrations of Two and Three Mutually Exclusive and Exhaustive Events.** The pictures in Figure 4.13 represent illustrations of the partitioning of the event space into two or three events.

■ EXAMPLE 4.19   A bag contains 200 balls of which sixty are white, fifty are red, forty are blue, thirty are yellow, and twenty are green. An experiment requires that a single ball be drawn from the bag. Let E_1 = drawing a white ball, E_2 = drawing a red ball, E_3 = drawing a blue ball, E_4 = drawing a yellow ball, and E_5 = drawing a green ball. What is the probability of drawing either a blue, red, or white ball?

Solution   Clearly events E_1 to E_5 constitute a partition of the event space; the events are therefore mutually exclusive. We can easily show that no ball is of any combination of colors, so that the probability of choosing a ball which is white, red, and blue is not possible, that is $P(E_1 \cap E_2 \cap E_3) = 0$; so also: $P(E_1 \cap E_2) = 0$; $P(E_1 \cap E_3) = 0$. The events E_1, E_2, and E_3 are mutually exclusive so that:

$$P(E_1 \text{ or } E_2 \text{ or } E_3) = P(E_1 \cup E_2 \cup E_3) = P(E_1) + P(E_2) + P(E_3)$$
$$= \frac{60}{200} + \frac{50}{200} + \frac{40}{200} = \frac{3}{4}$$

4.4   Bayes' Theorem

The simplest form of the Bayes' Theorem has to do with an event space S containing only two mutually exclusive and exhaustive events A_1 and A_2 each of which has some elements in common with another event B which is also in S as illustrated in Figure 4.14. Under these conditions then, Bayes' Theorem states as follows:

Given a probability space S, if A_1, and A_2 are mutually exclusive events in S, such that $E_1 \cup E_2 = S$ and B is any arbitrary event with respect to S, then:

$$P(A_1 \mid B) = \frac{P(A_1)P(B \mid A_1)}{P(A_1)P(B \mid A_1) + P(A_2)P(B \mid A_2)}.$$

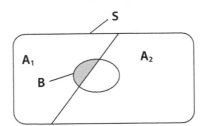

■ FIGURE 4.14   Two mutually exclusive and exhaustive events A_1 and A_2 both of which have some things in common with a third event B in S.

Consider the picture in Figure 4:14. B can be considered to be the union of two parts: The shaded part which it has in common with A_1, and the unshaded part which it has in common with A_2. The total probability of B can be written as:

$$P(B) = P(B \cap A_1) + P(B \cap A_2)$$

$$P(B \mid A_1) = \frac{P(B \cap A_1)}{P(A_1)} \Rightarrow P(A_1)P(B \mid A_1) = P(B \cap A_1)$$

$$P(B \mid A_2) = \frac{P(B \cap A_2)}{P(A_2)} \Rightarrow P(A_2)P(B \mid A_2) = P(B \cap A_2)$$

Thus, $P(B) = P(B \cap A_1) + P(B \cap A_2) = P(A_1)P(B \mid A_1) + P(A_2)P(B \mid A_2)$

Now consider $P(A_1 \mid B)$, by the foregoing, and by definition of conditional probability,

$$P(A_1 \mid B) = \frac{P(A_1 \cap B)}{P(B)}$$

So that $P(A_1 \mid B) = \dfrac{P(A_1 \cap B)}{P(B)} = \dfrac{P(A_1)P(B \mid A_1)}{P(A_1)P(B \mid A_1) + P(A_2)P(B \mid A_2)}$

Also $P(A_2 \mid B) = \dfrac{P(A_2 \cap B)}{P(B)} = \dfrac{P(A_2)P(B \mid A_2)}{P(A_1)P(B \mid A_1) + P(A_2)P(B \mid A_2)}$

We have just proved the Bayes' Theorem for two mutually exclusive and exhaustive events (A_1 and A_2), both of which have some things in common with a third event B, as shown in Figure 4.14.

Extension of Bayes' Theorem for Three Mutually Exclusive and Exhaustive Events

It is easy to extend this concept used in demonstrating Bayes' Theorem to three events. Consider Figure 4.15:

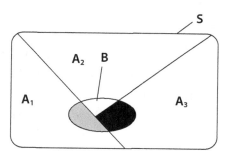

■ **FIGURE 4.15** Partition of event space S into three events with another event B in S which has some elements in common with all the partitions.

Following similar arguments used for two partitions or two mutually exclusive events, we can see that:

Total probability of $B = P(B) = P(A_1)P(B \mid A_1) + P(A_2)P(B \mid A_2) + P(A_3)PB \mid A_3)$

and $P(A_1 \mid B) = \dfrac{P(A_1 \cap B)}{P(B)} = \dfrac{P(A_1)P(B \mid A_1)}{P(A_1)P(B \mid A_1) + P(A_2)P(B \mid A_2) + P(A_3)P(B \mid A_3)}$

$P(A_2 \mid B) = \dfrac{P(A_2 \cap B)}{P(B)} = \dfrac{P(A_2)P(B \mid A_2)}{P(A_1)P(B \mid A_1) + P(A_2)P(B \mid A_2) + P(A_3)P(B \mid A_3)}$

$P(A_3 \mid B) = \dfrac{P(A_3 \cap B)}{P(B)} = \dfrac{P(A_3)P(B \mid A_3)}{P(A_1)P(B \mid A_1) + P(A_2)P(B \mid A_2) + P(A_3)P(B \mid A_3)}$

In fact it should be straightforward to extend this to k mutually exclusive events which are exhaustive, and which have some elements in common with any other event B in S.

■ EXAMPLE 4.20   A test for detecting a flu virus is known to detect the presence of the virus in those who have the flu 90% of the time. It also detects the absence of the flu in those who do not have the flu 99% of the time. If 10% of the people actually have the flu, find the following probabilities: (a) that a person who has been tested and declared to have the disease, actually has the disease, and (b) that a person who has been tested and declared to have the disease, actually does not have the disease.

Solution   This requires Bayes' Theorem for two partitions as shown below. In this case the event space S is partitioned into two mutually exclusive events, so that $S = \{D, D'\}$ where D = has the flu and D' = does not have the flu. Let T^+ denote that the test is positive and T^- denote that the test is negative.

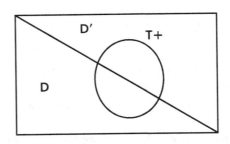

$$P(T^+) = P(D \cap T^+) + P(D' \cap T^+)$$

But, $P(T^+ \mid D) = \dfrac{P(D \cap T^+)}{P(D)} \Rightarrow P(D \cap T^+) = P(D)P(T^+ \mid D)$

Similarly, $P(T^+ \mid D) = \dfrac{P(D' \cap T^+)}{P(D')} \Rightarrow P(D' \cap T^+) = P(D')P(T^+ \mid D')$

$P(T^+) = P(D)P(T^+ \mid D) + P(D')P(T^+ \mid D')$

$P(D) = 0.10 \Rightarrow P(D') = 1 - 0.10 = 0.90$

$P(T^+ \mid D) = 0.90 \Rightarrow P(T^- \mid D) = 1 - 0.90 = 0.10$

$P(T^- \mid D') = 0.99 \Rightarrow P(T^+ \mid D') = 1 - 0.99 = 0.01$

$P(T^+) = 0.10 \times 0.90 + 0.01 \times 0.90 = 0.099 =$ total probability of T^+

$$P(D \mid T^+) = \frac{P(D)P(T^+ \mid D)}{P(D)P(T^+ \mid D) + P(D')P(T^+ \mid D')}$$

$$P(D' \mid T^+) = \frac{P(D')P(T^+ \mid D')}{P(D)P(T^+ \mid D) + P(D')P(T^+ \mid D')}$$

(a) $P(D \mid T^+) = \dfrac{0.9 \times 0.10}{0.9 \times 0.10 + 0.90 \times 0.01} = 0.90901$

(b) $P(D' \mid T^+) = \dfrac{0.90 \times 0.01}{0.9 \times 0.10 + 0.90 \times 0.01} = 0.090909$

■ EXAMPLE 4.21   A test for detecting a flu virus is known to detect the presence of the virus in those who have the flu 90% of the time. It also detects the absence of the flu in those who do not have the flu 92% of the time. If 10% of the people actually have the flu, find the following probabilities: (a) that a person who has been tested and declared not to have the disease, actually does not have the disease, and (b) that a person who has been tested and declared not to have the disease, actually has the disease.

Solution   This requires Bayes' Theorem for two partitions as shown below. In this case the event space S is partitioned into two mutually exclusive events, so that $S = \{D, D'\}$ where $D =$ has the flu and $D' =$ does not have the flu.

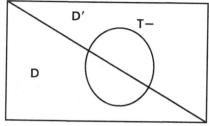

$$P(T^-) = P(D \cap T^-) + P(D' \cap T^-)$$

But, $P(T^- \mid D) = \dfrac{P(D \cap T^-)}{P(D)} \Rightarrow P(D \cap T^-) = P(D)P(T^- \mid D)$

Similarly, $P(T^- \mid D') = \dfrac{P(D' \cap T^-)}{P(D')} \Rightarrow P(D' \cap T^-) = P(D')P(T^- \mid D')$

$P(T^-) = P(D)P(T^- \mid D) + P(D')P(T^- \mid D')$

$P(D) = 0.10 \Rightarrow P(D') = 1 - 0.10 = 0.90$

$P(T^+ \mid D) = 0.10 \Rightarrow P(T^- \mid D) = 1 - 0.90 = 0.10$

$P(T^- \mid D') = 0.92 \Rightarrow P(T^+ \mid D) = 1 - 0.92 = 0.08$

$P(T^-) = 0.10 \times 0.10 + 0.90 \times 0.92 = 0.838 =$ total probability of T^-

$$P(D \mid T^-) = \dfrac{P(D)P(T^- \mid D)}{P(D)P(T^- \mid D) + P(D')P(T^- \mid D')}$$

$$P(D' \mid T^-) = \dfrac{P(D')P(T^- \mid D')}{P(D)P(T^- \mid D) + P(D')P(T^- \mid D')}$$

(a) $P(D \mid T^-) = \dfrac{0.10 \times 0.10}{0.9 \times 0.90 + 0.90 \times 0.92} = 0.0119$

(b) $P(D' \mid T^-) = \dfrac{0.90 \times 0.92}{0.9 \times 0.90 + 0.90 \times 0.92} = 0.9881$

■ **EXAMPLE 4.22**   Dr. Beck treats her migraine patients with only two medications D_1 and D_2. 60% of Dr. Beck's patients are treated with medication D_1 while the rest are treated with medication D_2. It is known from records that 90% of those who took medication D_1 fully recovered while 86% of those who took medication D_2 are known to have fully recovered.

(a) What is the total probability of full recovery if a migraine patient is treated by Dr. Beck?
(b) If a fully recovered patient of Dr. Beck is chosen at random, what is the probability that he/she was treated with (i) treatment D_1 (ii) treatment D_2.

Solution   Total probability is $P(F) = P(D_1 \cap F) + P(D_2 \cap F)$

But, $P(F \mid D_1) = \dfrac{P(D_1 \cap F)}{P(D_1)} \Rightarrow P(D_1 \cap F) = P(D_1)P(F \mid D_1)$

Similarly, $P(F \mid D_2) = \dfrac{P(D_2 \cap F)}{P(D_2)} \Rightarrow P(D_2 \cap F) = P(D_2)P(F \mid D_2)$

Total probability $= P(F) = P(D_1)P(F \mid D_1) + P(D_2)P(F \mid D_2)$

$$P(D_1) = 0.60 \Rightarrow P(D_2) = 1 - 0.60 = 0.40$$

$$P(F \mid D_1) = 0.90$$

$$P(F \mid D_2) = 0.86$$

$$\text{Total Probability} = P(F) = 0.90 \times 0.60 + 0.40 \times 0.86 = 0.884$$

$$P(D_1 \mid F) = \frac{P(D_1)P(F \mid D_1)}{P(D_1)P(F \mid D_1) + P(D_2)P(F \mid D_2)}$$

$$P(D_2 \mid F) = \frac{P(D_2)P(F \mid D_2)}{P(D_1)P(F \mid D_1) + P(D_2)P(F \mid D_2)}$$

(a) $P(D_1 \mid F) = \dfrac{0.60 \times 0.90}{0.60 \times 0.90 + 0.40 \times 0.86} = 0.61086$

(b) $P(D_2 \mid F) = \dfrac{0.40 \times 0.86}{0.60 \times 0.90 + 0.40 \times 0.86} = 0.38914$

4.5 Some Useful Mathematical Techniques

So far we have studied ways of computing the probabilities of events, and in the process, have restricted ourselves to those events in which it is very easy for us to enumerate all the simple events in an event. Some events are much more complex than those we have described in the previous sections. For such events, it is unlikely that we will be able to enumerate all the possible cases that arise. Even when we can it is tedious and labor intensive. It is therefore necessary at this stage that we study some mathematical principles which should enable us to avoid these tedious calculations.

4.5.1 Combinatorial Analysis

One principle which we shall find useful is **basic combinatorial analysis.**

Combinatorial analysis enables us to calculate the number of ways that a number of events can occur if order does not matter. For example, suppose that we have three bags; one contains four apples, the second contains ten oranges, while the third contains seven pears. We can easily work out the number of ways of choosing one apple, one orange and one pear from the three bags. Mathematically we can choose one apple from the first bag in $^4C_1$ ways (read as 4 combination 1 ways). Similarly, we choose the orange and pear from the second and third bags in $^{10}C_1$ and $^7C_1$ respective ways. The three events can happen in $^4C_1 \times {}^{10}C_1 \times {}^7C_1$ ways because choosing from the bags occurs independently. We shall define **combinations** and therefore work out explicitly how many ways in which the three events can happen. Before we define combinations, we need to define **factorial x, or x factorial.**

Definition of Factorial x

The quantity referred to as **x factorial, or factorial x,** is usually denoted mathematically by **$x!$** and is defined as:

$$x! = x(x - 1)(x - 2)(x - 3) \ldots (2)(1)$$

In terms of explicit numbers, we can write:

$$5! = 5 \times 4 \times 3 \times 2 \times 1$$
$$7! = 7 \times 6! = 7 \times 6 \times 5!$$
$$5!4! = (5 \times 4 \times 3 \times 2 \times 1) \times (4 \times 3 \times 2 \times 1)$$

In particular, we note that $1! = 1$ and $0! = 1$.

Definition of Combinations

The number of ways of choosing x things out of n distinct things without paying attention to order of arrangement is written as $\binom{n}{x}$ or sometimes as ^nC_x and can be defined as

$$\binom{n}{x} = \frac{n!}{x!(n - x)!}$$

Where:

$$n! = n(n - 1)(n - 2) \ldots (2)(1)$$
$$x! = x(x - 1)(x - 2) \ldots (2)(1)$$
$$(n - x)! = (n - x)(n - x - 1)(n - x - 2) \ldots (2)(1)$$

$\binom{n}{x}$ is often referred to as the number of combinations of n objects taking x at a time.

■ **EXAMPLE 4.23** Refer to the bags described in the above section, find the number of ways of choosing one apple, one orange, and one pear out of the three bags. Also find the number of ways of choosing two apples, four oranges, and three pears from the bags.

Solution We look for ways of choosing one apple from bag 1, one orange from bag 2, and one pear from bag 3 without regard to the order of the choice. By virtue of our earlier discussion the number of ways is $^4C_1 \times {}^{10}C_1 \times {}^7C_1$ ways, which is:

$$\frac{4!}{1!(4 - 1)!} \times \frac{10!}{1!(10 - 1)!} \times \frac{7!}{1!(7 - 1)!} = 4 \times 10 \times 7 = 280.$$

By similar arguments, the number of ways of choosing two apples from bag 1, four oranges from bag 2, and three pears from bag 3 is:

$$\frac{4!}{2!(4 - 2)!} \times \frac{10!}{4!(10 - 4)!} \times \frac{7!}{3!(7 - 3)!} = 6 \times 210 \times 35 = 44100.$$

■ **EXAMPLE 4.24** A bag contains ten oranges, seven pears, and four apples. Three fruits are chosen at random (without replacement) from the bag. What is the probability that one of each of the three kinds of fruits was chosen? Suppose that nine fruits were chosen at random from the bag (without replacement), what is the probability that four oranges, three pears, and two apples were chosen?

Solution We simply exploit our knowledge of combinations in the solutions of the problems. There are altogether twenty-one fruits in the bag. There are altogether $^{21}C_3$ ways of choosing three things out of twenty-one without regard to order. One orange can be chosen out of ten in $^{10}C_1$ ways, one pear can be chosen out of seven in $^7C_1$ ways, and one apple can be chosen out of four in $^4C_1$ ways, so that the three fruits can be chosen in $\binom{4}{1} \times \binom{10}{1} \times \binom{7}{1}$ ways.

$$P \text{ (one orange, one pear, and one apple were chosen)} = \frac{\binom{4}{1} \times \binom{10}{1} \times \binom{7}{1}}{\binom{21}{3}} =$$

$$\frac{280}{1330} = 0.2105.$$

Similarly, using the definitions of combinations, P (four oranges, three pears,

$$\text{and two apples were chosen)} = \frac{\binom{4}{2} \times \binom{10}{4} \times \binom{7}{3}}{\binom{21}{9}} = \frac{44100}{293930} = 0.150.$$

4.5.2 Permutations

Another mathematical concept which finds useful application in Statistics is **permutation**. An arrangement of n different objects taking x at a time, with attention paid to the order of the arrangement is called a permutation, and is often written as ^nP_x or $_nP_x$ where:

$$^nP_x = \frac{n!}{(n-x)!} = n(n-1)(n-2)(n-3)\ldots(n-x+1)$$

$$^nP_0 = \frac{n!}{(n-0)!} = 1$$

$$^nP_n = \frac{n!}{(n-n)!} = \frac{n!}{0!} = n(n-1)(n-2)(n-3)\ldots(2)(1)$$

If out of a objects in a group, a_1 are alike, a_2 are alike, . . . , a_k are alike, then the number of permutations of these a objects would be:

$$\frac{a!}{a_1!a_2!a_3!\ldots a_k!} \text{ where } \sum_{i-1}^{k} a_i = a$$

■ **EXAMPLE 4.25** Find the number of permutations of the letters A, B, C, and D taking two at a time.

Solution The number of permutations are: $^4P_2 = \frac{4!}{(4-2)!} = 12$

The arrangements are: AB, BA, AC, CA, AD, DA, BC, CB, BD, DB, CD, and DC

■ **EXAMPLE 4.26** What is the number of permutations in the words **classical, information,** and **association?**

(a) **classical** contains nine letters; 2 c's, 2 s's, 2 a's, 2 l's, and 1 i. The number of permutations is:

$$\frac{9!}{2!2!2!2!1!} = 22680.$$

(b) **information** contains eleven letters; 2 i's, 2 n's, 1 f, 2 o's, 1 r, 1 m, 1 a, and 1 t so that the number of permutations is:

$$\frac{11!}{2!2!2!1!1!1!1!1!1!} = 4989600.$$

(c) The word **association** is made up of eleven letters; 2 a's, 2 s's, 2 o's, 1 c, 2 i's, 1 t, and 1 n. The number of permutations is:

$$\frac{11!}{2!2!2!2!1!1!1!} = 2494800.$$

■ **EXAMPLE 4.27** Consider the following workers:

John	Linda	Marcus	Phil	Justin
Megan	Ray	Carol	Sue	Amanda

(a) If a committee of five is to be chosen from the workers, what is the probability that it must include at least two women?
(b) If Linda must be chosen as a member of the committee, what is the probability that the committee contains at least two females?

Solution There are ten workers (five males, five females). The number of ways of choosing five out of ten when order does not matter is $\binom{10}{5} = \frac{10!}{5!(5)!} = 252$ ways.

(a) The committee must be made up of: (i) two females and three males; (ii) three females and two males; (iii) four females and one male; or (iv) no males. The choices will have to be made in:

$$\binom{5}{2}\binom{5}{3} + \binom{5}{3}\binom{5}{2} + \binom{5}{4}\binom{5}{1} + \binom{5}{5}\binom{5}{0}$$
$$= 10 \times 10 + 10 \times 10 + 5 \times 5 + 1 \times 1 = 226 \text{ ways.}$$

P (choosing a committee of five with at least two females) $= \frac{226}{252} = 0.896825$

(b) The fact that Linda must be included, means that Linda is chosen with certainty or probability of 1. This reduces the number of women to be chosen or chosen from by 1. This means that the committee must be made up of either (i) Linda, one female and three males; (ii) Linda, two females, and two males; (iii) Linda three females and one male; (iv) Linda and no males. The choices will have to be made in:

$$\binom{1}{1}\binom{4}{1}\binom{5}{3} + \binom{1}{1}\binom{4}{2}\binom{5}{2} + \binom{1}{1}\binom{4}{3}\binom{5}{1} + \binom{1}{1}\binom{4}{4}\binom{5}{0}$$
$$= 1 \times 4 \times 10 + 1 \times 6 \times 10 + 1 \times 4 \times 5 + 1 \times 1 \times 1 = 121 \text{ ways.}$$

Also, since Linda is chosen with certainty, then we have only four out of nine to be chosen with uncertainty. This is done in $\binom{9}{4} = \dfrac{9!}{4!(5)!} = 126$ ways. (choosing a committee of five containing Linda and with at least two females) $= \dfrac{121}{126} = 0.960317$

■ **EXAMPLE 4.28**   An urn contains six yellow balls, five blue balls, and eight white balls. If balls are drawn one at a time, without replacement, four times; what is the probability that:

(a) A white ball is chosen in the first draw, a yellow ball is chosen second draw, a white ball is chosen in the third draw, and a blue ball is chosen in the fourth draw?
(b) white balls are chosen in the four draws?

Solution   The events of choosing the first, second, third, and fourth balls occur independently. The probabilities are multiplied to get the eventual probability that the choices will happen as stated.

(a) $P(w, y, w, b) = \dfrac{\binom{8}{1}}{\binom{19}{1}} \times \dfrac{\binom{6}{1}}{\binom{18}{1}} \times \dfrac{\binom{7}{1}}{\binom{17}{1}} \times \dfrac{\binom{5}{1}}{\binom{16}{1}} = \dfrac{8}{19} \times \dfrac{6}{18} \times \dfrac{7}{17} \times \dfrac{5}{16}$

$= \dfrac{1680}{93024} = 0.01806$

(b) $P(w, w, w, w) = \dfrac{\binom{8}{1}}{\binom{19}{1}} \times \dfrac{\binom{7}{1}}{\binom{18}{1}} \times \dfrac{\binom{6}{1}}{\binom{17}{1}} \times \dfrac{\binom{5}{1}}{\binom{16}{1}} = \dfrac{8}{19} \times \dfrac{7}{18} \times \dfrac{6}{17} \times \dfrac{5}{16}$

$= 0.01806$

References

Mood, A. M., Graybill, F. A., and Boes, D. C. (1974). Introduction to the theory of statistics. McGraw-Hill Co., NY

Exercises

1. Consider the following set of workers:

John	Jeremy	Linda	Marcus	Phil	Justin	Tyler
Megan	Anna	Ray	Carol	Sue	Amanda	Duncan

   (a) What is the probability of choosing a committee of five with at least two females?
   (b) If Anna must be chosen, what is the probability of choosing a committee of five with at least two females?

2. Using the set of workers in (1),
   (a) What is the probability of choosing a committee of six with at least three males?
   (b) If Jeremy must be chosen, what is the probability of choosing a committee of six with at least three males?

3. Using the set of workers in (1), if we consider only the men, and if only two of them are to be chosen to go to a course:
   (a) How many pairs are there? List all of them.
   (b) If all pairs of males chosen in (a) are equally likely, what is the probability that Phil is chosen?
   (c) If all pairs of males chosen in (a) are equally likely, what is the chance that either Justin or Phil is in the pair chosen?

4. Using the set of workers in (1), if we consider only the females, and if only two of them are to be chosen to go on a company cruise:
   (a) How many pairs are there? List them all.
   (b) If all pairs of females chosen in (a) are equally likely, what is the probability that Carol is chosen?
   (c) If all pairs of females chosen in (a) are equally likely, what is the chance that either Anna or Carol are in the pair chosen.
   (d) If all pairs of females chosen in (a) are equally likely, what is the chance that Anna, or Carol, or Sue is in the pair chosen?

5. Using the set of workers in (1), if only two of them are to be chosen to go for a special assignment:
   (a) Find the probability of choosing one male and one female.
   (b) Find the probability that the chosen pair is female.
   (c) Find the probability that the chosen pair is male.

6. Using all the set of workers in (1), if six workers are to be chosen:
   (a) What is the probability of choosing either more females or more males?
   (b) If Linda and Phil must be in the sample, what is the probability of choosing exactly two females?
   (c) What is the probability of choosing twice the number of males as females?
   (d) What is the probability of choosing twice the number of females as males?

Exercises on Events, Event Space and Probability

7. Two dice are rolled (assume outcomes are equally likely).
   (a) Obtain S the event space of all possible outcomes of the experiment.
   (b) Obtain event E_1 that face of die one is larger than face of die two and find its probability.
   (c) Obtain event E_2 that sum of face of die one and sum of face of die two is greater than 8, and find its probability,
   (d) Obtain event E_3 that the difference between the face of die one and the face of die two is less than 2, and find its probability.
   (e) Find (i) $E_1 \cap E_2$; (ii) $E_1 \cap E_3$ (iii) $E_2 \cap E_3$, and their probabilities.

8. Using the information from 7 find the following:
   (f) Find $E_1 \cup E_2$; (ii) $E_1 \cup E_3$ (iii) $E_2 \cup E_3$, and their probabilities.
   (g) Find the probability of $E_1 \cup E_2 \cup E_3$.

9. Consider a couple who gave birth to four children (assume a boy is equally likely as a girl):
   (a) Obtain S the event space of all possible outcomes if $B =$ boy and $G =$ girl.
   (b) Obtain event E_1 of no more than two girls in the family, and find its probability.
   (c) Obtain event E_2 of at least a boy in the family, and find its probability.
   (d) Obtain event E_3 for two or three girls.
   (e) Find $E_1 \cap E_2$, and its probability.
   (f) Find $E_1 \cup E_2$, and its probability.
   (g) Find $E_1 \cap E_3$, and its probability.

10. Using information of 9:
    (a) Find $E_1 \cup E_3$, and its probability.
    (b) Find $E_2 \cap E_3$, and its probability.
    (c) Find $E_2 \cup E_3$, and its probability.
    (d) Find $E_1 \cap E_2 \cap E_3$, and its probability.
    (e) Find the probability of $E_1 \cup E_3 \cup E_3$.

11. In a university, it is required that students of the Biological sciences take and pass Physics1(P1), Physics2(P2), or Physics3 (P3) before they graduate. 33% of biological sciences students take P1, 36% of them take P2 while the rest take P3. It is known that 80%, 85%, and 86% of students who take P1, P2, and P3, respectively, pass.
    (a) What is the total probability that a Biological sciences student will pass and graduate?
    (b) If a graduating biological science student is chosen at random, what is the probability that the student took and passed: (i) P1? (ii) P2? (iii) P3?

12. A bag contains eight red marbles and nine blue marbles. If two marbles are selected without replacement from the bag, what is the probability that:
    (a) Both marbles are different?
    (b) Both marbles are of the same color?

13. An urn contains eight yellow balls, seven blue balls, and nine white balls. If six balls are randomly chosen from the urn, what is the probability that:
    (a) Two white balls, two blue balls, and two yellow balls are chosen?
    (b) All balls are the same color?

14. An urn contains four green and nine yellow balls. Two balls are selected at random from the urn. Determine the probability that:
    (a) Two yellow balls are selected without replacement.
    (b) Balls of different colors are selected without replacement.
    (c) Balls of the same color are selected without replacement.

15. An urn contains four green and nine yellow balls. Four balls are chosen at random from the urn. Determine the probability that:
    (a) Two balls of different colors are selected.
    (b) At least two yellow balls are selected.

16. An urn contains five yellow and six blue balls.
    (a) If two balls are selected, without replacement, from this urn, what is the probability that the two balls selected are blue?
    (b) If two balls are selected, without replacement, from this urn, what is the probability that the two balls are of one color?

17. Three machines, A, B, and C, are used for producing items in a factory. Two of the machines (A and B) are prone to breaking down so often that they are rested more than machine C. Consequently A and B each produce 25% of the output while C produces the rest. It is known that A, B, and C produce 8%, 7%, and 4% defectives. What is the total probability that if an item produced by the factory is chosen at random and inspected, it is defective? What is the probability that it was produced by (a) machine A? (b) machine B? or (c) machine C?

18. A local grocer, Breadbasket, Inc.; sources 35% of its oranges from California, 30% from Florida, and the rest from Israel. Suppose that, due to logistics, it has been known that oranges from California are available at the shop 80% of the time, while those from Florida are available in the shop 87% of the time. Oranges from Israel can be found in the shop 78% of the time. Suppose the oranges are sold without stickers to indicate where they were produced. If you purchase a random orange from the shop, what is the probability that it is from: (a) California? (b) Florida? (c) Israel?

19. Insurance company AAMI has found some interesting statistics in a recent survey of drivers: 84% of men and 77% of women have crashed their cars. It is also known that 60.1% of a popular brand of cars is driven by women while the rest are driven by men. Assuming the percentages to be correct, if a random car of this popular brand was seen to have been crashed; what is the probability that it was crashed by: (a) a man? (b) a woman?

20. The event space for an experiment is $S = \{1, 2, 3, 4, 5, 6, 7, 8, 9, 10, 11, 12, 13, 14, 15\}$. We define the following events: $A_1 = \{2, 4, 6, 8, 10, 12, 14\}$; $A_2 = \{1, 5, 8, 9, 10, 13\}$; $A_3 = \{2, 3, 7, 13\}$; $A_4 = \{4, 7, 10, 13, 15\}$. Assume that all simple events in S are equally likely. Find the following probabilities:
    (a) $P(A_1 \cap A_2)$
    (b) $P(A_1 \cap A_3)$
    (c) $P(A_1 \cap A_4)$
    (d) $P(A_2 \cap A_3)$
    (e) $P(A_2 \cap A_4)$
    (f) $P(A_3 \cap A_4)$.

21. Using the information in 4.20, find the following probabilities:
    (a) $P(A_1 \cup A_2)$
    (b) $P(A_1 \cup A_3)$
    (c) $P(A_1 \cup A_4)$
    (d) $P(A_2 \cup A_3)$
    (e) $P(A_2 \cup A_4)$
    (f) $P(A_3 \cup A_4)$.

22. Using the information in 4.20, find the following probabilities:
    (a) $P(A_1 \cap A_2 \cap A_3)$
    (b) $P(A_1 \cap A_3 \cap A_4)$
    (c) $P(A_2 \cap A_3 \cap A_4)$
    (d) $P(A_1 \cup A_2 \cup A_3)$
    (e) $P(A_2 \cup A_3 \cup A_4)$
    (f) $P(A_1 \cup A_3 \cup A_4)$

23. The event space for an experiment is $S = \{2,3,4,5,6,7,8,9,10,11,12,14\}$. We define the following events:

    $A = \{2, 4, 6, 8, 10, 12, 14\}$; $B = \{2, 4, 7, 8, 9, 11, 12\}$;
    $C = \{5, 6, 8, 10, 11, 12\}$;

    Assume that all simple events in S are equally likely. Find the following probabilities:
    (a) $P(A \cap B)$
    (b) $P(A \cap C)$
    (c) $P(B \cap C)$
    (d) $P(A \cap B \cap C)$

24. Using the information in 4.23, find the following probabilities:
    (a) $P(A \cup B)$
    (b) $P(A \cup C)$
    (c) $P(B \cup C)$
    (d) $P(A \cup B \cup C)$

CHAPTER 5

Discrete Probability Distributions

5.0 Introduction

In chapter three, we constructed frequency polygons. The purpose was to help us get an idea of the shape of the kind of mathematical function that could be used to model the probabilities that the variable would take some values within its range or domain. In the frequency polygons the probabilities were represented by the frequencies of the classes. Most of what we got from the polygons was an idea of the shape of the function. We found that the shape could be symmetric, skewed to the left or right, unimodal, or multimodal. Such models are called probability distributions.

The kind of distribution used to model a variable would depend on the nature of the variable. A discrete variable is modeled by a discrete distribution, while continuous variables are modeled by continuous distribution. We discuss probability distributions in this section.

5.1 Probability Distributions for Discrete Random Variables

Some variables can only take point values such as . . . $-5, -4, -3, -2, -1, 0, 1, 2, 3, 4, 5, \ldots$ Such variables take no fractional values; for instance, number of people in a family, number of cars passing through a junction per unit time, number of babies born in a hospital per day, number of accidents on a highway per unit time, etc. None of these variables can take values such as 5.5, 6.3, or 2.1. They only take point values. When such a random variable takes values in its domain with some probabilities so that the sum of all the probabilities for all its values is 1.0, then the distribution of such a variable is a discrete distribution.

For instance, consider rolling a die, we can only have six faces (1, 2, 3, 4, 5, or 6) turn up. Let a random variable X represent the face of the die that can occur when we roll it. Never will a face of 3.5 or 4.5 or 2.5 occur when we roll the die. If we associate probabilities of 1/6 to each of the faces, then the sum of the probabilities for all the faces will be 1.0, and we have defined a discrete distribution. Thus, we present a formal definition of a discrete distribution.

If a **random variable, X,** can take point values, $x_1, x_2, \ldots, x_n$ with respective probabilities, $p_1, p_2, \ldots, p_n$ so that $\sum_{i=1}^{n} p_i = 1.0$, then we have defined a discrete probability distribution for X. Thus if we find a function $P(X)$ which takes values $p_1, p_2, \ldots, p_n$ when the values of X are $x_1, x_2, \ldots, x_n$ respectively, we say that X is a discrete random variable because it takes point values with some probabilities. It is usually required that $x_1, x_2, \ldots, x_n$ be finite, or countably infinite.

Consider Table 5.1:

■ **TABLE 5.1**

$X \rightarrow$	X_1	X_2	X_3	$\ldots$	X_{n-1}	X_n	**Total**
$P(x) \rightarrow$	P_1	P_2	P_3		P_{n-1}	P_n	**1.0**

X is a variable that can take discrete values $x_1, x_2, \ldots, x_n$ with associated probabilities $p_1, p_2, \ldots, p_n$ for the table to represent the discrete distribution for X, then $p_1 + p_2 + \ldots + p_n = 1.0$, provided that each p_i satisfies that $0 \leq p_i \leq 1.0$ for $i = 1, 2, 3, \ldots, n$.

We can generate a discrete distribution. Here is an example:

■ **EXAMPLE 5.1**   In a general hospital covering a catchment area, it is known that babies that are going to be born will be either male or female (boy or girl), in addition, it is known that approximately two in every three births is a female. If out of the antenatal register in this hospital, four deliveries were chosen at random, obtain the probability distributions for the random variable, X and Y which give the male and female composition of the four deliveries and their probabilities.

Solution   First, we list the sample space for the possible outcomes of the experiment with associated probabilities which was obtained as the product of probabilities of individual events due to the independence of the events. For instance, if we set B = boy, and G = girl, then $P(BBBB) = P(B) \cdot P(B) \cdot P(B) \cdot P(B) = \frac{1}{3} \cdot \frac{1}{3} \cdot \frac{1}{3} \cdot \frac{1}{3} = \frac{1}{81}$;

$P(BBBG) = P(B) \cdot P(B) \cdot P(B) \cdot P(G) = \frac{1}{3} \cdot \frac{1}{3} \cdot \frac{1}{3} \cdot \frac{2}{3} = \frac{2}{81}$ and the possible outcomes are:

List of All Outcomes (S)	Probability
BBBB	$1/81$
BBBG	$2/81$
BBGB	$2/81$
BBGG	$4/81$
BGBB	$2/81$

List of All Outcomes (S)	Probability
BGBG	$4/81$
BGGB	$4/81$
BGGG	$8/81$
GBBB	$2/81$
GBBG	$4/81$
GBGB	$4/81$
GBGG	$8/81$
GGBB	$4/81$
GGBG	$8/81$
GGGB	$8/81$
GGGG	$16/81$

Based on the above, we write the probability distribution of X and Y, the random variables which describe the possible gender-compositions:

No. of Boys (X)	Probability
0	$16/81$
1	$32/81$
2	$24/81$
3	$8/81$
4	$1/81$

No. of Girls (Y)	Probability
4	$16/81$
3	$32/81$
2	$24/81$
1	$8/81$
0	$1/81$

From the two distributions, we can clearly see that if out of the four births there was no girl, then there were four boys. Similarly, when three girls occurred in the sample, there was only one boy, and so forth, making sure that the total number of births is always four.

5.2 The Concept of Expectation, Mean, Variance, and Standard Deviation for a Discrete Distribution

A **random variable, X,** which takes the values, $x_1, x_2, \ldots, x_n$ with respective probabilities, $p_1, p_2, \ldots, p_n$ so that $\sum_{i=1}^{n} p_i = 1.0$, has a discrete distribution and that discrete distribution has a mean and variance. The mean denoted by μ, is defined as:

$$\mu = p_1 x_1 + p_2 x_2 + p_3 x_3 + \cdots + p_n x_n = \sum_{i=1}^{n} p_i x_i;$$

which is also called the expectation of X or the expected value of X, denoted by:

$$EX = p_1 x_1 + p_2 x_2 + p_3 x_3 + \cdots + p_n x_n = \sum_{i=1}^{n} p_i x_i = \mu, \text{ and we use mean}$$

and EX interchangeably.

It is easy to show that expectation actually measures the mean of a random variable. Using a crude example, if we recall that probability is defined in terms of relative frequencies, we can write $EX = \sum_{i=1}^{k} \left[\dfrac{f_i}{\sum_{i=1}^{k} f_i} \right] X_i$ from which we can easily see that this is the grouped frequency mean for X.

In order to find the variance of the variable X which has the discrete distribution described above, we need to find the **Expectation of X^2** denoted by EX^2. The Expectation of X^2 has the following definition:

$$EX^2 = p_1 x_1^2 + p_2 x_2^2 + p_3 x_3^2 + \cdots + p_n x_n^2 = \sum_{i=1}^{n} p_i x_i^2.$$

The variance of X is obtained as:

$$\sigma^2 = EX^2 - (EX)^2 = EX^2 - \mu^2$$
$$= \sum_{i=1}^{n} p_i x_i^2 - \mu^2.$$

The standard deviation of X is σ and is obtained from σ^2 as follows:

$$\sigma = \sqrt{EX^2 - (EX)^2} = \sqrt{EX^2 - \mu^2}$$
$$= \sqrt{\sum_{i=1}^{n} p_i x_i^2 - \mu^2}.$$

■ **EXAMPLE 5.2**   In Example 5.1, we found the distributions of X and Y. Find the mean, variance, and standard deviations for the distributions of: (a) X and (b) Y.

Solution   (a)

No. of Boys (X)	Probability (p_i)	$p_i x_i$	$p_i x_i^2$
0	$^{16}/_{81}$	$0 \times \dfrac{16}{81} = 0$	$0^2 \times \dfrac{16}{81} = 0$
1	$^{32}/_{81}$	$1 \times \dfrac{32}{81} = \dfrac{32}{81}$	$1^2 \times \dfrac{32}{81} = \dfrac{32}{81}$
2	$^{24}/_{81}$	$2 \times \dfrac{24}{81} = \dfrac{48}{81}$	$2^2 \times \dfrac{24}{81} = \dfrac{96}{81}$
3	$^{8}/_{81}$	$3 \times \dfrac{8}{81} = \dfrac{24}{81}$	$3^2 \times \dfrac{8}{81} = \dfrac{72}{81}$
4	$^{1}/_{81}$	$4 \times \dfrac{1}{81} = \dfrac{4}{81}$	$4^2 \times \dfrac{1}{81} = \dfrac{16}{81}$
Total→	1.00	$EX = \dfrac{108}{81}$	$EX^2 = \dfrac{216}{81}$

Thus,

$$EX = \frac{108}{81}; \; EX^2 = \frac{216}{81}$$

$$\sigma^2_{boys} = EX^2 - (EX)^2 = \frac{216}{81} - \left(\frac{108}{81}\right)^2 = \frac{216(81) - 108^2}{81^2} = 0.88889$$

$$\sigma_{boys} = \sqrt{0.88889} = 0.94281.$$

(b)

No. of Girls (Y)	Probability (p_i)	$p_i x_i$	$p_i x_i^2$
4	$^{16}/_{81}$	$4 \times \dfrac{16}{81} = \dfrac{64}{81}$	$4^2 \times \dfrac{16}{81} = \dfrac{256}{81}$
3	$^{32}/_{81}$	$3 \times \dfrac{32}{81} = \dfrac{96}{81}$	$3^2 \times \dfrac{32}{81} = \dfrac{288}{81}$
2	$^{24}/_{81}$	$2 \times \dfrac{24}{81} = \dfrac{48}{81}$	$2^2 \times \dfrac{24}{81} = \dfrac{96}{81}$
1	$^{8}/_{81}$	$1 \times \dfrac{8}{81} = \dfrac{8}{81}$	$1^2 \times \dfrac{8}{81} = \dfrac{8}{81}$
0	$^{1}/_{81}$	$0 \times \dfrac{1}{81} = 0$	$0^2 \times \dfrac{1}{81} = 0$
Total→	1.00	$EX = \dfrac{216}{81}$	$EX^2 = \dfrac{648}{81}$

$$EY = \frac{216}{81}; \; EY^2 = \frac{648}{81}$$

$$\sigma^2_{girls} = EY^2 - (EY)^2 = \frac{648}{81} - \left(\frac{216}{81}\right)^2 = \frac{648(81) - 216^2}{81^2} = 0.88889$$

$$\sigma_{girls} = \sqrt{0.88889} = 0.94281.$$

■ **EXAMPLE 5.3**   Mr. Zales has been a car salesman with a car sales company on Division Street, St. Cloud, MN for many years. An examination of his books shows that the number of cars he sells daily has never exceeded five and sometimes it can be zero. The following was the distribution of his daily car sales as obtained from his records:

X	0	1	2	3	4	5
P(x)	0.083	0.215	0.35	0.176	0.112	0.064

Obtain the mean, variance and standard deviation for Mr. Zales daily car sales.

Solution

$X \rightarrow$	0	1	2	3	4	5	Total
$P(x) \rightarrow$	0.083	0.215	0.35	0.176	0.112	0.064	1.0
$x \cdot p(x) \rightarrow$	0	0.215	0.63	0.528	0.448	0.32	2.211
$X^2 \cdot p(x) \rightarrow$	0	0.215	1.26	1.584	1.792	1.6	6.591

$EX = 2.211; EX^2 = 6.591$

$\sigma^2 = EX^2 - (EX)^2 = 6.591 - (2.211)^2 = 1.702479$

$\sigma = \sqrt{1.702479} = 1.304791$

The mean, variance and standard deviation are 2.211, 1.702479, and 1.304791. We can obtain the discrete distribution for a variable based on relative frequency or count data, and then determine the mean, variance, and standard deviation as described previously. We show another example.

■ **EXAMPLE 5.4**   A study of families recorded the number of families that had Y number of children:

Y	0	1	2	3	4	5	6	7
Count	6	14	220	112	40	9	2	1

Obtain the discrete distribution for Y by finding the relative frequencies for each value of Y. Find the mean, variance, and standard deviation of the distribution of children per family based on the relative frequencies obtained for each value of Y.

Solution

Y	0	1	2	3	4	5	6	7	Total
Count	6	14	220	112	40	9	2	1	404
Probability or Relative Frequency	0.0149	0.03465	0.54455	0.2772	0.099	0.0223	0.00495	0.0025	1.0
$y \cdot p(y)$	0	0.03465	1.08911	0.8316	0.396	0.1115	0.0297	0.0175	2.5101
$y^2 \cdot p(y)$	0	0.03465	2.17822	2.4948	1.584	0.5575	0.1782	0.1225	7.1499

$$EY = 2.5101; \, EY^2 = 7.1499$$
$$\sigma^2 = EY^2 - (EY)^2 = 7.1499 - (2.5101)^2 = 0.8495$$
$$\sigma = \sqrt{0.8495} = 0.92167$$

The mean, variance, and standard deviation are 2.5101, 0.8495, and 0.92167.

5.3   Other Discrete Distributions

In the previous sections we discussed the general discrete distribution values and used them to model different phenomena which have discrete outcomes; such as Mr. Zales car sales, or number of children per family. There are many families of distributions which fall into the category of discrete distributions. These distributions have been given names, such as the Bernoulli, the binomial, the Poisson, the geometric, the hypergeometric, and the negative binomial. We will treat three of these distributions: the Bernoulli, the Binomial, and the Poisson.

5.3.1 The Bernoulli Distribution

A coin toss experiment leads to only two possible outcomes, $S = \{\text{head, tail}\}$. This type of experiment has many real life equivalents which have only two possible outcomes. Here are some examples:

- In pregnancy we expect the outcome to be a boy or a girl.
- If we plant a seed, we expect it to germinate or not germinate.
- A person is either married or unmarried.
- A rocket flies when launched or fails.
- If an issue crops up we are either satisfied or not satisfied.
- The result of sitting for an exam is pass or fail.
- Manufactured items are either defective or non-defective.

We have only given a few. Since the outcomes occur by chance, they can be modeled by the Bernoulli distribution. We can write the functional form of the Bernoulli distribution as:

$$p(x) = p^x(1 - p)^{1-x} \, ; x = 0, 1$$

where the parameter, p, is such that $0 \leq p \leq 1.0$. Generally X is regarded as the variable, with possible outcomes of 0 or 1 corresponding to the two possible outcomes; such as head or tail for the coin-flipping experiment. If head is the focus, then the occurrence of a head is regarded as a "success," while occurrence of a tail is regarded as a "failure." In this model p is the probability of success while $1 - p$ is the probability of failure. Sometimes $1 - p$ is replaced by q. Like other discrete distributions we described previously, the Bernoulli distribution has a mean, a variance, and a standard deviation. They are p, pq, and $\sqrt{pq}$, respectively.

The mean, variance, and standard deviation of this distribution can be found using formulas we described earlier as follows:

$mean = \mu = EX = 1 \times p + (1 - p) \times 0 = p$. The mean of the Bernoulli distribution is p.

$variance = \sigma^2 = E(X^2) - \mu^2 = E(X^2) - p^2$

$E(X^2) = 1^2 \times p + 0^2 \times (1 - p) = p$

$\sigma^2 = p - p^2 = p(1 - p) = pq$ if $q = 1 - p$.

■ **EXAMPLE 5.5** Suppose that in human pregnancy, a boy is as equally likely as a girl.

(a) Find the mean, variance, and standard deviation of the Bernoulli distribution for human births.
(b) What are the mean, variance, and standard deviation for the distribution of human births if probability of a boy is 0.45?

Solution In this case, if we focus on either boy or girl as success, probability of "success" would be equal to probability of failure.

(a) The applicable Bernoulli distribution would be:

$$p(x) = \begin{cases} \dfrac{1}{2} \text{ if } x = 1 \\ \dfrac{1}{2} \text{ if } x = 0 \end{cases}$$

The mean is $\mu = \dfrac{1}{2}$ and the variance is $\sigma^2 = \dfrac{1}{2} - \left(\dfrac{1}{2}\right)^2 = \dfrac{1}{4} \Rightarrow \sigma = \sqrt{\dfrac{1}{4}} = \dfrac{1}{2}$.

Thus the variance for this distribution is $\dfrac{1}{4}$ while the standard deviation is $\dfrac{1}{2}$.

(b) In this case, if we focus on boy as "success," probability of "success" is 0.45, while 0.55 would be the probability of failure. The applicable Bernoulli distribution would be:

$$p(x) = \begin{cases} 0.45 \text{ if } x = 1 \\ 0.55 \text{ if } x = 0 \end{cases}.$$

The mean is $\mu = 0.45$ and the variance is $\sigma^2 = 0.45 - (0.45)^2 = 0.2475 \Rightarrow \sigma = \sqrt{0.2475} = 0.497494$. Thus the variance for this distribution is 0.2475 while the standard deviation is 0.497494.

5.3.2 The Binomial Distribution

When a series of n Bernoulli trials have taken place, the outcome of that series usually gives rise to a Binomial variable, X. Thus, a binomial variable X is the sum of n Bernoulli variables, so that if $x_1, x_2, \ldots, x_n$ are such Bernoulli variables, then $X = x_1 + x_2 + \ldots, + x_n$. Often the Binomial distribution is a more popular vehicle for studying Bernoulli trials. If X is a binomial variable, the distribution of X can be written functionally as:

$$p(x) = \binom{n}{x} p^x (1 - p)^{n-x}; \; x = 0, 1, 2, \ldots, n,$$

and called the binomial distribution. In chapter four, we defined the quantity $\binom{n}{x}$ when we treated combinatorial analysis. The quantity is now an integral part of the definition of the binomial distribution and is evaluated as part of the computation of the probabilities that X would assume some discrete values within its domain.

In this case, it is assumed that n Bernoulli trials have taken place and x successes have been observed, which implies that $n - x$ failures have occurred. As in the Bernoulli case, $0 \leq p \leq 1.0$. A random variable, X, which is the outcome of a series of experiments; the outcome of each of which can be classed into two broad categories, "success" and "failure," with probability of success given by p, and probability of failure given by $q = 1 - p$, a **binomial random variable.** It can easily be seen that if we set $n = 1$ in the above binomial distribution, we obtain the Bernoulli distribution.

The binomial distribution can be used to work out practical problems related to management decisions. In the next example binomial distribution was used by a horticultural marketing store to arrive at a rough approximation of the proportion of space to devote to annuals and perennials in their shop. They already know that the probability that any of their customers would buy an annual is 3/5.

■ **EXAMPLE 5.6**   CCT flower/plants stores took a sample of 1800 customers (typical weekly clientele) who bought four flower/plants from them, and wished to calculate the expected number of customers who buy:

    (i) no perennial
    (ii) two or three perennials
    (iii) three annuals
    (iv) no annuals

Solution   This is a typical case in which the binomial distribution can be applied. Roughly, we can define annual = "success," and perennial = "failure" and model the problem as arising from a binomial random variable, X, with $n = 4$, $p = 3/5$ and $q = 2/5$. With this, we can work out the probabilities of a customer buying a combination of annuals and perennials, and hence the expected number of customers.

The specific binomial distribution which applies to this problem is:

$$p(x) = \binom{4}{x}\left(\frac{3}{5}\right)^x \left(\frac{2}{5}\right)^{4-x}; \; x = 0, 1, 2, 3, 4.$$

In each case, the expected number = probability $\times$ 1800

(i) $P(\text{no perennial}) = P(4 \text{ annuals}) = \binom{4}{0}\left(\frac{3}{5}\right)^4\left(\frac{2}{5}\right)^0 = \dfrac{81}{625} = 0.1296$

$$\text{Expected number} = \frac{81}{625} \times 1800 = 233.28 \approx 233$$

(ii) P(2 or 3 perennials) = P(2 or 1 annuals)

$$= \binom{4}{2}\left(\frac{3}{5}\right)^2\left(\frac{2}{5}\right)^2 + \binom{4}{1}\left(\frac{3}{5}\right)\left(\frac{2}{5}\right)^3 = \frac{216}{625} + \frac{96}{625} = \frac{312}{625} = 0.4992$$

$$\text{Expected number} = \frac{312}{625} \times 1800 = 898.56 \approx 899$$

(iii) P(3 annuals) = $\binom{4}{3}\left(\frac{3}{5}\right)^3\left(\frac{2}{5}\right)^2 = \frac{216}{625} = 0.3456$

$$\text{Expected number} = \frac{216}{625} \times 1800 = 622.08 \approx 622$$

(iv) P(no annual) = $\binom{4}{0}\left(\frac{3}{5}\right)^0\left(\frac{2}{5}\right)^4 = \frac{16}{625} = 0.0256$

$$\text{Expected number} = \frac{16}{625} \times 1800 = 46.08 \approx 46$$

■ **EXAMPLE 5.7**   A company which transports tourists to a wild-life resort deliberately over-books its buses because it knows from previous experience that only 75% of the passengers turn up. For a particular stop for which she normally reserves ten seats, she booked thirteen passengers. (i) What is the probability that all the ten seats will be occupied? (ii) What is the probability that only 60% of the reserved seats are occupied? (iii) What is the probability that they would have to pay compensation to some passengers for not finding space for them?

Solution   Because only 75% of passengers turn up, the probability that any customer will turn up is 0.75. We use the binomial distribution, and assume that $p = 0.75, n = 13$ and for the first question, $x = 10$ and for the second question, $x = 60\%$ of $10 = 6$. The specific binomial distribution which applies to this situation:

$$p(x) = \binom{13}{x}(0.75)^x(0.25)^{13-x}; x = 0, 1, 2, 3, \dots, 13$$

(i) P(10 *turn up*) = $P(X = 10) = \binom{13}{10}(0.75)^{10}(0.25)^{13-10} = 0.25165$

(ii) P(60% *is occupied*) = $P(X = 6) = \binom{13}{6}(0.75)^6(0.25)^{13-6} = 0.01864$

(iii) P(they would have to pay compensation to some persengers)
$= P$(11, 12, or 13 passengers turn up) = $P(11) + P(12) + P(13)$

$$= \binom{13}{11}(0.75)^{11}(0.25)^{13-11} + \binom{13}{12}(0.75)^{12}(0.25)^{13-12}$$

$$+ \binom{13}{13}(0.75)^{13}(0.25)^{13-13}$$

$$= 0.205896 + 0.102948 + 0.0237573 = 0.3326012$$

This decision to book thirteen passengers for ten seats is likely to lead to compensating at least one, two, or three passengers about a third of the time.

■ **EXAMPLE 5.8**   If two unbiased dice are thrown together eight times, find the probability that the sum of the two faces which turn up (i) is exactly 10, five times; and (ii) is greater than or equal to 10, four times.

We can use the binomial distribution to solve this problem. For (i) we need to work out how many times the sum of two faces can be exactly 10. There are thirty-six possible outcomes for the two dice, such as (1, 1), (1, 2), . . ., (6, 6), as illustrated in the figure below:

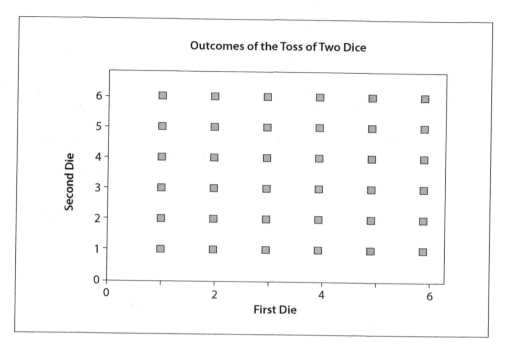

However, the sum of two faces can only be 10 in a few of these cases, namely, (4, 6), (5, 5), (6, 4). This means that the probability of getting exactly two faces whose sum is 10 is 3/36=1/12. Using the binomial distribution we can easily show that the probability of any of these combinations of faces turning up five out of eight tosses would be modeled by the following specific binomial distribution:

$$p(x) = \binom{8}{x}\left(\frac{1}{12}\right)^{x}\left(\frac{11}{12}\right)^{8-x} ; x = 0, 1, 2, 3, \ldots, 8$$

P (sum of two faces being 10, five times out of eight) is $p(5) = \binom{8}{5}\left(\frac{1}{12}\right)^{5}\left(\frac{11}{12}\right)^{8-5} = 0.0001733$

(ii) The sum of two faces equal or exceed 10 when the combination of faces that turn up are (4, 6), (5, 5), (6, 4), (5, 6), (6, 5) and (6, 6). Hence in a single roll of the two dice, the probability of the sum of the values of two faces being equal or exceeding 10 is 6/36, or 1/6. The binomial distribution which applies to this situation:

$$p(x) = \binom{8}{x}\left(\frac{1}{6}\right)^{x}\left(\frac{5}{6}\right)^{8-x} ; x = 0, 1, 2, 3, \ldots, 8$$

Hence, p (sum of two faces equal or exceed 10, four times out of eight) =
$p(4) = \binom{8}{4}\left(\frac{1}{6}\right)^{4}\left(\frac{5}{6}\right)^{8-4} = 0.026048$

■ **EXAMPLE 5.9**   The probability that the gestation period of a woman will exceed nine months is 0.314. In ten human births, what is the probability that the number of pregnancies in which gestation period exceeds nine months is:

(a) between three and five inclusive?
(b) no more than three?
(c) exactly six?
(d) greater than three, but no more than eight?

For the above random variable, X, which is the distribution of gestation periods which exceed nine months; obtain the mean, variance, and standard deviation.

Solution   This is a binomial problem because a pregnancy can either exceed or not exceed nine months. Ten pregnancies represent ten trials. The relevant binomial distribution which applies to the gestation period is:

$$p(x) = \binom{10}{x}(0.314)^x(1 - 0.314)^{10-x}; \; x = 0, 1, 2, \ldots, 10$$

(a) $P(3 \leq x \leq 5) = \binom{10}{3}(0.314)^3(0.686)^{10-3} + \binom{10}{4}(0.314)^4(0.686)^{10-4}$

$\qquad\qquad\qquad + \binom{10}{5}(0.314)^5(0.686)^{10-5}$

$\qquad\qquad = 0.265607 + 0.212756 + 0.116861 = 0.595224.$

(b) $P(0 \leq x \leq 3) = p(0) + p(1) + p(2) + p(3) = \binom{10}{0}(0.314)^0(0.686)^{10-0}$

$\qquad\qquad\qquad + \binom{10}{1}(0.314)^1(0.686)^{10-1} + \binom{10}{2}(0.314)^2(0.686)^{10-2}$

$\qquad\qquad\qquad + \binom{10}{3}(0.314)^3(0.686)^{10-3}$

$\qquad\qquad = 0.02308 + 0.105644 + 0.217603 + 0.265607 = 0.611934.$

(c) $P(X = 6) = p(6) = \binom{10}{6}(0.314)^6(0.686)^{10-6} = 0.044575.$

(d) $P(3 < x \leq 8) = p(4) + p(5) + p(6) + p(7) + p(8)$

$\qquad = \binom{10}{4}(0.314)^4(0.686)^{10-4} + \binom{10}{5}(0.314)^5(0.686)^{10-5}$

$\qquad\qquad + \binom{10}{6}(0.314)^6(0.686)^{10-6} + \binom{10}{7}(0.314)^7(0.686)^{10-7}$

$\qquad\qquad + \binom{10}{8}(0.314)^8(0.686)^{10-8}$

$\qquad = 0.212756 + 0.116861 + 0.044575 + 0.011659 + 0.002001$

$\qquad = 0.387853.$

■ **EXAMPLE 5.10**   An electronic component is mass-produced and then tested unit by unit on an automatic testing machine which classifies the unit as good or defective. However, there is a probability of 0.2 that the machine will misclassify the unit. Each unit is therefore tested six times and the classification is accepted if so classified three times or more. After six tests, what is the probability of misclassification of a unit?

Solution Here again, we can use the binomial distribution to find the probability that an item is misclassified after six tests. Since probability of a classification being right is $1 - 0.2 = 0.8$, we note that if an item is given a classification 0, 1, or 2 times, it is not accepted; that is, the classification is regarded as incorrect.

Let X be the probability of correct classification. Hence, this is a binomial distribution, $b(n, p)$ with $p = 0.8$, $n = 6$ with the specific binomial distribution of:

$$p(x) = \binom{6}{x}(0.80)^x(1 - 0.80)^{6-x}; x = 0, 1, 2, \ldots, 6.$$

$$P \text{ (classification being rejected)} = P(x = 0, 1, \text{ or } 2) = P(0) + P(1) + p(2)$$

$$= p(x \le 2) = \binom{6}{0}(0.8)^0(1 - 0.8)^{6-0}$$

$$+ \binom{6}{1}(0.8)^1(1 - 0.8)^{6-1}$$

$$+ \binom{6}{2}(0.8)^2(1 - 0.8)^{6-2}$$

$$= 0.01696$$

This means that instead of 80% of the items which used to be classified correctly after six classifications, $100(1 - 0.01696) = 98.3\%$ are now properly classified.

■ **EXAMPLE 5.11** It is known that only 30% of a type of electrical component fitted on a model of a car manufactured by DCC Company will last more than one year. If twelve cars of that model are chosen at random and this component inspected, what is the probability that: (i) four or more are functioning (after one year of use)? (ii) the number functioning lies between three and five inclusive? (iii) none is in good order?

Solution Let X be the number of functioning cars. Using the binomial, with $p = 0.30$, $q = 0.7$ and $n = 12$, which leads to the specific binomial distribution of $p(x) = \binom{12}{x}(0.3)^x(1 - 0.7)^{12-x}$ we solve the problems:

(i) $p(X \ge 4) = 1 - p(X = 0, X = 1, X = 2, \text{ or } X = 3)$
$$= 1 - \{P(1) + P(2) + P(3) + P(4)\} = 1 - p(X \le 3)$$
$$= 1 - \binom{12}{0}(0.3)^0(0.7)^{12-0} - \binom{12}{1}(0.3)^1(0.7)^{12-1}$$
$$- \binom{12}{2}(0.3)^2(0.7)^{12-2} + \binom{12}{3}(0.3)^3(0.7)^{12-3}$$
$$= 1 - 0.0138 - 0.0712 - 0.1678 - 0.2397 = 1 - 0.4925$$
$$= 0.5075$$

(ii) $P(3 \le X \le 5) = \binom{12}{3}(0.3)^3(0.7)^{12-3} + \binom{12}{4}(0.3)^4(0.7)^{12-4}$
$$+ \binom{12}{5}(0.3)^5(0.7)^{12-5}$$
$$= 0.2397 + 0.2311 + 0.1585 = 0.6293$$

(iii) $P(X = 0) = p(0) = \binom{12}{0}(0.3)^0(0.7)^{12-0} = 0.0138$

5.3.3 The Poisson Distribution

The Poisson distribution was discovered by a Frenchman named Simeon Poisson (1781–1840). This distribution is used to model the number of occurrences of events per unit time, unit area, or unit volume. It is extensively used in the Telecommunications industry. It is one of the tools used for studying queues at service points with a view to eliminating congestion. This distribution can therefore be used to model things like number of telephone calls passing through an exchange per unit time, number of births at a maternity ward per unit time, number of customers arriving at a service point (supermarket check-outs, gas stations, emergency room, doctor's outpatient clinic if patients are allowed to show up without appointment, etc.), number of rare snails/acre in a rainforest, the incidence of a disease per unit area, number of organisms by unit volume of pond water, etc.

The distribution is written as follows:

$$p(x) = \frac{\lambda^x e^{-\lambda}}{x!}; x = 0, 1, 2, \ldots, \infty;$$

where x is the number of occurrences; λ is the mean number of occurrences per unit time, per unit area, or per unit volume; and where $e = 2.71828\ldots$, the base of natural logarithm. The domain of x is $0, 1, 2 \ldots$ to infinity.

It can be shown that the mean, variance, and standard deviation of the Poisson distribution are: λ, λ and $\sqrt{\lambda}$ respectively, but showing it is beyond the scope of this text.

■ **EXAMPLE 5.12** It is expected that, on average, four snails of a rare species would be found per acre in a rainforest. A biologist searches a random acre in the forest for the snails. What is the probability that she finds:

    (a) exactly five snails?
    (b) between three and five snails inclusive?
    (c) no snails?
    (d) more than three snails?

Solution Here the rate of occurrence per unit area (acre) is four. Therefore $\lambda = 4$. Thus the Poisson distribution modeling the distribution of the snails in this forest is:

$$p(x) = \frac{4^x e^{-4}}{x!}; x = 0, 1, 2, 3, \ldots$$

(a) $P(5) = \dfrac{4^5 e^{-4}}{5!} = 0.156293$

(b) $P(3 \le x \le 5) = P(3) + P(4) + P(5) = \dfrac{4^3 e^{-4}}{3!} + \dfrac{4^4 e^{-4}}{4!} + 0.156293$

                $= 0.195367 + 0.195367 + 0.156293 = 0.547027$

(c) $P(0) = \dfrac{4^0 e^{-4}}{0!} = e^{-4} = 0.018316$

(d) $P(X > 3) = 1 - P(X \le 3) = 1 - P(0) - P(1) - P(2) - P(3)$

$$= 1 - 0.018316 - \frac{4^1 e^{-4}}{1!} - \frac{4^2 e^{-4}}{2!} - 0.195367$$

$$= 1 - 0.018316 - 0.073263 - 0.146525 - 0.195367$$

$$= 1 - 0.43347 = 0.56653$$

■ **EXAMPLE 5.13** On average four children are born daily at a local maternity in Heathsville. If a day is chosen at random, what is the probability that: (i) no more than two children would be born? (ii) more than two, but less than five, children would be born? (iii) five or more children would be born at Heathsville maternity?

Solution The number born daily at the Maternity can be modeled by a Poisson distribution, whose mean is the number of births per unit time (day). The specific distribution that applies is:

$$p(x) = \frac{4^x e^{-4}}{x!}; x = 0, 1, 2, \ldots,$$

(i) $P(0) + P(1) + P(2) = \dfrac{4^0 e^{-4}}{0!} + \dfrac{4^1 e^{-4}}{1!} + \dfrac{4^2 e^{-4}}{2!}$

$$= 0.018316 + 0.073263 + 0.146525 = 0.238103$$

(ii) $P(2 \le X < 5) = P(2) + P(3) + P(4) = \dfrac{4^2 e^{-4}}{2!} + \dfrac{4^3 e^{-4}}{3!} + \dfrac{4^4 e^{-4}}{4!}$

$$= 0.146525 + 0.195367 + 0.195367 = 0.537259$$

(iii) $P(X \ge 5) = 1 - P(X \le 4)$

$$= 1 - \left(\frac{4^0 e^{-4}}{0!} + \frac{4^1 e^{-4}}{1!} + \frac{4^2 e^{-4}}{2!} + \frac{4^3 e^{-4}}{3!} + \frac{4^4 e^{-4}}{4!} \right)$$

$$= 1 - (0.018316 + 0.073263 + 0.146525 + 0.195367$$

$$+ 0.195367) = 0.371163$$

■ **EXAMPLE 5.14** At an outpatient clinic, the number of patients arriving for treatment per hour follows the Poisson distribution, with mean 3.5. If an hour of the day is chosen at random, what is the probability that:

  (i) at most three patients arrived during the hour?
  (ii) between two and four patients, inclusive, arrived at the facility in an hour?
  (iii) more than two but less than six?
  (iv) What are the mean, variance, and standard deviation of the distribution of arriving?

Solution (i) $P(0) + P(1) + P(2) + P(3) = \dfrac{3.5^0 e^{-3.5}}{0!} + \dfrac{3.5^1 e^{-3.5}}{1!} + \dfrac{3.5^2 e^{-3.5}}{2!}$

$$+ \frac{3.5^3 e^{-3.5}}{3!}$$

$$= 0.03020 + 0.105691 + 0.184959 + 0.215785$$

$$= 0.536635$$

(ii) $P(2 \le X \le 4) = P(2) + P(3) + P(5) = \dfrac{3.5^2 e^{-3.5}}{2!} + \dfrac{3.5^3 e^{-3.5}}{3!} + \dfrac{3.5^4 e^{-3.5}}{4!}$

$$= 0.184959 + 0.215785 + 0.188812 = 0.589557$$

(iii) $P(2 < X < 6) = p(3) + p(4) + p(5) = \dfrac{3.5^3 e^{-3.5}}{3!} + \dfrac{3.5^4 e^{-3.5}}{4!} + \dfrac{3.5^5 e^{-3.5}}{5!}$

$= 0.215785 + 0.188812 + 0.132169 = 0.536766$

(iv) $mean = 3.5$, $variance = 3.5$, $STD = \sqrt{3.5} = 1.870829$

■ **EXAMPLE 5.15**   On average, the number of a particular organism found by biologists in a 1 ml of water from the research pond at XYZ university is four. If 1ml of water is scooped randomly from this pond, what is the probability that the number of the organisms found: (i) is less than or equal to three? (ii) is greater than four?

Solution   The mean number per unit volume is four, so the Poisson distribution which applies is:

$$P(x) = \dfrac{4^x e^{-4}}{x!}; x = 0, 1, 2, \ldots$$

$$p(x) = \dfrac{4^x e^{-4}}{x!}; x = 0, 1, 2, \ldots, \infty$$

(i) $P(0) + P(1) + P(2) + P(3) = \dfrac{4^0 e^{-4}}{0!} + \dfrac{4^1 e^{-4}}{1!} + \dfrac{4^2 e^{-4}}{2!} + \dfrac{4^3 e^{-4}}{3!}$

$= 0.018316 + 0.073263 + 0.146525 + 0.195367$

$= 0.43347$

(ii) $P(X > 4) = 1 - P(X \le 4) = 1 - P(X \le 3) - P(4)$

$= 1 - 0.43347 - \dfrac{4^4 e^{-4}}{4!}$

$= 1 - 0.43347 - 0.195367 = 1 - 0.628837 = 0.371163$

5.4   A Different Form for Poisson Distribution

Sometimes, this distribution is written in the form:

$$p(x) = \dfrac{(\lambda t)^x e^{-\lambda t}}{x!}; x = 0, 1, 2, \ldots$$

0 *elsewhere*

In this second form, the use of Poisson distribution for representing time-dependence is emphasized. In this we can use it to model the number of occurrences of an event in a time interval t. Thus, we can use this distribution to model the number of telephone calls arriving at an exchange in a time interval of t, number of cars passing through a junction in a time interval of t, number of babies born in a city hospital in a given period t, number of customers arriving at a service point in a time interval t (where λ is the number of observations per unit time), etc. Thus, $p(x)$ is the probability that x events occur in the interval t. Under this form of the distribution, occurrences in two different intervals are independent. It can be shown that λt is both the mean and variance of this version of the Poisson distribution. This is shown in the next example.

■ **EXAMPLE 5.16**   Telephone calls arrive at an exchange at the rate of five per minute. What is the probability that three calls arrive in fifteen seconds? Find the probability of three or more calls in fifteen seconds.

Solution   In this case, it is better to use the Poisson distribution in the form:

$$p(x) = \frac{(\lambda t)^x e^{-\lambda t}}{x!}; \; x = 0, 1, 2, \ldots$$

Since there are sixty seconds in a minute, then the rate of arrival of five per minute is equivalent to $\frac{5}{60} = \frac{1}{12} =$ one call per twelve seconds. Hence $\lambda = \frac{1}{12}$. In fifteen seconds, we have $\lambda t = \frac{1}{12}(15) = \frac{5}{4}$. Let X be the number of calls arriving in fifteen seconds, then the distribution for these calls is:

$$p(x) = \frac{\left(\frac{5}{4}\right)^x e^{-5/4}}{x!}; \; x = 0, 1, 2, \ldots$$

$$P(3 \text{ calls in 15 seconds}) = P(X = 3) = \frac{\left(\frac{5}{4}\right)^3 e^{-5/4}}{3!} = 0.0933$$

$$P(3 \text{ or more calls in 15 seconds}) = P(X \geq 3) = 1 - P(X < 3)$$
$$= 1 - P(X = 2) - P(X = 1) - P(X = 0)$$
$$= 1 - \frac{\left(\frac{5}{4}\right)^2 e^{-5/4}}{2!} - \frac{\left(\frac{5}{4}\right)^1 e^{-5/4}}{1!} - \frac{\left(\frac{5}{4}\right)^0 e^{-5/4}}{0!}$$
$$= 1 - 0.2238 - 0.3581 - 0.2865 = 0.1316$$

■ **EXAMPLE 5.17**   Twenty calls per hour arrive at a Central Computer, in accordance with a Poisson process. Find the probability that in an hour five calls arrive in the first twenty minutes, and four calls arrive in the last fifteen minutes.

Solution   Twenty calls arrive per hour $\Rightarrow \lambda = 20/60 = 1/3$ call arrives per minute. The underlying distribution which is Poisson is:

$$p(x) = \frac{\left(\frac{1}{3}t\right)^x e^{-\left(\frac{1}{3}t\right)}}{x!}; \; x = 0, 1, 2, \ldots$$

Here, we apply the independent increment and stationary properties of a Poisson process. We obtain the probability of occurrence of the numbers of events in each of the two non-overlapping intervals and multiply them because of the independence property to obtain the desired probability:

$P(5 \text{ calls in first 20 minutes and 4 calls in last 15 minutes})$
  $= P(5 \text{ calls in the first 20 minutes}).$

$$P(4 \text{ calls in the last 15 minutes}) = \frac{\left(\frac{1}{3} \cdot 20\right)^5 e^{-\left(\frac{1}{3} \cdot 20\right)}}{5!} \times \frac{\left(\frac{1}{3} \cdot 15\right)^4 e^{-\left(\frac{1}{3} \cdot 15\right)}}{4!}$$
$$= 0.1397 \times 0.1755 = 0.0245$$

5.5   The Use of MINITAB to Calculate Probabilities for Some Discrete Distributions

5.5.1 Calculating Binomial Probabilities using MINITAB

■ **EXAMPLE 5.18**   In the production of a certain type of paper, a process used is known to produce 20% defectives. If ten papers are produced, what is the probability that: (a) exactly nine are not defective, (b) greater than five are not defective, and (c) between four and eight papers are not defective.

Solution   This is a binomial problem in which P(paper not defective), $p = 0.8$; while P(paper is defective), $q = 0.20$; $n = 10$. To find the probabilities, we use MINITAB as follows:

```
MTB > set c1
DATA> 0 1 2 3 4 5 6 7 8 9 10
DATA> end
MTB > pdf c1 c2;
SUBC> binomial 10 0.8.
```

The output is the table below:

x	p(x)
0	1.02E-07
1	4.1E-06
2	7.37E-05
3	0.000786
4	0.005505
5	0.026424
6	0.08808
7	0.201327
8	0.30199
9	0.268435
10	0.107374

With this output, we see that the solutions for the questions above are:

(a) $P(X = 9) = 0.268435$

(b) $P(X > 5) = 1 - p(X < 5)$
$$= 1 - (0.005505 - 0.000786 - 7.37E - 05 - 4.1E - 06$$
$$- 1.02E - 07)$$
$$= 1 - 0.006369 = 0.993631$$

(c) $P(4 \leq X \leq 8) = p(4) + p(5) + p(6) + p(7) + p(8)$
$$= 0.005505 + 0.026424 + 0.08808 + 0.201327 + 0.30199$$
$$= 0.623336$$

EXAMPLE 5.19   We re-do Example 5.11 using MINITAB:

It is known that only 30% of a type of electrical component fitted on a model of a car manufactured by DCC Company last more than one year. If twelve cars of that model are chosen at random and this component inspected, what is the probability that: (i) four or more are functioning? (ii) the number functioning lies between three and five inclusive? (iii) none is in good order?

Solution $p = 0.30$; $q = 0.7$; $n = 12$. We carry out the analysis in MINITAB as follows:

```
MTB > set c1
DATA> 0 1 2 3 4 5 6 7 8 9 10 11 12
DATA> end
MTB > pdf c1 c2;
SUBC> binomial 12 0.3.
```

The output:

x	p(x)
0	0.013841
1	0.071184
2	0.16779
3	0.2397
4	0.23114
5	0.158496
6	0.079248
7	0.029111
8	0.007798
9	0.001485
10	0.000191
11	1.49E-05
12	5.31E-07

(i) $P(X > 4) = 1 - P(X \leq 3)$
$= 1 - 0.013841 - 0.071184 - 0.16779 - 0.2397$
$= 1 - 0.49252 = 0.50758$

(ii) $P(3 \leq X \leq 5) = 0.2397 + 0.23114 + 0.158496 = 0.629336$

(iii) $P(X = 0) = p(0) = 0.013841$

5.5.2 Calculating Poisson Probabilities using MINITAB

EXAMPLE 5.20   Accidents occur on a particular busy state road at the rate of 3.5 per month. What is the probability that, for a month chosen at random, (a) less than three accidents, (b) more than three accidents, and (c) between two and five accidents inclusive, occur on the road.

Solution This is a Poisson problem as it occurs per unit time. Here $\lambda = 3.5$ and we can use MINITAB to find the solutions. Because the domain of the variable X is countably infinite, we can only enter in MINITAB just enough values from 0, 1, . . . to cover the values we need to evaluate their probabilities and use the idea that the sum of the probabilities is 1.0 to find the rest.

Here, the highest value we need to enter is 5, since it is the highest discrete value, whose probability we need to find. We present a MINITAB solution:

```
MTB > set c1
DATA> 0 1 2 3 4 5
DATA> end
MTB > pdf c1 c2;
SUBC> Poisson 3.2.
```

Output:

x	p(x)
0	0.040762
1	0.130439
2	0.208702
3	0.222616
4	0.178093
5	0.113979

(a) $P(X < 3) = p(0) + p(1) + p(2) = 0.040762 + 0.130439 + 0.208702$
$= 0.379904$
(b) $P(X > 3) = 1 - p(X \leq 3) = 1 - 0.379904 - 0.222616 = 0.39748$
(c) $P(2 \leq X \leq 5) = p(2) + p(3) + p(4) + p(5)$
$= 0.208702 + 0.222616 + 0.178093 + 0.113979 = 0.723391$

■ **EXAMPLE 5.21** **We redo Example 5.13 using MINITAB.**
On average four children are born daily at a local maternity ward in Heathsville. If a day is chosen at random, what is the probability that: (i) no more than two children would be born; (ii) two, but less than five children would be born; or (iii) five or more children would be born at Heathsville maternity?

Solution The number of babies born daily at the Maternity can be modeled by a Poisson distribution whose mean is the number of births per unit time (day). Using MINITAB for the specific distribution that applies we have:

```
MTB > set c1
DATA> 0 1 2 3 4 5
DATA> end
MTB > pdf c1 c2;
SUBC> Poisson 4.
```

Output:

x	p(x)
0	0.018316
1	0.073263
2	0.146525
3	0.195367
4	0.195367
5	0.156293

(i) $P(0) + P(1) + P(2) = 0.018316 + 0.073263 + 0.146525 = 0.238103$

(ii) $P(2 \leq X < 5) = P(2) + P(3) + P(4)$
$$= 0.146525 + 0.195367 + 0.195367 = 0.537259$$

(iii) $P(X \geq 5) = 1 - P(X \leq 4)$
$$= 1 - (0.238103 + 0.195367 + 0.195367) = 0.378579$$

Exercises

5.1 An examination of the record for Mr. Jorgenson, another car salesman at Division Street in St. Cloud, MN, shows that the following is the distribution of X, his daily car sales.

X	0	1	2	3	4	5
P(x)	0.15	0.20	0.25	0.18	0.14	0.08

Obtain the mean, variance, and standard deviation for his daily car sales.

5.2 Let a fair die be rolled, and let X be the face of the die that turns up. The distribution of the faces of the dies would be:

X	1	2	3	4	5	6
P(x)	1/6	1/6	1/6	1/6	1/6	1/6

Obtain the mean, variance, and standard deviation for faces of the die in this experiment.

5.3 Research biologists surveyed a rainforest for rare species of snails. They divided the survey region into acres and thoroughly searched each acre for the snails. Let X be the number of snails/acre that they found; the distribution of their findings are shown in the table below.

X	0	1	2	3	4	5	6	7
P(x)	0.08	0.17	0.22	0.24	0.12	0.08	0.05	0.04

Obtain the mean, variance, and standard deviation for snails in this study.

5.4 A conservationist found the following distribution for number of bird nests/square mile whacked by a woodpecker:

X	0	1	2	3	4	5
P(x)	0.28	0.32	0.22	0.10	0.04	0.04

Obtain the mean, variance, and standard deviation for bird nests whacked by a woodpecker in this study.

5.5 Here is the distribution of girls (X) in families with five children:

X	0	1	2	3	4	5
$P(x)$	0.03125	0.15625	0.3125	0.3125	0.15625	0.03125

Obtain the mean, variance, and standard deviation for number of girls in the family.

5.6 Evaluate the following:

(a) $\binom{15}{4}$

(b) $\binom{10}{3}$

(c) $\binom{14}{6}$

5.7 Evaluate the following:

(a) $\binom{13}{5}$

(b) $\binom{9}{2}$

(c) $\binom{11}{4}$

5.8 Evaluate the following:
(a) $(0.6)^5(0.4)^8$
(b) $(0.35)^2(0.65)^7$
(c) $(0.41)^4(0.59)^7$

5.9 Evaluate the following:
(a) $(0.36)^4(0.4)^{11}$
(b) $(0.55)^3(0.45)^7$
(c) $(0.341)^6(0.659)^8$

5.10 Evaluate the following:

(a) $\binom{15}{4}(0.36)^4(0.4)^{11}$

(b) $\binom{10}{3}(0.55)^3(0.45)^7$

(c) $\binom{14}{6}(0.341)^6(0.659)^8$

5.11 Evaluate the following:

(a) $\binom{13}{5}(0.6)^5(0.4)^8$

(b) $\binom{9}{2}(0.35)^2(0.65)^7$

(c) $\binom{11}{4}(0.41)^4(0.59)^7$

5.12 It is known that 67% of the people in a local county support the proposal to reserve a piece of land for conservation. If eleven people are chosen at random from the county what is the probability that:
(a) between three and six, inclusive, support the proposal?
(b) less than five support the proposal?
(c) exactly seven support the proposal?

5.13 Gophers are a menace on Mr. Jones' farm. From the records kept by him, he observes, on average, four gophers per acre. If an acre is chosen at random at Mr. Jones' farm, what is the probability of observing:
(a) exactly four gophers?
(b) three or more gophers?
(c) between two and five gophers?

5.14 In a search for a rare butterfly species, a part of the Amazon was divided into acres and the researcher found on average 3.5 butterflies per acre. Assume butterflies are spread in the area according to a Poisson distribution. If one acre is chosen at random, what is the probability that the researcher finds:
(a) exactly three butterflies?
(b) no more than four butterflies?
(c) between four and six butterflies, inclusive?
(d) greater or equal to three, but less than seven butterflies?

5.15 A laboratory population of fruit-flies is made up of 45% gray flies and 55% black flies. If a sample of twelve flies is chosen at random from the population. What is the probability that:
(a) more than four flies are gray?
(b) no fly is gray?
(c) between three and six flies are black, inclusive?

5.16 Seventy percent of rockets made by RPC company are known to fly when fired. If ten rockets are test fired, find the probability that:
(a) exactly five will fly?
(b) at least two, but no more than five, will fly?
(c) no more than four will fly?

5.17 A process for producing computer hard disks is known to yield 8% defectives. If fifteen computer hard disks are randomly taken at the end of the production line; what is the probability that:
(a) no more than three disks are defective?
(b) exactly five are defective?
(c) What is the mean, variance, and standard deviation of the number of defectives for the fifteen chosen?

5.18 It is known that on average the number of accidents occurring at a very busy highway intersection is 3.4 per month. If any month is chosen at random, what is the probability of observing:
(a) exactly three accidents?
(b) between three and six accidents, inclusive?
(c) at least four accidents at the intersection?

5.19 For each hour of production the mean number of defective items yielded by a production process is known to be two. If an hour of production is chosen at random, what is the probability that:
(a) no defective item,
(b) more than two defective items, and
(c) exactly four defective items, were produced?

5.20   The emergency room of a large hospital receives on average five emergency cases per hour. What is the probability that, in a random hour,
   (a) exactly five,
   (b) between four and seven, and
   (c) less than three cases arrive?

5.21   The number of bottles arriving at a service point for corking per five seconds is on average 3.9. If a five seconds is chosen at random, what is the probability that:
   (a) exactly three bottles?
   (b) no more than four bottles?
   (c) between four and six bottles, inclusive?
   (d) greater than or equal to three bottles arrive at the same service point?

5.22   Grand Metropolitan Hospital keeps records of emergency room visits due to respiratory problems. The record reveals that the average number of patients in this category arriving per month is 6.9. If a month is chosen randomly from one of the evenings, what is the probability that the number of patients who arrive is:
   (a) exactly four?
   (b) at most three?
   (c) between four and eight, inclusive?
   (d) greater than four?

5.23   During the day, cars are known to arrive at the rate of six/minute at a toll gate. If any minute of the day is chosen at random, what is the probability that:
   (a) no car arrives?
   (b) less than three cars arrive?
   (c) four or more cars arrive at the toll gate?

5.24   On average, 5.2 people are caught by cops making u-turns at unauthorized places on a highway per year. If a year is chosen randomly, what is the probability that people who are caught making u-turns in the wrong places on the highway is:
   (a) exactly five?
   (b) less than five?
   (c) between three and six, inclusive?

5.25   A human–computer interface software which produces documents by listening to human voices is reputed to make only 2.6 errors on average per page of document. If a page of document produced by this software is chosen at random and examined, what is the probability of finding:
   (a) no more than three errors?
   (b) between two and five errors, inclusive?
   (c) no error?

5.26   Ninety percent of those who go to church say that they go because they love God. Suppose that twelve church-going individuals are surveyed, what is probability that:
   (a) exactly seven,
   (b) greater than six,
   (c) between five and nine, inclusive, and
   (d) eight or less, will say they go to church because they love God?

CHAPTER 6

Continuous Distributions

6.0   Introduction

Continuous distributions are used for modeling variables which are continuous. The families of continuous distributions which are often encountered in statistical analysis include: the uniform, the normal, the exponential, the gamma, the Beta, the chisquare, the t, and the F, among many others. In this section, we will only discuss two of them, namely the uniform (sometimes called the rectangular) and the normal. The discussion of the rest is beyond the scope of this text. We will treat the properties of these distributions, and explore how we can use them in practical statistical analysis.

6.1   The Uniform Distribution

A random variable X is said to have a uniform or **rectangular** distribution if its distribution can be represented by the function:

$$f(x) = \frac{1}{b - a}; a \leq x \leq b; -\infty < a < b < \infty.$$

The random variable X is defined to be uniformly distributed over the finite interval $[a, b]$. This means that the probability of the random variable is uniformly distributed over the interval, $[a, b]$. It provides a useful way of representing some important random phenomena. It means that probability of the variable X taking values anywhere in the interval between a and b, inclusive, is the same.

The uniform distribution can be used to model events that can happen at any time in the interval $[a, b]$ with equal probability. For instance a company whose business involves the shipment of parcels from a conflict zone to their headquarters in the USA may find that there is a minimum time, *a*, and a maximum time, *b*, between arrival of shipments following their being dispatched; the arrivals may be considered equally probable in the period between *a* and *b*. The arrivals of the parcels can be modeled by a uniform distribution as described. We illustrate the Uniform distribution in Figure 6.1.

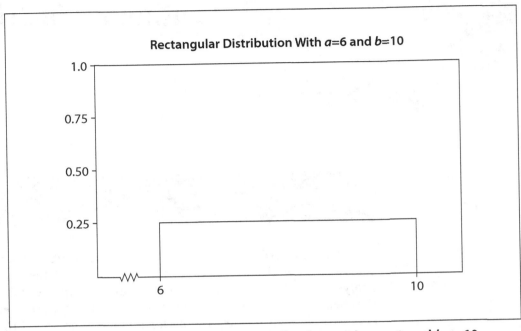

■ FIGURE 6.1   Rectangular Distribution with *a* = 6 and *b* = 10

As we can see from Figure 6.1, the sketch of the curve of *X* when the distribution is deemed to be Uniform in [*a*, *b*] is rectangular. The area under the curve which is used as a density (continuous distribution) pertaining to any interval between two values (a_1 and b_1, say, $a \leq a_1 \leq b_1 \leq b$), that is within the domain of the variable *X*, represents the probabilities of realizing values between them. In this case the area under the entire uniform "curve" is a rectangle and as such, we can calculate the probabilities for any sub-interval within its range by using the methods for evaluating areas under a rectangle, determined as base × height (or width × length). Since the uniform "curve" on [*a*, *b*] is a distribution, then its area should be 1.0. The height of the rectangle is $\dfrac{1}{b-a}$, arising from its being a distribution and the fact that its base is $b - a$. From this, evaluation of probabilities for given intervals in its domain of width, will involve the calculations of sub-rectangles of the distribution.

The uniform distribution is symmetric about its mean. This means that its mean and median are the same.

6.1.1 The Mean, Variance, and Standard Deviation of the Uniform Distribution

The mean, variance, and standard deviation of a variable, *X*, which has the uniform distribution as stated above, can be obtained from first principles; however, this involves calculus, which is not used in this text, so we will only state the values as follows:

$$\text{Mean of } X = \mu = \frac{b + a}{2}$$

$$\text{Variance of } X = \sigma^2 = \frac{(b - a)^2}{12}$$

$$\text{Therefore, standard deviation of } X = \sigma = \sqrt{\frac{(b - a)^2}{12}} = \frac{b - a}{\sqrt{12}}$$

■ **EXAMPLE 6.1**   John observed that the 8:00 A.M. commuter train he takes to work never arrives on time and arrives late with same probability anywhere in the interval between 8:05 A.M. and 8:20 A.M. On a randomly chosen day, if John stands on the platform to catch the train, what is the probability that the train will arrive: (a) before 8:00 A.M.? (b) between 8:06 A.M. and 8:13 A.M.? (c) after 8:08 A.M.? (d) Obtain the (i) mean, (ii) variance, and (iii) standard deviation for the value of late arrival times of the train.

Solution   This can be modeled by the Uniform distribution as in Figure 6.2.

The routine late arrivals occur between 8:05 and 8:20; if we use minutes as our measure of time, the $a = 5$ and $b = 20$. The height of the rectangle is $\dfrac{1}{20 - 5} = \dfrac{1}{15}$

(a) P(train arrives before 8:00 A.M.) $= P(X < 0) = 0$, because the train never arrives on time.

(b) P(train arrives between 8:06 A.M. and 8:13 A.M.) $= P(6 \leq X \leq 13) =$
$(13 - 6) \times \dfrac{1}{15} = \dfrac{7}{15} = 0.4667.$
Here we found the interval $(13 - 6) = 7$ as the base of the sub-rectangle, and the height is $\dfrac{1}{15}$, so that the area under the distribution is obtained by using base $\times$ height.

(c) Similarly, P(train arrives after 8:08 A.M.) $= P(8 \leq X \leq 20) =$
$(20 - 8) \times \dfrac{1}{15} = \dfrac{12}{15} = 0.8.$

(d) (i) mean $= \mu = \dfrac{20 + 5}{2} = 12.5$ minutes $\Rightarrow$ mean arrival occurs at 8:12:30 A.M.

    (ii) Variance $= \sigma^2 = \dfrac{(20 - 5)^2}{12} = \dfrac{225}{12} = 18.75$

    (iii) Standard deviation $= \sigma = \sqrt{\dfrac{(20 - 5)^2}{12}} = 4.3301$

■ **EXAMPLE 6.2**   Now consider a company whose business involves shipment of parcels from a conflict zone to their headquarters in the USA. They have noticed that it takes the parcels anywhere between five days and twenty-six days to arrive after dispatch. Find the probability that following dispatch a parcel will: (a) arrive in the first week? (b) in the second week? (c) in the third week? (d) in the fourth week? (e) Obtain the (i) mean, (ii) variance, and (iii) standard deviation for the value of the arrivals of the parcels.

Solution   This can also be modeled by the Uniform distribution as we discussed previously. The arrivals of the parcels occur between five and twenty-six days after dispatch; this means that $a = 5$ and $b = 26$. The height of the rectangle is $\dfrac{1}{26-5} = \dfrac{1}{21}$.

(a) P(a parcel arrives in the first week) $= P(X < 7) =$

$(7-5) \times \dfrac{1}{21} = \dfrac{2}{21} = 0.095238.$

(b) P(a parcel arrives in the second week) $= P(7 \le X \le 14) =$

$(14 - 7) \times \dfrac{1}{21} = \dfrac{7}{21} = 0.3333.$

(c) Similarly, P(a parcel arrives in the third week) $= P(14 \le X \le 21) =$

$(21 - 14) \times \dfrac{1}{21} = \dfrac{7}{21} = 0.3333.$

(d) Similarly, P(a parcel arrives in the fourth week) $= P(21 \le X \le 26) =$

$(26 - 21) \times \dfrac{1}{21} = \dfrac{5}{21} = 0.2381.$

(e) (i) mean $= \mu = \dfrac{26+5}{2} = 15.5$ days.

   (ii) Variance $= \sigma^2 = \dfrac{(26-5)^2}{12} = \dfrac{441}{12} = 36.75$ days^2.

   (iii) Standard deviation $= \sigma = \sqrt{\dfrac{(26-5)^2}{12}} = 6.0622$ days.

■ **EXAMPLE 6.3**   A random variable, X, is uniformly distributed over the interval $[-b, a]$.

(a) Write the specific uniform distribution that applies.
(b) Find its mean, variance, and standard deviation.
(c) If $11 > -b$, find $P[X \le 11]$.

Solution   (a) The specific uniform distribution which applies is:

$$f(x) = \frac{1}{a+b}; \quad -b \le x \le a$$

(b) The mean is:

$$\mu = \frac{(a + (-b))}{2}$$

The variance is:

$$\text{Var } X = \sigma^2 = \frac{(a+b)^2}{12}$$

The standard deviation $= \sigma = \sqrt{\dfrac{(a+b)^2}{12}}$

(c) Height in this case is $= \dfrac{1}{a+b}$

Thus $P(X \le 11) = (11 - (-b))\dfrac{1}{a+b} = \dfrac{11+b}{a+b}$

EXAMPLE 6.4 The commute times to work of Mr. Fridley is uniformly distributed between twenty-nine and forty-five minutes.

    (a) Obtain the five number summary for the distribution of his commute times.
    (b) Obtain the upper 27th percentile of his commute times.
    (c) Obtain the lower 66th percentile of his commute times.

Solution

    (a) The specific uniform distribution which applies to Mr. Fridley's commute times is:

$$f(x) = \begin{cases} \dfrac{1}{45-29} = \dfrac{1}{16}; & 29 \le x \le 45 \\ 0 \text{ elsewhere} \end{cases}$$

    (b) The median is

$$\mu = \dfrac{45+29}{2} = 37$$

The Q_1 is the lower 25th percentile

$$\dfrac{29+Q_1}{16} = 0.25 \Rightarrow 29 + Q_1 = 16(0.25) \Rightarrow Q_1 = 29 + 4 = 33$$

The Q_3 is the upper 25th percentile

$$\dfrac{45 - Q_3}{16} = 0.25 \Rightarrow 45 - Q_3 = 16(0.25) \Rightarrow Q_3 = 45 - 4 = 41$$

The 5-number summary is minimum $= 29$ $Q_1 = 33$ median $= 37$ $Q_3 = 41$ maximum $= 45$

    (c) The upper 27th percentile is $P_{1-0.27} = P_{0.73} =$ upper 73rd percentile

$$\dfrac{P_{0.73} - 29}{16} = 0.73 \Rightarrow P_{0.73} - 29 = 16(0.73) \Rightarrow P_{0.73} = 29 + 11.68$$
$$= 40.68$$

    (d) The lower 66th percentile is $P_{.66}$

$$\dfrac{P_{0.66} - 29}{16} = 0.66 \Rightarrow P_{0.66} - 29 = 16(0.66) \Rightarrow P_{0.66} = 29 + 10.56$$
$$= 39.56$$

6.2   The Normal Distribution

The normal distribution or (more exactly) the family of normal distributions, is one of the very commonly encountered density functions used by statisticians for modeling a continuous variable. The normal distribution plays a central role in statistical endeavors. This is because virtually every statistical distribution can be approximated by the normal distribution, provided that a large enough sample is taken. The normal distribution also provides an important bridge between the discrete and continuous distributions, because the normal distribution can be used to provide approximate models for the binomial and Poisson distributions.

The normal distribution is represented by the function:

$$f(x) = \frac{1}{\sigma\sqrt{2\pi}} e^{-\frac{1}{2}\left(\frac{x-\mu}{\sigma}\right)^2} \quad -\infty < x < \infty.$$

This statement can be put in a short-hand as $X \sim N(\mu, \sigma^2)$. It is a model for a continuous variable X whose domain is the real line $(-\infty, \infty)$. The mean of the normal distribution is μ, while the variance is σ^2. The relative frequency curve of X (or the distribution of X) is represented by the area under the curve of this function which, when sketched, is bell-shaped and symmetric about the mean of the variable, μ as shown in Figure 6.2. On account of this symmetry, the mean and the median are the same. The curve also peaks at the mean, μ, indicating that the mean has the highest frequency of occurrence of all other values that X can take. Thus, the mean is also the mode for the normal distribution.

When this curve is sketched, the total area under it for its domain (the real line $(-\infty, \infty)$) can be shown to be 1.0. This satisfies the rule that, for a function which is a statistical distribution, the total probability for all values of the random variable which it is modeling should be unity. Thus in order to find the probability of any subsection of the domain, all we need do is calculate the area under the curve for that section.

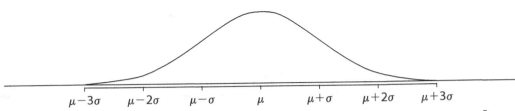

■ **FIGURE 6.2   Normal Distribution with Mean, μ and Variance, σ^2**

6.2.1 The Standard Normal Distribution

Consider the variable, $X \sim N(\mu, \sigma^2)$: that is a variable which is normally distributed with mean μ and variance σ^2. Consider another variable, $Z = \dfrac{X - \mu}{\sigma}$: this is another normal random variable, called the **standard normal variable,** which has **the standard normal distribution. This variable has** a normal distribution, with a mean of zero and a variance of one. If a value of X is converted to Z, then the Z is the **standardized value** of the X. Thus $Z \sim (0, 1)$. The density function for the random variable, Z, is:

$$f(z) = \frac{1}{\sqrt{2\pi}}e - \frac{1}{2}z^2; -\infty < z < \infty.$$

Due to the fact that the standard normal variate is more mathematically tractable than other members of this parametric family (the normal distributions) it has been extensively tabulated. The probabilities for other members of this normal distribution family are read via the table of the standard normal random variable. This table can have different formats. A very popular format shows the cumulative probabilities for different values of Z. By converting other normal distributions with means and variances other than 0 and 1, to Z, we can calculate their probabilities via Z. This obviates the need to directly calculate the probabilities for other combinations of parameters of other members of the normal parametric family of distributions.

In Figure 6.3 we show the sketch of the standard normal distribution and indicate the values of Z for 1, 2, and 3 standard deviations away from its mean of zero.

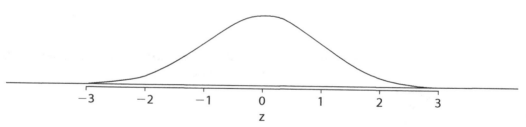

■ **FIGURE 6.3**   **The Standard Normal Distribution Showing Values of Z Up to 3 Standard Deviations Away from the Mean**

6.2.2 The 68–95–99 Rule, or the Empirical Rule

If we use the normal curve to model a variable, it is known that for an interval that spans one standard deviation of that variable from its mean, on both sides, the area covered is 68.27%, which means we would have exhausted 68.27% of the values of that variable. For an interval which spans two standard deviations from the mean, the area covered is 95.45% and about 95.45% of the values of the variable will lie between two standard deviations away from its mean. The percentage for an interval that spans three standard deviations from the mean is 99.73%. In Figures 6.4 (a, b, and c), we illustrate the areas covered for the standard normal variable Z.

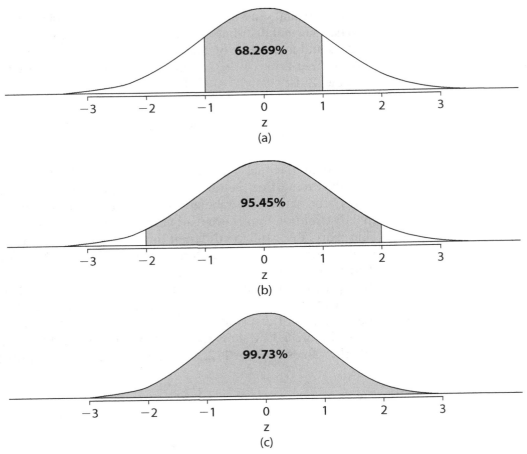

FIGURE 6.4   The 68–95–99 Rule for Areas Covered Under the Standard Normal Distribution

6.2.3 Application of Empirical Rule to Normal Distribution with Mean $= \mu$ and Variance $= \sigma^2$

The same rule applies for a typical normal distribution with mean μ and variance σ^2. Consider the variable which is normally distributed so that $X \sim N(80, 49)$. Applying the rule, we can determine the range for which one would expect 68.27% of the values of X to lie. Similar ranges can be found for 95.45%, and 99.73% of the values of X. It turns out in this case that the range for 68.27% is (73, 87). For 95.45% and 99.73% respectively, the ranges are (66, 94) and (59, 101) as shown in Figure 6.5 (a, b, and c).

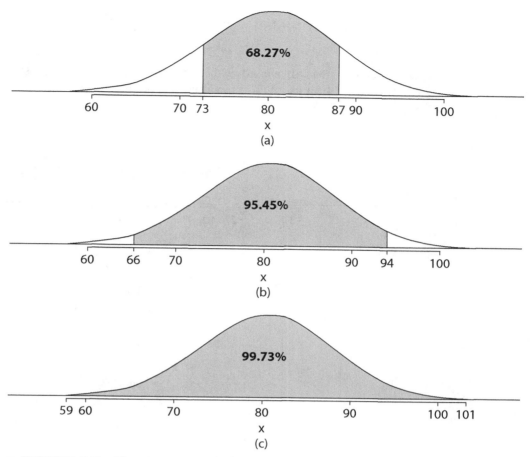

■ **FIGURE 6.5**   **The 68–95–99 Rule Applied to a Normal Distribution with Mean = 80 and Variance = 49**

■ **EXAMPLE 6.5**   The heights of mothers X are known to be normally distributed with mean 65.78 inches and variance of 13.69 inches^2. What is the probability that the height of a mother chosen at random is:

(a) greater than 61.2 inches?
(b) lies between 63.5 and 69 inches?
(c) less than 64.3?

Solution $$Z = \frac{X - \mu}{\sigma}$$

Using this, we calculate the Z for any value of X, and read the table of standard normal distribution (Appendix II, pages 195 and 415) to obtain the associated probabilities:

$$\mu = 65.78; \sigma^2 = 13.69 \Rightarrow \sigma = 3.7$$

(a) $P(X > 61.2) = P\left[Z > \dfrac{61.2 - 65.78}{3.7}\right]$

$= P(Z > -1.24) = 1 - P(Z \leq 1.24) = 1 - 0.1075 = 0.8925$

It is always advisable to draw the picture of the standard normal distribution as an aid in evaluating the probabilities.

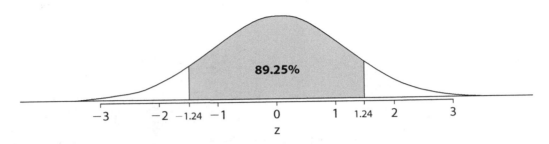

(b) $P(63.5 < X < 69) = P\left[\dfrac{63.5 - 65.78}{3.7} < Z < \dfrac{69 - 65.78}{3.7}\right]$

$= P(-0.62 < Z < 0.87) = 0.8079 - 0.2676 = 0.5403$

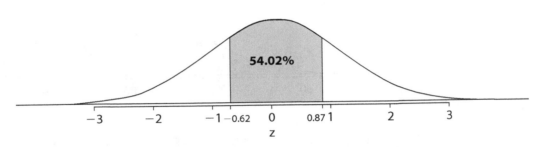

(c) $P(X < 64.3) = P\left[Z < \dfrac{64.3 - 65.78}{3.7}\right] = P(Z < -0.40) = 0.3446$

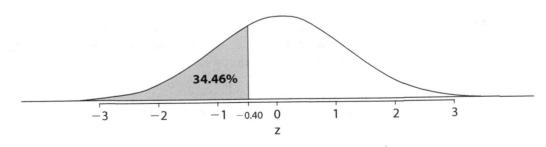

■ **EXAMPLE 6.6** Y, the gestation period (the period of time between human conception and labor) is believed to be normally distributed, with a mean of 280 days and a variance of 49 days^2. What proportion of pregnancies:

(a) will exceed 290 days?
(b) will lie between 272 and 294 days?
(c) will take less than 270 days?

Solution $Z = \dfrac{Y - \mu}{\sigma}$

Again, using this, we calculate the Z for any value of Y, and read the table of standard normal distribution (Appendix II, page 197) to obtain the associated probabilities:

$$\mu = 280; \sigma^2 = 49 \Rightarrow \sigma = 7$$

(a) $P(Y > 290) = P\left[Z > \dfrac{290 - 280}{7}\right]$

$\qquad = P(Z > 1.43) = 1 - P(Z \leq 1.43) = 1 - 0.9236 = 0.0764$

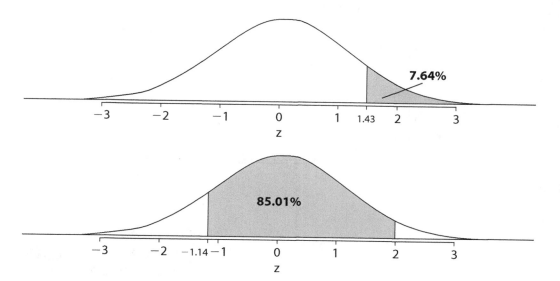

(b) $P(272 < Y < 294) = P\left[\dfrac{272 - 280}{7} < Z < \dfrac{294 - 280}{7}\right]$

$\qquad = P(-1.14 < Z < 2.0) = 0.9772 - 0.1271 = 0.8501$

(c) $P(Y < 270) = P\left[Z < \dfrac{270 - 280}{7}\right] = P(Z < -1.43) = 0.0764$

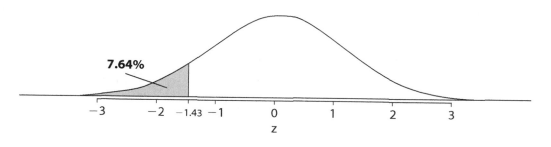

■ **EXAMPLE 6.7** X, the weights of newborn babies, are believed to be normally distributed with mean of 3.1 kg and a variance of 0.114 kg². What proportion of newborn babies have weight that:

   (a) exceeds 3.80 kg?
   (b) lies between 2.90 and 3.8 kg?
   (c) takes less than 2.83 kg?

Solution $$Z = \frac{X - \mu}{\sigma}$$

Again, using this, we calculate the Z for any value of X, and read the table of standard normal distribution (Appendix II, pages 195 and 415) to obtain the associated probabilities:

$$\mu = 3.1; \sigma^2 = 0.114 \Rightarrow \sigma = 0.337639$$

(a) $P(X > 3.8) = P\left[Z > \dfrac{3.8 - 3.1}{0.337639}\right]$
$$= P(Z > 2.07) = 1 - P(Z \leq 2.07) = 1 - 0.9808 = 0.0192$$

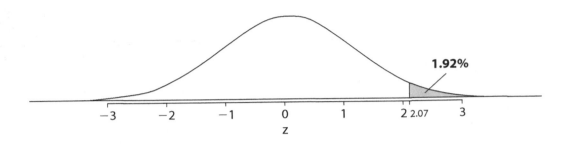

(b) $P(2.83 < X < 3.8) = P\left[\dfrac{2.83 - 3.1}{0.337639} < Z < \dfrac{3.8 - 3.1}{0.337639}\right]$
$$= P(-0.80 < Z < 2.07) = 0.9808 - 0.2119 = 0.7689$$

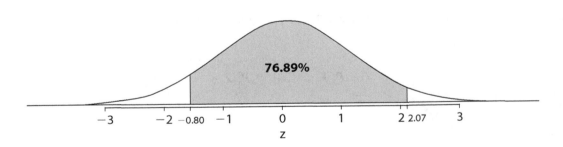

(c) $P(X < 2.83) = P\left[Z < \dfrac{2.83 - 3.1}{0.337639}\right] = P(Z < -0.80)$
$$= 0.2119$$

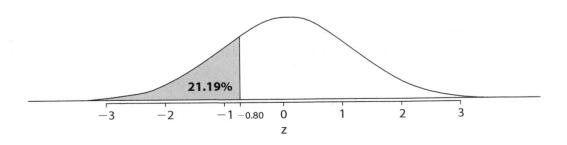

EXAMPLE 6.8 The ACT composite scores, Y, in 2009 are believed to be normally distributed with a mean score of 21.1 and variance of 26.01. What is the probability that any student who took the ACT composite in 2009 scored:

(a) between 12.8 and 30.5?
(b) greater than 16.5
(c) less than 23.9?

Solution $Z = \dfrac{Y - \mu}{\sigma}$

As usual, we calculate the Z for any value of Y, and read the table of standard normal distribution for the associated probabilities:

$$\mu = 21.1; \sigma^2 = 26.01 \Rightarrow \sigma = 5.1$$

(a) $P(12.8 < Y < 30.5) = P\left[\dfrac{12.8 - 21.1}{5.1}\right] < Z < \left[\dfrac{30.5 - 21.1}{5.1}\right]$
$= P(-1.63 < Z < 1.84)$
$= P(Z < 1.84) - P(Z < -1.63) = 0.9671 - 0.0516$
$= 0.9155$

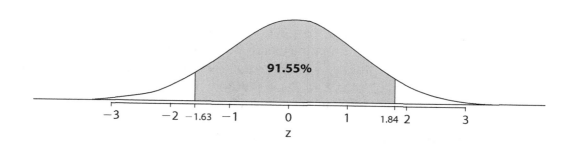

(b) $P(Y > 16.5) = P\left[Z > \dfrac{16.5 - 21.1}{5.1}\right] = p(Z > -0.9)$
$= 1 - P(Z < -0.9) = 1 - 0.1841 = 0.8159$

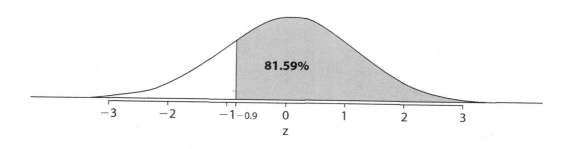

(c) $P(Y < 23.9) = P\left[Z < \dfrac{23.9 - 21.1}{5.1}\right] = P(Z < 0.55)$
$= 0.7088$

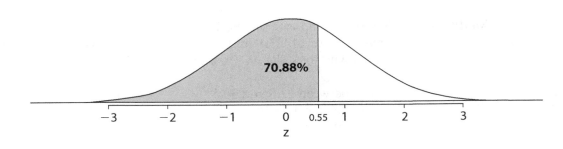

70.88%

■ **EXAMPLE 6.9** Find the values of Z_1 if the following are true:

$\quad$ (i) $P(-Z_1 < Z < Z_1) = 0.8764$
$\quad$ (ii) $P(-Z_1 < Z < 2.02) = 0.8404$
$\quad$ (iii) $P(-1.77 < Z < Z_1) = 0.9280$

Solution $\quad$ (i) $P(-Z_1 < Z < Z_1) = 0.8764 \Rightarrow P(Z < -Z_1) + P(Z > Z_1)$
$\qquad\qquad\qquad = 1 - 0.8764 = 0.1336$
$\qquad$ But $P(Z < -Z_1) = P(Z > Z_1) \Rightarrow 2P(Z < -Z_1) = 0.1336$
$\qquad P(Z < -Z_1) = 0.0668 \Rightarrow Z_1 = 1.5.$
$\qquad$ We can see from the figures that: $P(-1.5 < Z < 1.5) = 0.8764$

$\quad$ (ii) $P(-Z_1 < Z < 2.02) = 0.8404 \Rightarrow P(Z < 2.02_1) - P(Z < -Z_1) = 0.8404$
$\qquad P(Z < 2.02) = 0.9783$
$\qquad P(Z < 2.02) + P(Z < -Z_1) = 0.8404$
$\qquad P(Z < -Z_1) = 0.9783 - 0.8404 = 0.1379$
$\qquad$ From the tables, $Z_1 = 1.09$ or $-Z_1 = -1.09$

$\quad$ (iii) $P(-1.77 < Z < Z_1) = 0.9280 \Rightarrow P(Z < -Z_1) = 0.9280 + P(Z < -1.77)$
$\qquad P(Z < Z_1) = 0.9280 + 0.0384 = 0.9664$
$\qquad$ From the tables, $Z_1 = 1.83$

■ **EXAMPLE 6.10** The IQs of twelve year olds, X, are normally distributed, so that: $X \sim N(100, 117)$, with a mean of 100 and variance of 117. Find the probability that:

$\quad$ (i) $P(X > 97)$
$\quad$ (ii) $P(103 < X < 110)$

Solution

$\quad$ (i) $P(X > 97) = P\left[Z > \dfrac{97 - 100}{\sqrt{117}}\right] = P(Z > -0.28)$
$\qquad\qquad = 1 - P(Z < -0.28) = 1 - 0.3987 = 0.6013$

$\quad$ (ii) $P(103 < X < 110) = P\left[\dfrac{103 - 100}{\sqrt{117}} < Z < \dfrac{110 - 100}{\sqrt{117}}\right]$
$\qquad\qquad = P(0.28 < Z < 0.92) = 0.8212 - 0.6013 = 0.2199$

■ **EXAMPLE 6.11** The weights of individual bars of bio-fuel made by certain manufacturing process are known to be approximately normally distributed with mean of 127 and standard deviation 4.

$\quad$ (a) Find $P[118 \leq X \leq 137]$ where X represents weight.
$\quad$ (b) What proportion of the bars were of weight greater than 130?

Solution

(a) $P(118 < X < 137) = P(-2.25 < Z < 2.5) = 0.9938 - 0.0122 = 0.9816$

(b) $P(X > 130) = P\left(Z > \dfrac{130 - 127}{4}\right) = P(Z > 0.75)$

$$= 1 - 0.7734 = 0.2266$$

■ EXAMPLE 6.12 The weights of boxes of shoes made of leather produced by Burtons Ranch are known to have weights which are normally distributed with a mean of 115 kg, and standard deviation of 3 kg. How many of these boxes would have weights:

(a) between 115 and 118.5 kg?
(b) less than 124.7 kg?
(c) greater than 122.4 kg?

Solution

(a) $P(115 < X < 118) = P\left(\dfrac{115 - 115}{3} \le Z \le \dfrac{118.5 - 115}{3}\right)$

$$= P(0 \le Z \le 1.17) = 0.8790 - 0.5000 = 0.3790$$

(b) $P(X < 124.7) = P\left(Z \le \dfrac{124.7 - 115}{3}\right) = P(Z \le 3.23) = 0.9994$

(c) $P(X < 124.7) = P\left(Z \le \dfrac{122.4 - 115}{3}\right) = P(Z > 2.47)$

$$= 1 - P(Z \le 2.47) = 1 - 0.9932 = 0.0068$$

6.2.4 Percentiles of the Normal Distribution

We are familiar with the idea that in reporting ACT or SAT scores, a student is told that his or her score is in the upper 5th percentile or 2nd percentile, and so on. These percentiles are largely based on the idea that the scores of all the several thousands of students who took the test are normally distributed with a given variance and standard deviation. Often it is based on the assumption of normality for the data, due to the large number of candidates who took the examinations.

Thus, if we know the mean and variance of a variable which is normally distributed, we can find the value for any specified percentile of the distribution.

For instance, say that $X \sim N(\mu, \sigma^2)$ and we want to find the value of X at the lower p^{th} percentile. We can denote the value of X at the lower p^{th} percentile as X_p. It is important to know that the lower p^{th} percentile is the same as upper $(100 - p)^{th}$ percentile. The foregoing means that the cumulative probability up to the pth percentile is p%. This is illustrated in Figure 6.6:

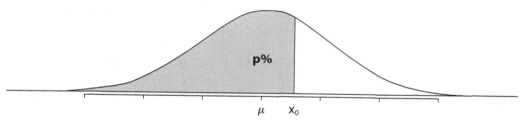

■ FIGURE 6.6   **The Normal Distribution Showing the Lower p^{th} Percentile**

We want to obtain the value of X_p. This can be done by using the definition of Z. It is clear from the definition of Z that:

$$Z_p = \frac{X_p - \mu}{\sigma}.$$

From this we see that:

$$\sigma Z_p = X_p - \mu$$
$$\mu + \sigma Z_p = X_p \text{ or } X_p = \mu + \sigma Z_p;$$

where Z_p is the value of the standard normal variable in the pth percentile position. With the above equation, we can solve for X_p once we know μ and σ. We do this by reading the standard normal distribution table of Z to find the value of Z_P.

■ **EXAMPLE 6.13** Suppose that $X \sim N(80, 49)$, obtain the:

(a) upper 10th percentile,
(b) lower 78th percentile,
(c) lower 40th percentile,

for X.

Solution

(a) The upper 10th percentile = the lower 90th percentile, so we need to find the lower 90th percentile.

$$X \sim N(80, 49) \Rightarrow \mu = 80; \sigma = \sqrt{49} = 7$$
$$X_{0.90} = \mu + \sigma Z_{0.90} = 80 + 7Z_{0.90}$$

From the table of the standard normal distribution, the nearest tabulated value to 0.90 is 0.8997, with a corresponding $Z = 1.28$. We use this as an approximate $Z_{0.90}$, so that:

$$X_{0.90} = 80 + 7Z_{0.90} = 80 + 7(1.28) = 88.96.$$

(b) Similarly, the lower 78th percentile is obtained as:

$$X_{0.78} = 80 + 7Z_{0.78}$$

From the table of the standard normal distribution, the nearest tabulated value to 0.78 is 0.7794, with a corresponding $Z = 0.77$. We use this as an approximate $Z_{0.78}$, so that:

$$X_{0.78} = 80 + 7Z_{0.78} = 80 + 7(0.77) = 85.39.$$

(c) Again, the lower 40th percentile is obtained as:

$$X_{0.40} = 80 + 7Z_{0.40}$$

From the table of the standard normal distribution, the nearest tabulated value to 0.40 is 0.4013, with a corresponding $Z = -0.25$. We use this as an approximate $Z_{0.40}$, so that:

$$X_{0.40} = 80 + 7Z_{40} = 80 + 7(-0.25) = 78.25.$$

■ **EXAMPLE 6.14** Y, the number of days of deployment of a soldier in the warfront, is believed to be normally distributed with a mean 275 days and variance of 85 days^2. Obtain the number of days of deployment a soldier would stay in the warfront to be in the:

   (a) lower 75th percentile.
   (b) lower 25th percentile.
   (c) lower 95th percentile.

Solution   (a) The lower 75th percentile:

$$Y \sim N(275, 85) \Rightarrow \mu = 275; \sigma = \sqrt{85} = 9.21944$$
$$X_{0.75} = \mu + \sigma Z_{0.75} = 275 + 9.21944(Z_{0.75})$$

From the table of the standard normal distribution, the nearest tabulated value to 0.75 is 0.7486, with a corresponding $Z = 0.67$. We use this as an approximate $Z_{0.75}$, so that:

$$X_{0.75} = 275 + 9.219544(Z_{0.75}) = 275 + 9.219544(0.67) = 281.1771$$

(b) Similarly, the lower 25th percentile is obtained as:

$$X_{0.25} = 275 + 9.219544(Z_{0.25})$$

From the table of the standard normal distribution, the nearest tabulated value to 0.25 is 0.2514, with a corresponding $Z = -0.67$. We approximate $Z_{0.25}$, as -0.67, so that:

$$X_{0.25} = 275 + 9.219544(Z_{0.25}) = 275 + 9.219544(-0.67) = 268.8229.$$

(c) Again, the upper 5th percentile = the lower 95th percentile. It is obtained as:

$$X_{0.95} = 275 + 9.219544(Z_{0.95})$$

From the table of the standard normal distribution, the nearest tabulated value to 0.95 is between 0.9505 and 0.9495, with a corresponding $Z = 1.645$ – obtained by adding the Z's corresponding to 0.9495 and 0.9505, which are 1.64 and 1.65, and dividing by 2. With this as an approximate value of $Z_{0.95}$, then:

$$X_{0.95} = 275 + 9.219544(Z_{0.95}) = 275 + 9.219544(1.645) = 290.1662.$$

■ **EXAMPLE 6.15** X, the weight of individual bars of bio-fuel made by a certain manufacturing process, is known to be approximately normally distributed with mean of 127 and standard deviation 4.

(a) A potential buyer requires at least that 95% of all the bars to be of greater weight than 122. Do these bars meet the specification?

(b) What weight, X, does a bio-fuel bar need to have for 80% of the weights of other bio-fuel bars to be above it?

Solution

(a) Under this distribution, the weight of a bar has to be above 122 to be within specification:

$$P(X > 122) = \frac{122 - 127}{4} = P(X > -1.25) = 1 - 0.1056 = 0.8944.$$

We can see that only 89.44% of the bars would be of weight 122 or higher.

We can approach the problem from the point of view of finding the lower 5th percentile for the above distribution of the bio-fuels. If it is less than 122, then it means that the bio-fuels will not meet the specifications.

From the table $Z_{0.05} = -1.645$. If X is the 5th percentile, then 95% of the bars would be of greater weight than X, $\frac{X - 127}{4} = -1.645 \Rightarrow X =$

$127 - 4(1.645) = 120.42$

Clearly this does not meet the specification that 95% should be greater than 122. Alternatively, if the buyer were to re-specify a weight of $X = 120.42$, then 95% of the bars would be of weight X and above.

(b) We need the lower 20th percentile of X, which is the upper 80th percentile. From the table, the nearest value to 0.2 is 0.1977 with a Z value of -0.85. So the 20th percentile is:

$$X_{0.20} = \mu + \sigma Z_{0.20} = 127 + (-0.85)(4) = 123.60.$$

The weight X which is less than the weights of 80% of the bio-fuels is 123.60.

6.3 Relationship Between the Binomial and the Normal Distributions

The binomial distribution has the functional form given in chapter five (see section 5.3.2). Using that function, suppose that the probability that a seed will germinate is 0.6. If 1000 seeds were planted, we may need to evaluate the probability that between 590 and 630 would germinate. The probability can be written as:

$$P(590 \leq X \leq 630) = \binom{1000}{590}(0.6)^{590}(1 - 0.6)^{1000-590}$$
$$+ \binom{1000}{590}(0.6)^{591}(1 - 0.6)^{1000-591} + \ldots + \binom{1000}{630}(0.6)^{630}(1 - 0.6)^{1000-630}$$

This probability is very tedious to evaluate and will consume a lot of time. We need an easier and less arduous way for evaluating it.

It turns out that when n, the number of trials in a binomial experiment, is large and neither p nor q is close to zero, then we can approximate the binomial distribution by a normal distribution. When n, the number of trials in a binomial experiment, becomes large, the shape of the binomial approaches the shape of the normal distribution; that is the bell-shaped symmetric form.

We can illustrate this by simulating the binomial experiment of flipping a fair coin for some values of n, the number of trials. Here, we present the shapes of the binomial distributions for $n = 10, 20, 50, 100,$ and 1100.

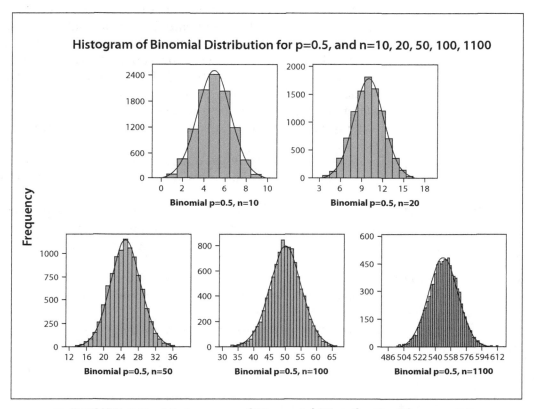

■ **FIGURE 6.7**   Histogram of Binomial Distribution for $p = 0.5$, and $n = 10, 20, 5, 100, 1100$

It is clear from the results that as n becomes larger, the binomial distribution approaches the normal distribution.

In such a situation, the distribution of X, the total number of trials, is normal with mean, np, and variance, npq. The resulting normal distribution is:

$$f(x) = \frac{1}{\sqrt{2\pi npq}} e - \frac{(x - np)^2}{2npq}; \; -\infty < x < +\infty.$$

In the above normal distribution, the usual mean μ is replaced by np, while the variance σ^2 is replaced by npq. As usual, we can easily obtain the probability of X, the binomial variable, using:

$$Z = \frac{X - np}{\sqrt{npq}}.$$

Because we are approximating a discrete distribution (the binomial) by a continuous distribution (the normal), we need to correct for continuity.

6.4 Correcting for Continuity

In the use of the normal to approximate the binomial, we have to pay particular attention to the framing of the question. For instance, if we want to evaluate the Z for $X > 291$, we need to replace the X in Z by 291.5 and not 291. If we need to evaluate the Z for $X < 291$, we replace X in the formula for Z by 290.5. Similarly, if we need to evaluate probability for $X \leq 291$, we replace 291 in the formula for Z by 291.5. These rules we observe are similar to rules observed in rounding up or rounding down the values by 0.5. For instance, under the rounding rules 1.5 is rounded up to 2, anything less than 1.5 but higher than 1 is rounded down to 1. By the same token, 0.5 can be rounded up to 1, but anything less than 0.5 would have to be rounded down to zero. Thus $X = 1$ will span everything from 0.5 up to but not including 1.5.

Here are some examples of applications of the normal approximation to the binomial distribution and the use of continuity correction in the process.

EXAMPLE 6.16   Suppose that 1000 oak seeds were planted and the probability of germination is 0.60, what is the probability that: (a) greater than 612 germinate? (b) greater than 590 and up to 630 germinate? (c) less than 615 germinate?

Solution $n = 1000; p = 0.6; q = 0.4; np = 0.6 \times 1000 = 600;$
$npq = 0.6 \times 0.4 \times 1000 = 240; \sqrt{npq} = \sqrt{240} = 15.49193$

We apply correction for continuity here:

(a) $P(X > 612) = P\left(Z > \dfrac{612.5 - 600}{15.4913}\right) = P(Z > 0.81)$
$$= 1 - 0.7910 = 0.2090$$

Again, we carefully apply the continuity correction in the following solutions to (b) and (c):

(b) $P(590 < X \leq 530) = \left(\dfrac{590.5 - 600}{15.4913} < Z \leq \dfrac{630.5 - 600}{15.4913}\right)$
$$= P(-0.61 < Z \leq 1.97) = 0.9756 - 0.2709 = 0.7047$$

(c) $P(X < 615) = P\left(Z < \dfrac{614.5 - 600}{15.4913}\right) = P(Z < 0.94) = 0.8264$

EXAMPLE 6.17   A production line for shoes is known to turn up 8% defectives. What is the probability that out of 500 shoes produced: (i) between 35 and 45 shoes, inclusive (ii) more than 48 shoes, and (iii) less than 38 shoes, will be defective?

Solution   Here again, we use normal approximation to the binomial and apply continuity correction of 0.5 as required, because we are converting discrete numbers to continuous numbers. We note that numbers 34.5 and above, but less than 35, are rounded up to 35; similarly numbers less than 45.5, but greater than 45, are rounded down to 45. This is similar to the use of class boundaries we discussed in chapter one when assigning data to classes in frequency table.

By using the normal distribution to approximate the binomial distribution, we note that the mean is $np = 0.08(500) = 40$ and variance is $npq = 0.08(0.92)(500) = 36.8$ so that $\sqrt{npq} = \sqrt{36.8} = 6.0663$

We now calculate the probabilities:

(a) $P(35 \le X > 45) = P\left(\dfrac{34.5 - 40}{6.0663} \le Z \le \dfrac{45.5 - 40}{6.0663}\right)$

$= P(-0.91 \le Z \le 0.91) = 0.8186 - 0.1814 = 0.6372$

(b) $P(X > 48) = P\left(Z > \dfrac{48.5 - 40}{6.0663}\right) = P(Z > 1.40) = 1 - 0.9192 = 0.0808$

(c) $P(X < 38) = P\left(Z < \dfrac{37.5 - 40}{6.0663}\right) = P(Z < -0.41) = 0.3409$

■ **EXAMPLE 6.18** It is known at the CTZ University Medical Services that women treated by in-vitro fertilization who became pregnant with twins have equal probability of giving birth to one gender only twins or mixed gender twins. Of 500 women treated by CTZ University Medical Services who became pregnant with twins, what is the probability that mixed gender twins differ from the mean by: (a) ten twins? (b) thirty twins? (c) What is the chance that 255 or more one gender only twins are born?

Solution Again, here we use a continuity correction of 0.5, to convert discrete numbers to continuous numbers, so that numbers 239.5 and above, but less than 240 are rounded up to 240; similarly, numbers less than 260.5 but greater than 260 are rounded down to 260 as we described in Example 3.16.

We can use the normal approximation for the binomial to solve this problem.

$$P(\text{mixed gender twins}) = p = \frac{1}{2} = P(\text{one gender twins}) = q;\ n = 500;$$

$$np = 500 \times \frac{1}{2} = 250$$

$$npq = \frac{1}{2} \times \frac{1}{2} \times 500 = 125 \Rightarrow \sqrt{npq} = \sqrt{125} = 11.18$$

(a) Let X be the number of mixed gender twins. $P(X$ does not differ from 250 by more than 10 twins)

$$P(240 \le X \le 260) = P\left(\frac{239.5 - 250}{11.18} \le Z \le \frac{260.5 - 250}{11.18}\right)$$
$$= P(-0.94 \le Z \le 0.94) = 0.8264 - 0.1736 = 0.6528$$

(b) Let X be the number of mixed gender twins. $P(X$ does not differ from 250 by more than 30 twins)

$$P(220 \le X \le 280) = P\left(\frac{219.5 - 250}{11.18} \le Z \le \frac{280.5 - 250}{11.18}\right)$$
$$= P(-2.73 \le Z \le 2.73) = 0.9968 - 0.0032 = 0.9936$$

(c) $P(X > 255) = P\left(Z > \dfrac{255.5 - 250}{11.18}\right)$
$$= P(Z > 0.49) = 1 - 0.6879 = 0.3121$$

■ **EXAMPLE 6.19** If it is known that a pregnant cow is equally is likely to give birth to a female calf as it is to give birth to a male calf; if 300 cows calve in Hartersley's farm, what is the probability that the number of bulls born: (a) lie between 120 and 180, inclusive? (b) is less than 135? (c) What is the probability that the number of female cows born is less than 128 or more than 158?

Solution The total number of births is 300. The probability of a bull = probability of a cow $= \frac{1}{2}$. Using the normal approximation to the binomial distribution, we note that the mean is $np = 300 \times \frac{1}{2} = 150$ and the variance, $\sigma^2 = npq = 300 \times \frac{1}{2} \times \frac{1}{2} = 75 \Rightarrow \sigma = \sqrt{75} = 8.6603$.

(a) Let X be the number of bulls, then P(between 120 and 180 bulls inclusive)

$$P(120 \le X \le 180) = P\left(\frac{119.5 - 150}{8.6603} \le Z \le \frac{180.5 - 150}{8.6603}\right)$$
$$= P(-3.52 \le Z \le 3.52) = 0.9998 - 0.0002$$
$$= 0.9996.$$

(b) Let X be the number of bulls, then P[number of bulls is less than 135] is $P[X < 135]$

$$P(X \le 135) = P\left(Z \le \frac{134.5 - 150}{8.6603}\right) = P(Z \le -1.79) = 0.0367.$$

(c) Let X be the number of cows. Here the keyword "or" means addition of the two probabilities as shown below. P[less than 128 or more than 158 cows] is:

$$P(X < 128) + P(X > 158) = P\left(Z < \frac{127.5 - 150}{8.6603}\right) + P\left(Z > \frac{158.5 - 150}{8.6603}\right)$$
$$= P(Z < -2.60) + P(Z > 0.98)$$
$$= 0.0047 + (1 - 0.8365) = 0.0047 + 0.1635$$
$$= 0.1682.$$

6.5 Normal Approximation of the Poisson Distribution

The normal distribution can be used to approximate the Poisson distribution if λ, the mean, is sufficiently large. The approximation is considered good if $\lambda \ge 10$. As was the case with the Binomial distribution, we need to use a continuity correction.

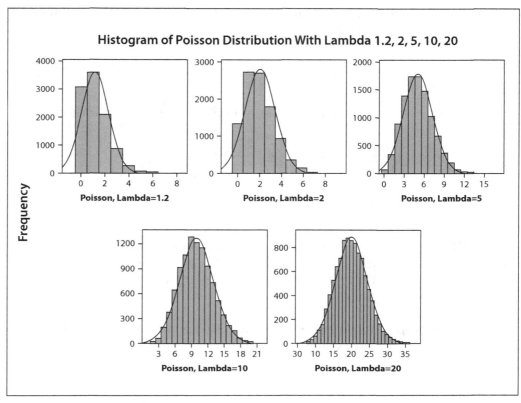

FIGURE 6.8 Histograms of Poisson Distributions with Lambda = 1.2, 2, 5, 10, 20

We can see from the results represented by the histograms of Figure 6.8 that as λ becomes large, the Poisson distribution approaches the normal distribution.

The distribution of X, the total number of occurrences per unit time, is normal with mean, λ and variance, λ. The resulting normal distribution is:

$$f(x) = \begin{cases} \dfrac{1}{\sqrt{2\pi\lambda}}e - \dfrac{(x-\lambda)^2}{2\lambda}; & -\infty < x < +\infty \\ 0 \text{ elsewhere} \end{cases}$$

In the above normal distribution, the usual mean μ is replaced by λ, while the variance σ^2 is replaced by λ. As usual, we can easily obtain the probability of X, the Poisson variable, using:

$$Z = \frac{X - \lambda}{\sqrt{\lambda}}.$$

Again, because we are approximating a discrete distribution (the Poisson) by a continuous distribution (the normal), we need to correct for continuity.

EXAMPLE 6.20   Twenty calls per hour arrive at a Central Computer in accordance with a Poisson process. Find the probability that in an hour (a) more than 30 calls, (b) less than 15, and (c) between 10 and 35 inclusive calls, would arrive at the computer.

Solutions   We use the normal approximation to the Poisson and as usual, we use continuity correction:

(a) $P(X > 30) = P\left(Z > \dfrac{30.5 - 20}{\sqrt{20}}\right) = P(Z > 2.35) = 1 - 0.9906 = 0.0094$

(b) $P(X < 15) = P\left(Z < \dfrac{14.5 - 20}{\sqrt{20}}\right) = P(Z < -1.23) = 0.1094$

(c) $P(10 < X < 35) = P\left(\dfrac{9.5 - 20}{\sqrt{20}} < Z < \dfrac{35.5 - 20}{\sqrt{20}}\right)$
$= P(-2.35 < Z < 3.47) = 0.9997 - 0.0094 = 0.9903$

■ **EXAMPLE 6.21**   The mean number of snails found per acre in a research in a rain forest was 15. If any of the acres in the forest is chosen at random, what is the probability that (a) at least 11 snails (b) less than 23 snails (c) between 9 and 20 snails, would be found.

Solutions   Event occurs per unit area. We approximate the Poisson by the normal and adjust values with continuity correction:

(a) $P(X > 11) = P\left(Z > \dfrac{10.5 - 15}{\sqrt{15}}\right)$
$= P(Z > -1.16) = 1 - 0.1230 = 0.9770$

(b) $P(X < 15) = P\left(Z < \dfrac{22.5 - 15}{\sqrt{15}}\right) = P(Z < 1.94) = 0.9738$

(c) $P(10 < X < 35) = P\left(\dfrac{8.5 - 15}{\sqrt{15}} < Z < \dfrac{20.5 - 15}{\sqrt{15}}\right)$
$= P(-1.68 < Z < 1.42) = 0.9222 - 0.0465 = 0.8757$

■ **EXAMPLE 6.22**   Machines arrive randomly for repair at a repair shop at the rate of 18 per day. If a day is chosen at random, what is the probability that (a) less than 28 machines (b) greater than 10 but less than 26 (c) at least 15 machines arrive at the repair shop.

Solution   Again, we approximate the Poisson by the normal and adjust values with continuity correction:

(a) $P(X < 28) = P\left(Z < \dfrac{27.5 - 18}{\sqrt{18}}\right) = P(Z < 2.24) = 0.9875$

(b) $P(10 < X < 26) = P\left(\dfrac{10.5 - 18}{\sqrt{18}} < Z < \dfrac{25.5 - 18}{\sqrt{18}}\right)$
$= P(-1.77 < Z < 1.77) = 0.9616 - 0.0384 = 0.9232$

(c) $P(X \geq 15) = P\left(Z \leq \dfrac{14.5 - 18}{\sqrt{18}}\right)$
$= 1 - P(Z < -0.82) = 1 - 0.2061 = 0.7939$

6.6	**Use of MINITAB to Evaluate Normal and Rectangular Distributions**

We can use the MINITAB to evaluate the probabilities for continuous distributions just like we did for discrete distributions. In using MINITAB, we have to invoke the cdf (cumulative distribution function) in order to evaluate probabilities for continuous distributions. We will illustrate the use of MINITAB by doing some examples.

6.6.1 Use of MINITAB to Evaluate Probabilities for Rectangular Distribution

EXAMPLE 6.23 A variable, X, is distributed such that $X \sim U(10, 45)$. Obtain the probabilities that: (a) $X > 23$, (b) $17 < X < 36$, and (c) $X < 32$. Use MINITAB to find these probabilities.

Solution

```
MTB > set c1
DATA> 10 17 23 32 36 45
DATA> end
MTB > cdf c1 c2;
SUBC> uniform 10 45.
```

Output:

x	cdf(x)
10	0.0
17	0.2
23	0.371429
32	0.628571
36	0.742857
45	1.0

(a) $P(X > 23) = 1 - P(X < 23) = 1 - 0.371429 = 0.628571$
(b) $P(17 < X < 36) = 0.742857 - 0.20 = 0.542857$
(c) $P(X < 32) = 0.628571$

EXAMPLE 6.24 A variable Y is distributed as:

$$f(y) = \frac{1}{12}; \; -4 \le y \le 8;$$

What is the probability that: (a) $Y < 2$, (b) $1 < Y < 6$, and (c) $Y > 0$?

```
MTB > set c1
DATA> -4 0 1 2 6 8
DATA> end
MTB > cdf c1 c2;
SUBC> uniform -4 8.
```

Output:

x	cdf(x)
−4	0.0
0	0.333333
1	0.416667
2	0.5
6	0.833333
8	1.0

(a) $P(Y < 2) = 0.50$
(b) $P(1 < Y < 6) = 0.833333 - 0.416667 = 0.416667$
(c) $P(X > 0) = 1 - 0.33333 = 0.66667$

6.6.2 Use of MINITAB to Evaluate Probabilities for Normal Distribution

■ **EXAMPLE 6.25**   A variable, X, is distributed such that $X \sim N(160, 81)$. Obtain the probabilities that: (a) $X > 143$, (b) $157 < X < 176$, and (c) $X < 192$. Use MINITAB to find these probabilities.

Solution   Here $\mu = 160$ and $\sigma^2 = 81$ so that $\sigma = \sqrt{81} = 9$

```
MTB > set c1
DATA> 143 157 176 192
DATA> end
MTB > cdf c1 c2;
SUBC> normal 160 9.
```

Output:

x	cdf(x)
143	0.029453
157	0.369441
176	0.96228
192	0.999811

(a) $P(X > 143) = 1 - P(X < 143) = 1 - 0.029453 = 0.970547$
(b) $P(157 < X < 176) = 0.96228 - 0.36944 = 0.592838$
(c) $P(X < 192) = 0.999811$

■ **EXAMPLE 6.26**   A variable, Y, is distributed so that $Y \sim N(79, 44)$;
What is the probability that (a) $Y < 69.6$, (b) $67.3 < Y < 85.3$, and (c) $Y > 75$?

Solution   Here $\mu = 79$ and $\sigma^2 = 44$ so that $\sigma = \sqrt{44} = 6.63325$

```
MTB > set c1
DATA> 67.3 69.6 85.3 75
DATA> end
MTB > cdf c1 c2;
SUBC> normal 79 6.63325.
```

Output:

x	cdf(x)
67.3	0.038879
69.6	0.078226
75.0	0.273247
85.3	0.828883

(a) $P(Y < 69.6) = 0.078226$
(b) $P(67.3 < Y < 85.3) = 0.828883 - 0.038879 = 0.790004$
(c) $P(X > 75) = 1 - 0.273247 = 0.726753$

Exercises

6.1 For human pregnancies, the World Health Organization defines normal term for delivery as between 259 days and 294 days following the inception of the pregnancy. Assume that normal deliveries in human pregnancies occur uniformly in this interval. Obtain the probability that a baby is born: (i) by day 277 or more, (ii) between 266 and 280 days, and (iii) under 270 days following the inception of the pregnancy.

6.2 A four year College noticed that students graduate for the BS degree in seven to seventeen semesters. Assume that graduation of students occurs uniformly in this interval of semesters.
  (a) Obtain the five number summary for graduation period for the college, in semesters. What is the proportion of students that graduate:
  (b) in less than or eight semesters?
  (c) between eight and twelve semesters?
  (d) What is the number of semesters required for exactly 50% of students to graduate?
  (e) Find the interquartile range for the distribution of graduation in semesters.

6.3 A factory that produces steel rods that are supposed to be of lengths of twenty inches, find that the lengths are spread uniformly between 19.80 and 20.2 inches.
  (a) Find the five number summary for the distribution of the rods.
  (b) What is the proportion of rods whose lengths are above 20 inches?
  (c) Suppose it is decided to reject the rods which are in the lower 10th percentile and in the upper 10th percentile; what are the cut-off values?

6.4 A variable X is uniformly distributed between $[-a, 2a]$.
  (a) Write down the particular form of the Uniform distribution which applies.
  (b) Obtain the mean, variance, and standard deviation for this distribution in terms of a.

6.5 The waiting times between filing a State tax refund and getting a refund is between two and thirty-six days. Assume refunds are released uniformly from the State Revenue office.
  (a) State the specific form of the uniform distribution that applies in this case.
  (b) Obtain the probability that one waits for a state refund for more than four weeks.
  (c) What is the number of days one has to wait to be among the 30% that waited longest?
  (d) What is the number of days one has to wait to be among the first 40% that received their refunds?

6.6   Consider the following uniform distribution:

$$f(x) = \frac{1}{5}; \; -3 \le x \le 2;$$

(a) Find the five number summary for the distribution.
(b) Obtain the lower 60th percentile for this distribution.
(c) Obtain the upper 34th percentile for this distribution.
(d) Obtain the variance and standard deviation for the distribution.

Exercises on the Normal Distribution

6.7   If Y is normally distributed with mean 80 and variance 20; obtain the standardized values (Z) for the following Y values:
(a) 87,
(b) 88.95,
(c) 72,
(d) 68, and
(e) 93.

6.8   A certain animal that lives in seawater is believed to contain calcium in a concentration which is normally distributed with mean 30 mmole/kg and variance of 6.6 mmole^2/kg. What proportions of the animal contain calcium in the concentrations which are:
(a) above 26.6 mmole/kg?
(b) between 25 and 34 mmole/kg?
(c) less than 27 mmole/kg?

6.9   With $Z \sim N(0, 1)$, obtain the values of Z_1:
(a) $P(-Z_1 \le Z \le Z_1) = 0.78$
(b) $P(-1.25 \le Z \le Z_1) = 0.8403$
(c) $P(-Z_1 \le Z \le 1.68) = 0.5367$

6.10   The scores in STAT 193 are believed to follow the normal distribution with a mean of 72 and variance of 90. What is the proportion of students whose scores:
(a) lie between 55 and 81?
(b) are less than 50?
(c) are greater than 95?

6.11   The scores in an aptitude test are believed to follow the normal distribution with a mean of 66 and a variance of 53. What is the proportion of candidates whose scores:
(a) lie between 50 and 77?
(b) are less than 47?
(c) are greater than 82?

6.12   Using the table of the standard normal distribution, find the following probabilities:
(a) $P(-1.23 < Z < 2.05)$
(b) $P(Z < -1.84)$
(c) $P(Z > 1.68)$
(d) $P(Z > -1.33)$
(e) $P(Z < 2.41)$

6.13   The heights of eighteen-year-old students in County Durham are normally distributed with a mean of 64 inches and a variance (σ^2) of 16 inches^2. Find the probability that a randomly selected eighteen-year-old student in County Durham has a height of:
(a) less than 70?
(b) between 60 and 72?
(c) greater than 57?

6.14   The diameters of 3-year-old oak trees are normally distributed as Y with mean 30.2 and variance of 15.29. What is the probability that an oak tree has diameter:
(a) $Y > 36.7$
(b) $26.1 < Y < 37.8$
(c) $Y < 27.4$

6.15 X, the IQs of twelve year olds, are normally distributed with a mean of 100 and a variance of 112. Find the probability that the IQ of a twelve-year-old is:
(a) $X > 97$
(b) $94 < X < 107$
(c) $104 < X < 118$
(d) $X < 89$

6.16 Y is normally distributed with mean of 110 and variance of 104.
(a) $P(97 \le Y \le 118)$
(b) $P(Y > 104)$
(c) $P(Y < 123)$

6.17   The following pictures show a variable X which is normally distributed with a mean of 80 and a variance of 36. In each case, obtain the probability represented by the shaded area.

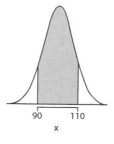
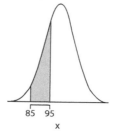
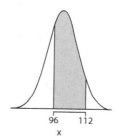
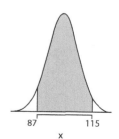

6.18   The following pictures show a variable X which is normally distributed with a mean of 100 and a variance of 49. Obtain the probability represented by the shaded area.

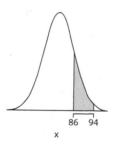
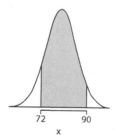
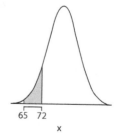
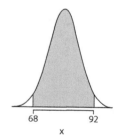

6.19   The following pictures show a variable *X* which is normally distributed with a mean of 90 and a variance of 42.25. Obtain the probability represented by the shaded area.

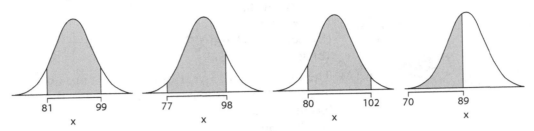

6.20   The following pictures show a variable *X* which is normally distributed with a mean of 78 and a variance of 25. Obtain the probability represented by the shaded area.

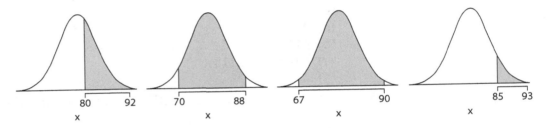

Exercises on Percentiles of the Normal Distribution

The following data show the 2009 SAT scores for participants from Minnesota:

Critical Thinking		Mathematics		Writing	
Mean	Variance	Mean	Variance	Mean	Variance
595	11025	609	10201	578	10609

6.21   Assume that the SAT scores in Critical Thinking are normally distributed; obtain the: (i) lower 96th percentile, and (ii) upper 66th percentile.

6.22   Assume that the SAT scores in Mathematics are normally distributed; obtain the: (i) upper 28th percentile, (ii) lower 53rd percentile, and (iii) lower 26th percentile.

6.23   Assume that the SAT scores in Writing are normally distributed; obtain the: (i) upper 20th percentile, (ii) lower 70th percentile, and (iii) upper 5th percentile.

6.24   In IQs of adults, *X* are assumed to be normally distributed, so that $X \sim N(100, 225)$. What is the IQ of an adult in the:
(a) lower 99th percentile?
(b) lower 25th percentile?
(c) lower 75th percentile?
(d) What is the median?

6.25 Weekly TV watching by elementary school children is normally distributed with a mean of nineteen hours and a variance of 12.24 hours^2. How many hours of TV does an elementary school child need to watch to be in the:
(a) upper 4th percentile?
(b) median?
(c) lower 85th percentile?

6.26 The heights of American men are normally distributed with a mean of 69 inches and a variance of 8.5. Obtain the height of an American man in the:
(a) lower 92nd percentile,
(b) upper 55th percentile, and
(c) upper 3rd percentile.

6.27 The heights of American women are normally distributed with a mean of 64 inches and a variance of 7.9. Obtain the height of an American woman in the:
(a) lower 90th percentile,
(b) upper 6th percentile, and
(c) lower 80th percentile.

6.28 The weekly number of hours spent by professors on teaching, advising, research, and other aspects of the job are normally distributed with a mean of 60.3 hours and a variance of 27.8. How many hours does a professor have to work weekly to be in the:
(a) upper 2nd percentile?
(b) median?
(c) lower 90th percentile?

6.29 Blood health is generally important for our well-being. Hematologists use the whole blood hemoglobin concentration in the diagnosis and medical management of patients with hematologic disease. The normal whole blood hemoglobin counts for adult males are believed to be normally distributed with mean = 16 g/dL and a variance of 1.33 g^2/dL. Obtain the:
(a) upper 30th percentile,
(b) upper 5th percentile, and
(c) lower 88th percentile for hemoglobin concentrations of adult males.

6.30 A variable $Y \sim N(190, 90)$ what is the:
(a) 75th percentile,
(b) 90th percentile, and
(c) 95th percentile for Y?

Exercises on Normal Approximation to the Binomial Distribution

6.31 In a local county it is known that 66% of the women support the proposed reservation of a piece of land for conservation. If 730 of the women from this county were surveyed, what is the probability that the number of women who support the land conservation will be:
(a) more than 500?
(b) greater than 460 but no more than 490? and
(c) less than 464?

6.32 In 2010 elections, 56% voted against Proposition 19, which would legalize recreational marijuana in California. If 820 Californians who voted were surveyed, determine the probability that the number of people who opposed Proposition 19 is:
(a) between 417 and 489 inclusive,
(b) less than 416, and
(c) would differ from 450 by 30.

6.33 A loaded coin is such that the head turns up only 43% of the time. The coin was flipped 500 times. What is the probability that the number of times the head turns up is:
(a) greater than 182?
(b) less than 233?
(c) lies between 185 and 235?

6.34 Among online tax filers last year, 72% chose to receive their refund via direct deposit. If 1200 of the online tax filers for last year are surveyed what is the probability that:
(a) more than 840,
(b) less than 890, and
(c) between 850 and 900 inclusive, received their refund via direct deposit.

6.35 Binge drinking among American college students is on the rise, along with its consequences, according to survey research by *Health Day News*. The same survey showed that 51% of all college males have binged on alcohol. Based on this finding, if 1500 male students were interviewed, what is the probability that:
(a) between 720 and 805,
(b) less than 790, and
(c) over 767, have binged on alcohol?

6.36 Canada's Boreal Forest is the world's largest intact forest and is home to more than 300 bird species, including 63% finches. If 1000 finches are caught at random from a mixture which migrated from different parts of the world, what is the probability that the number of finches from Canada's Boreal Forest:
(a) differs from 630 by exactly 35?
(b) is less than 600?
(c) is greater than 640?

6.37 An opinion poll by Gallup found in 2007 that 88% were saying they would vote for a well-qualified woman for president of the USA. In that year, if 900 voters were surveyed, what is the probability that:
(a) over 770,
(b) between 790 and 822 inclusive, and
(c) less than 820, would support a female president?

6.38 Eighty percent of breast lumps are **not** cancerous, but benign growths, such as fibrocystic breast disease. If 1020 women found lumps in their breasts, what is the probability that:
(a) more than 778,
(b) less than 840, and
(c) between 785 and 835, are not cancerous?

6.39 The probability that the gestation period of a woman will exceed nine months is 0.314. In 1000 human births, what is the probability that the number in which gestation period exceeds nine months is:
(a) less than 287?
(b) greater than 340?
(c) between 286 and 336?

6.40 If it is known that the probability a pregnant Gebu cow gives birth to a male calf is 0.45, if 300 Gebu cows calve in Mr. Jones' farm, what is the probability that the number of bulls born:
(a) lies between 120 and 158?
(b) is less than 149?
(c) What is the probability that the number of female calves born is less than 122 but no more than 155?

Exercises on Normal Approximation for the Poisson

6.41 A forest officer observed on average forty bird nests per square mile of a managed forest. If a square mile of this forest is chosen at random, what is the probability that:
(a) no more than 55,
(b) more than 33 but less than 51, and
(c) more than 35 bird nests, would be found in it?

6.42 A trained eye of the manufacturer observed on average seventy defects per piece of textile manufactured by a pilot plant. A piece is made up of approximately 18×3 square feet. If a piece of the cloth is chosen at random, what is the probability that:
(a) less than 58,
(b) more than 50 but less than 88,
(c) more than 79 defects, would be found in it.

6.43 The number of patients arriving at an outpatients' clinic serving a large conurbation is fifty-two per hour for every working day. If an hour of a working day is chosen at random, what is the probability that:
(a) less than 40,
(b) more than 55 but less than 68, and
(c) more than 37 patients would be seen?

6.44 On average in recent years, in Minnesota, seventy-five fatalities have occurred due to drivers who were texting, using cell phones, or were otherwise distracted by similar technology while driving. If such a year is chosen at random, what is the probability that:
(a) less than 48,
(b) more than 53 but less than 86, and
(c) more than 83 fatalities occurred due to this type of distraction of the driver.

6.45 In the USA, gun-related accidental deaths occur at the rate of 849 per year. If a year is chosen at random, what is the probability that the number of gun-related accidental deaths is:
(a) less than 878,
(b) between 800 and 900 inclusive, and
(c) more than 790 would have occurred.

6.46   A variable Y is normally distributed with a mean of 75 and a variance of
41. What is the probability that:
(a) $Y > 80.2$?
(b) $70.4 < Y < 83.1$?
(c) $73.7 < Y < 78.4$?

Table of the Standard Normal Distribution

Z	0	0.01	0.02	0.03	0.04	0.05	0.06	0.07	0.08	0.09	Z
−3.7	0.0001	0.0001	0.0001	0.0001	0.0001	0.0001	0.0001	0.0001	0.0001	0.0001	−3.7
−3.6	0.0002	0.0002	0.0002	0.0001	0.0001	0.0001	0.0001	0.0001	0.0001	0.0001	−3.6
−3.5	0.0002	0.0002	0.0002	0.0002	0.0002	0.0002	0.0002	0.0002	0.0002	0.0002	−3.5
−3.4	0.0003	0.0003	0.0003	0.0003	0.0003	0.0003	0.0003	0.0003	0.0003	0.0002	−3.4
−3.3	0.0005	0.0005	0.0005	0.0004	0.0004	0.0004	0.0004	0.0004	0.0004	0.0004	−3.3
−3.2	0.0007	0.0007	0.0006	0.0006	0.0006	0.0006	0.0006	0.0005	0.0005	0.0005	−3.2
−3.1	0.0010	0.0009	0.0009	0.0009	0.0008	0.0008	0.0008	0.0008	0.0007	0.0007	−3.1
−3.0	0.0014	0.0013	0.0013	0.0012	0.0012	0.0011	0.0011	0.0011	0.0010	0.0010	−3.0
−2.9	0.0019	0.0018	0.0018	0.0017	0.0016	0.0016	0.0015	0.0015	0.0014	0.0014	−2.9
−2.8	0.0026	0.0025	0.0024	0.0023	0.0023	0.0022	0.0021	0.0021	0.0020	0.0019	−2.8
−2.7	0.0035	0.0034	0.0033	0.0032	0.0031	0.0030	0.0029	0.0028	0.0027	0.0026	−2.7
−2.6	0.0047	0.0045	0.0044	0.0043	0.0042	0.0040	0.0039	0.0038	0.0037	0.0036	−2.6
−2.5	0.0062	0.0060	0.0059	0.0057	0.0055	0.0054	0.0052	0.0051	0.0049	0.0048	−2.5
−2.4	0.0082	0.0080	0.0078	0.0076	0.0073	0.0071	0.0070	0.0068	0.0066	0.0064	−2.4
−2.3	0.0107	0.0104	0.0102	0.0099	0.0096	0.0094	0.0091	0.0089	0.0087	0.0084	−2.3
−2.2	0.0139	0.0136	0.0132	0.0129	0.0126	0.0122	0.0119	0.0116	0.0113	0.0110	−2.2
−2.1	0.0179	0.0174	0.0170	0.0166	0.0162	0.0158	0.0154	0.0150	0.0146	0.0143	−2.1
−2.0	0.0228	0.0222	0.0217	0.0212	0.0207	0.0202	0.0197	0.0192	0.0188	0.0183	−2.0
−1.9	0.0287	0.0281	0.0274	0.0268	0.0262	0.0256	0.0250	0.0244	0.0239	0.0233	−1.9
−1.8	0.0359	0.0352	0.0344	0.0336	0.0329	0.0322	0.0314	0.0307	0.0301	0.0294	−1.8
−1.7	0.0446	0.0436	0.0427	0.0418	0.0409	0.0401	0.0392	0.0384	0.0375	0.0367	−1.7
−1.6	0.0548	0.0537	0.0526	0.0516	0.0505	0.0495	0.0485	0.0475	0.0465	0.0455	−1.6
−1.5	0.0668	0.0655	0.0643	0.0630	0.0618	0.0606	0.0594	0.0582	0.0571	0.0559	−1.5
−1.4	0.0808	0.0793	0.0778	0.0764	0.0749	0.0735	0.0722	0.0708	0.0694	0.0681	−1.4
−1.3	0.0968	0.0951	0.0934	0.0918	0.0901	0.0885	0.0869	0.0853	0.0838	0.0823	−1.3
−1.2	0.1151	0.1131	0.1112	0.1094	0.1075	0.1057	0.1038	0.1020	0.1003	0.0985	−1.2
−1.1	0.1357	0.1335	0.1314	0.1292	0.1271	0.1251	0.1230	0.1210	0.1190	0.1170	−1.1
−1.0	0.1587	0.1563	0.1539	0.1515	0.1492	0.1469	0.1446	0.1423	0.1401	0.1379	−1.0

Continued.

■ Table of the Standard Normal Distribution—cont'd

Z	0	0.01	0.02	0.03	0.04	0.05	0.06	0.07	0.08	0.09	Z
−0.9	0.1841	0.1814	0.1788	0.1762	0.1736	0.1711	0.1685	0.1660	0.1635	0.1611	−0.9
−0.8	0.2119	0.2090	0.2061	0.2033	0.2005	0.1977	0.1949	0.1922	0.1894	0.1867	−0.8
−0.7	0.2420	0.2389	0.2358	0.2327	0.2297	0.2266	0.2236	0.2207	0.2177	0.2148	−0.7
−0.6	0.2743	0.2709	0.2676	0.2644	0.2611	0.2579	0.2546	0.2514	0.2483	0.2451	−0.6
−0.5	0.3085	0.3050	0.3015	0.2981	0.2946	0.2912	0.2877	0.2843	0.2810	0.2776	−0.5
−0.4	0.3446	0.3409	0.3372	0.3336	0.3300	0.3264	0.3228	0.3192	0.3156	0.3121	−0.4
−0.3	0.3821	0.3783	0.3745	0.3707	0.3669	0.3632	0.3594	0.3557	0.3520	0.3483	−0.3
−0.2	0.4207	0.4168	0.4129	0.4091	0.4052	0.4013	0.3974	0.3936	0.3897	0.3859	−0.2
−0.1	0.4602	0.4562	0.4522	0.4483	0.4443	0.4404	0.4364	0.4325	0.4286	0.4247	−0.1
0.0	0.5000	0.4960	0.4920	0.4880	0.4840	0.4801	0.4761	0.4721	0.4681	0.4641	0.0
0.1	0.5398	0.5438	0.5478	0.5517	0.5557	0.5596	0.5636	0.5675	0.5714	0.5754	0.1
0.2	0.5793	0.5832	0.5871	0.5910	0.5948	0.5987	0.6026	0.6064	0.6103	0.6141	0.2
0.3	0.6179	0.6217	0.6255	0.6293	0.6331	0.6368	0.6406	0.6443	0.6480	0.6517	0.3
0.4	0.6554	0.6591	0.6628	0.6664	0.6700	0.6736	0.6772	0.6808	0.6844	0.6879	0.4
0.5	0.6915	0.6950	0.6985	0.7019	0.7054	0.7088	0.7123	0.7157	0.7190	0.7224	0.5
0.6	0.7258	0.7291	0.7324	0.7357	0.7389	0.7422	0.7454	0.7486	0.7518	0.7549	0.6
0.7	0.7580	0.7612	0.7642	0.7673	0.7704	0.7734	0.7764	0.7794	0.7823	0.7852	0.7
0.8	0.7881	0.7910	0.7939	0.7967	0.7996	0.8023	0.8051	0.8079	0.8106	0.8133	0.8
0.9	0.8159	0.8186	0.8212	0.8238	0.8264	0.8289	0.8315	0.8340	0.8365	0.8389	0.9
1.0	0.8413	0.8438	0.8461	0.8485	0.8508	0.8531	0.8554	0.8577	0.8599	0.8621	1.0
1.1	0.8643	0.8665	0.8686	0.8708	0.8729	0.8749	0.8770	0.8790	0.8810	0.8830	1.1
1.2	0.8849	0.8869	0.8888	0.8907	0.8925	0.8944	0.8962	0.8980	0.8997	0.9015	1.2
1.3	0.9032	0.9049	0.9066	0.9082	0.9099	0.9115	0.9131	0.9147	0.9162	0.9177	1.3
1.4	0.9192	0.9207	0.9222	0.9236	0.9251	0.9265	0.9279	0.9292	0.9306	0.9319	1.4
1.5	0.9332	0.9345	0.9357	0.9370	0.9382	0.9394	0.9406	0.9418	0.9430	0.9441	1.5
1.6	0.9452	0.9463	0.9474	0.9485	0.9495	0.9505	0.9515	0.9525	0.9535	0.9545	1.6
1.7	0.9554	0.9564	0.9573	0.9582	0.9591	0.9599	0.9608	0.9616	0.9625	0.9633	1.7
1.8	0.9641	0.9649	0.9656	0.9664	0.9671	0.9678	0.9686	0.9693	0.9700	0.9706	1.8

■ Table of the Standard Normal Distribution—cont'd

Z	0	0.01	0.02	0.03	0.04	0.05	0.06	0.07	0.08	0.09	Z
1.9	0.9713	0.9719	0.9726	0.9732	0.9738	0.9744	0.9750	0.9756	0.9762	0.9767	1.9
2.0	0.9773	0.9778	0.9783	0.9788	0.9793	0.9798	0.9803	0.9808	0.9812	0.9817	2.0
2.1	0.9821	0.9826	0.9830	0.9834	0.9838	0.9842	0.9846	0.9850	0.9854	0.9857	2.1
2.2	0.9861	0.9865	0.9868	0.9871	0.9875	0.9878	0.9881	0.9884	0.9887	0.9890	2.2
2.3	0.9893	0.9896	0.9898	0.9901	0.9904	0.9906	0.9909	0.9911	0.9913	0.9916	2.3
2.4	0.9918	0.9920	0.9922	0.9925	0.9927	0.9929	0.9931	0.9932	0.9934	0.9936	2.4
2.5	0.9938	0.9940	0.9941	0.9943	0.9945	0.9946	0.9948	0.9949	0.9951	0.9952	2.5
2.6	0.9953	0.9955	0.9956	0.9957	0.9959	0.9960	0.9961	0.9962	0.9963	0.9964	2.6
2.7	0.9965	0.9966	0.9967	0.9968	0.9969	0.9970	0.9971	0.9972	0.9973	0.9974	2.7
2.8	0.9974	0.9975	0.9976	0.9977	0.9977	0.9978	0.9979	0.9980	0.9980	0.9981	2.8
2.9	0.9981	0.9982	0.9983	0.9983	0.9984	0.9984	0.9985	0.9985	0.9986	0.9986	2.9
3.0	0.9987	0.9987	0.9987	0.9988	0.9988	0.9989	0.9989	0.9989	0.9990	0.9990	3.0
3.1	0.9990	0.9991	0.9991	0.9991	0.9992	0.9992	0.9992	0.9992	0.9993	0.9993	3.1
3.2	0.9993	0.9993	0.9994	0.9994	0.9994	0.9994	0.9994	0.9995	0.9995	0.9995	3.2
3.3	0.9995	0.9995	0.9996	0.9996	0.9996	0.9996	0.9996	0.9996	0.9996	0.9997	3.3
3.4	0.9997	0.9997	0.9997	0.9997	0.9997	0.9997	0.9997	0.9997	0.9998	0.9998	3.4
3.5	0.9998	0.9998	0.9998	0.9998	0.9998	0.9998	0.9998	0.9998	0.9998	0.9998	3.5
3.6	0.9998	0.9999	0.9999	0.9999	0.9999	0.9999	0.9999	0.9999	0.9999	0.9999	3.6
3.7	0.9999	0.9999	0.9999	0.9999	0.9999	0.9999	0.9999	0.9999	0.9999	0.9999	3.7

CHAPTER 7

Sampling Distributions of the Mean, the Proportion, and the Central Limit Theorem

The population Y is such that $Y = \{2, 3, 5, 7, 10, 12\}$. So the population is made up of only six values. Let us select a simple random sample of size four from this population using the principles we learned in chapter one. Following what we learned in chapter one, we label all the members of the population, as in Table 7.1 below:

TABLE 7.1 Assigning Labels to the Values of Y

Y	2	3	5	7	10	12
Label	0	1	2	3	4	5

Here $N = 6$, a one-digit number, so we label all the units in the population as 0, 1, 2, 3, 4, 5. From the Table of random digits (Appendix I), let us arbitrarily plan to select samples as follows:

Sample 1: Start from row 7, column 13
Sample 2: Start from row 5, column 21
Sample 3: Start from row 3, column 15
Sample 4: Start from row 10, column 18
Sample 5: Start from row 12, column 20

We select samples of sizes $n = 4$, proceeding row-wise and counting one digit at a time, discarding any number outside the range of the labels, and taking any label that appears once and only once. Please see the samples on the following page.

■ **TABLE 7.2** Chosen Simple Random Samples of Sizes Four

	Chosen Labels	Chosen Y Values				Means	S.D.
Sample 1	1, 3, 0, 2	3	7	2	5	4.25	2.217356
Sample 2	0, 3, 5, 4	2	7	12	10	7.75	4.349329
Sample 3	4, 5, 1, 3	10	12	3	7	8	3.91578
Sample 4	5, 3, 2, 4	12	7	5	10	8.5	3.109126
Sample 5	0, 2, 4, 1	2	5	10	3	5	3.559026

Looking at the five samples of sizes four chosen, we see that they are all different. They represent five out of many samples of sizes four that we could have chosen from the population Y. Samples, sample means, and variances vary and so does the standard deviation. They all vary according to sample.

7.1 Sampling Distribution of the Sample Mean

The reason we collect sample data is in order to reach conclusions (make inference) about the population from which the sample was drawn. In this section we want to explore the relationship between the mean of a population and the mean of a sample taken from the population. We already know, especially from the demonstration above, that when sampling from the same population, taking samples of the same size n, we often get different observations in the different samples. These differences in observations in the different samples indicate that there is variability among random samples of the same size that can be selected from the same population. This variability is exemplified by the values of S.D. in Table 7.2. The differences in the observations in the samples lead to different sample means. Therefore, the sample mean $\overline{X}$ is a random variable, taking different values for the different random samples of same size from same population. Just like we need to know the distribution of a random variable X so that we can obtain the probability with which it takes different values, so also we need to know the distribution of the means of the different samples.

Besides that, it turns out that $\overline{X}$ is used for making inference about population mean, μ. It is therefore important that we know the behavior of $\overline{X}$ and its distribution. These are important for inference. This need leads us to study the sampling distribution of the sample mean.

7.1.1 Mean, Variance, and Standard Deviation of a Sample Mean

We state without proof that if X is a random variable, and we take a random sample of size n from X, and compute the sample mean, $\overline{X}$, then the value of $\overline{X}$ is a random variable, since it changes for every sample of size n from X. Thus if X has distribution, so does $\overline{X}$. We discuss this distribution in this section.

It can be shown that if X is a random variable with mean, μ, and variance σ^2:

(a) **If we keep taking samples of size n from X, and X is an infinite population, or we sample with replacement from X, and X is a finite population, then the mean of the sample means is μ, and its variance is $\dfrac{\sigma^2}{n}$, and its standard deviation (called the standard error) is $\sqrt{\dfrac{\sigma^2}{n}}$.**

(b) **If we keep taking samples of size n from a finite population without replacement, the mean of the sample means will be μ, while the variance of the sample means would be $\dfrac{\sigma^2}{n} \cdot \dfrac{N-n}{N-1}$. The standard deviation of the sample means (which is called a *the standard error of the mean*) for a sample of size n, and population size of N, is $\sqrt{\dfrac{s^2}{n} \cdot \dfrac{N-n}{N-1}}$.**

7.1.2 Sampling Distribution of the Sample Mean when the Variable is Normally Distributed

If we sample from a variable X which is normally distributed so that $X \sim N(\mu, \sigma^2)$, then $\overline{X}$ is normally distributed with the same mean as X, but the variance, is as stated previously, $\dfrac{\sigma^2}{n}$. In other words, $\overline{X} \sim N\left(\mu, \dfrac{\sigma^2}{n}\right)$. This can be proved, but the proof is beyond the scope of this text. However, we can use simulation to generate samples from a normal distribution and show that it is normally distributed. Figure 7.1 shows the histogram of means of 2000 samples which were generated from the normal variable $X \sim N(80, 196)$. Here we are sampling from the normal distribution, and the shape of the density overlaid on the histogram is normal.

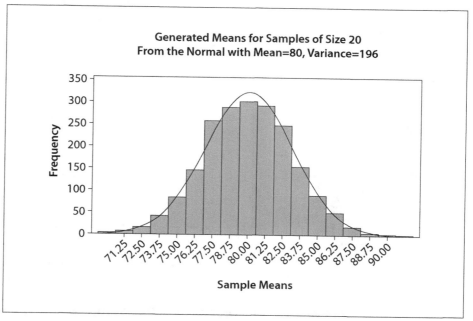

FIGURE 7.1   Generated Means for Samples of Size 20 from the Normal with Mean = 80, Variance = 196

We repeat the same for three more sample sizes and present their histograms together.

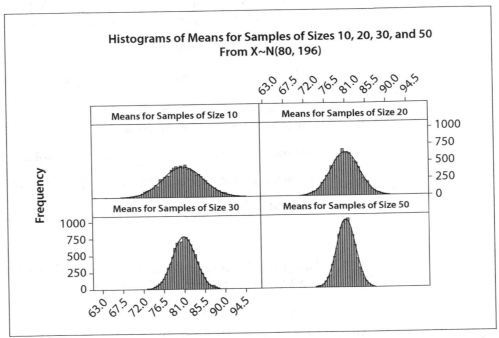

■ **FIGURE 7.2**   **Histograms of Means for Samples of Sizes 10, 20, 30 and 50 from** $X \sim N(80, 196)$

Next, we present the densities for the means of 10,000 samples of sizes 50, 30, 20, and 10 from $X \sim N(80, 196)$.

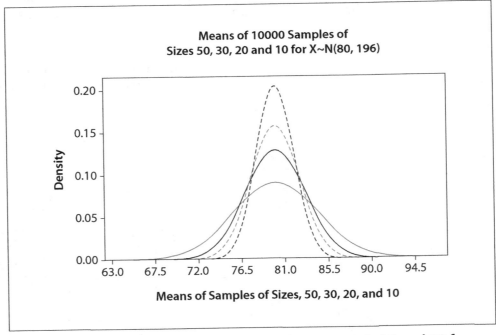

■ **FIGURE 7.3**   **Means of 10000 Samples of Sizes 50, 30, 20 and 10 for** $X \sim N(80, 196)$

Examination of the figures show that the variance of $\overline{X}$ is smaller, when n is larger, leading to a more peaked and much narrower normal distribution. In short with n larger, the means are closer to the true mean of the population. The smaller the sample size, the further the value of $\overline{X}$ is from the true population mean, μ. This reflects the fact that $\overline{X} \sim N(\mu, \sigma^2/n)$, showing that the variance of $\overline{X}$ is affected by the size of the sample.

We can infer from the shapes of the densities that the sample means are normally distributed with their means being 80. The variances change, as do the n, but they depend on the original variance of 196. In fact, the variances are proportional to the 196 and the proportionality constant has to do with the sample size. The pictures so far generally agree with the fact that $\overline{X} \sim N(\mu, \sigma^2/n)$. Thus, we may be able to evaluate the probability that $\overline{X}$ will take on different values by converting values of $\overline{X}$ to Z, the standard normal variable so that:

$$Z = \frac{\overline{X} - \mu}{\sigma/\sqrt{n}} \text{ or } Z = \frac{\sqrt{n}(\overline{X} - \mu)}{\sigma} \tag{7.1}$$

Let us apply the formula (7.1) and use it to determine the probabilities that the sample will take different values when the variable is normally distributed.

■ **EXAMPLE 7.1**   Suppose that samples of size eighteen are taken from the variable $X \sim N(80, 49)$; what is the proportion of samples for which $\overline{X}$'s would: (a) be greater than 83? (b) lie between 76 and 85? (c) be less than 82?

Solution   We use the idea that $Z = \dfrac{\overline{X} - \mu}{\sigma/\sqrt{n}}$ if X is normally distributed with mean, μ and variance σ^2

$\mu = 80; \sigma = \sqrt{49} = 7, n = 18.$

(a) $P(\overline{X} > 83) = P\left(Z > \dfrac{83 - 80}{7/\sqrt{18}}\right) = P(Z > 1.82) = 1 - 0.9656 = 0.0344$

(b) $P(76 < \overline{X} < 85) = P\left(\dfrac{76 - 80}{7/\sqrt{18}} < Z < \dfrac{85 - 80}{7/\sqrt{18}}\right)$
$= P(-2.42 < Z < 3.03) = 0.9988 - 0.0078 = 0.9910$

(c) $P(\overline{X} < 82) = P\left(Z < \dfrac{82 - 80}{7/\sqrt{18}}\right) = P(Z < 1.21) = 0.8869$

■ **EXAMPLE 7.2**   The diameters X of trees of the same age grown by a forest conservation company are believed to be normally distributed with mean 30.2 *cm*, and variance 15.29 *cm*. If a sample of fourteen trees is taken from this group, what is the probability that:

(a) $\overline{X} < 27.9$ *cm*.
(b) $27 \leq \overline{X} \leq 33$ *cm*.
(c) $\overline{X} > 29$ *cm*.

Solutions   We use the fact that if $X \sim N(30.2, 15.29)$ then $\overline{X} \sim N\left(30.2, \frac{15.29}{14}\right)$

(a) $P(\overline{X} < 27.9) = P\left(Z < \frac{27.9 - 30.2}{\sqrt{15.29/14}}\right) = P(Z < -2.2) = 0.0139$

(b) $P(27 < \overline{X} < 33) = P\left(\frac{27 - 30.2}{\sqrt{15.29/14}} < Z < \frac{33 - 30.2}{\sqrt{15.29/14}}\right)$
$$= P(-3.06 < Z < 2.68) = 0.9963 - 0.0011 = 0.9952$$

(c) $P(\overline{X} > 29) = P\left(Z > \frac{29 - 30.2}{\sqrt{15.29/14}}\right) = P(Z > -1.15)$
$$= 1 - 0.1251 = 0.8749$$

■ **EXAMPLE 7.3**   A junior high school in a school district has 655 students. In this school, the height is normally distributed with mean 66 inches and variance of 30 inches^2. Suppose we draw a random sample of twenty-five students from this junior high school; what is the probability that the average height of the sample would be: (a) less than 63 inches? (b) between 62 and 69? (c) greater than 68.6?

Solutions   (a) $P(\overline{X} < 63) = P\left(Z < \frac{63 - 66}{\sqrt{30/5}}\right) = P(Z < -2.74) = 0.0031$

(b) $P(62 < \overline{X} < 69) = P\left(\frac{62 - 66}{\sqrt{30/5}} < Z < \frac{69 - 66}{\sqrt{30/5}}\right)$
$$= P(-3.65 < Z < 2.74) = 0.9969 - 0.0001 = 0.9968$$

(c) $P(\overline{X} > 68.6) = P\left(Z > \frac{68.6 - 66}{\sqrt{30/5}}\right) = P(Z > 2.37) = 1 - 0.9911$
$$= 0.0089$$

7.2   Sampling Distribution of the Sample Mean when Population is Not Normal

So far, with the pictures above (Figures 7.1, 7.2, and 7.3) we have been able to show (without proving it) that if we draw samples of different sizes from a variable X which is normally distributed so that $X \sim N(\mu, \sigma^2)$, the means of the samples will be normally distributed with the same mean as X, and a variance which is proportionate to the variance of X. It turns out that the proportionality constant is the inverse of the sample size n. The proof of this is beyond the scope of this text.

7.2.1   Sampling Without Replacement

Consider the following data, the heights (in centimeters) of the six members of a volleyball team, whom we name A, B, C, D, E, and F:

Population	A	B	C	D	E	F
Heights	184	185	176	175	172	183

The population size is $N = 6$; and the mean height for this population of players is given as:

$$\mu = \frac{\sum_{i=1}^{N} X_i}{N} = \frac{\sum_{i=1}^{6} X_i}{6} = \frac{1075}{6} = 179.16667 \text{ cm.}$$

Also, the population variance and standard deviation of the heights of the team are obtained as:

$$\sigma^2 = \frac{\sum_{i=1}^{N} X_i^2}{N} - \mu^2 = \frac{\sum_{i=1}^{6} X_i^2}{6} - 179.16667^2 = \frac{192755}{6} - 179.16667^2$$

$$= 25.13889 \Rightarrow \sigma = \sqrt{25.13889} = 5.01387$$

To illustrate the sampling distribution of the mean, our goal is to estimate the mean height of the volleyball team by sampling from the team. We decide to take a sample of size four and compute the mean. We can obtain all the possible samples of size 4 for $N = 6$. There are only fifteen of them, which agrees with the number of ways of choosing 4 things out of 6, that is $\binom{6}{4} = 15$.

This population is not normally distributed. By sampling from this population without replacement, we will show that the mean of the sample means is the same as the population mean, the variance of the sample means is $\frac{\sigma^2}{n} \cdot \frac{N-n}{N-1}$, and its standard deviation is $\sqrt{\frac{\sigma^2}{n} \cdot \frac{N-n}{N-1}}$. To do so, we take samples of size four from the population of the volleyball team without replacement. We can list all the samples as in the table below. There are altogether fifteen samples, whose compositions are shown in the table below:

	A	B	C	D	E	F	Sample Means
Heights	184	185	176	175	172	183	
Samples							
A,B,C,D	184	185	176	175			180
A,B,C,E	184	185	176		172		179.25
A,B,C,F	184	185	176			183	182
A,B,D,E	184	185		175	172		179
A,B,D,F	184	185		175		183	181.75

Continued.

	A	B	C	D	E	F	Sample Means
Heights	184	185	176	175	172	183	
Samples							
A,B,E,F	184	185			172	183	181
A,C,D,E	184		176	175	172		176.75
A,C,D,F	184		176	175		183	179.5
A,C,E,F	184		176		172	183	178.75
A,D,E,F	184			175	172	183	178.5
B,C,D,E		185	176	175	172		177
B,C,D,F		185	176	175		183	179.75
B,D,E,F		185		175	172	183	178.75
C,D,E,F			176	175	172	183	176.5
B,C,E,F		185	176		172	183	179
							179.16667

The mean of the fifteen sample means agrees with the population mean of 179.16667. On the other hand the variance of the means is a proportion of the variance of X.

The variance and standard deviation of the sample means is:

$$\sigma_{\overline{X}}^2 = \frac{\sum_{i=1}^{15}(\overline{X}_i - \mu)^2}{15}$$

$$= \frac{(180 - 179.16667)^2 + (180 - 179.16667)^2 + \ldots + (179 - 179.16667)^2}{15}$$

$$= 2.513889$$

$$\sigma_{\overline{x}} = \sqrt{\frac{\sum_{i=1}^{15}(\overline{X}_i - \mu)^2}{15}} = \sqrt{2.513889} = 1.585525$$

Thus with samples of size $n = 4$ chosen without replacement from this population of $N = 6$, the variance is $\frac{\sigma^2}{4} \cdot \frac{6-4}{6-1} = \frac{\sigma^2}{10}$. In this case, we are taking samples of size $n = 4$, with $N = 6$, the standard error would be $\sqrt{\frac{\sigma^2}{10}}$. The results above agree.

We can see that for the above example that $\sigma^2 = 25.13889$ and that the variance of $\overline{X}$ is 2.513889, agreeing with the above findings. We have therefore demonstrated (without proof) that if we sample without replacement from a finite population, then the mean of $\overline{X}$ is still μ, but the variance of $\overline{X}$ is not the same as

$\sigma_{\bar{x}}^2 = \dfrac{\sigma^2}{n}$, rather we have to use a correction factor. In that case, after application of

a correction factor, we have: $\sigma_{\bar{x}}^2 = \dfrac{N-n}{N-1} \cdot \dfrac{\sigma^2}{n}$ and $\sigma_{\bar{x}} = \sqrt{\dfrac{N-n}{N-1}} \cdot \dfrac{\sigma}{\sqrt{n}}$.

If we are sampling from an infinite population or sampling with replacement from a finite population, the mean of the sample mean $\bar{X}$, is μ (same as the population mean) while its variance is $\dfrac{\sigma^2}{n}$. We demonstrate this by sampling with replacement from a finite population.

7.2.2 Sampling With Replacement from a Finite Population

Consider sampling from a population X which is $X = \{2, 3, 5, 7, 10, 12\}$. The mean of X is $\mu = 6.5$ and its variance $\sigma^2 = 12.92$. If we take samples of size 2, then all the possible samples we can obtain by sampling with replacement are given in Table 7.3:

■ **TABLE 7.3**

Sample			Mean	Sample			Means
1	2	2	2	19	7	2	4.5
2	2	3	2.5	20	7	3	5
3	2	5	3.5	21	7	5	6
4	2	7	4.5	22	7	7	7
5	2	10	6	23	7	10	8.5
6	2	12	7	24	7	12	9.5
7	3	2	2.5	25	10	2	6
8	3	3	3	26	10	3	6.5
9	3	5	4	27	10	5	7.5
10	3	7	5	28	10	7	8.5
11	3	10	6.5	29	10	10	10.0
12	3	12	7.5	30	10	12	11
13	5	2	3.5	31	12	2	7
14	5	3	4	32	12	3	7.5
15	5	5	5	33	12	5	8.5
16	5	7	6	34	12	7	9.5
17	5	10	7.5	35	12	10	11
18	5	12	8.5	36	12	12	12
19	7	2	4.5	overall			6.5

There are 36 samples, or 6^2 samples, 2 being the sample size and 6 the population size. So if we take samples of size n from a population of size N with replacement, the total number of samples we can take is $(N)^n$. To use sample of size 3, with Y above, there will be 216 samples.

If we took samples of size 3 with replacement from the same population, we can arrive at the same conclusion: the mean for the means of the 216 samples would be the same as the population mean, that is 6.5; its variance will be different, weighted by the inverse of 3 instead of the inverse of 2 which applied when we took samples of size 2. It is clear that the means of the samples gravitate to the mean of the population; in this case 6.5. We can see that the overall mean of the samples is the same as the population mean of 6.5.

If we draw similar samples from a population X which is not normal, what is the distribution of the sample mean, $\overline{X}$? Does $\overline{X}$ behave in a similar way as X above? Let us use the same method we employed for the case when the population X is normal. Here we present the histograms for the sample means, $\overline{X}$, for 10,000 samples drawn from X, for sample sizes $n = 2, 3, 4,$ and 5.

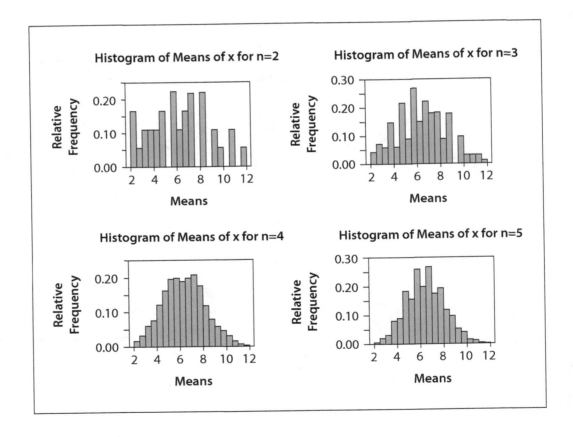

Even from this small population, when sampling for sample-sizes of $n = 2, 3, 4,$ and 5, the histograms show that the means are approximately normally distributed. The approximation gets better for the larger sample sizes. It can be seen that the histograms are centered at the mean of Y, which is at $\mu = 6.5$.

7.3   The Central Limit Theorem:

Again, using the results of these simulations for X, which is not normally distributed, we get the same result. We conclude that the sample mean of X is approximately normally distributed with mean, $\mu_{\bar{X}} = \mu$, and its variance $\sigma_{\bar{x}}^2 = \dfrac{\sigma^2}{n}$. The standard deviation of $\bar{X}$ is $\sigma_{\bar{x}} = \dfrac{\sigma}{\sqrt{n}}$. The implication of these figures is that for n which is large enough, the mean of the means of the samples will eventually be equal to the mean of the population, but its variance is weighted by the inverse of the sample size. Also, for large enough sample sizes the distribution of the sample means will be normally distributed with mean, $\mu_{\bar{x}} = \mu$, and variance $\sigma_{\bar{x}}^2 = \dfrac{\sigma^2}{n}$.

If we sample from a population which is relatively large compared to the sample size, the difference between sampling with replacement and sampling without replacement is minimal.

The central limit theorem has something to do with the behavior of the sampling distribution of the mean, so we state it here, and show how it applies:

Central limit theorem states that if a large enough sample is drawn from a population X, then $\bar{X}$, the sample mean, is approximately normally distributed no matter what was the original distribution of X. As a rule of the thumb, statisticians agree that a sample size $n \geq 30$, is large enough.

The implication of central limit theorem is that for a large sample of size n drawn from X with mean μ, and variance σ^2, $\bar{X} \sim N(\mu, \sigma^2/n)$. This means that a population does not have to be normal for this to be true. In fact, if we know the mean and variance of a population we draw large samples from, the distribution of their means will be approximately normal as described.

From the point of view of the central limit theorem, let us revisit the population $Y = \{2, 3, 5, 7, 10, 12\}$ from Table 7.1. Looking at the results of the simulation presented in the histograms above, we see that the distributions of the sample means for the different sample sizes are approximately normal with a mean of 6.5.

Let us apply the central limit theorem to large samples from some of the distributions that we have treated so far.

■ **EXAMPLE 7.4**   If a large sample of size n is taken from X which is Poisson distributed with a parameter λ, find the distribution of $\bar{X}$.

Solution   The mean of X is λ, and the variance of X is λ. By central limit theorem, $\bar{X} \sim N(\lambda, \lambda/n)$ approximately. This means that we can convert $\bar{X}$ to Z, where $Z = \dfrac{\bar{X} - \lambda}{\sqrt{(\lambda/n)}}$.

■ **EXAMPLE 7.5**   If X is uniformly distributed in the interval $[a, b]$, find the distribution of $\bar{X}$ if a large sample is taken from it.

Solution   The mean of X is $\dfrac{a + b}{2}$, and the variance of X is $\dfrac{(b - a)^2}{12}$. By central limit theorem,

$$\overline{X} \sim N\left[\frac{a + b}{2}, \frac{(b - a)^2}{12n}\right], \text{ approximately. The equivalent } Z = \frac{\overline{X} - (a + b)/2}{\sqrt{(b - a)^2/12n}}.$$

■ **EXAMPLE 7.6**   If a large sample of size m is taken from a binomial population with probability of success p, and number of trials n, find the distribution of $\overline{X}$.

Solution   The mean of X is np, and the variance of X is npq. By central limit theorem, the sample mean $\overline{X} \sim N(np, npq/m)$, approximately, so that $Z = \dfrac{\overline{X} - np}{\sqrt{npq/m}}$.

■ **EXAMPLE 7.7**   This year, the mean age at first marriage for men in the USA is 26.6, with a variance of 31.47. If a sample of one hundred men who got married this year is surveyed, what is the probability that the mean age would be: (a) greater than 25.2? (b) lie between 25 and 27.9? (c) less than 28?

Solution   In this case, data may or may not be normally distributed, but we can appeal to the central limit theorem in order to obtain the desired probabilities. Thus, by central limit theorem we assume that approximately $\overline{X} \sim N(26.6, 31.47)$, so that $\mu = 26.6$ and $\sigma = 5.61$. The desired probabilities are:

(a) $P(\overline{X} > 25.2) = P\left(Z > \dfrac{25.2 - 26.6}{5.61/\sqrt{100}}\right) = P(Z > -2.50) = 1 - 0.0062$
$$= 0.9938$$

(b) $P(25 < \overline{X} < 27.9) = P\left(\dfrac{25 - 26.6}{5.61/\sqrt{100}} < Z < \dfrac{27.9 - 26.6}{5.61/\sqrt{100}}\right)$
$$= P(-2.85 < Z < 2.32) = 0.9898 - 0.0022 = 0.9876$$

(c) $P(\overline{X} < 28) = P\left(Z < \dfrac{28 - 26.6}{5.61/\sqrt{100}}\right) = P(Z < 2.50) = 0.9938$

■ **EXAMPLE 7.8**   Calls arrive at a Customer Service center of a large corporation on average five calls per minute. If a sample of fifty hours is taken from their hours of service, what is the probability that the mean number of calls per hour is: (a) between 293 and 308? (b) at least 296? (c) less than 305?

Solution   In this case, the mean number of calls per hour is $5 \times 60 = 300$. Since these calls are arriving from a large population of customers, they therefore follow the Poisson distribution with mean 300, and variance 300. However, by central limit theorem, $\overline{X} \sim (\lambda, \lambda/n)$. In this case, $\overline{X} \sim N(300, 6)$.

For $\overline{X}$, $\mu = 300$, $\sigma_{\overline{X}} = \sqrt{6} = 2.44949$.

(a) $P(293 < \overline{X} < 308) = P\left(\dfrac{293 - 300}{2.44949} < Z < \dfrac{308 - 300}{2.44949}\right)$
$$= P(-2.86 < Z < 3.27) = 0.9995 - 0.0021 = 0.9974$$

(b) $P(\overline{X} > 296) = P\left(Z > \dfrac{296 - 300}{2.44949}\right) = P(Z > -1.63) = 1 - 0.05155$
$$= 0.94855$$

(c) $P(\overline{X} < 305) = P\left(Z < \dfrac{305 - 300}{2.44949}\right) = P(Z < 2.04) = 0.9793$

7.4 Sampling Distribution of the Proportion

Recall that trials have outcomes. The outcomes of the trial are the event space for the trial. Usually experiments are made up of one or more trials. If out of n trials, an attribute of primary interest in the trial (an outcome or a subset of outcomes in the space of outcomes of the trial) is observed x times, then the proportion of occurrence of the attribute of primary interest is estimated as:

$$\hat{p} = \frac{x}{n} \tag{7.2}$$

It is clear from lessons we learned in sampling that as we repeat the experiment another n times, we are very likely to obtain a different value for $\hat{p}$ than we obtained in the previous set of trials. Thus, $\hat{p}$ is a variable, and varies from one sample to another. Just like the mean, we can use exploratory analysis to try to establish the distribution of all the $\hat{p}$'s that we obtain by repeating the experiments. One easy way to explore its distribution would be to plot the histogram of the values of $\hat{p}$ and examine the underlying shape that the histogram suggests. Is it symmetric, bell-shaped, skewed to the left or to the right? Is it humped in more than one place?

When we divide the set of outcomes of a trial into two; (a) attribute of primary interest and (b) other attributes within the trial space, then (a) will lead to the observation of $\hat{p}$, while (b) will lead to the observation of $\hat{q}$ where:

$$\hat{q} = \frac{n - x}{n} \tag{7.3}$$

$\hat{q}$ is the proportion of outcomes not of primary interest. Essentially due to the randomness of the occurrence of $\hat{p}$ and so $\hat{q}$, $\hat{p}$ is a binomial variable. We can therefore simulate its behavior in the long run. Statisticians have long established that in the long run (that is, for large number of trials), the distribution of $\hat{p}$ is approximately normal, with a mean of p and a variance of $\dfrac{pq}{n}$. p is the true population proportion and $q = 1 - p$.

In Figure 7.4, we show the results of simulation for $\hat{p}$ when the true population proportion is 0.55.

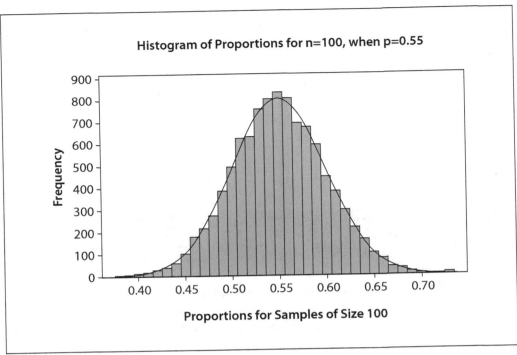

■ **FIGURE 7.4**   Histogram of Proportions for $n = 100$, when $p = 0.55$

It clearly shows that the mean of the proportions is 0.55.

Next, we present the histograms for different values of n when the true population proportion is 0.40.

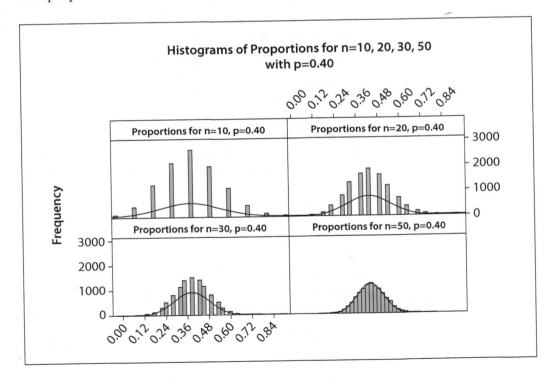

We can see that at $n = 30$, the distribution is barely normal. However at $n = 50$, there is no doubt that the underlying distribution of the proportions is normal.

The implication is that if n is large enough, then

$$\hat{p} \sim N\left(p, \frac{p(1-p)}{n}\right) \tag{7.4}$$

With this we can obtain the probability, for a fixed n which is large enough of observing different values of $\hat{p}$ via the standard normal distribution. Thus,

$$Z = \frac{\hat{p} - p}{\sqrt{\dfrac{p(1-p)}{n}}} \tag{7.5}$$

■ EXAMPLE 7.9 It is known that 62% of high school students surf the NET for uses other than research and home work. If a sample of 120 students is taken from this group, what is the probability that

(a) $\hat{p}$ is greater than 0.68 (b) Between 0.58 and 0.66 (c) less than 0.55

Solution We use the fact that $Z = \dfrac{\hat{p} - p}{\sqrt{\dfrac{p(1-p)}{n}}}$; We first obtain the value of $\sqrt{\dfrac{p(1-p)}{n}} =$

$$\sqrt{\frac{0.62(1-0.62)}{120}} = 0.04431$$

(a) $P(\hat{p} > 0.68) = P\left(Z > \dfrac{0.68 - 0.62}{0.04431}\right) = P(Z > 1.35) = 1 - 0.9115$
$$= 0.0885$$

(b) $P(0.58 < \hat{p} < 0.66) = P\left(\dfrac{0.58 - 0.62}{0.04431} < Z < \dfrac{0.66 - 0.62}{0.04431}\right)$
$$= P(-0.90 < Z < 0.90) = 0.8159 - 0.1841 = 0.6318$$

(c) $P(\hat{p} < 0.55) = P\left(Z < \dfrac{0.55 - 0.62}{0.04431}\right) = P(Z < -1.58) = 0.0571$

■ EXAMPLE 7.10 Ash.Org reports that 38% of male students in the USA use tobacco in one form or another. Based on this, if 200 male students from the USA are chosen at random and surveyed for tobacco use, what is the probability that the proportion of tobacco users would: (a) be less than 0.48? (b) lie between 0.30 and 0.44? (c) be greater than 0.34?

Solution Here, again we use the fact that $Z = \dfrac{\hat{p} - p}{\sqrt{\dfrac{p(1-p)}{n}}}$; Again, we obtain the value of

$$\sqrt{\frac{p(1-p)}{n}} = \sqrt{\frac{0.38(1-0.38)}{200}} = 0.034322$$

(a) $P(\hat{p} < 0.48) = P\left(Z < \dfrac{0.48 - 0.38}{0.034322}\right) = P(Z < 2.91) = 0.9982$

(b) $P(0.30 < \hat{p} < 0.44) = P\left(\dfrac{0.30 - 0.38}{0.034322} < Z < \dfrac{0.44 - 0.38}{0.034322}\right)$

$= P(-2.33 < Z < 1.75) = 0.9599 - 0.0099 = 0.95$

(c) $P(\hat{p} > 0.34) = P\left(Z > \dfrac{0.34 - 0.38}{0.034322}\right) = P(Z < -1.17) = 1 - 0.1210$

$= 0.879$

■ **EXAMPLE 7.11** The Center for Disease Control (CDC) reports that the incidence of anemia occurs in the third trimester of pregnancy for low income women. If 254 low income women are pregnant, what is the probability that the proportion that would suffer from anemia would be (a) more than 0.35 (b) between 0.24 and 0.36 and (c) less than 0.27.

Solution

$$\sqrt{\frac{p(1-p)}{n}} = \sqrt{\frac{0.29(1-0.29)}{254}} = 0.028472$$

(a) $P(\hat{p} > 0.35) = P\left(Z > \dfrac{0.35 - 0.29}{0.028472}\right) = P(Z > 2.11) = 1 - 0.9826$

$= 0.0174$

(b) $P(0.24 < \hat{p} < 0.36) = P\left(\dfrac{0.24 - 0.29}{0.028472} < Z < \dfrac{0.36 - 0.29}{0.028472}\right)$

$= P(-1.76 < Z < 2.46) = 0.9931 - 0.0392 = 0.9539$

(c) $P(\hat{p} < 0.27) = P\left(Z > \dfrac{0.27 - 0.29}{0.028472}\right) = P(Z < -0.70) = 0.2420$

■ **EXAMPLE 7.12** It is reported that, nationally, 64% of women who get married before age twenty-four, get divorced. If 350 women who married before age twenty-four are surveyed, what is the probability that the proportion who get divorced is: (a) less than 0.57? (b) lies between 0.59 and 0.69? (c) greater than 0.67?

Solution

$$\sqrt{\frac{p(1-p)}{n}} = \sqrt{\frac{0.64(1-0.64)}{350}} = 0.025657$$

(a) $P(\hat{p} < 0.57) = P\left(Z < \dfrac{0.57 - 0.64}{0.025657}\right) = P(Z < -2.73) = 0.0032$

(b) $P(0.59 < \hat{p} < 0.69) = P\left(\dfrac{0.59 - 0.64}{0.025657} < Z < \dfrac{0.69 - 0.64}{0.025657}\right)$

$= P(-1.95 < Z < 1.95) = 0.9744 - 0.0256 = 0.9488$

(c) $P(\hat{p} > 0.67) = P\left(Z > \dfrac{0.67 - 0.64}{0.025657}\right) = P(Z > 1.17) = 1 - 0.8790$

$= 0.1210$

Exercises

For the distributions mentioned below, assume that large samples of n have been chosen from them. Apply the central limit theorem and determine: (a) the distribution of the sample mean $\overline{Y}$, (b) the variance, and (c) the standard deviation, of the ensuing distribution.

7.1 Y is uniformly distributed in the interval (7, 14); $n = 35$.

7.2 Y is uniformly distributed in the interval (10, 36); $n = 40$.

7.3 Y is Poisson distributed with $\lambda = 44$; $n = 50$.

7.4 Y is Poisson distributed with $\lambda = 300$; $n = 60$.

7.5 Y is Poisson distributed with $\lambda = 35$; $n = 36$.

7.6 Y is Binomially distributed with $p = 0.44$; $n = 50$; take a sample of size 100.

7.7 Y is Binomially distributed with $p = 0.3$; $n = 40$; take a sample of size 89.

7.8 Y is Binomially distributed with $p = 0.35$; $n = 35$; take a sample of size, 120.

7.9 In the St. Cloud area, mean cost for plumbers to tune-up a typical home heating system for winter is 100 dollars, with a variance of 560 dollars^2. If a sample of costs of tune-ups for fifty households in the area is taken, what is the probability that the mean is:
(a) greater than 92?
(b) between 95 and 108?
(c) less than 110 dollars?

7.10 The mean life time of the fruit fly, Drosophila melanogaster in a laboratory, is sixty days with a variance of 235 days^2. If a sample of forty-five flies were taken from this population, what is the probability that the mean life time is:
(a) less than 55 days?
(b) between 58 and 66 days?
(c) greater than 65 days?

7.11 Repeat Exercise 7.10 with a sample size of fifty.

7.12 The following data show the 2009 SAT scores for participants from Minnesota:

Critical Thinking		Mathematics		Writing	
Mean	Variance	Mean	Variance	Mean	Variance
595	11025	609	10201	578	10609

In 2009, the mean score of participants from Minnesota in SAT Critical Thinking test is 595 with variance of 11025. If a sample of sixty participants were to be taken from this group, what is the probability that the mean is:
(a) less than 572?
(b) between 577 and 590?
(c) greater than 610?

7.13   In 2009, the mean score of participants from Minnesota in the SAT Mathematics test is 609 with variance of 10201. If a sample of one hundred participants were to be taken from this group, what is the probability that the mean is:
(a) less than 580?
(b) between 612 and 620?
(c) greater than 605?

7.14   In 2009, the mean score of participants from Minnesota in the SAT Writing is 578 with variance of 10609. If a sample of eighty participants were to be taken from this group, what is the probability that the mean is:
(a) less than 560?
(b) between 570 and 590?
(c) greater than 582.

7.15   The owner of a hotel chain with hundreds of rooms claims that their rooms are occupied, on average, 300 days a year, with a variance of 7680 days^2. If a sample of one hundred rooms is taken from the chain, what is the probability that the mean occupancy of the rooms is:
(a) less than 280?
(b) between 290 and 315?
(c) greater than 308 days?

7.16   A manufacturer of hybrid cars states that the mean life of the battery they used for their hybrid cars is 100,000 miles with a variance of $(100,000)^2$ miles. If the claim is true, and a sample of one hundred of their hybrid cars are selected and the lives of the installed batteries are studied, what is the probability that the mean is:
(a) less than 90,000?
(b) between 93,000 and 107,000?
(c) at least 105,000?

7.17   A state claims that the average monthly cost of home care for a category of patients is 4290.00 dollars with a variance of 1732800 dollars^2. Assuming this is true and a sample of forty such patients are taken, what is the probability that the mean cost of looking after the patients is:
(a) less than 3650?
(b) between 4000 and 4600?
(c) greater than 3900?

7.18   Over the years, a city has kept a record of the daily amount of suspended particulate matter in the air. It is known that the daily mean amount of suspended particulate matter in the city is 68 $\mu g/m^2$ with a variance of 102.82 $\mu g^2/m^3$. A sample of air quality in the city is taken for thirty-seven days . What is the probability that the mean will:
(a) exceed 71.5 $\mu g/m^2$?
(b) be between 63.5 and 72.3 $\mu g/m^2$?
(c) be less than 64 $\mu g/m^2$?

7.19 A city states that the mean water consumption for its residents per household is 10,569 gallons/month, with a variance of 1760700 gallons^2/month. If forty households are surveyed, what is the probability that mean water consumption would:
(a) differ from the mean by 500 gallons?
(b) not exceed the mean by 450 gallons?
(c) be less than the mean by 245 gallons?

7.20 $X \sim N(63, 139)$. If a sample of size 100 is taken from X, find the probability that:
(a) $\overline{X} \geq 65.5$
(b) $63.7 \leq \overline{X} \leq 66.2$
(c) $\overline{X} < 60$

7.21 The Bureau of Transportation Statistics, classifies Weather related delays or cancelations of flights are as NAS. It reports that 31% of all flights delays or cancelations are NAS. If 260 delayed or canceled flights are surveyed, what is the probability that the proportion of NAS
(a) less than 0.24?
(b) between 0.27 and 0.35?
(c) is greater than 0.36?

7.22 The UNICEF reports that 7% of children that are born in USA are underweight. If 700 births in USA are surveyed, what is the probability that the proportion of babies born underweight
(a) lies between 0.05 and 0.09?
(b) is greater than 0.06?
(c) is lower than 0.085?

7.23 *USA Today* reports, quoting the Department of Education, that the proportion of four-year college students who graduate in four years is 0.53. If 1000 students of four-year colleges are surveyed, what is the probability that
(a) more than .50,
(b) between 0.49 and 0.57,
(c) less than 0.565, would graduate in four years?

7.24 The US Department of Education states that students at for-profit institutions represent 11% of all students, and their graduates represent 43% of all education **loan defaulters.** If 920 students of for-profit institutions who took education loans are surveyed, what is the probability that the proportion of defaulters:
(a) is more than 0.39?
(b) between 0.38 and 0.46?
(c) is less than 0.45?

7.25 The National Institute on Drug Abuse reports that 17% of all fraternity members in colleges abuse painkillers. If 500 fraternity members are surveyed, what is the probability that the proportion of abusers of painkillers:
(a) is greater than 0.145?
(b) less than 0.20?
(c) is between 0.15 and 0.205?

7.26   The University of Western Ontario found that for junior nurses working across Ontario, a whopping 66% were experiencing "symptoms of burnout, including emotional exhaustion and depression." Based on this, if a sample of 150 junior nurses in Ontario is taken, what is the probability that the proportion experiencing symptoms of burnout, including emotional exhaustion and depression, will:
(a) exceed 0.72?
(b) lie between 0.6 and 0.75?
(c) be no more than 0.57?

7.27   The Food and Drug Administration (FDA) consumer magazine reports that 14% of cases of TB are resistant to the antibiotics (isoniazid and rifampin) which are currently used for treating the disease. If 407 cases of TB are surveyed, what is the probability that the proportion of resistance to the current treatment is:
(a) less than 0.10?
(b) between 0.09 and 0.12?
(c) greater 0.17?

7.28   An actuary is a financial expert who applies mathematical and statistical methods to assess financial and other risks relating to various contingent events and then arrives at the rate or premium that a person should pay to get a particular insurance coverage. There is an acute shortage of qualified actuaries so that only 30% of the people who do the job worldwide are qualified, with some consequences. If a survey of 400 people who do the job of an actuary are surveyed, what is the probability that the proportion who are qualified:
(a) exceeds 0.24?
(b) is between 0.26 and 0.36?
(c) is less than 0.35?

7.29   Minnesota House of Representative research shows that the proportion of first time DWI (driving while intoxicated) offenders in Minnesota who recidivate is 25%. If 250 first time DWI offenders are surveyed, what is the probability that the proportion of recidivism would be:
(a) greater than 0.28?
(b) less than 0.20?
(c) lies between 0.21 and 0.32?

7.30   National Young Driver Survey (NYDS) reported that 20% of fourteen- to eighteen-year-old drivers involved in fatal crashes in 2006 did not have a license. This means unlicensed teens are significantly over-represented in fatal crashes. If 450 fourteen to eighteen year olds who had driven without a license in 2006 are surveyed, what is the probability that the proportion involved in fatal crashes:
(a) is less than 0.16?
(b) is greater than 0.25?
(c) is between 0.17 and 0.23?

CHAPTER 8

Confidence Interval for the Single Population Mean and Proportion

The goal of sampling is mostly to be able to reach some conclusion (or make an inference) about the population. It means that when we sample, we intend to use some attributes of the sample to make some inference about its population equivalent. The process of making some inference starts with estimation. We use a sample value (called a statistic) to estimate (make an informed guess of) a population value (called a parameter). In this section, we discuss the estimation of parameters, particularly, confidence interval estimates.

8.1   Estimation

An estimate is an informed guess because it depends on actual data from the population and a mathematical formula which ensures that such a guess has some acceptable science behind it. **Estimation is the process of using a mathematical formula, or some mathematical formulae, to calculate from observed data, a value, or some values, for an unknown parameter, or a set of unknown parameters.** Estimation is an important part of the general field of statistics called Statistical inference. Statistical inference is made up of the body of theory and methods used for drawing conclusions about a population based on information obtained from a sample.

A parameter is a constant which appears in a probability mass function or the probability density function of a family of distributions. The values assigned to the parameter(s) determine the shape of the distribution. The parameter is a population value. Its true value is calculated from information or data obtained from the entire population. Often, however, the information for the entire population is unavailable or impossible to collect. Because of this, it is often estimated from sample data taken from the population.

We have already encountered some parameters in the course of our treatment of some statistical distributions which we have considered in this text.

For example, the parameters of a Binomial distribution with probability mass function:

$$p(x) = \binom{n}{x} p^x (1-p)^{n-x}; \, x = 0, 1, 2, \ldots, n \qquad (8.1)$$

are n and p.

In the case of the Poisson distribution, with probability mass function:

$$p(x) = \frac{\lambda^x e^\lambda}{x!}; \, x = 0, 1, 2, \ldots \qquad (8.2)$$

the sole parameter is λ; λ is both its mean and its variance.

On the other hand the normal distribution with probability density function:

$$f(x) = \frac{1}{\sigma\sqrt{2\pi}} e - \frac{(x-\mu)^2}{2\sigma^2}; \, -\infty < x < +\infty \qquad (8.3)$$

has two parameters, μ and σ^2, which correspond to its mean and its variance.

In order to estimate a parameter, we often need a statistic. A statistic or some function of a statistic is needed for hypothesis testing. Hypothesis testing is another aspect of statistical inference which we shall discuss in chapter 9. In hypothesis testing, we use both the value and distribution of a statistic to reach a decision about a hypothesized value of a parameter.

A **statistic** is a quantity that is computed from a sample. A statistic is a sample value. When there is more than one, we refer to them as statistics. For instance, two well-known statistics are the sample mean, $\overline{X}$, and the sample variance, s^2.

A well-known statistic is the sample mean, $\overline{X}$. Its value, which is obtained from a sample, is used to estimate the population mean, μ. Such an estimate is called a point estimate. A **point estimate** is a single statistic obtained from sample values and used as an approximation, a "best guess," for a parameter. $\overline{X}$, is a point estimate because it gives a single value as an estimate of μ and does not give a range of values within which the parameter μ could be found.

8.2 Estimating the Population Mean, μ, When the Variable is Normally Distributed

If the random variable X is normally distributed, then $X \sim N(\mu, \sigma^2)$. Usually neither μ, nor σ^2, is known. The task is to estimate the mean, μ. Sometimes, in the past, several samples have been taken from the population so that some guess can be made about the variance, σ^2. This does not happen often; however, it is possible to build a theory for estimating the population mean by assuming that σ^2 is known. The more realistic and practical situation that we often encounter is that both μ and σ^2 are unknown and we seek to estimate μ. In the next sections, we treat the estimation of the mean in both cases.

8.2.1 Confidence Interval for the Population Mean, μ, When the Variable is Normally Distributed, Assuming σ^2 is Known

If $X \sim N(\mu, \sigma^2)$, the sample mean $\overline{X}$ is the best candidate among other sample statistics for estimation of the population mean. This is based on the sampling distribution of the sample mean. We discovered in the previous chapter that with repeated sampling from a normal distribution, the mean of the sample means is μ and its variance is σ^2/n. Under the 68–95–99 rule, some of the properties of the sampling distribution of the mean $\overline{X}$ can be shown in the Figure 8.1 (a, b, c).

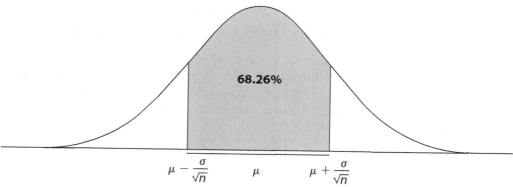

■ FIGURE 8.1 (A)   68.26% of sample means lie between 1 standard error from the mean.

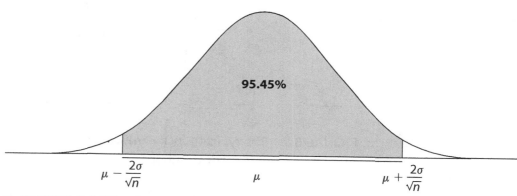

■ FIGURE 8.1 (B)   95.45% of sample means lie 2 standard errors from the mean.

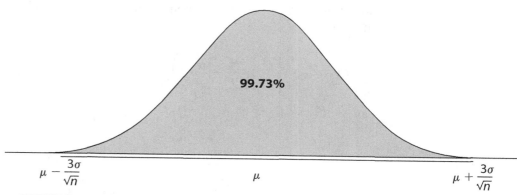

■ FIGURE 8.1 (C)   99.73% of sample means lie 3 standard errors from the mean.

We have demonstrated that the relationships between sample means and the population mean, shown in Figure 8.1 (a, b, c). In Figure 8.1 (c), the fact that 99.73% of $\overline{X}$ would lie in the interval $\left[\mu - \dfrac{3\sigma}{\sqrt{n}}, \mu + \dfrac{3\sigma}{\sqrt{n}}\right]$ indicates that μ would lie between $\left[\overline{X} - \dfrac{3\sigma}{\sqrt{n}}, \overline{X} + \dfrac{3\sigma}{\sqrt{n}}\right]$ with the same probability of 0.9973 or 99.73% of the time. Since we used the same sampling method and selected several samples and obtained the distribution of the mean of the sample means, and we based our interval on the confidence that the sample mean would lie in the interval 99.73% of the time, so will the true population mean. Similar statements can be made for Figures 8.1a and 8.1b which lead to 68.26% and 95.45% confidence, respectively.

Thus, the fact that $\overline{X} \sim N(\mu, \sigma^2/n)$, means that $Z = \dfrac{\overline{X} - \mu}{\sigma/\sqrt{n}}$.

To construct a typical $100(1 - \alpha)$% confidence interval for μ, when data is normal, with σ known, the upper and lower limits on the distribution of Z (the standard normal variable) would be equidistant from zero (the mean of Z) and therefore of the same magnitude and of opposite signs, so that we get a picture like Figure 8.2.

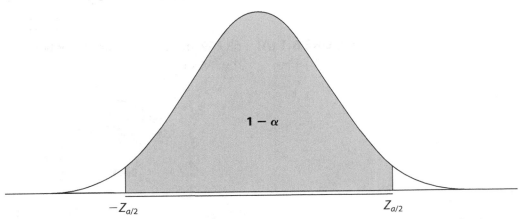

$1 - \alpha$

$-Z_{\alpha/2}$ $Z_{\alpha/2}$

■ **FIGURE 8.2**   The typical $100(1 - \alpha)$% confidence interval on the standard normal distribution.

For us to be $100(1 - \alpha)$% confident that we will find the mean, μ, within the interval on the distribution of $\overline{X}$ to be similar to the interval $[-Z_{\alpha/2}, Z_{\alpha/2}]$ that is shown in Figure 8.2, then by converting $\overline{X}$ to Z, the following must be true:

$$P\left(-Z_{\alpha/2} \leq \frac{\overline{X} - \mu}{\sigma/\sqrt{n}} \leq Z_{\alpha/2}\right) = 1 - \alpha$$

$$\Rightarrow P\left(-Z_{\alpha/2} \cdot \frac{\sigma}{\sqrt{n}} \leq \overline{X} - \mu \leq Z_{\alpha/2} \cdot \frac{\sigma}{\sqrt{n}}\right) = 1 - \alpha$$

$$\Rightarrow P\left(-\overline{X} - Z_{\alpha/2} \cdot \frac{\sigma}{\sqrt{n}} \leq -\mu \leq -\overline{X} + Z_{\alpha/2} \cdot \frac{\sigma}{\sqrt{n}}\right) = 1 - \alpha$$

$$\Rightarrow P\left(\overline{X} - Z_{\alpha/2} \cdot \frac{\sigma}{\sqrt{n}} \leq \mu \leq \overline{X} + Z_{\alpha/2} \cdot \frac{\sigma}{\sqrt{n}}\right) = 1 - \alpha$$

It follows that the interval that will guarantee us a $100(1 - \alpha)\%$ confidence that the mean, μ, is within it, is:

$$100(1 - \alpha)\% \text{ CI for } \mu = \left[\overline{X} - Z_{\alpha/2} \cdot \frac{\sigma}{\sqrt{n}}, \overline{X} + Z_{\alpha/2} \cdot \frac{\sigma}{\sqrt{n}}\right] \qquad (8.4)$$

When we compute the values, we have the lower limit of the interval as $\overline{X} - Z_{\alpha/2} \cdot \frac{\sigma}{\sqrt{n}}$ and the upper limit of the interval as $\overline{X} + Z_{\alpha/2} \cdot \frac{\sigma}{\sqrt{n}}$. Sometimes for ease of remembrance, this interval is stated as: $100(1 - \alpha)\%$ CI for $\mu = \overline{X} \pm Z_{\alpha/2} \cdot \frac{\sigma}{\sqrt{n}}$. In this form, we can state that a typical $100(1 - \alpha)\%$ confidence interval is made up of two parts:

Sample-Based Point Estimate $\pm$ Margin of Error

In the $100(1 - \alpha)\%$ in 8.4, $\overline{X}$ is the sample-based point estimate of μ and the margin of error $= Z_{\alpha/2} \cdot \frac{\sigma}{\sqrt{n}}$. Confidence intervals are preferred over point estimates, or intervals with no confidence attached, because the confidence interval provides us with the precision attached to the estimate.

For the purpose of confidence interval estimation, we will read the table of the normal distribution by finding the probability in the table and reading the Z that corresponds to it.

■ **EXAMPLE 8.1**   Suppose that the following data was taken from a population which is normal with variance 91 and unknown mean. Establish the (a) 96%, and (b) 93% confidence limits for the mean of the population. What are the margins of error in each case?

X: 40 26 39 14 42 18 25 43 46 27 19 47 19 26 35 34 15

Solution $\overline{X} = \dfrac{515}{17} = 30.29$

$96\% = 100(1 - 0.04)\% \Rightarrow \alpha = 0.04;$

$\alpha/2 = 0.02 \Rightarrow 96\% \text{ CI} = 30.29 \pm Z_{0.02}\sqrt{\dfrac{91}{17}}$

(a) $\left(30.29 - 2.05\sqrt{\dfrac{91}{17}}, 30.29 + 2.05\sqrt{\dfrac{91}{17}}\right)$

(25.547, 35.033)

margin of error = 4.743

$93\% = 100(1 - 0.07)\% \Rightarrow \alpha = 0.07 \Rightarrow \alpha/2 = 0.035;$

$93\% \text{ CI} = 30.29 \pm Z_{0.035}\sqrt{\dfrac{91}{17}}$

(b) $\left(30.29 - 1.81\sqrt{\dfrac{91}{17}}, 30.29 + 1.81\sqrt{\dfrac{91}{17}}\right)$

(26.102, 34.478)

margin of error = 4.188

■ **EXAMPLE 8.2** The heights of eighteen year-olds males in North Jade County are known to be normally distributed with variance 0.076 m^2 while heights of eighteen-year-old females from the county are also normally distributed with variance of 0.065. Samples of 260 male, and 200 female 18-year-olds were obtained. The sample mean for height for the males was 1.68 m the sample mean height for the females was 1.57 m.

   (a) Obtain the 94% confidence limits for the mean height of eighteen year old males in the county. Obtain the margin of error.
   (b) Obtain the 96% confidence limits for the mean height of eighteen year old females in the county. Obtain the margin of error.

	Sample Size	Mean	Variance	SD
Males	260	1.68	0.076	0.275680975
Females	200	1.57	0.065	0.254950976

$$\overline{X}_1 = 1.68; \sigma_1^2 = 0.076 \Rightarrow \frac{\sigma_1}{\sqrt{n_1}} = \frac{0.275681}{\sqrt{260}} = 0.017097$$

(i) 94% CI $\Rightarrow 100(1 - 0.06)\% \Rightarrow \alpha = 0.06, \alpha/2 = 0.03$

$$94\% \text{ CI for } \mu_1 = 1.68 \pm Z_{0.03} \times \frac{\sigma_1}{\sqrt{n_1}} = 1.68 \pm 1.88 \times 0.017097$$
$$= (1.6479, 1.7121)$$
margin of error $= 1.88 \times 0.017097 = 0.03214$

$$\overline{X}_2 = 1.57; \sigma_2^2 = 0.065 \Rightarrow \frac{\sigma_2}{\sqrt{n_2}} = \frac{0.065}{\sqrt{200}} = 0.018028$$

(ii) 96% CI $\Rightarrow 100(1 - 0.04)\% \Rightarrow \alpha = 0.04, \alpha/2 = 0.02$

$$96\% \text{ CI for } \mu_2 = 1.57 \pm Z_{0.02} \times \frac{\sigma_2}{\sqrt{n_2}} = 1.57 \pm 2.05 \times 0.018028$$
$$= (1.5330, 1.6070)$$
margin of error $= 2.05 \times 0.018028 = 0.0369574$

■ **EXAMPLE 8.3** Clean air is an important part our daily concerns. The Environmental Protection Agency (EPA) monitors ozone concentrations at different locations. The following data represent a sample from eight-hour daily maximum ozone concentrations as measured by the Clean Air Status and Trends Network (CASTNET) at a location. Assume that the eight-hour ozone concentration measurements are normally distributed with variance of 242.39.

X: 68 68 73 71 71 67 62 67 62 59 63 65 65 67 72 66 61 76 75 58 64 61 76 68 63 47

Obtain the (a) 90%, and (b) 97% mean maximum ozone concentration at this location. Find the margins of error in each case.

Solution $\overline{X} = \dfrac{1715}{26} = 65.96; \sigma^2 = 242.39 \Rightarrow 15.569; \dfrac{\sigma}{\sqrt{n}} = \dfrac{15.569}{\sqrt{26}} = 3.0533$

(a) 90% CI $\Rightarrow 100(1 - 0.10)\% \Rightarrow \alpha = 0.10,\ \alpha/2 = 0.05$

$$\text{90\% CI for } \mu = 65.96 \pm Z_{0.05} \times \frac{\sigma}{\sqrt{n}} = 65.96 \pm 1.645 \times 3.0533$$

$$= (60.9373,\ 70.9827)$$

$$\text{margin of error} = 1.645 \times 3.0533 = 5.0227$$

(b) 97% CI $\Rightarrow 100(1 - 0.03)\% \Rightarrow \alpha = 0.03,\ \alpha/2 = 0.015$

$$\text{97\% CI for } \mu = 2.17 \pm Z_{0.015} \times \frac{\sigma}{\sqrt{n}} = 65.96 \pm 2.17 \times 3.0533$$

$$= (59.3343,\ 72.5857)$$

$$\text{margin of error} = 2.17 \times 3.0533 = 6.6257$$

■ **EXAMPLE 8.4** The number of students enrolled in Public Universities in a given year in the USA is believed to be normally distributed with a variance $\sigma^2 = (9572)^2$. A sample of enrollments is given below:

21756 25000 6000 35000 15090 7880 19659
11113 12418 8388 2000 19237 12589 10852

Obtain a (a) 95%, and (b) 99% confidence intervals for mean number enrolled for the year. Obtain the margins of error in each case.

Solution

$$\overline{X} = \frac{206982}{14} = 14784;\ \sigma^2 = (9572)^2 \Rightarrow \sigma = 9572;\ \frac{\sigma}{\sqrt{n}} = \frac{9572}{\sqrt{14}} = 2558.225$$

(a) 95% CI $\Rightarrow 100(1 - 0.05)\% \Rightarrow \alpha = 0.05,\ \alpha/2 = 0.025$

$$\text{95\% CI for } \mu = 14784 \pm Z_{0.025} \times \frac{\sigma}{\sqrt{n}} = 14784 \pm 1.96 \times 2558.225$$

$$= (9769.88,\ 19798.12)$$

$$\text{margin of error} = 1.96 \times 2558.225 = 5014.12$$

(b) 99% CI $\Rightarrow 100(1 - 0.01)\% \Rightarrow \alpha = 0.01,\ \alpha/2 = 0.005$

$$\text{99\% CI for } \mu = 2.575 \pm Z_{0.005} \times \frac{\sigma}{\sqrt{n}} = 14784 \pm 2.575 \times 2558.225$$

$$= (8196.57,\ 21371.43)$$

$$\text{margin of error} = 2.575 \times 2558.225 = 6587.428$$

8.2.2 Finding Confidence Intervals for the Mean When σ^2 is Unknown

If we draw a sample from a population which is normal with unknown mean and unknown variance, then we have to estimate the variance by calculating s^2. In that case we cannot use the normal distribution to obtain the confidence interval for the mean. When we draw a sample of size n from a population that is normally distributed, with unknown mean and variance, then in this case the underlying distribution of the sample mean is t with $n - 1$ degrees of freedom. As we learned earlier, a sample size n is small if $n < 30$. This means that

$$T = \frac{\bar{x} - \mu}{s/\sqrt{n}} \sim t(n-1) \tag{8.5}$$

$$\text{where } s^2 = \frac{\sum_{i=1}^{n} x_i^2 - n\bar{x}^2}{n-1}$$

Since the statistic T is distributed as t with $n-1$ degrees of freedom, we can only set up the confidence interval for the mean of this population via the t-distribution with $n-1$ degrees of freedom. In this case the $100(1-\alpha)\%$ confidence interval for the mean is obtained by using the formula:

$$100(1-\alpha)\% \text{ CI for } \mu = \left[\bar{x} - t(n-1, \alpha/2) \times \frac{s}{\sqrt{n}}, \bar{x} + t(n-1, \alpha/2) \times \frac{s}{\sqrt{n}} \right]$$

■ **EXAMPLE 8.5** A random sample of sixty married couples who have been married for seven years, were asked the number of children they have. The results of the survey are as follows:

0 0 0 3 3 3 1 3 2 2 3 1 3 2 4 0 3 3 3 1 0 2 3 3 1 4 2 3 1 3
3 5 0 2 3 0 4 4 2 2 3 2 2 2 2 3 4 3 2 2 1 4 3 2 4 2 1 2 3 2

Obtain a 95% confidence interval for the mean number of children that couples have after seven years of marriage.

Solution

$$\bar{x} = \frac{136}{60} = 2.26667$$

$$s^2 = \frac{\sum_{i=1}^{60} x_i^2 - n\bar{x}^2}{n-1} = \frac{396 - 60(2.26667)^2}{60-1} = 1.487$$

$$s = \sqrt{1.487} = 1.2194$$

$$100(1-\alpha)\% \text{ CI} = \left[\bar{x} - Z_{\alpha/2} \times \frac{s}{\sqrt{n}}, \bar{x} + Z_{\alpha/2} \times \frac{s}{\sqrt{n}} \right]$$

From the tables (Appendix III) we see that $t_{(59, 0.025)} = 2.001$, so that:

$$\left[2.2667 - 2.001 \times \frac{1.2194}{\sqrt{60}}, 2.26667 + 2.001 \times \frac{1.2194}{\sqrt{60}} \right] = [1.9517, 2.5817]$$

■ **EXAMPLE 8.6** During a time of economic recession, the data below represent a sample of the number of hours some people worked per week in a city in USA.

X: 40 26 39 14 42 18 25 43 46 27 19 47 19 26 35 34 15 44 40 38 31 46
  52 25 35 35 33 29 34 41 49 28 52 47 35 48 22 33 41 51 27 14 54 45

Using the data, obtain the (a) 95% (b) 98% confidence interval for the mean number of hours people in this city worked during the period. What are the margins of error.

Solution

$$\bar{x} = \frac{1544}{44} = 39.0909; \quad s^2 = \frac{\sum_{i=1}^{60} x_i^2 - n\bar{x}^2}{n-1} = \frac{59564 - 44(35.0909)^2}{44-1} = 125.201;$$

$$s = \sqrt{125.201} = 11.1893$$

To find the 95% CI, from the tables we see that $t_{(44-1, 0.025)} = 2.017$, so that

$$\bar{x} \pm t_{(44-1, 0.025)} \times \frac{s}{\sqrt{n}} = 35.0909 \pm 2.017 \times \frac{11.1893}{\sqrt{44}} = [31.6885, 38.4933].$$

$$\text{margin of error} = 2.017 \times \frac{11.1893}{\sqrt{44}} = 3.4024$$

To find the 98% CI, from the tables we see that $t_{(44-1, 0.01)} = 2.416$, so that

$$35.0909 \pm 2.416 \times \frac{11.1893}{\sqrt{44}} = [31.0155, 39.1664].$$

$$\text{margin of error} = 2.416 \times \frac{11.1893}{\sqrt{44}} = 4.0754$$

■ EXAMPLE 8.7 The following data is a sample of NFL career passing touchdowns for NFL professionals:

> 88 67 87 86 86 85 85 85 128 126 125 84 84 44 83 80 79 79 45
> 33 32 31 94 30 82 79 79 25 33 88 87 86 100 96 158 96 94 31

(a) Obtain a (a) 95%, and (b) 99% confidence interval for the mean of the number of career passing touchdowns in the NFL.

Solution

$$\bar{x} = \frac{2980}{38} = 78.42; \quad s^2 = \frac{\sum_{i=1}^{60} x_i^2 - n\bar{x}^2}{n-1} = \frac{267630 - 38(78.42)^2}{38-1} = 917.1693;$$

$$s = \sqrt{917.1693} = 30.2848$$

$$100(1-\alpha)\% \text{ CI} = \left[\bar{x} - t_{(n-1, \alpha/2)} \times \frac{s}{\sqrt{n}}, \bar{x} + t_{(n-1, \alpha/2)} \times \frac{s}{\sqrt{n}} \right]$$

(a) To find the 95% $= 100(1 - 0.05)\%$ CI $\Rightarrow \alpha = 0.05; \alpha/2 = 0.025.$
From the tables we see that $t_{(38-1, 0.025)} = 2.026$, so that

$$78.42 \pm 2.026 \times \frac{30.2848}{\sqrt{38}} = [68.4666, 88.3734].$$

$$\text{margin of error} = 2.026 \times \frac{30.2848}{\sqrt{38}} = 9.9534$$

(b) To find the 99% $= 100(1 - 0.01)\%$ CI $\Rightarrow \alpha = 0.01; \alpha/2 = 0.005.$
From the tables we see that $t_{(38-1, 0.005)} = 2.715$, so that

$$78.42 \pm 2.715 \times \frac{30.2848}{\sqrt{38}} = [65.0816, 91.7584].$$

$$\text{margin of error} = 2.715 \times \frac{30.2848}{\sqrt{38}} = 13.3384$$

■ **EXAMPLE 8.8**   Assume that temperature in August at St. Cloud is normally distributed. The following data show the temperatures for a sample of ten days in August:

$$67 \quad 71 \quad 62 \quad 76 \quad 66 \quad 71 \quad 63 \quad 82 \quad 65 \quad 77$$

Establish a (i) 95%, and (ii) 99% confidence interval for the mean temperature in August at St. Cloud.

Solution   We note that $n = 10$, and the degree of freedom of t must be $10 - 1 = 9$, $t(9, 0.025) = 2.262$, $t(9, 0.005) = 3.250$. We use t here because the underlying population is normal, but the sample size is small.

$$\overline{x} = \frac{700}{10} = 70.0; \ s^2 = \frac{49394 - 10(70)^2}{10 - 1} = \frac{394}{9} = 43.7777 \Rightarrow s = \sqrt{43.7777} = 6.6165$$

The 95% CI $= 100(1 - 0.05)\%$ CI for mean temperature is obtained as:

$$\left(70 - t(9, 0.025) \times \frac{6.6165}{\sqrt{10}}, \ 70 + t(9, 0.025) \times \frac{6.6165}{\sqrt{10}}\right)$$

$$= \left(70 - 2.262 \times \frac{6.6165}{\sqrt{10}}, \ 70 + 2.262 \times \frac{6.6165}{\sqrt{10}}\right) = (65.267, 74.733)$$

The 99% CI $= 100(1 - 0.01)\%$ CI for mean temperature is obtained as:

$$\left(70 - t(9, 0.005) \times \frac{6.6165}{\sqrt{10}}, \ 70 + t(9, 0.005) \times \frac{6.6165}{\sqrt{10}}\right)$$

$$= \left(70 - 3.250 \times \frac{6.6165}{\sqrt{10}}, \ 70 + 3.250 \times \frac{6.6165}{\sqrt{10}}\right) = (63.20, 76.80)$$

■ **EXAMPLE 8.9**   The following data show a random sample of points that an NBA top basketball player scored in fifteen games:

$$16 \quad 24 \quad 27 \quad 27 \quad 25 \quad 30 \quad 27 \quad 25 \quad 25 \quad 18 \quad 18 \quad 28 \quad 29 \quad 33 \quad 26$$

Assuming that the points scored by this player per game are normally distributed, obtain the (a) 95%, and (b) 98% confidence interval for his mean score per game.

Solution   We use the t-distribution because sample size is small, and data is normal.
$100(1 - \alpha)\%$ CI for mean score per game:

$$\overline{x} = \frac{378}{15} = 25.2; \ s^2 = \frac{9832 - 15(25.2)^2}{15 - 1} = 21.88571 \Rightarrow s = \sqrt{21.88571} = 4.6782$$

(a) 95% CI $= \left(25.2 - t(14, 0.025) \times \frac{4.6782}{\sqrt{15}}, \ 25.2 + t(14, 0.025) \times \frac{4.6782}{\sqrt{15}}\right)$

$$= \left(25.2 - 2.1448 \times \frac{4.6782}{\sqrt{15}}, \ 25.2 + 2.1448 \times \frac{4.6782}{\sqrt{15}}\right)$$

$$= (22.61, 27.79)$$

(b) 98% CI $= \left(25.2 - t(14, 0.01) \times \dfrac{4.6782}{\sqrt{15}}, 25.2 + t(14, 0.01) \times \dfrac{4.6782}{\sqrt{15}}\right)$

$= \left(25.2 - 2.6245 \times \dfrac{4.6782}{\sqrt{15}}, 25.2 + 2.6245 \times \dfrac{4.6782}{\sqrt{15}}\right)$

$= (22.03, 28.37)$

■ **EXAMPLE 8.10** The data below represent a sample of weights for nineteen female bears obtained from those living in a conservation area. Assume weights of female bears are normally distributed.

166 220 204 144 140 105 26 125 132 116
182 356 316 202 62 236 76 48 29

Obtain the (i) 99%, and (ii) 90% confidence intervals for the mean weight of the female bears.

$$\bar{x} = \frac{\sum\limits_{i=1}^{19} x_i}{19} = \frac{2885}{19} = 151.8421$$

$$s^2 = \frac{\sum\limits_{i=1}^{19} x_i^2 - n\bar{x}^2}{19 - 1} = \frac{585095 - 19(151.8421)^2}{19 - 1} = 8168.363 \Rightarrow s = 90.37899$$

We use the t-distribution because sample size is small, and data is normal.
$100(1 - \alpha)$% CI for mean weight:
The **99% CI** for the mean is:

$$\left(\bar{X} - t(n - 1, \alpha/2)\frac{s}{\sqrt{n}}, \bar{X} + t(n - 1, \alpha/2)\frac{s}{\sqrt{n}}\right)$$

$$= \left(151.8421 - t(18, 0.005) \times \frac{90.37899}{\sqrt{19}}, 151.8421 + t(18, 0.005) \times \frac{90.378991}{\sqrt{19}}\right)$$

$$= \left(151.8421 - 2.8784 \times \frac{90.37899}{\sqrt{19}}, 151.8421 + 2.8784 \times \frac{90.378991}{\sqrt{19}}\right)$$

$$= (92.16032, 211.5239)$$

90% CI for the mean is:

$$\left(\bar{X} - t(n - 1, \alpha/2)\frac{s}{\sqrt{n}}, \bar{X} + t(n - 1, \alpha/2)\frac{s}{\sqrt{n}}\right)$$

$$= \left(151.8421 - t(18, 0.05) \times \frac{90.37899}{\sqrt{19}}, 151.8421 + t(18, 0.05) \times \frac{90.378991}{\sqrt{19}}\right)$$

$$= \left(151.8421 - 1.7341 \times \frac{90.37899}{\sqrt{19}}, 151.8421 + 1.7341 \times \frac{90.378991}{\sqrt{19}}\right)$$

$$= (115,8866, 187.7976)$$

8.3   Confidence Intervals for the Population Proportion

In the previous sections we have used the sample mean to estimate the population mean. We have used the sample mean to estimate the value of a characteristic which represents the central value of that characteristic for the population. We already know that the mean is very important, and the sample mean helps us get an idea of the value of the characteristic for the population even when we cannot carry out a census which requires us to find out all the values of the attribute for the entire population.

8.3.1   The Need to Know a Population Proportion

Just like sample means are important in statistical inquiry, the sample proportion is important. We often want to know what proportion of a population has an attribute that is of interest to us. Again hampered by the inability to carry out a census, we use sample proportions to estimate population proportions.

Proportions feature in the important questions of real life that we ask. We ask questions such as:

- What proportion of students graduating from high school goes on to obtain a college degree?
- What proportion of our society lives below poverty levels?
- What proportion of retired adults avails itself of flu shots for protection during the winter?
- What proportion of African American males wind up incarcerated or tied up in the penal system?
- What proportion of the work-force is unemployed?
- What proportion of a segment of society is likely to suffer from one disease or another?
- What proportion of the electorate will vote for candidate John?

Many researches, especially social research, focus on proportions, although the results get reported in percentages. We can see that whenever an inquiry is looking for the percentages, it is the same thing as looking for the proportions. For instance, opinion polls are reported in percentages.

8.3.2   Estimating Population Proportion—Point and Confidence Interval Estimates

Our focus in this section is on how to estimate the population proportion, p. In particular, we want to go beyond getting the point estimate for p, and obtain an interval for p for which we are $100(1 - \alpha)\%$ confident that the population proportion would be found in it. To do this, we hop back to our discussion of the sampling distribution of the sample proportion.

At section 7.4, Equation 7.2, we see that if we take a sample of size n, and out of the n, x have the attribute which is of primary interest to us, the point estimate for the population proportion p is:

$$\hat{p} = \frac{x}{n}$$

(8.6)

But we know from Chapter 7 that the sampling distribution of $\hat{p}$ is normal so that $\hat{p} \sim N\left(p, \dfrac{p(1-p)}{n}\right)$. This means that the standard error of $\hat{p}$ is $\sqrt{\dfrac{p(1-p)}{n}}$. From this, we know that $Z = \dfrac{\hat{p} - p}{\sqrt{\dfrac{p(1-p)}{n}}}$. Some implications of the distribution of the sample proportion as described are depicted in Figure 8.3 (a, b, c). The figures shown as Figure 8.3 (a, b, c) depict the idea that 68.26% of the sample proportions would lie 1 standard error from population proportion, p; 95.45% of the sample proportions would lie 2 standard errors from population proportion, p; and 99.73% of the sample proportions would lie 3 standard errors from population proportion, p.

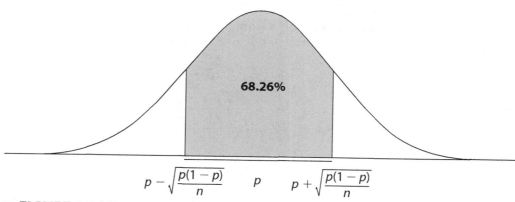

■ **FIGURE 8.3 (A)**   68.26% of the sample proportions lie 1 standard error from the population proportion p.

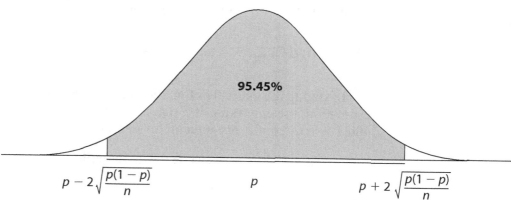

■ **FIGURE 8.3 (B)**   95.45% of the sample proportions lie within 2 standard errors from p.

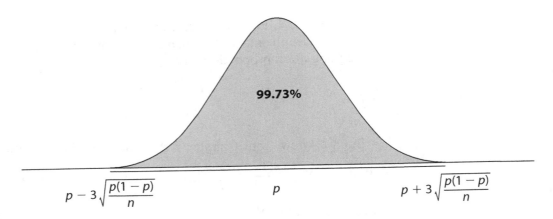

$$p - 3\sqrt{\frac{p(1-p)}{n}} \qquad\qquad p \qquad\qquad p + 3\sqrt{\frac{p(1-p)}{n}}$$

■ **FIGURE 8.3 (C)**   99.73% of the sample proportions lie with 3 standard errors from p.

Following similar arguments as we used for the sample mean, regarding the typical $100(1 - \alpha)\%$ confidence interval for the mean, for us to be $100(1 - \alpha)\%$ confident that we will find the population proportion, p, within the interval on the distribution of $\hat{p}$ similar to the interval $[-Z_{\alpha/2}, Z_{\alpha/2}]$ that is shown in Figure 8.2, then by converting $\hat{p}$ to Z, the following must be true:

$$P\left(-Z_{\alpha/2} \le Z = \frac{\hat{p} - p}{\sqrt{p(1-p)/n}} \le Z_{\alpha/2}\right) = 1 - \alpha$$

$$\Rightarrow P\left(-Z_{\alpha/2} \cdot \sqrt{\frac{p(1-p)}{n}} \le \hat{p} - p \le Z_{\alpha/2} \cdot \sqrt{\frac{p(1-p)}{n}}\right) = 1 - \alpha$$

$$\Rightarrow P\left(-\hat{p} - Z_{\alpha/2} \cdot \sqrt{\frac{p(1-p)}{n}} \le -p \le -\hat{p} + Z_{\alpha/2} \cdot \sqrt{\frac{p(1-p)}{n}}\right) = 1 - \alpha$$

$$\Rightarrow P\left(\hat{p} - Z_{\alpha/2} \cdot \sqrt{\frac{p(1-p)}{n}} \le p \le \hat{p} + Z_{\alpha/2} \cdot \sqrt{\frac{p(1-p)}{n}}\right) = 1 - \alpha$$

It follows that the interval that will guarantee us a $100(1 - \alpha)\%$ confidence that the population proportion p is within it, is:

$$100(1 - \alpha)\% \text{ CI for } p = \left[\hat{p} - Z_{\alpha/2} \cdot \sqrt{\frac{\hat{p}(1 - \hat{p})}{n}}, \hat{p} + Z_{\alpha/2} \cdot \sqrt{\frac{\hat{p}(1 - \hat{p})}{n}}\right] \quad (8.7)$$

In stating the interval in Equation 8.7, we replaced p by $\hat{p}$ because at the time of estimation, the only value of p that is available is the $\hat{p}$ from the sample. To find the $100(1 - \alpha)\%$ confidence interval for p, we have the lower limit of the interval for p as $\hat{p} - Z_{\alpha/2} \cdot \sqrt{\frac{\hat{p}(1 - \hat{p})}{n}}$, while the upper limit is $\hat{p} + Z_{\alpha/2} \cdot \sqrt{\frac{\hat{p}(1 - \hat{p})}{n}}$. The margin of error is $Z_{\alpha/2} \cdot \sqrt{\frac{\hat{p}(1 - \hat{p})}{n}}$.

To obtain a $100(1 - \alpha)\%$ confidence interval for a population proportion, p, we need to:

(a) Take a sample of size n individuals from the population of interest

(b) x out of n individuals in the sample have an attribute that is of interest to us while $n - x$ of the sample do not have the attribute of interest.

(c) We calculate $\hat{p} = \dfrac{x}{n}$

(d) When the level of confidence is $100(1 - \alpha)\%$, we determine α, and obtain $Z_{\alpha/2}$ from the tables of standard normal distribution (Appendix II)

(e) The $100(1 - \alpha)\%$ confidence interval for p, is $\hat{p} \pm z_{\alpha/2} \times se$ where

$$se = \sqrt{\frac{\hat{p}(1 - \hat{p})}{n}}$$

(f) We interpret the confidence interval.

Let us do some examples:

■ EXAMPLE 8.11 A survey of 200 pregnant women from a city found that seventy-six are smokers. Obtain the (a) 93%, and (b) 95% confidence interval, for the proportion of pregnant women in the city who smoke. Obtain the margin of error in each case.

Solution First we obtain a point estimate for p, that is $\hat{p} = \dfrac{76}{200} = 0.38$.

This is an estimate of the proportion of pregnant women in that city who are smokers.

$$se = \sqrt{\frac{\hat{p}(1 - \hat{p})}{n}} = \sqrt{\frac{0.38(1 - 0.38)}{200}} = 0.034322$$

(i) confidence level = 93%
93% = $100(1 - 0.07)\% \Rightarrow \alpha = 0.07 \Rightarrow \alpha/2 = 0.035$
93% CI: $\hat{p} \pm Z_{0.035} \times se$; $Z_{0.035} = 1.81$
$0.38 \pm 1.81 \times 0.034322 = (0.317877, 0.442123)$
margin of error = $1.81 \times 0.034322 = 0.062123$

(ii) confidence level = 95%
95% = $100(1 - 0.05)\% \Rightarrow \alpha = 0.05 \Rightarrow \alpha/2 = 0.025$
95% CI: $\hat{p} \pm Z_{0.025} \times se$; $Z_{0.025} = 1.96$
$0.38 \pm 1.96 \times 0.034322 = (0.312729, 0.447271)$
margin of error = $1.96 \times 0.034322 = 0.067271$

■ EXAMPLE 8.12 In a survey of 1020 doe in a wildlife reserve, 377 were found to be pregnant. Establish a (i) 90%, and (ii) 98% confidence interval for the proportion of doe that are pregnant.

Solution $x = 377, n = 1020, \hat{p} = \dfrac{377}{1020} = 0.3696078,$

$$se = \sqrt{\frac{\hat{p}(1 - \hat{p})}{n}} = \sqrt{\frac{0.3696078(1 - 0.3696078)}{1020}} = 0.01511388$$

(i) confidence level = 90%
90% = $100(1 - \alpha)\% \Rightarrow \alpha = 0.1 \Rightarrow \alpha/2 = 0.05$
90% CI: $\hat{p} \pm Z_{0.05} \times se$; $Z_{0.05} = 1.645$
$0.3696078 \pm 1.645 \times 0.01511388 = (0.34447455, 0.39444701)$

(ii) confidence level = 98%
$$98\% = 100(1 - \alpha)\% \Rightarrow \alpha = 0.02 \Rightarrow \alpha/2 = 0.01$$
$$98\% \text{ CI: } \hat{p} \pm Z_{0.01} \times se; Z_{0.01} = 2.33$$
$$0.3696078 \pm 2.33 \times 0.01511388 = (0.33439246, 0.40482314)$$

■ EXAMPLE 8.13 In a sample of 450 ex-prisoners, it was found that 197 re-offended. Obtain a (i) 97%, and (ii) 99% confidence intervals for the proportion of ex-prisoners who re-offended in the population of ex-prisoners.

Solution $x = 197, n = 450, \hat{p} = \dfrac{197}{450} = 0.437778,$

$$se = \sqrt{\frac{\hat{p}(1 - \hat{p})}{n}} = \sqrt{\frac{0.437778(1 - 0.437778)}{450}} = 0.023387$$

(i) confidence level = 97%
$$97\% = 100(1 - \alpha)\% \Rightarrow \alpha = 0.03 \Rightarrow \alpha/2 = 0.015$$
$$97\% \text{ CI: } \hat{p} \pm Z_{0.015} \times se; Z_{0.015} = 2.17$$
$$0.437778 \pm 2.17 \times 0.023387 = (0.387028, 0.488528)$$

(ii) confidence level = 99%
$$99\% = 100(1 - \alpha)\% \Rightarrow \alpha = 0.01 \Rightarrow \alpha/2 = 0.005$$
$$99\% \text{ CI: } \hat{p} \pm Z_{0.005} \times se; Z_{0.005} = 2.575$$
$$0.437778 \pm 2.575 \times 0.023387 = (0.377556, 0.498)$$

■ EXAMPLE 8.14 In developing a flu vaccine, of 600 people who were surveyed about whether they were confident that the vaccine will keep them and their family safe from flu, 402 were confident that it would. Obtain (a) 94%, and (b) 98% confidence interval for the population proportion who were confident that it would. Obtain the margin of error in each case.

Solution $\hat{p} = \dfrac{405}{600} = 0.675; se = \sqrt{\dfrac{\hat{p}(1 - \hat{p})}{n}} = \sqrt{\dfrac{0.675(1 - 0.675)}{600}} = 0.019121$

(i) confidence level = 94%
$$94\% = 100(1 - 0.06)\% \Rightarrow \alpha = 0.06 \Rightarrow \alpha/2 = 0.03$$
$$94\% \text{ CI: } \hat{p} \pm Z_{0.03} \times se; Z_{0.03} = 1.88$$
$$0.675 \pm 1.88 \times 0.019121 = (0.639052, 0.710948)$$
margin of error = $1.88 \times 0.019121 = 0.035948$

(ii) confidence level = 98%
$$98\% = 100(1 - 0.02)\% \Rightarrow \alpha = 0.02 \Rightarrow \alpha/2 = 0.01$$
$$98\% \text{ CI: } \hat{p} \pm Z_{0.01} \times se; Z_{0.01} = 2.33$$
$$0.675 \pm 2.33 \times 0.019121 = (0.630447, 0.719553)$$
margin of error = $2.33 \times 0.019121 = 0.044553$

■ EXAMPLE 8.15 Out of 1000 people who were surveyed about whether the US medical bill which was signed into law in 2010 should be repealed, 570 said it should not. Obtain the (a) 99%, and (b) 97% confidence interval for proportion of people who do not want the law repealed. What is the margin of error in each case?

Solution $\hat{p} = \dfrac{570}{1000} = 0.57; se = \sqrt{\dfrac{\hat{p}(1 - \hat{p})}{n}} = \sqrt{\dfrac{0.57(1 - 0.57)}{1000}} = 0.015656$

(a) confidence level = 99%
$99\% = 100(1 - 0.01)\% \Rightarrow \alpha = 0.01 \Rightarrow \alpha/2 = 0.005$
99% CI: $\hat{p} \pm Z_{0.005} \times se; Z_{0.005} = 2.575$
$0.57 \pm 2.575 \times 0.015656 = (0.529687, 0.610313)$
margin of error = $2.575 \times 0.015656 = 0.040313$

(b) confidence level = 97%
$97\% = 100(1 - 0.03)\% \Rightarrow \alpha = 0.03 \Rightarrow \alpha/2 = 0.015$
97% CI: $\hat{p} \pm Z_{0.015} \times se; Z_{0.015} = 2.17$
$0.57 \pm 2.17 \times 0.015656 = (0.536027, 0.603973)$
margin of error = $2.17 \times 0.015656 = 0.033973$

8.4 Finding the Sample Size for Attaining a Confidence Level and a Margin of Error in Estimating μ, the Population Mean

Often before taking a sample from a population in order to make inference about a parameter, we determine in advance that we want to attain a level of confidence in our interval estimates of $100(1 - \alpha)\%$, and a margin of error, E. This is a good practice as it enables us to achieve the goals of making our confidence level as high as we want and to minimize the margin of error as much as we desire. In order to attain these goals we have to compensate in our sampling. Our desire to attain high confidence and low or small margin of error would inevitably lead us to take a large sample.

We recall that sampling from the normal distribution with known variance, the margin of error for a $100(1 - \alpha)\%$ confidence for the mean, μ is:

$$E = Z_{\alpha/2} \cdot \frac{\sigma}{\sqrt{n}} \tag{8.8}$$

A little algebra here will show that the sample size we need to attain the margin of error of E at $100(1 - \alpha)\%$ is:

$$n = \frac{(Z_{\alpha/2})^2 \sigma^2}{E^2} \tag{8.9}$$

Determination of sample size, n, in this way is useful for sampling generally with a view to attaining some confidence with a set margin of error; it is particularly useful in industry where the quality control officer (usually a statistician) sets up the process to produce within certain margins of error. When setting up that margin of error, the quality control officer needs to be $100(1 - \alpha)\%$ confident that production will continue within the upper and lower limits of the interval. To do so, the officer chooses the margin of error and the confidence level in advance and determines how many items to sample. If the process follows the normal distribution with some known variance, then Equation 8.9 is used to determine n.

■ **EXAMPLE 8.16** A factory produces steel rods that should be 20 cm long. If it is known that the variance is 0.0184 cm^2, to be 95% confident at a margin of error of 0.04, what sample size should the quality control officer choose?

Solution $\sigma^2 = 0.0184;\ 95\%\ \text{CI} \Rightarrow 100(1 - 0.05)\% \Rightarrow \alpha = 0.05 \Rightarrow \alpha/2 = 0.025$

$Z_{0.025} = 1.96$

$$\therefore n = \frac{(Z_{\alpha/2})^2\,\sigma^2}{E^2} = \frac{(1.96)^2(0.0184)}{0.04^2} = 44.1784 \approx 45$$

The quality control officer should choose a sample of size 45. Note that the sample size is not rounded down to 44, rather it is rounded up to 45. The reason is that a sample of size 44 will not allow us to attain the targeted margin of error $E = 0.04$, it will give a slightly higher margin, that is an $E > 0.04$. We therefore chose 45, which allows us to attain a margin of error smaller than targeted. So taking a sample of size 45, the margin of error attained $E < 0.04$, which is good, as a lower margin of error is desirable.

We can demonstrate that the formula for n in Equation 8.9 is correct by applying it to Example 8.1 questions.

■ **EXAMPLE 8.17**

(a) In Example 8.1 (a) we found a 96% confidence interval for the mean, μ, when $n = 17$, $\sigma^2 = 91$. We found a margin of error of 4.743; show that the sample size, n, is 17 for the confidence and the margin of error.

(b) in Example 8.1 (b) we found a 93% confidence interval for the mean, μ, when $n = 17$, $\sigma^2 = 91$. We found a margin of error of 4.188; show that the sample size, n, is 17 for the confidence and the margin of error.

Solution

(a) $E = 4.743;\ \sigma^2 = 91;\ 96\%\ \text{CI} \Rightarrow 100(1 - 0.04)\% \Rightarrow \alpha = 0.04;\ \alpha/2 = 0.02$

$Z_{0.02} = 2.05$

$$n = \frac{(Z_{\alpha/2})\,\sigma^2}{E^2} = \frac{2.05^2(91)}{4.743^2} = 16.9976 \approx 17$$

(b) $E = 4.188;\ \sigma^2 = 91;\ 93\%\ \text{CI} \Rightarrow 100(1 - 0.07)\% \Rightarrow \alpha = 0.07;\ \alpha/2 = 0.035$

$Z_{0.035} = 1.81$

$$n = \frac{(Z_{\alpha/2})\,\sigma^2}{E^2} = \frac{1.81^2(91)}{4.188^2} = 16.9971 \approx 17$$

■ **EXAMPLE 8.18**   Consider Example 8.3 where EPA determines the eight-hour ozone concentrations. Suppose that in that study, the EPA wants to be 98% confident that the estimate of the mean will not differ by more than 2.5, what is the number of measurements they need to attain these goals?

Solution $E = 2.5;\ \sigma^2 = 242.39;\ 98\%\ \text{CI} \Rightarrow 100(1 - 0.02)\% \Rightarrow \alpha = 0.02;\ \alpha/2 = 0.01$

$Z_{0.01} = 2.33$

$$n = \frac{(Z_{\alpha/2})\,\sigma^2}{E^2} = \frac{2.33^2(242.39)}{2.5^2} = 210.5458 \approx 211$$

■ **EXAMPLE 8.19**   Consider Example 8.4, if the aim is to keep the estimate of mean number of students to within a quarter of the population standard deviation from the mean at 96% confidence level, what is the size of the sample needed (that is, number of universities needed).

Solution $\sigma^2 = (9572)^2 \Rightarrow \sigma = 9572 \Rightarrow E = \dfrac{\sigma}{4} = \dfrac{9572}{4} = 2393;$

(a) 96% CI $\Rightarrow 100(1 - 0.04)\% \Rightarrow \alpha = 0.04$, $\alpha/2 = 0.02$; $Z_{0.02} = 2.05$

$$n = \frac{(Z_{\alpha/2})\,\sigma^2}{E^2} = \frac{2.05^2(9572)^2}{2393^2} = 67.24 \approx 68$$

8.5 Finding the Sample Size for Attaining a Confidence Level and a Set Margin of Error in Estimating p, the Population Proportion

We recall that sampling for the purpose of estimation of the population proportion p, we were able to find the $100(1 - \alpha)\%$ confidence for the mean, p, as:

$$E = Z_{\alpha/2} \cdot \sqrt{\frac{\hat{p}(1 - \hat{p})}{n}} \tag{8.10}$$

Again, with a little algebra we can show that the sample size we need to attain the margin of error of E at $100(1 - \alpha)\%$ is:

$$n = \frac{(Z_{\alpha/2})^2\,\hat{p}(1 - \hat{p})}{E^2} \tag{8.11}$$

We use this formula to determine the sample size n. There is sometimes the rule of thumb where there is no estimate of p known, in which case $\hat{p} = 0.5$, for the purpose of choosing a rough sample size. However if some information from which $\hat{p}$ can be obtained is available, it should be determined using that sample. In short, to use the $\hat{p}$ in the formula, we need a prior study which had determined the value of $\hat{p}$, or else we use $\hat{p} = 0.5$.

■ **EXAMPLE 8.20**   A sample of one hundred married couples who are over fifty showed that fifty-five were grandparents. What sample size do we need to take in order to conduct a survey in which we are 99% confident, and the margin of error is no greater than 0.015?

Solution $\hat{p} = \dfrac{55}{100} = 0.55$; 99% CI $\Rightarrow 100(1 - 0.01)\% \Rightarrow \alpha = 0.01 \Rightarrow \alpha/2 = 0.005$;

$Z_{0.005} = 2.575$; $E = 0.015$;

$\therefore n = \dfrac{(Z_{\alpha/2})^2\,\hat{p}(1 - \hat{p})}{E^2} = \dfrac{(2.575)^2(0.55)(0.45)}{0.015^2} = 7293.688 \approx 7294$

We can demonstrate that the formula for n in Equation 8.11 is correct by applying it to Example 8.11 questions.

■ **EXAMPLE 8.21**   A survey of two hundred pregnant women from a city found that seventy-six are smokers. Obtain the (a) 93%, and (b) 95% confidence interval for the proportion of pregnant women in the city who smoke. Obtain the margin of error in each case.

(a) In Example 8.11 (a) if we want to be 93% confident with a margin of error of 2%, what sample size need we take?

(b) in Example 8.11 (b) if we want to be 95% confident with a margin of error of 2%, what sample size should we use in our study?

Solution $\hat{p} = \dfrac{76}{200} = 0.38$

(a) confidence level = 93%

$93\% = 100(1 - 0.07)\% \Rightarrow \alpha = 0.07 \Rightarrow \alpha/2 = 0.035$

$Z_{0.035} = 1.81$

$\therefore n = \dfrac{(Z_{\alpha/2})^2\, \hat{p}(1 - \hat{p})}{E^2} = \dfrac{(1.81)^2(0.38)(0.62)}{0.02^2} = 1929.623 \approx 1930$

(b) confidence level = 95%

$95\% = 100(1 - 0.05)\% \Rightarrow \alpha = 0.05 \Rightarrow \alpha/2 = 0.025$

$Z_{0.025} = 1.96$

$\therefore n = \dfrac{(Z_{\alpha/2})^2\, \hat{p}(1 - \hat{p})}{E^2} = \hat{p} = 0.38;$

$95\% \text{ CI} \Rightarrow 100(1 - 0.01)\% \Rightarrow \alpha = 0.01 \Rightarrow \alpha/2 = 0.005;$

$Z_{0.025} = 1.96; E = 0.02;$

$\therefore n = \dfrac{(Z_{\alpha/2})^2\, \hat{p}(1 - \hat{p})}{E^2} = \dfrac{(1.96)^2(0.38)(0.62)}{0.02^2} = 2262.702 \approx 2263$

EXAMPLE 8.22   Consider Example 8.12 how many doe need to be sampled in order to be 99% confident and within a margin of error of 1.25%?

Solution $x = 377, n = 1020, \hat{p} = \dfrac{377}{1020} = 0.3696078,$

confidence level = 99%

$99\% = 100(1 - \alpha)\% \Rightarrow \alpha = 0.01 \Rightarrow \alpha/2 = 0.005$

$Z_{0.005} = 2.575$

$n = \dfrac{(Z_{\alpha/2})^2\, \hat{p}(1 - \hat{p})}{E^2} = \dfrac{(2.575)^2(0.3696078)(1 - 0.3696078)}{0.0125^2}$

$= 9887.498 \approx 9888.$

EXAMPLE 8.23   Office of National Statistics of Great Britain (General Household Survey (Longitudinal)) reports a survey outcome that showed that 75% of households have access to more than one car. What sample size is needed for this survey to have a margin of error of 2.5% at 98% confidence level?

Solution $\hat{p} = 0.75; E = 0.025$

$98\% \text{ CI} \Rightarrow 100(1 - 0.02)\% \Rightarrow \alpha = 0.02, \alpha/2 = 0.01; Z_{0.01} = 2.33$

$n = \dfrac{(Z_{\alpha/2})^2\, \sigma^2}{E^2} = \dfrac{2.33^2(0.75)(0.25)}{0.025^2} = 1628.67 \approx 1629$

8.6	**Large Samples—Using Central Limit Theorem (CLT) to Find Confidence Intervals for the Mean**

If we draw a sample from a population which is not necessarily normal with unknown mean and unknown variance, then we have to estimate the variance by calculating s^2. In that case we can use a normal distribution (if n is large, usually thirty and above) to find the $100(1 - \alpha)\%$ confidence interval for the mean. The confidence interval found this way is only approximate. This is an appeal to the central limit theorem. Under the CLT, the resulting approximate Z based on sample mean would be written as:

$$Z = \frac{\bar{x} - \mu}{s/\sqrt{n}} \sim N(0,1)$$

$$\text{where } s^2 = \frac{\sum_{i=1}^{n} x_i^2 - n\bar{x}^2}{n - 1}$$

We use this as a basis for establishing a $100(1 - \alpha)\%$ confidence interval for the mean, using the sample data. We want the $100(1 - \alpha)\%$ confidence interval for the mean μ. The formula is:

$$\left[\bar{x} - Z_{\alpha/2} \times \frac{s}{\sqrt{n}}, \bar{x} + Z_{\alpha/2} \times \frac{s}{\sqrt{n}} \right]$$

So that for 95% confidence limits for the mean, we need the formula:

$$\left[\bar{x} - Z_{0.025} \times \frac{s}{\sqrt{n}}, \bar{x} + Z_{0.025} \times \frac{s}{\sqrt{n}} \right]$$

The approximation looks very good when compared to the use of t-distribution for the same confidence intervals. Here we revisit Examples 8.5–8.7, which we did earlier, but this time we use CLT.

■ **EXAMPLE 8.24** A random sample of 60 married couples who have been married for seven years, were asked the number of children they have. The results of the survey are as follows:

0 0 0 3 3 3 1 3 2 2 3 1 3 2 4 0 3 3 3 1 0 2 3 3 1 4 2 3 1 3
3 5 0 2 3 0 4 4 2 2 3 2 2 2 2 3 4 3 2 2 1 4 3 2 4 2 1 2 3 2

Obtain a 95% confidence interval for the mean number of children that couples have after seven years of marriage.

Solution

$$\bar{x} = \frac{136}{60} = 2.26667$$

$$s^2 = \frac{\sum_{i=1}^{60} x_i^2 - n\bar{x}^2}{n - 1} = \frac{396 - 60(2.6667)^2}{60 - 1} = 1.487$$

$$s = \sqrt{1.487} = 1.2194$$

$$100(1 - \alpha)\% \text{ CI} = \left[\bar{x} - Z_{\alpha/2} \times \frac{s}{\sqrt{n}}, \bar{x} + Z_{\alpha/2} \times \frac{s}{\sqrt{n}} \right]$$

From the tables we see that $Z_{0.025} = 1.96$, so that:

$$\left[2.2667 - 1.96 \times \frac{1.2194}{\sqrt{60}}, 2.26667 + 1.96 \times \frac{1.2194}{\sqrt{60}} \right] = [1.9581, 2.5752]$$

Recall that with t-distribution, we obtained **[1.9517, 2.5817]** which is slightly wider than we obtained by using CLT.

■ **EXAMPLE 8.25**   During a time of economic recession, the data below represent a sample of the number of hours some people worked per week in a city in USA.

X: 40 26 39 14 42 18 25 43 46 27 19 47 19 26 35 34 15 44 40 38 31 46
52 25 35 35 33 29 34 41 49 28 52 47 35 48 22 33 41 51 27 14 54 45

Using the data, obtain the (a) 95% (b) 98% confidence interval for the mean number of hours people in this city worked during the period. What are the margins of error.

Solution

$$\bar{x} = \frac{1544}{44} = 35.0909; \quad s^2 = \frac{\sum_{i=1}^{60} x_i^2 - n\bar{x}^2}{n - 1} = \frac{59564 - 44(35.0909)^2}{44 - 1} = 125.201;$$

$$s = \sqrt{125.201} = 11.1893$$

To find the 95% CI, from the tables we see that $Z_{0.025} = 1.96$, so that

$$\bar{x} \pm Z_{0.025} \times \frac{s}{\sqrt{n}} = 35.0909 \pm 1.96 \times \frac{11.1893}{\sqrt{44}} = [31.7847, 38.3971].$$

Margin of error $= 1.96 \times \dfrac{11.1893}{\sqrt{44}} = 3.3062$

To find the 98% CI, from the tables we see that $Z_{0.01} = 2.33$, so that

$$35.0909 \pm 2.33 \times \frac{11.1893}{\sqrt{44}} = [31.1605, 39.0213].$$

Margin of error $= 2.33 \times \dfrac{11.1893}{\sqrt{44}} = 3.9304.$

Recall that under the t-distribution, the 95% and 98% CIs were **(31.6885, 38.4933)** and **(31.0155, 39.1664)** respectively.

■ **EXAMPLE 8.26**   The following data is a sample of NFL career passing touchdowns for NFL professionals.

88 67 87 86 86 85 85 85 128 126 125 84 84 44 83 80 79 79 45
33 32 31 94 30 82 79 79 25 33 88 87 86 100 96 158 96 94 31

Obtain a (a) 95%, and (b) 99% confidence interval for the mean of the number of career passing touchdowns in NFL.

Solution $\bar{x} = \dfrac{2980}{38} = 78.42$; $s^2 = \dfrac{\sum\limits_{i=1}^{60} x_i^2 - n\bar{x}^2}{n-1} = \dfrac{267630 - 38(78.42)^2}{38-1} = 917.1693$;

$s = \sqrt{917.1693} = 30.2848$

$$100(1 - \alpha)\% \ CI = \left[\bar{x} - Z_{\alpha/2} \times \frac{s}{\sqrt{n}}, \bar{x} + Z_{\alpha/2} \times \frac{s}{\sqrt{n}}\right]$$

(a) To find the 95% = 100(1 − 0.05)% CI ⟹ $\alpha = 0.05$; $\alpha/2 = 0.025$.
From the tables we see that $Z_{0.025} = 1.96$, so that:

$$78.42 \pm 1.96 \times \frac{30.2848}{\sqrt{38}} = [68.7908, 88.0492].$$

Margin of error = $1.96 \times \dfrac{30.2848}{\sqrt{38}} = 9.6292$.

(b) To find the 99% = 100(1 − 0.01)% CI ⟹ $\alpha = 0.01$; $\alpha/2 = 0.005$.
From the tables we see that $Z_{0.005} = 2.575$, so that:

$$78.42 \pm 2.575 \times \frac{30.2848}{\sqrt{38}} = [65.7694, 91.0706].$$

Margin of error = $2.575 \times \dfrac{30.2848}{\sqrt{38}} = 12.6506$

Recall that under the *t*-distribution, the 95% and 99% CIs were **(68.4666, 88.3734)** and **(65.0816, 91.7584)** respectively.

8.7   Confidence Intervals Using MINITAB

8.7.1 Use of MINITAB for Confidence Intervals for the Single Population Mean When Variance is Known and Data is Normally Distributed or Sample Size is Large

When the mean is unknown and the variance is known, we can construct the $100(1 - \alpha)\%$ confidence interval for the population mean. We illustrate this by using some of the examples we have done by manual methods.

EXAMPLE 8.27   Recall Example 8.1. Suppose that the following data was taken from a population which is normal with variance 91 and unknown mean. Using MINITAB obtain the (a) 96%, and (b) 93% confidence limits for the mean of the population.

X: 40 26 39 14 42 18 25 43 46 27 19 47 19 26 35 34 15

Solution   96% Confidence Interval

```
MTB > set c1
DATA> 40 26 39 14 42 18 25 43 46 27 19 47 19 26 35 34 15
DATA> END
```

```
DATA> name c1 'X'
MTB > let k1=sqrt(91)
MTB > oneZ c1;
SUBC> sigma k1;
SUBC> confidence 96.
```

One-Sample Z: X

```
The assumed standard deviation = 9.53939

    Variable   N   Mean   StDev   SE Mean      96% CI
       X      17  30.29  11.21     2.31    (25.54, 35.05)
```

93% Confidence Interval

```
MTB > name c1 'X'
MTB > let k1=sqrt(91)
MTB > oneZ c1;
SUBC> sigma k1;
SUBC> confidence 93.
```

One-Sample Z: X

```
The assumed standard deviation = 9.53939

    Variable   N   Mean   StDev   SE Mean      93% CI
       X      17  30.29  11.21     2.31    (26.10, 34.49)
```

■ **EXAMPLE 8.28** Y follows the normal distribution, whose variance is 66. The following is a sample taken from Y: 70 76 86 96 88 100 74 75 79 77 81 85 86 79 87

Using MINITAB and the data, find the (a) 98%, and (b) 99% confidence intervals for the mean of Y.

Solution   98% confidence interval for μ:

```
set c2
DATA> 70 76 86 96 88 100 74 75 79 77 81 85 86 79 87
DATA> end
MTB > name c2 'Y'
MTB > let k2=sqrt(66)
MTB > oneZ c2;
SUBC> sigma k2;
SUBC> confidence 98.
```

One-Sample Z: Y

```
The assumed standard deviation = 8.12404

    Variable   N   Mean   StDev   SE Mean      98% CI
       Y      15  82.60   8.25     2.10    (77.72, 87.48)
```

The 99% confidence interval

```
MTB > oneZ c2;
SUBC> sigma k2;
SUBC> confidence 99.
```

One-Sample Z: Y

```
The assumed standard deviation = 8.12404

Variable   N    Mean   StDev   SE Mean       99% CI
       Y   15   82.60   8.25     2.10    (77.20, 88.00)
```

■ **EXAMPLE 8.29** The following is a sample of measurements for noise made by twenty aircraft, in decibels: 79 80 82 79 74 91 87 74 88 86 84 84 81 77 77 89 76 85 86 78. The city engineer's measurements show that the noise has a variance of 43. Using MINITAB, construct the (a) 90%, and (b) 95% confidence interval, if the noise of the aircraft is normally distributed.

Solution 90% confidence interval for mean of the aircraft noise, μ:

```
MTB > let k3=sqrt(43)
MTB > set c3
DATA> 79 80 82 79 74 91 87 74 88 86 84 84 81 77 77 89 76 85 86 78
DATA> end
MTB > oneZ c2;
SUBC> sigma k3;
SUBC> confidence 90.
```

One-Sample Z: Y

```
The assumed standard deviation = 6.55744

Variable   N    Mean   StDev   SE Mean      90% CI
       Y   15   82.60   8.25     1.69    (79.82, 85.38)
```

95% confidence interval for μ:

```
MTB > oneZ c2;
SUBC> sigma k3;
SUBC> confidence 95.
```

One-Sample Z: Y

```
The assumed standard deviation = 6.55744

Variable   N    Mean   StDev   SE Mean      95% CI
       Y   15   82.60   8.25     1.69    (79.28, 85.92)
```

■ **EXAMPLE 8.30** A random sample of sixty married couples who have been married for seven years, were asked the number of children they have. The results of the survey are as follows:

```
0 0 0 3 3 3 1 3 2 2 3 1 3 2 4 0 3 3 3 1 0 2 3 3 1 4 2 3 1 3
3 5 0 2 3 0 4 4 2 2 3 2 2 2 2 3 4 3 2 2 1 4 3 2 4 2 1 2 3 2
```

Obtain the (a) 95%, and (b) 99% confidence interval for the mean number of children that couples have after seven years of marriage.

Solution   95% confidence interval for mean number of children:

```
MTB > name c4 'children'
MTB > describe c4
```

Descriptive Statistics: Children

```
Variable N  N* Mean  SE Mean StDev Minimum   Q1   Median  Q3   Maximum
children 60 0  2.267 0.157   1.219 0.000    2.000 2.000  3.000 5.000


MTB > oneZ c4;
SUBC> sigma 1.219;
SUBC> confidence 95.
```

One-Sample *Z*: Children
```
The assumed standard deviation = 1.219


Variable   N    Mean   StDev   SE Mean      95% CI
children   60   2.267  1.219    0.157    (1.958, 2.575)
```

99% confidence interval for the mean number of children

```
MTB > oneZ c4;
SUBC> sigma 1.219;
SUBC> confidence 99.
```

One-Sample *Z*: Children
```
The assumed standard deviation = 1.219


Variable   N    Mean   StDev   SE Mean      99% CI
children   60   2.267  1.219    0.157    (1.861, 2.672)
```

8.7.2 Confidence Interval for a Single Population Mean Using MINITAB When Data are Normally Distributed and Variance is Unknown

When data are normally distributed and variance is unknown, the *t*-distribution is used for constructing the confidence interval. We now use some examples to illustrate the use of MINITAB to construct the confidence interval for the mean.

■ **EXAMPLE 8.31**   During a time of economic recession, the data below represents a sample of the number of hours some people worked per week in a city in the USA.

X: 40 26 39 14 42 18 25 43 46 27 19 47 19 26 35 34 15 44 40 38 31 46
52 25 35 35 33 29 34 41 49 28 52 47 35 48 22 33 41 51 27 14 54 45

Using MINITAB and the data, obtain the (a) 95%, and (b) 92% confidence interval for the mean number of hours people in this city worked during the period.

Solution   95% Confidence interval for mean number of hours worked

```
> onet c1;
SUBC> confidence 95.
```

One-Sample *T*: C1

```
Variable   N   Mean   StDev   SE Mean      95% CI
   C1     44  35.09   11.19    1.69    (31.69, 38.49)
```

92% Confidence interval for mean number of hours worked

```
MTB > onet c1;
SUBC> confidence 92.
```

One-Sample *T*: C1

```
Variable   N   Mean   StDev   SE Mean      92% CI
   C1     44  35.09   11.19    1.69    (32.07, 38.12)
```

■ **EXAMPLE 8.32**   A sample of the weekly number of patients without any medical insurance who were treated by the Community Dental services of St. Cloud Technical College is given below:

$$30\ 27\ 33\ 20\ 19\ 24\ 15\ 29\ 21\ 16\ 17\ 25\ 29$$

Assume the data are normally distributed. Use MINITAB to obtain the (a) 90%, and (b) 95% confidence interval for the mean number treated weekly.

Solution   90% confidence interval for the mean

```
MTB > set c6
DATA> 30 27 33 20 19 24 15 29 21 16 17 25 29
DATA> end
MTB > name c6 'patients'
MTB > onet c6;
SUBC> confidence 90.
```

One-Sample *T*: Patients

```
Variable    N   Mean   StDev   SE Mean      90% CI
patients   13  23.46   5.90     1.64    (20.55, 26.38)
```

96% confidence interval for the mean

```
MTB > name c6 'patients'
MTB > onet c6;
SUBC> confidence 96.
```

One-Sample *T*: Patients

```
Variable    N   Mean   StDev   SE Mean      96%        CI
patients   13  23.46   5.90     1.64    (19.70, 27.23)
```

■ EXAMPLE 8.33   The weekly attendance of a church in St. Cloud, MN is believed to be normally distributed. A sample of attendances for some weeks were selected as follows:

278 301 330 290 304 288 293 287 361 298 332 316 326 289 304 320

Using MINITAB, obtain the (a) 92%, and (b) 97% confidence intervals for the mean church attendance.

Solution   92% confidence interval for the mean

```
MTB > name c7 'attendance'
MTB > set c7
DATA> 278 301 330 290 304 288 293 287 361 298 332 316 326 289 304 320
DATA> end
MTB > onet c7;
SUBC> confidence 92.
```

One-Sample *T*: Attendance
```
     Variable    N    Mean    StDev   SE Mean       92% CI
    attendance   16  307.31   21.95    5.49    (297.01, 317.62)
```

97% confidence interval for the mean

```
MTB > onet c7;
SUBC> confidence 97.
```

One-Sample *T*: Attendance
```
     Variable    N    Mean    StDev   SE Mean       97% CI
    attendance   16  307.31   21.95    5.49    (294.16, 320.46)
```

8.7.3 Confidence Interval for a Single Population Proportion Using MINITAB

Again, in MINITAB we use *OneProportion* to obtain the $100(1 - \alpha)\%$ confidence interval for the true population proportion. We use an example to illustrate.

■ EXAMPLE 8.34   For a fee, the TotalGym company guarantees its gym equipment would last for four years or they will replace it free of charge. If the proportion of equipment which breaks down is less than 10% in four years, they make a profit. They surveyed 5567 equipment and found that 457 had broken down in four years. Obtain the (a) 91%, and (b) 97% confidence interval for proportion which break down within four years.

Solution   91% confidence interval

```
MTB > POne 5567 457;
SUBC> Confidence 91;
SUBC> UseZ.
```

Test and CI for One Proportion

```
Sample      X       N     Sample p            91% CI
   1       457    5567    0.082091    (0.075853, 0.088328)
```

Using the normal approximation.

97% confidence interval

```
MTB > POne 5567 457;
SUBC> Confidence 97;
SUBC> UseZ.
```

Test and CI for One Proportion

```
Sample      X       N     Sample p            97% CI
   1       457    5567    0.082091    (0.074107, 0.090075)
```

Using the normal approximation.

■ **EXAMPLE 8.35**   The proportion of births to unmarried mothers in the United States has risen steeply over the past few decades, the Center for Disease Control of the USA reported in May 2009. A survey of 42,866 births show that 17,146 were born to unwed mothers. Use MINITAB to obtain the (a) 94%, and (b) 98% confidence interval for the true proportion of unwed mothers in the USA.

Solution   94% confidence interval

```
MTB > POne 42866 17146;
SUBC> Confidence 94;
SUBC> UseZ.
```

Test and CI for One Proportion

```
Sample      X       N      Sample p            94% CI
   1      17146   42866    0.0399991    (0.395540, 0.404441)
```

Using the normal approximation.

98% confidence interval

```
MTB > POne 42866 17146;
SUBC> Confidence 98;
SUBC> UseZ.
```

Test and CI for One Proportion

```
Sample      X       N      Sample p            98%            CI
   1      17146   42866    0.0399991    (0.394486, 0.405495)
```

Using the normal approximation.

Exercises

Confidence intervals when σ^2 is unknown and data are normally distributed

8.1 Preservation of wild life is good for our environment and for us. The following data are the number of bird-nests, Y, which were found per hectare that was surveyed in a wild life reserve:

4, 10, 11, 4, 6, 8, 6, 3, 5, 8, 8, 0, 8, 8, 10, 10, 2, 2, 0, 10,
14, 7, 0, 10, 4, 0, 10, 8, 10, 5, 8, 0, 1, 8, 3, 5, 7, 2, 11, 10

Obtain the (a) 95%, and (b) 99% confidence interval for mean number of bird nests found in a hectare of the reserve. Obtain the margin of error in each case.

8.2 We are all accustomed to late take-offs of flights. Some are necessary, others are not. The following data show a sample of the number of minutes by which flights of a well-known airline were late before they took off.

15 25 33 45 67 48 60 33 46 33 21 24 38 40 43
26 64 43 65 34 23 14 35 44 57 54 34 33 30 49
14 17 34 23 20 25 37 34 23 43 45 61 34

Obtain the (a) 90%, and (b) 95% confidence interval for the mean number of minutes of lateness.

8.3 Snow is important for agriculture, winter sports, and business. The presence of snow sometimes slows or disrupts some daily activities. The following data show a sample of the number of days that measurable snow was on the ground in a Mid-Western US city before it melted:

18 10 31 40 53 50 30 29 26 45 10 13 30 19 16 21 33 30
9 19 27 39 61 42 54 27 40 27 15 18 32 34 37 20 58 37

What are the (a) 90%, and (b) 98% confidence intervals for the mean number of days measurable snow was on the ground?

8.4 A production process was being designed to complete the production of an item every ten minutes. This kind of design is often used in order to control quality, cost, or other factors. In a pilot study, a sample of measurements of times of production of the items were obtained as recorded below:

10.1 10.7 10.5 9.8 9.0 11.0 11.5 10.9 8.9 10.0 9.7 10.2
11.1 10.3 10.1 10.6 10.7 9.5 9.4 9.6 10.1 10.7 10.5
9.8 9.0 11.0 11.5 10.9 8.9 10.0 9.7 10.2 11.1 10.3
10.1 10.6 10.7 9.5 9.4 9.6 9.6 10.1 10.2 9.7 10.2

What are the (a) 96%, and (b) 98% confidence intervals for the mean time it took to produce an item?

8.5   Some of the important things we need for modern life are connected to mining. A sample of hourly wages (in $) paid in the mining industry is presented below:

12.83 14.21 16.53 19.34 17.17 18.89 17.62 18.31 17.64
18.85 18.58 13.59 18.10 17.44 18.70 14.03 12.31 18.53
14.03 12.31 18.53 17.82 20.47 16.93 19.03 15.21 19.80
21.44 20.90 22.79 16.76 15.88 16.38 18.49 21.44

Obtain the (a) 95%, and (b) 99% confidence intervals for hourly wages paid in the mining industry.

8.6   Nutritionists often need to know how much of a nutrient is in a quantity of food in order to determine what a reasonable daily intake amount is. Here is a sample of determinations of the percentage of nutrient content of a fruit, made by a research food scientist.

3.56 6.40 5.60 4.56 4.55 5.33 5.56 6.12 3.45 6.34 4.33
5.40 5.70 4.89 6.20 5.70 4.56 5.33 5.34 5.10 6.00 5.77
5.45 6.10 5.41 6.30 6.40 5.40 5.33 4.98 4.88 4.12 5.14
4.13 6.00 4.87 4.56 4.10 5.20

Obtain the (a) 90%, and (b) 98% confidence interval for the mean percentage nutrient content of the fruit.

8.7   Pulse rates between sixty and one hundred when at rest is considered normal. The following is a sample of pulses obtained for some individuals when at rest:

67 89 70 77 68 90 76 81 76 78 84 83 90 94 76
88 79 78 85 82 77 91 86 83 78 91 76 69 65 74
73 78 77 79 76 66 78

What are the (a) 95%, and (b) 99% confidence intervals for the mean heart rate?

8.8   The following are the times (in minutes) used by a sample of students at a high school, to complete a titration experiment in a chemistry class.

16.70 19.70 22.50 26.90 18.12 19.70 23.50 23.10 24.70
34.00 23.50 23.60 25.00 23.10 24.70 34.00 23.50 23.60
25.80 25.70 29.70 32.50 36.50 28.20 29.70 23.50 20.80
24.00 34.00 28.90 27.90 28.90 27.90 31.30 32.10 34.50

What are the (a) 90%, and (b) 95% confidence intervals for the mean titration time?

8.9   Doctors use blood glucose (A1c) measurements to determine whether one is potentially at risk of developing diabetes. Determinations of A1c levels of a sample of adults aged thirty who underwent medical exams yielded the following:

4.5 5.4 4.7 5.0 5.7 4.9 4.4 5.8 4.6 4.9 5.2 5.4 5.2
5.6 5.3 5.4 4.8 4.7 4.5 5.5 5.4 4.8 4.8 5.4 4.6 5.4
5.6 4.7 5.8 4.8 4.7 5.9 5.3

Obtain the (a) 95%, and (b) 98% confidence interval for the mean A1c level for adults of age thirty.

8.10 The following data show the number of times a sample of students cut their hair in a year:

7   9 20 18   7 10 12 12 20 18   6   8   9 12 20
18 16 12   6   7 10 12   4   6   9 12 12 18 12 18
6   9   7   9 15 20 18   6 12 15   6

Obtain the (a) 95%, and (b) 99% confidence interval for the mean number of times the students cut their hair in a year.

8.11 A police study of traffic violations at a busy intersection in a big US city, observed the following number of "failure to yield" violations per hour:

24, 26, 36, 20, 14, 36, 40, 21, 24, 26, 28, 40,14, 16, 31, 23

If the occurrence of "failure to yield" violations are normally distributed, obtain the (a) 95%, and (b) 98% confidence interval for the mean number of "failure to yield" violations occurring at this intersection. Obtain the margin of errors in each case.

8.12 A sample of weekly working hours of a sample of young resident doctors showed that they had worked:

70 65 76 78 80 60 67 70 65 65 70 72 75 79 71 68 80 74

If the weekly hours they worked are normally distributed, obtain the (a) 90%, and (b) 99% confidence interval for the mean number of hours they worked a week. Obtain the margin of error in each case.

8.13 The following show the number of hours a sample of high school students from a local high school studied a week:

21 15 18 24 15 18 20 23 24 12 10 20 12 18 20 14 17

Assume that the weekly study hours are normally distributed; find the (a) 95%, and (b) 90% confidence intervals for the mean number of study hours, and the margin of error in each case.

8.14 A sample of determinations of the concentration of a compound made by students during titration, gave the following values:

4.4 3.4 4.0 3.8 4.1 4.4 3.7 3.9 3.5 4.5 5.0 5.1 3.8 4.4 5.1 5.5 5.6 4.5 5.2 4.9 5.2

Assume that the determinations are normally distributed; find the (a) 95%, and (b) 90% confidence intervals for the mean concentration of the compound, and the margin of error in each case.

8.15 The number of calories in a hamburger depends on what ingredients are included. The following are determinations of calorific values of a sample of homemade large hamburgers:

350 375 400 356 378 379 340 320 330 324 360 390

Assume that the calorific values for the hamburgers are normally distributed. Obtain the (a) 80%, and (b) 95% confidence intervals for the mean calorific values of the burgers.

8.16 Young people consume media mostly online. Edison Research, a research outfit in NJ, reports that American young adults spend more time online than consuming other forms of media. The following are the times in minutes that a sample of young adults reported that they spent online the day before they were surveyed:

201 172 180 159 188 205 136 149 220 137 140 186 210 195

If time young adults spend on the internet is normally distributed, obtain the (a) 90%, and (b) 99% confidence interval for the mean time spent by them on the internet.

8.17 An experiment was performed to determine the lifetime of a brand of electric light bulbs. A sample of eleven bulbs was used and the lifetimes were recorded to nearest hour as follows:

805 925 860 851 871 830 871 850 849 860 901

Assume that the lifetimes of the bulbs are normally distributed. Obtain the (a) 95%, and (b) 98% confidence intervals for the mean lifetimes of the bulbs.

8.18 A sample of items, each of which is supposed to weigh 120 grams, were taken off the production line and weighed. The following are their weights:

126 120 124 124 120 115 126 115 122 126 122 123 116 122 122 123 116 118 119

Obtain the (i) 99%, and (ii) 90% confidence intervals for the mean weight of items.

8.19 A sample of the wind gusts experienced in a city and its vicinity were measured at the same time for some days in miles per hour as follows:

10 25 50 35 21 25 32 42 27 24 26 15 18

Obtain the (a) 80%, and (b) 95% confidence interval for mean wind gust in that city, at that time of the day.

8.20 In a certain city the time it takes, in minutes, for a sample of individuals to pedal from home to work are recorded below:

22 19 24 31 29 29 21 15 27 23 37
31 30 26 16 26 12 23 48 22 29 28

If the cycling times are known to be normally distributed with an unknown mean, find the (i) 90%, and (ii) 80% confidence intervals for the mean of the cycling times.

8.21 A sample of the number of days it took for patients who had the flu to recover are:

5 7 9 10 7 8 10 11 8 10 9 6 8 10 9 10 11

Assume recovery time is normally distributed. Obtain the (a) 90%, and (b) 98% confidence interval for the mean recovery time.

Exercises on Confidence Intervals for Proportions

8.22 The New York Times, on August 30, 2006, reported that Census figures show scant improvement in city poverty rate. The poverty rate among households headed by a woman with children is of interest to government, researchers, community, and social groups. A survey of 800 families in this category showed that 392 were living below poverty levels. Obtain the (a) 96%, and (b) 98% confidence intervals for the proportion of such families which live below poverty level. Obtain the margin of error in each case.

8.23 There is a decline in reading the print versions of newspapers. Pew Research center media consumption survey in 2008 showed an increase in those who read newspapers online. Another survey showed that out of 1030 adults surveyed, 720 reported reading newspaper only online. Find the (a) 90%, and (b) 97% confidence interval for the proportion who read the newspapers online. Obtain the margin of error in each case.

8.24 Jewels are important and are often purchased for birthdays, weddings, anniversaries, and expressions of affection, among other things. A survey of 1150 households with annual income of $50,000 and above showed that 529 purchased some jewelry in the previous year. Find the (a) 92%, and (b) 95% confidence interval for the proportion of families with annual income of $50,000 and above who purchased some jewels in the previous year. Obtain the margin of error in each case.

8.25   Births that occur before thirty-seven completed weeks of gestation are defined as preterm. A full-term pregnancy is forty weeks. A survey of births in a US state found that 128 out of 895 births were preterm. Obtain the (a) 80%, and (b) 85% confidence intervals for proportion of babies who were born preterm in this state. Obtain the margin of error in each case.

8.26   The Washington Post, on November 25, 2009, reports: "Fewer Americans believe in global warming, poll shows." Global warming is a current issue in our public discourse, and is debated all over the world. In a survey of 1080 individuals, 802 believe that global warming is happening. Obtain the (a) 99%, and (b) 94% confidence intervals for the proportion of the population that believe that global warming is happening. Obtain the margin of error in each case.

8.27   High school graduation has received a lot of attention in recent years. The national high school graduation rate has slipped in recent years according to an analysis of the most recent data in "Diplomas Count 2010." Another survey found that out 1500 students that were eligible to graduate last year, 1032 actually graduated. Find the (a) 91%, and (b) 95% confidence intervals for the proportion of eligible high school students who actually graduated last year. Obtain the margin of error in each case.

8.28   Arguments continue as some politicians insist that tax cuts for the rich create jobs, and others do not believe it. A major news organization in the USA (CNN) on August 20, 2010, released an opinion poll finding in which 69% of Americans want to let the tax cuts for the rich expire. If the number in the survey was 1200, obtain the (a) 96%, and (b) 98% confidence intervals for proportion of Americans who want to let the tax cuts for the rich expire. Obtain the margin of error in each case.

8.29   A survey of 1006 Americans showed that 804 believe that Congress should consider reallocating federal subsidies from fossil fuels to solar. Find the (a) 92%, and (b) 94% confidence interval for the proportion of Americans who believe that Congress should so reallocate subsidies. Obtain the margin of error in each case.

8.30   It is reported that for patients who suffer from one of the cancers, if the disease is diagnosed early and treated, a very high percent will recover fully. A sample of 850 was chosen from the database for those whose cancer was diagnosed early, and 678 were found to have fully recovered following treatment. Obtain the (a) 90%, and (b) 95% confidence interval for those whose cancer was diagnosed early and treated, that fully recovered. Obtain the margin of error in each case.

8.31   There has been a concern that in boxing, most of the injuries are sustained in the head region leading to very debilitating diseases later in life. A sample of 107 injuries which occurred due to participation in boxing fights, ninety-four were found to be located in the head. Obtain (a) 92%, and (b) 99% confidence interval for the proportion of boxing injuries which are on the head region. Obtain the margin of error in each case.

■ Table of the *t*-Distribution

df	t.10	t.05	t.025	t.01	t.005	df
1	3.078	6.314	12.706	31.821	63.657	1
2	1.886	2.920	4.303	6.965	9.925	2
3	1.638	2.353	3.182	4.541	5.841	3
4	1.533	2.132	2.776	3.747	4.604	4
5	1.476	2.015	2.571	3.365	4.032	5
6	1.440	1.943	2.447	3.143	3.707	6
7	1.415	1.895	2.365	2.998	3.499	7
8	1.397	1.860	2.306	2.896	3.355	8
9	1.383	1.833	2.262	2.821	3.250	9
10	1.372	1.812	2.228	2.764	3.169	10
11	1.363	1.796	2.201	2.718	3.106	11
12	1.356	1.782	2.179	2.681	3.055	12
13	1.350	1.771	2.160	2.650	3.012	13
14	1.345	1.761	2.145	2.624	2.977	14
15	1.341	1.753	2.131	2.602	2.947	15
16	1.337	1.746	2.120	2.583	2.921	16
17	1.333	1.740	2.110	2.567	2.898	17
18	1.330	1.734	2.101	2.552	2.878	18
19	1.328	1.729	2.093	2.539	2.861	19
20	1.325	1.725	2.086	2.528	2.845	20
21	1.323	1.721	2.080	2.518	2.831	21
22	1.321	1.717	2.074	2.508	2.819	22
23	1.319	1.714	2.069	2.500	2.807	23
24	1.318	1.711	2.064	2.492	2.797	24
25	1.316	1.708	2.060	2.485	2.787	25
26	1.315	1.706	2.056	2.479	2.779	26
27	1.314	1.703	2.052	2.473	2.771	27
28	1.313	1.701	2.048	2.467	2.763	28

Continued.

■ Table of the *t*-Distribution—cont'd

df	t.10	t.05	t.025	t.01	t.005	df
29	1.311	1.699	2.045	2.462	2.756	29
30	1.310	1.697	2.042	2.457	2.750	30
31	1.309	1.696	2.040	2.453	2.744	31
32	1.309	1.694	2.037	2.449	2.738	32
33	1.308	1.692	2.035	2.445	2.733	33
34	1.307	1.691	2.032	2.441	2.728	34
35	1.306	1.690	2.030	2.438	2.724	35
36	1.306	1.688	2.028	2.434	2.719	36
37	1.305	1.687	2.026	2.431	2.715	37
38	1.304	1.686	2.024	2.429	2.712	38
39	1.304	1.685	2.023	2.426	2.708	39
40	1.303	1.684	2.021	2.423	2.704	40
41	1.303	1.683	2.020	2.421	2.701	41
42	1.302	1.682	2.018	2.418	2.698	42
43	1.302	1.681	2.017	2.416	2.695	43
44	1.301	1.680	2.015	2.414	2.692	44
45	1.301	1.679	2.014	2.412	2.690	45
46	1.300	1.679	2.013	2.410	2.687	46
47	1.300	1.678	2.012	2.408	2.685	47
48	1.299	1.677	2.011	2.407	2.682	48
49	1.299	1.677	2.010	2.405	2.680	49
50	1.299	1.676	2.009	2.403	2.678	50
51	1.298	1.675	2.008	2.402	2.676	51
52	1.298	1.675	2.007	2.400	2.674	52
53	1.298	1.674	2.006	2.399	2.672	53
54	1.297	1.674	2.005	2.397	2.670	54
55	1.297	1.673	2.004	2.396	2.668	55
56	1.297	1.673	2.003	2.395	2.667	56

◼ Table of the *t*-Distribution—cont'd

df	t.10	t.05	t.025	t.01	t.005	df
57	1.297	1.672	2.002	2.394	2.665	57
58	1.296	1.672	2.002	2.392	2.663	58
59	1.296	1.671	2.001	2.391	2.662	59
60	1.296	1.671	2.000	2.390	2.660	60
61	1.296	1.670	2.000	2.389	2.659	61
62	1.295	1.670	1.999	2.388	2.657	62
63	1.295	1.669	1.998	2.387	2.656	63
64	1.295	1.669	1.998	2.386	2.655	64
65	1.295	1.669	1.997	2.385	2.654	65
66	1.295	1.668	1.997	2.384	2.652	66
67	1.294	1.668	1.996	2.383	2.651	67
68	1.294	1.668	1.995	2.382	2.650	68
69	1.294	1.667	1.995	2.382	2.649	69
70	1.294	1.667	1.994	2.381	2.648	70
71	1.294	1.667	1.994	2.380	2.647	71
72	1.293	1.666	1.993	2.379	2.646	72
73	1.293	1.666	1.993	2.379	2.645	73
74	1.293	1.666	1.993	2.378	2.644	74
75	1.293	1.665	1.992	2.377	2.643	75
76	1.293	1.665	1.992	2.376	2.642	76
77	1.293	1.665	1.991	2.376	2.641	77
80	1.292	1.664	1.990	2.374	2.639	80
85	1.292	1.663	1.988	2.371	2.635	85
95	1.291	1.661	1.985	2.366	2.629	95
100	1.290	1.660	1.984	2.364	2.626	100
150	1.287	1.655	1.976	2.351	2.609	150
200	1.286	1.653	1.972	2.345	2.601	200
250	1.285	1.651	1.969	2.341	2.596	250

Continued.

■ **Table of the *t*-Distribution—cont'd**

df	t.10	t.05	t.025	t.01	t.005	df
500	1.283	1.648	1.965	2.334	2.586	500
600	1.283	1.647	1.964	2.333	2.584	600
700	1.283	1.647	1.963	2.332	2.583	700
800	1.283	1.647	1.963	2.331	2.582	800
900	1.282	1.647	1.963	2.330	2.581	900
1000	1.282	1.646	1.962	2.330	2.581	1000
2000	1.282	1.646	1.961	2.328	2.578	2000

CHAPTER 9

Tests of Hypotheses

9.0   Introduction

Suppose someone makes a statement of belief: "The mean height of men and women in the USA are equal." There are many who would agree with this statement. But there are others who would not agree with it. Among those who do not agree with it would be those who would say, for instance: "The mean height of men and women in the USA are not equal." Others among those who disagree would say: "The mean height of men is greater than the mean height of women in the USA." The first thing to note is that each of these statements is referring to a population; in this case, the USA population. Thus, to further elucidate, **a hypothesis** is a postulate or claim we make, or a belief we hold, or an idea we have about a population or a set of populations.

Each of the statements made is a postulate, an idea, a claim, or a hypothesis about a population. The fact that there is stark disagreement between the statements about an aspect of the population, namely the mean heights for males and females, lead to a research situation. This research situation created by competing hypotheses is the basis for hypotheses testing which is the subject of this section.

9.1   The Null and Alternative Hypotheses

The first statement described in the introductory section, which asserts equality of the means, states a claim that, if not challenged by the differing claim, would have remained and would have been accepted. By the fact that it is a statement that can be either challenged or not challenged, it is the status quo belief. Since it triggers disagreement that could lead to research that may find that it is not true, leads statisticians to refer to such a statement as the **null hypothesis**—a hypothesis that is to be nullified. Either of the other statements which challenge the status quo belief, is the research idea that could be accepted in order to nullify the null hypothesis. Statisticians refer to such postulates that can be used to nullify the null hypothesis by the name, the **alternative hypothesis.**

In statistical decision theory we usually have these two hypotheses; that is, two competing ideas about a population or some populations. One of these hypotheses is called the null hypothesis and is usually denoted by H_0. The other, which is called the alternative hypothesis, is denoted by H_1 or H_A. In reality, most of the hypotheses which are tested are usually about an unknown parameter (population value) such as population mean μ, and population variance σ^2. There are other hypotheses tested in the field of nonparametric statistics which will not be treated in this text. Usually the alternative hypothesis is the one that would be true if the null hypothesis turns out to be untrue. Thus, the tests are primarily set up to reject or accept the null hypothesis—in short to nullify the null hypothesis. This is why the current belief, the statement of claim, or the status quo idea is assigned to the null hypothesis. That is also the reason the research, new idea, or the challenging claim being tested is put in the alternative hypothesis. Once the null hypothesis is rejected (nullified), we will accept the alternative.

9.1.1 Definitions

Null hypothesis (denoted by H_0): It states the status quo postulate, belief, or claim about a population.

Alternative hypothesis (denoted by H_1): It states the research idea, the new postulate, or the opposing claim about the population that we wish to investigate. Normally, if the null hypothesis is not true, the alternative hypothesis should be true.

Test statistic: In practice, we usually take a sample from the target population and use it to estimate the **parameter.** The sample value which is used to estimate the parameter is called a **statistic.** In testing a hypothesis, we usually obtain a relevant statistic (a value from the sample related to the parameter about which the hypothesis is being tested) whose distribution is known. This value is then used to test the hypothesis about the parameter. Such a statistic is called a **test statistic.**

Tests of significance: Sometimes, tests of hypotheses are called **tests of significance.** This is because the statistician uses a sample value (based on data) to make a decision as to whether the idea in the null hypothesis is supported. When the data does not support H_o, the null hypothesis is rejected, signaling that the data are **significant;** presumably signaling that the data are significantly different from what would be expected if the null hypothesis was true. Thus, **tests of hypotheses** are the same as **tests of significance.**

9.2 Errors Associated with Hypotheses Tests

In carrying out tests of hypotheses, we usually make some decisions. Decision making is fraught with errors. Thus, in hypothesis testing we can make errors. Making errors is inevitable if we must research and innovate. At some point in history, the only medication available for pain was aspirin. If we were afraid to make errors, no one would have tried to research pain medications and we would not have the variety that exists today, some perhaps better than aspirin. This is because no one would have been able to postulate that there is a better alternative for curing pain than aspirin, due to fear of making an error. There are two possible errors that can be made in hypothesis testing. These errors are named Type I and Type II errors. Statisticians are aware of these errors and also recognize that they have consequences. To minimize their consequences, statisticians fix or control the levels of the risks associated with each error in advance, before testing the hypotheses, thus minimizing their consequences.

Type I Error

This error occurs when we reject H_0 when it is true (or should have been accepted). In testing hypotheses, it is possible that we make the error of rejecting the null hypothesis when it is true. This is identified as making a Type I error. We make provision for making this error by setting α, the level of significance of the test, in advance.

Type II Error

This type of error is made when we accept the null hypothesis when it is not true (should have been rejected).

The following table shows the decisions we could make in hypothesis testing and the errors that may be associated to them.

Decisions	H_0 is true	H_0 is untrue
Accept H_0	No error made	Type II error made
Reject H_0	Type I error made	No error made

9.2.1 Sizes of Types I and II Errors

The size of a Type I error is denoted as α, where α = Probability of rejecting H_0 when H_0 is true. This is a conditional probability which can be written as:

$\alpha = P(\text{rejecting } H_0 \mid H_0 \text{ is true})$. This probability is equal to the level of significance (l.o.s) of the test.

The size of a Type II error is denoted as β, where β = Probability of accepting H_0 when H_0 is untrue or H_1 is true. This is also a conditional probability which can be written as:

$\beta = P(\text{accepting } H_0 \mid H_0 \text{ is untrue}) = P(\text{accepting } H_0 \mid H_1 \text{ is true})$.

Both α and β (the sizes of Types I and II errors) represent the risks of making Types I and II errors respectively. Often we fix one of them, particularly, α, knowing that β would then depend on it.

9.2.2 Level of Significance

Along with setting up the null hypothesis we set up a level of significance usually denoted by α, which could be expressed in percentage. It represents the level of error we make if we reject the null hypothesis when it should have been accepted. The level of significance represents the maximum probability with which we are prepared to risk making a Type I error. Since this error could be serious, we make it as small as possible. For this reason, values of α ranging from a fraction of a percent to 5% are frequently used. Sometimes some use values higher than these, up to 10%. However, above 5% the risk begins to be progressively less acceptable. The value of β, the risk of Type II error follows from the value of α.

It should be noted that α represents the following:

$\alpha = P(\text{rejecting } H_0 \mid H_0 \text{ is true});$

$\alpha = \text{size of type I error};$

$\alpha = \text{size of the rejection region in classical approaches to the tests of hypothesis.}$

9.2.3 Finding the Values of α the Size of Type I Error, β the Size of Type II Error, and the Power of Test

Suppose that we believe of people who have regular systolic blood pressure (SBP), that their SBPs are normally distributed with mean of 120 and variance of 25. If a person with SBP of 132 is diagnosed as having high SBP, what is the size of type I error?

Usually for the above situation where we intend to declare a person's SBP high, we will generally set up a hypothesis which looks like:

$H_0 : \mu_{SBP} = 120$ vs. $H_1 : \mu_{SBP} > 120;$

and progressively calculate the risk of type I error for values of SBP greater than 120.

$\alpha = P(\text{rejecting } H_0 \mid H_0 \text{ is true}).$

In the above case, we have to assume that H_0 is true, and under the distribution of H_0, we determine the probability of deciding that someone with SBP of 132 has high SBP.

$\alpha = P(\text{saying somebody with } SBP = 132 \text{ has high } SBP \text{ when } H_0 \text{ is true}) =$
$$P\left(Z > \frac{132 - 120}{\sqrt{25}}\right) = P(Z > 2.4) = 0.0082.$$

The probability of making Type I error in the above circumstance is 0.0082. We can use this idea to determine what value of SBP we will declare abnormal by setting up the value of the size of Type I error we are happy to allow in advance. Suppose that we are happy to allow a Type I error of size 1.5%, we see from the table of the standard normal distribution that $Z = 2.17$, and $P(Z > 2.17) = 0.015$. This means that the value we want is 2.17 standard deviation units from the mean of 120, that is $SBP = 120 + 2.17\sqrt{25} = 130.9$. Thus under these assumptions people with SBP > 130.9 should have their SBP declared abnormal. We have established that null hypothesis is accepted only when $\overline{X} \leq 130.9$.

In the above hypotheses set up, how do we determine the value of β, the size of Type II error?

Suppose that we made a Type II error under the above conditions, that means that we accepted H_0 when it is untrue. Assuming $\alpha = 0.015$, the area for which H_0 is accepted is $P(\overline{X} \leq 130.9)$. If the actual mean is $\mu = 135$ and $\sigma^2 = 25$, then:

$$\beta = P(\overline{X} \leq 130.9) \text{ when } \mu = 135 \text{ and}$$
$$\sigma = 5 \Rightarrow \beta = P\left[Z < \frac{130.9 - 135}{5}\right] = P(Z < -0.82) = 0.2061.$$

The knowledge of β allows us to calculate the **power of the test.**

9.2.4 Power of a Test

The power of a test is the probability of the test to lead us to reject the null hypothesis when it is false; quantitatively, the power of the test $= 1 - \beta$.

In the above example, power of the test $= 1 - \beta = 1 - 0.2061 = 0.7939$. This is the probability that we would reject the null hypothesis when it is untrue. Later we will explore the power of the test further in an example problem (see Example 9.2).

9.3   Tests of Hypotheses

Based on sampling theory and the underlying distributions of some sample statistics, we can calculate the value of an appropriate test statistic for a typical hypothesis that we want to test. To use a statistic for testing a hypothesis about a parameter, the statistic must be an estimate of the parameter, and we must know the underlying sampling distribution of the statistic. The statistic is used for constructing a test statistic. In an actual test, the value of the test statistic is compared to some tabulated values related to the underlying distribution of the test statistic. Usually, the tabulated values obtained for the test are based on the distribution of the test statistics under the assumption that the H_0 is true and on the level of significance, α. Using this we can decide whether the findings from our sample data differ **significantly** from what is postulated for the population(s). The process by which we arrive at this decision is called a test of hypothesis, or a test of significance. Hypothesis testing is part of a subject area in Statistics called Inferential Statistics, or Statistical Inference.

9.3.1 Tests for a Single Population Mean Based on Normal Distribution

As we stated earlier, in statistical theory, the value of a **parameter** (a constant) may be estimated by a **statistic (a variable from a sample).** Some of these statistics are known to follow the normal distribution. In such a case, if the statistic is X_s, it will be known that its mean under the assumption that it is normally distributed will be μ_s, and its variance will be σ_s^2. If the variance of the normal population σ^2 is known, we use it to obtain σ_s^2. A sample of size n is taken from the population X, which is normally distributed, to test a hypothesis about its mean. In order to test that the mean is μ_0 against an alternative that it is not, the hypotheses are set up as follows:

$$H_0{:}\mu = \mu_0 \text{ vs. } H_1{:}\mu \neq \mu_0.$$

The statistic which is associated with the tests about the mean of a population is the sample mean, $\overline{X}$. It is with this that we obtain the **test statistic, Z_c,** which is used in the test. The above test has a two-sided or two-tailed alternative. The test statistic obtained for this test is compared with appropriate tabulated values of Z. Z has the standard normal distribution, and the value read depends on the chosen level of significance. We know that Z is normally distributed with mean zero and variance 1:

$$Z_c = \frac{\overline{X} - \mu}{\sigma_{\overline{x}}} \text{ or } Z_c = \frac{\overline{X} - \mu}{\sigma/\sqrt{n}} \text{ or } Z_c = \frac{\sqrt{n}(\overline{X} - \mu)}{\sigma} \qquad (9.1)$$

provided that we know the value of σ^2, the variance. If σ^2 is unknown, and we take a sample $X_1, X_2, \ldots, X_n$, then a statistic which has a t-distribution must be used. We discuss the use of such a statistic later.

There are actually three versions of possible hypotheses which can be stated for the above problem and the three sets of hypotheses can be tested using the same test statistic, Z_c. The choice of alternative hypotheses determines the differences in the sets of hypothesis that can be tested. Two one-sided alternatives are available along with the single two-sided alternative stated earlier. The choice is made according to the interest of the researcher. Although the same test statistic would do for the three versions of the test, the decision rules are different for the three versions of the hypotheses:

Case 1: One-sided to the right: $H_o{:}\mu = \mu_o$ vs. $H_1{:}\mu > \mu_{so}$
Case 2: One-sided to the left: $H_o{:}\mu = \mu_o$ vs. $H_1{:}\mu < \mu_o$
Case 3: Two-sided: $H_o{:}\mu = \mu_o$ vs. $H_1{:}\mu \neq \mu_o$

9.3.2 Classical Approach to Tests of Hypotheses

Usually, in the classical approach to hypothesis testing, we choose α the level of significance in advance in order to minimize the risk of Type I error. Choosing α enables us to delineate the area of the distribution of the test statistic under the assumption that H_0 is true, where, if the value of the computed test statistic falls, we should reject the null hypothesis. We usually do this by choosing the **critical value(s)** on the basis of α, and using it, or them, to demarcate between **acceptance** and **rejection** regions. There is only one critical value and one rejection region in one-sided tests. In a two-sided test, there are two critical values, leading to two rejection regions. Each of the rejection regions in a two-sided test is of size $\alpha/2$. In Figure 9.1, we present the rejection and acceptance regions for testing hypothesis in Case 1. The critical value is Z_α so that the rejection region comprise the area $\{Z > Z_\alpha\}$, ensuring that $P(Z > Z_\alpha) = \alpha$. Any value of the test statistic which falls into this region or exceeds Z_α leads to the rejection of the null hypothesis. Similarly, Figure 9.2 shows the regions for testing hypothesis of Case 2. The rejection region comprises the area $\{Z < -Z_\alpha\}$, so that $P(Z < -Z_\alpha) = \alpha$. For this set of hypotheses, any value of the test statistic which falls into this region or is less than $-Z_\alpha$ leads to the rejection of the null hypothesis. In Figure 9.3, we illustrate the rejection and acceptance region for the two-sided test. There are two rejection regions to the right and left, each of size $\alpha/2$. The corresponding critical values are $-Z_{\alpha/2}$ and $+Z_{\alpha/2}$ so that $P(Z < -Z_{\alpha/2}) = \alpha/2$ and $P(Z < +Z_{\alpha/2}) = \alpha/2$.

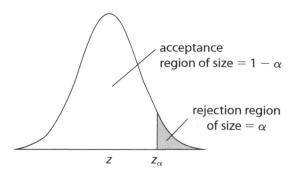

■ **FIGURE 9.1**　**Rejection and Acceptance Regions for One-Sided Test (Case 1)**

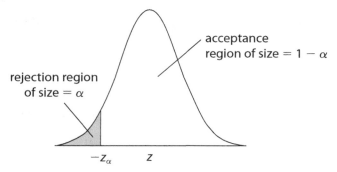

■ **FIGURE 9.2**　**Rejection and Acceptance Regions for the One-Sided Test (Case 2)**

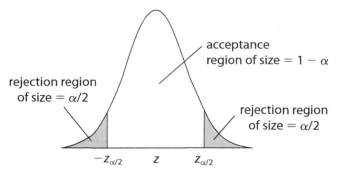

■ **FIGURE 9.3**　**Rejection and Acceptance Regions for the Two-Sided Test (Case 3)**

We use an example to illustrate the test of hypothesis under the normal distribution.

■ **EXAMPLE 9.1**   It is known that the heart rates of a large number of students after a given period on a tread mill are normally distributed with a variance of 88.92. To test the hypothesis that the mean rate is 77 per minute, the following random sample of thirty-five students were chosen and their heart rates after the specified period of working at a given rate on a tread mill were measured. The heart rates obtained were:

$$66 \ 75 \ 92 \ 72 \ 72 \ 92 \ 68 \ 62 \ 87 \ 75 \ 62 \ 76 \ 75 \ 90 \ 79 \ 76 \ 80 \ 60$$
$$83 \ 75 \ 92 \ 65 \ 73 \ 65 \ 74 \ 83 \ 68 \ 65 \ 70 \ 88 \ 80 \ 68 \ 66 \ 62 \ 68$$

Carry out this test against an alternative that it is less than 77 using a level of significance, $\alpha = 5\%$.

Solution   First, we set up the null hypothesis and the alternative:

$$H_0{:}\mu = 77 \text{ vs. } H_1{:}\mu < 77$$

We know from statistical theory, that $\overline{X} \sim N\left(\mu, \dfrac{\sigma^2}{n}\right)$. The required test statistic, Z_c, is distributed as standard normal with mean zero and variance 1. $Z_c = \dfrac{\sqrt{n}(\overline{X} - \mu)}{\sigma} \sim N(0, 1)$ and can be read from the tables for a given level of significance, α. Now,

$$\sum_{i=1}^{35} X_i = 2604 \Rightarrow \overline{X} = \frac{1}{35}\sum_{i=1}^{35} X_i = \frac{2604}{35} = 74.40$$

$$Z_c = \frac{\sqrt{35}(74.40 - 77)}{\sqrt{88.92}} = -1.63.$$

Using $\alpha = 5\%$, we read from the table of $Z \sim N(0, 1)$ that $Z_\alpha = 1.645$ so that $-Z_\alpha = -1.645$.

Since our test corresponds to Case 2 above, we compare Z_c with $-Z_{0.05}$. Clearly, $Z_c = -1.63 > -1.645$ so we accept the null hypothesis and reject the alternative that the mean heart rate is less than 77 per minute.

Alternatively, we can make our decision by drawing the figure of the normal distribution and identifying the rejection and acceptance regions, so that if Z_c falls into the rejection region we reject the null hypothesis; otherwise, we accept it. This is illustrated in Figure 9.4:

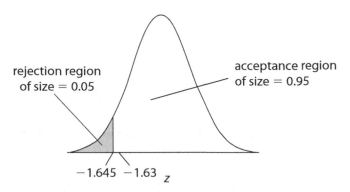

■ **FIGURE 9.4   Rejection and Acceptance Regions When $\alpha = 0.05$**

Clearly, -1.63 falls into the acceptance region, so we accept the null hypothesis that $\mu = 77$. In the above decision the size of Type I error is 0.05, or the probability of us rejecting H_0 when H_0 is true is 0.05. It should be noted that this probability applied only up until we made the error. Once the error is made, it is certain that we have made the error. The probability of making the error at that point is 1.0.

In the next example, we illustrate the calculation of β, the size of Type II error, and we determine the power of the test.

■ EXAMPLE 9.2   Suppose that in Example 9.1, the true mean of the heartbeats is really 72 with a variance of 111.72. What was the probability of Type II error? Obtain the power of the test.

Solution   Since we tested the hypothesis in Example 9.1 at $\alpha = 0.05$, with a critical value of -1.645. This means that the value we want is -1.645 standard deviation units from the mean of 77; that is, mean heart-rate $(\overline{X}) = 77 - 1.645\sqrt{(111.72/35)} = 74.38$. Thus under these assumptions we would accept values of $\overline{X} \geq 74.38$. We accepted H_0 and made the Type II error if indeed the true mean of the heartbeats is 72 with a variance of 111.72. $\beta = P(\text{accepting } H_0 \mid H_0 \text{ is untrue or } H_1 \text{ is true})$. Now:

$$\beta = P\,(\overline{X} > 74.38) \text{ when } \mu = 72 \text{ and}$$
$$\sigma^2 = 111.72 \Rightarrow \beta = P\left[Z > \frac{74.38 - 72}{\sqrt{111.72/35}}\right] = P(Z > 1.33) = 0.0918$$

The power of the test is $1 - \beta = 1 - 0.0918 = 0.9082$.

■ EXAMPLE 9.3   A Biotech company claims that the time it takes for a biomass to degrade is normally distributed with mean of 1500 hrs. and variance of 157.25 hrs². This is disputed by a rival company which believes that it is not equal to 1500 hours. The lifetimes of a sample of twenty of these biomasses were obtained as follows:

1493 1494 1496 1508 1504 1483 1502 1486 1493 1484
1506 1508 1478 1496 1501 1486 1485 1491 1488 1498

Can we support the claim of the producer at a 5% level of significance?

Solution   First, we set up the null hypothesis and the alternative. In this case, the alternative should be two-sided as the dispute states that it is not equal. The disputed idea is put in the alternative hypothesis:

$$H_0{:}\mu = 1500 \text{ vs. } H_1{:}\mu \neq 1500.$$

The test statistic is obtained as:

$$\sum_{i=1}^{20} X_i = 29880 \Rightarrow \overline{X} = \frac{1}{20}\sum_{i=1}^{20} X_i = \frac{29880}{20} = 1494$$
$$Z_c = \frac{\sqrt{20}(1494.0 - 1500)}{\sqrt{157.25/20}} = -2.14.$$

Using $\alpha = 5\%$, we need two rejection regions, each of size $\alpha/2$. We read from the table of $Z \sim N(0, 1)$ that $-Z_{\alpha/2} = -Z_{0.025} = -1.96$; $+Z_{\alpha/2} = Z_{0.025} = 1.96$.

These are the two critical values needed to set up the rejection and acceptance regions. The regions are presented pictorially in Figure 9.5:

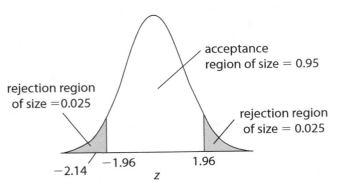

■ **FIGURE 9.5   The Rejection and Acceptance Regions for Example 9.3**

It is clear that the value of the test statistic (-2.14) is such that it falls into the rejection region. So we reject the null hypothesis and accept the alternative. Therefore the counter-claim of the rival company has prevailed at the 5% level of significance.

It should be noted that the counter-claim may not prevail at a different level of significance. For instance at $\alpha = 1\%$, the critical values would have been -2.575 and $+2.575$, and the test statistic would have fallen into the acceptance region, leading to the acceptance of the null hypothesis at the 1% level of significance. The reason for this is that we have an increased probability of making the right decision to accept H_0 when H_0 is true from 95% to 99%. We have to widen the interval where we can be right, commensurate to the increased probability. The issue that arises here points out the fact that a relationship exists between two-sided hypothesis testing and confidence intervals. We will pursue this later.

9.4   The *p*-Value Approach to Testing Hypothesis

In general, *p*-value is the probability of observing the values you observed (the data you collected or some function of it) and every other value that is more extreme, under the assumption that H_0 is true. The smaller the *p*-value, the stronger is the evidence that the data you collected supports the alternative hypothesis. For instance, a test statistic depends on the sample observation; if its probability under the assumption that H_0 is true is relatively small compared to its probability under the assumption that H_1 is true, then the test statistic supports the alternative hypothesis and is an extreme value. To get a better understanding of *p*-value, we need to understand the concept which leads us to consider data to be extreme or as extreme values.

9.4.1   Direction of Extreme and Extreme Values

Values which are less likely when their probabilities are computed under the assumption that the null hypothesis is true, but which are more likely when their probabilities are computed under the alternative hypothesis, generally support the alternative hypothesis and are said to be extreme values. The position on the domain of a variable corresponding to values which are extreme is the direction of extreme. If larger values of the variable tend to lend support to alternative hypothesis, we say that the direction of extreme is to the right. On the other hand, if smaller values tend to lend support to the alternative hypothesis then the direction of extreme is to the left. It is generally true that for the set of hypotheses described in Case 1 the direction of extreme would be to the right, while the direction of extreme will be to the left for hypotheses of Case 2. The direction of extreme for two-sided tests (Case 3) is both to the right and to the left.

9.4.2 Using p-Value to Make Decisions When Testing Hypotheses

The p-value is the probability attained by the test statistic we obtained from our sample, or something more extreme, evaluated under the distribution specified under the null hypothesis, H_0. A value is considered extreme if it is more probable to occur under H_1 than under H_0. We can easily read off the p-values, by reference to the appropriate table of the distribution of the test statistic under H_0. Sometimes the arrangement of the table in a text may not make it useful for obtaining p-values. In the alternate we can always use software to generate p-values. Among the software that can be used are Microsoft EXCEL, MINITAB, SAS, S, R, JMP, etc. To make decisions whether to reject or accept a null hypothesis, the p-value is usually compared to the α, the level of significance.

Decision Rule
If p-value $< \alpha$, we reject H_0 and accept H_1. If p-value $\geq \alpha$, accept H_0.

When the extreme values are to the right, and the value of the test statistic is Z_C, then the p-value is depicted in Figure 9.6. This applies to hypotheses under Case 1. The p-value is calculated as p-value $= P(Z > Z_C)$.

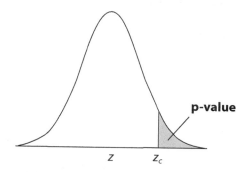

■ **FIGURE 9.6** The p-Value for Test Statistic Z_C When Extreme Values are to the Right

Also, when the extreme values are to the left, and the value of the test statistic is Z_C, negative or not, then the p-value is as depicted in Figure 9.7. This applies to hypotheses under Case 2. The p-value is calculated as p-value $= P(Z < -Z_C)$.

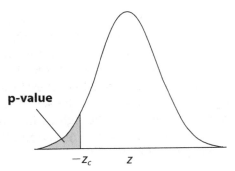

■ **FIGURE 9.7**   *p*-Value for Statistic Z_C When Extreme Values are to the Left

Similarly when the extreme values are to the left and to the right, and the value of the test statistic is Z_C, the value, with the sign changed, is used to demarcate the *p*-value on the other side as shown in Figure 9.8. This applies to hypotheses under Case 3. The *p*-value is calculated as: *p*-value $= P(Z > Z_C) + P(Z < -Z_C)$.

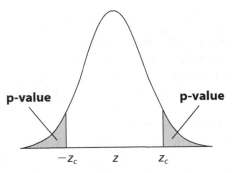

■ **FIGURE 9.8**   *p*-Value for Statistic Z_C With Extreme Direction Both to the Left and to the Right

We illustrate the use of *p*-value by revisiting the examples we treated under the classical approach to hypothesis testing.

■ **EXAMPLE 9.4**   Calculate the *p*-value for the test of Example 9.1, and use it to make the decision for the test of Example 9.1

Solution

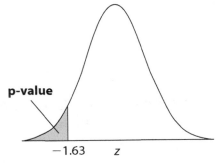

■ **FIGURE 9.9**   *P*-Value for Example 9.1

We see that the test statistic has a value of -1.63, and that the direction of extreme is to the left. We evaluate the *p*-value as: $P - \text{value} = P(Z < -1.63) = 0.0516$.

The level of significance for this test is $\alpha = 0.05$, implying that p-value $> \alpha$; so we accept H_0. The conclusion we reached under the p-value approach is exactly the same as we did by the classical approach.

■ **EXAMPLE 9.5**   Calculate the p-value for the test of Example 9.3, and use it to make the decision for the test of Example 9.3

Solution

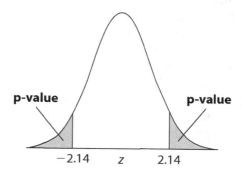

■ **FIGURE 9.10**   *P*-Value for Example 9.3

The test statistic had a value of -2.14, this value is located on the left, so we assign the opposite sign to it, and use it to demarcate for the p-value on the right. Both areas make up the p-value, as shown in Figure 9.10. So now, the p-value is obtained as: p-value $= P(Z < -2.14) + P(Z > 2.14) = 0.0162 + (1 - 0.9838) = 0.0324$. The level of significance for this test is $\alpha = 0.05$. Since p-value < 0.05, we reject the null hypothesis and accept the alternative—the same conclusion we reached by classical approach.

Although we have discussed p-value in the context of the standard normal distribution $Z \sim N(0, 1)$, because the test statistic has a standard normal distribution, it applies in the same way when the test statistic is distributed otherwise. It is difficult however to find a table for other distributions which are configured for us to read the p-value as in the table of $Z \sim N(0, 1)$. Software will be used in those cases. Most software when used to carry out the test will, as a matter of course, supply the p-value along with other test characteristics as part of the output.

The p-value approach reaches the same conclusion as the classical approach, but the p-value approach is more advantageous because we can use the p-value and compare it to any other level of significance we have in mind different to the one stipulated. We can read journal articles and, based on the p-values supplied by the researchers, determine our own level of significance and see whether we agree with their conclusions or not.

9.5 Tests About Single Population Mean Drawn from Normal Populations, When Variance is Unknown—Use of *t*-Distribution

When samples are taken from the normal distribution without knowing either the mean or the variance, then we cannot use a test statistic that is based on the standard normal distribution, $Z \sim N(0, 1)$. Instead of using the normal distribution for such samples, statisticians have found that the t-distribution gives better results when used to represent the distributions of statistics calculated when mean and variance are unknown.

When the data is believed to have come from a normal distribution, and the population variance is unknown, regardless of the sample size the best test statistic to be used for the tests discussed in the previous section has a t-distribution with $n - 1$ degrees of freedom where n is the sample size. Further, if the sample size is small, then the t-statistic is the best to use. It is also possible to conduct approximate tests based on central limit theorem (CLT) and therefore the normal distribution, provided that large samples ($n \geq 30$) are taken from the populations. We present the analysis based on CLT shortly.

A sample of size n may be taken from a population which is believed to be normal with unknown standard deviation, the sample standard deviation with some modifications could then be used in place of the population standard deviation in the calculation of the test statistic. In order to test the hypotheses (represented by Cases 1–3):

Case 1: $H_o: \mu = \mu_o$ vs. $H_1: \mu > \mu_o$

or

Case 2: $H_o: \mu = \mu_o$ vs. $H_1: \mu < \mu_o$

or

Case 3: $H_o: \mu = \mu_o$ vs. $H_1: \mu \neq \mu_o$

We use:

$$t_c = \frac{(\overline{X} - \mu_0)}{s/\sqrt{n}} \text{ or } t_c = \frac{\sqrt{n}(\overline{X} - \mu_0)}{s}$$

where $s^2 = \dfrac{\sum\limits_{i=1}^{n} X_i^2 - n\overline{X}^2}{n - 1}$;

t_c is t distributed with $n - 1$ degrees of freedom. This will be used as the test statistic for testing the two one-sided, and the one two-sided tests stated above in Cases 1–3. In the examples we present, we shall only discuss the classical tests. The p-value approaches are conducted in a similar way as described earlier for hypotheses that can be tested under the normality assumption. To obtain the p-values for t-tests we need a calculator, or computer software such as EXCEL, MINITAB, SAS, SPSS, etc.

Decision Rules

Let $\alpha\%$ be the level of significance for the tests. Then here are the decision rules for this test:

Case 1: If $t_c > t(n - 1, \alpha)$ reject null hypothesis, accept the alternative
Case 2: If $t_c < -t(n - 1, \alpha)$ reject null hypothesis, accept the alternative
Case 3: If $t_c > t(n - 1, \alpha/2)$ or $t_c < -t(n - 1, \alpha/2)$ reject null hypothesis, accept the alternative

In Figure 9.11, we present the rejection and acceptance regions for testing hypotheses in Case 1 above. The critical value is $t_{(n-1, \alpha)}$ so that the rejection region comprise the area $\{t_{(n-1)} > t_{(n-1, \alpha)}\}$, ensuring that $P(t_{(n-1)} > t_{(n-1, \alpha)}) = \alpha$. Any value of the test statistic which falls into this region or exceeds $t_{(n-1, \alpha)}$ leads to the rejection of the null hypothesis. Similarly, Figure 9.12 shows the regions for testing hypothesis of Case 2. The rejection region comprises $\{t_{(n-1)} < -t_{(n-1, \alpha)}\}$, so that $P(t_{(n-1)} < -t_{(n-1, \alpha)}) = \alpha$. For this set of hypotheses, any value of the test statistic which falls into this region or is less than $-t_{(n-1, \alpha)}$, leads to the rejection of the null hypothesis. In Figure 9.13, we illustrate the rejection and acceptance regions for the two-sided test. There are two rejection regions to the right and left, each of size $\alpha/2$. The corresponding critical values are $-t_{(n-1, \alpha/2)}$ and $t_{(n-1, \alpha/2)}$ so that $P(t_{(n-1)} > t_{(n-1, \alpha/2)}) = \alpha/2$ and $P(t_{(n-1)} < -t_{(n-1, \alpha/2)}) = \alpha/2$.

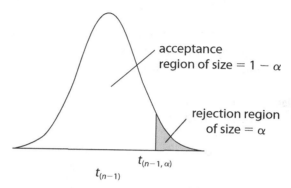

■ FIGURE 9.11 **Rejection and Acceptance Regions for Testing Hypothesis for a Single Population Mean Using *t*-Distribution (Case 1)**

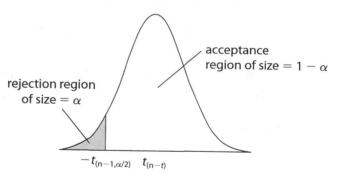

■ FIGURE 9.12 **Rejection and Acceptance Regions for Testing Hypothesis about a Single Population Mean Using *t*-Distribution (Case 2)**

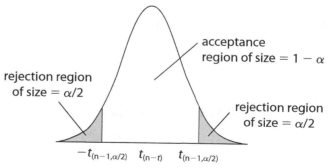

■ FIGURE 9.13 **Rejection and Acceptance Regions for Testing Hypothesis about a Single Population Mean Using the *t*-Distribution (Case 3)**

We use an example to illustrate the tests of hypothesis when the distribution is normal, and sample size small with unknown variance.

EXAMPLE 9.6 (Referring to data of Example 8.7)

The following data is a sample of NFL career passing touchdowns for NFL professionals:

88 67 87 86 86 85 85 85 128 126 125 84 84 44 83 80 79 79 45
33 32 31 94 30 82 79 79 25 33 88 87 86 100 96 158 96 94 31

Using classical and *p*-value approaches, test the hypothesis that the mean career passing touchdowns is 70 against that it is not, at the 2% level of significance. Assume that NFL touchdowns are normally distributed.

Solution Here variance is unknown and touchdowns are normally distributed. We carry out the required test using the *t*-distribution.

$$H_0 : \mu = 70 \text{ vs. } H_1 : \mu \ne 70;$$

$$\bar{x} = \frac{2980}{38} = 78.42; \quad s^2 = \frac{\sum_{i=1}^{38} x_i^2 - n\bar{x}^2}{n-1} = \frac{267630 - 38(78.42)^2}{38-1} = 917.1693;$$

$$s = \sqrt{917.1693} = 30.2848$$

$$t_c = \frac{\sqrt{38}(78.42 - 70)}{30.2848} = 1.71$$

With $\alpha = 0.02$, $\Rightarrow \alpha/2 = 0.01$. From the table of *t*-distribution, we obtain the critical values, based on $t_{(n-1, 0.01)}$, that is $t_{(37, 0.01)} = 2.431$ and -2.431.

Classical Approach

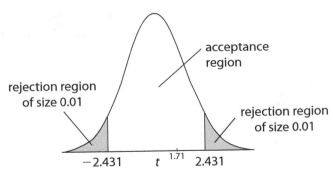

■ **FIGURE 9.14**   **Critical Values, Rejection and Acceptance Regions for Example 9.6**

We accept the null hypothesis and conclude that the mean number of NFL career touchdowns is 70.

p-Value Approach

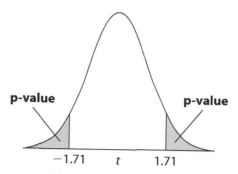

p-value   p-value

-1.71 t 1.71

■ **FIGURE 9.15** *p*-Value for Example 9.6

To obtain the *p*-value, we use EXCEL: In EXCEL, we enter in one of the cells = **TDIST(1.71,37,2)**. In this entry in EXCEL, 1.71 is the value of the test statistic, $37 = n - 1$ is the number of degrees of freedom, while 2 indicates that this is a two-sided test. The output of EXCEL is 0.095641, that is EXCEL computed the probabilities:

$$p\text{-value} = P(t_{(37)} < -1.71) + P(t_{(37)} > 1.71) = 0.04782 + 0.04782 = 0.095641$$

The *p*-value is greater than $\alpha = 0.02$, so we accept the null hypothesis and conclude that mean career NFL touchdowns is 70.

■ **EXAMPLE 9.7** The amount people spend on eating out affects businesses such as restaurants and related establishments. The following is a sample of how much people spend on eating out every week (in dollars):

10 10   8 19   20   18 15 21 25 23 25 38 33 35 40 38 40 45 59
65 58 98 96 102 103 87 87 68 56 43 65 80 84 33 26 28 30

It is believed that an individual spends on average $50 a week eating out, but a restaurant owner disputes it, thinking that it is less. Test this claim at 5% level of significance. Assume that people's weekly expenses on eating out are normally distributed.

Solution   Again, here variance is unknown and no assumption of normality was made for the data. We could use central limit theorem, and carry out a test assuming normality because of the large sample of size 37. The t-test assumes that the means of the different samples are normally distributed; it does not assume that the population is normally distributed. However, here we present a test based on *t*-distribution.

$$H_0: \mu = 50 \text{ vs. } H_1: \mu < 50;$$

$$\bar{x} = \frac{1731}{37} = 46.784; s^2 = \frac{\sum_{i=1}^{37} x_i^2 - n\bar{x}^2}{n-1} = \frac{111001 - 37(46.784)^2}{37-1} = 833.8408;$$

$$s = \sqrt{833.8408} = 28.8763$$

$$t_c = \frac{\sqrt{37}(46.784 - 50)}{28.8763} = -0.68$$

With $\alpha = 0.05$. From the table of t-distribution, we obtain the critical value, based on $t_{(n-1, 0.05)}$, that is $t_{(36, 0.05)} = -1.688$.

Classical Approach

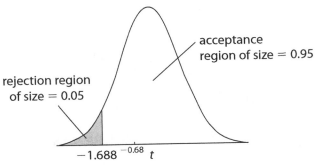

FIGURE 9.16 Rejection and Acceptance Regions, Example 9.7

We accept the null hypothesis and conclude that the mean weekly amount a person spends eating out is $50.00 at a 5% level of significance.

p-Value Approach

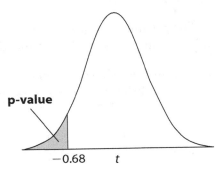

FIGURE 9.17 p-Value for Example 9.7

In a cell in EXCEL, we enter = **TDIST(0.67, 36, 1)**. The output of EXCEL is 0.253567. EXCEL computed:

$$p\text{-value} = P(t_{(36)} < -0.67) = 0.253567$$

The p-value is greater than $\alpha = 0.05$, so we do not reject the null hypothesis and conclude that mean weekly expenditure for eating out is $50.00 at a 5% level of significance.

■ **EXAMPLE 9.8** A manager of a market garden chain was planning for inventory purposes and believed that a particular shop in the chain sold, on average, sixty plants a day. He took a random sample of plant sales for twenty-one days and obtained the following figures:

71 50 76 62 43 22 63 66 93 68 63 10 81 25 61 61 66 32 59 20 67

Assuming that daily sales of plants is normally distributed, test the claim of the manager using a level of significance of 5% against that it is less.

Solution   We state the hypothesis to be tested as follows:

$$H_0\!:\!\mu = 60 \text{ vs. } H_1\!:\!\mu < 60.$$

Now we calculate all the statistics required for the test: $\overline{X}$, s^2 and t_c.

$$\overline{X} = \frac{\sum\limits_{i=1}^{21} X_i}{21} = \frac{1159}{21} = 55.19948$$

$$s^2 = \frac{\sum\limits_{i=1}^{21} X_i^2 - 21 \times \overline{X}^2}{21 - 1} = \frac{73539 - 21(55.19948)^2}{21 - 1} = 478.6619 \Rightarrow s = 21.87834$$

$$t_c = \frac{\sqrt{n}(\overline{X} - \mu_0)}{s} = \frac{\sqrt{21}(55.19948 - 60)}{21.87834} = -1.00739.$$

From the tables (See Appendix III) we see that $t_{(n-1, \alpha)} = t_{(20, 0.05)} = 1.725$. Since the $t_c > -t_{(20, 0.05)}$, then we accept the null hypothesis. The shop probably sells on average, sixty plants per day.

p-Value Approach
In any cell in EXCEL, we enter = **TDIST(1.00739, 20, 1)**. The output from EXCEL is 0.16289, the p-value. The p-value is greater than the level of significance, $\alpha = 0.05$, so we do not reject the null hypothesis; we conclude that the mean number of plants sold per day is sixty.

■ **EXAMPLE 9.9**   The following is a sample of how much people paid for a four oz. scoop of ice cream (in cents):

120 199 156 175 145 180 175 145 159 159 135 139 149

It is believed that an individual spends, on average, 150 cents per four oz. scoop of ice cream. Assume payments for four oz. scoops are normally distributed. Test this claim against that it is not equal at both 5% and 1% levels of significance.

Solution $H_0\!:\!\mu = 150 \text{ vs. } H_1\!:\!\mu \neq 150;$

$$\overline{x} = \frac{2036}{13} = 156.6154; s^2 = \frac{\sum\limits_{i=1}^{13} x_i^2 - n\overline{x}^2}{n - 1} = \frac{324346 - 13(156.6154)^2}{13 - 1} = 456.4231;$$

$$s = \sqrt{456.4231} = 21.36406$$

$$t_c = \frac{\sqrt{13}(156.6154 - 150)}{21.36406} = 1.1165$$

From the tables, we read $t_{(12, 0.025)} = 2.179$, $t_{(12, 0.005)} = 3.055$. In both cases, we accept the null hypothesis, and conclude the average cost of a four oz. scoop of ice cream is 150 cents.

p-Value Approach
In a cell in EXCEL, we enter = **TDIST(1.1165,12,2)**. The output from EXCEL is 0.286066, which is the p-value. The p-value is greater than the level of significance, $\alpha = 0.05$, so we do not reject the null hypothesis; we conclude that the mean amount spent per scoop is 150 cents.

■ **EXAMPLE 9.10**   It is claimed that the weights of fuel made from agricultural waste produced by JTC Company is 1900 lbs. A manager suspects that one of the shifts of workers is not producing according to specification and took a sample of twenty fuels produced by the shift, and measured their weights. The results were as follows:

> 1895 1924 1891 1889 1915 1910 1904 1894 1889 1906
> 1903 1913 1922 1886 1893 1993 1914 1897 1895 1912.

Assume the weights are normally distributed. Test the hypothesis that the shift is working within specifications using a level of significance of (i) 1% and (ii) 5%.

Solution   In this case, we can carry out a two-sided t-test because the data are normally distributed and the variance is unknown. We set up the hypothesis to be tested as follows:

$$H_0: \mu = 1900 \text{ vs. } H_1: \mu \neq 1900$$

Now we calculate all the statistics required for the test:

$\overline{X}, s^2$ and t_c.

$$\overline{X} = \frac{\sum_{i=1}^{20} X_i}{20} = \frac{36250}{20} = 1907.25$$

$$s^2 = \frac{\sum_{i=1}^{20} X_i^2 - 20 \times \overline{X}^2}{20 - 1} = \frac{72762267 - 20(1907.25)^2}{20 - 1} = 537.6711 \Rightarrow s = 23.1877$$

$$t_c = \frac{\sqrt{n}(\overline{X} - \mu_0)}{s} = \frac{\sqrt{20}(55.19948) - 1907.25}{23.1877} = 1.3983$$

From the tables, we read $t_{(19, 0.025)} = 2.093$, $t_{(19, 0.005)} = 2.861$. In both cases, we accept the null hypothesis, and conclude that the mean is 1900 lbs., and the shift is working according to specification.

p-Value Approach
In a cell in EXCEL, we enter = **TDIST(1.13983, 19, 2)**. The output from EXCEL is 0.178133, which is the p-value. The p-value is greater than the level of significance, $\alpha = 0.05$, so we do not reject the null hypothesis; we conclude that the mean weight is 1900.

9.5.1 Large Sample Approach Based on Central Limit Theorem

As stated earlier, if sample size is large, for the following test statistic:

$$Z_c = \frac{\sqrt{n}(\overline{X} - \mu)^2}{s} \text{ where } s^2 = \frac{\sum\limits_{i=1}^{n} X_i - n(\overline{X})^2}{n - 1};$$

we can assume that it has standard normal distribution and use it to test the hypotheses regarding the population mean, μ.

We will carry out the tests of hypotheses which were called for in Examples 9.6 and 9.7 using this method:

■ **EXAMPLE 9.11**   **Doing Example 9.6 Using Central Limit Theorem**

The following data is a sample of NFL career passing touchdowns for NFL professionals:

88 67 87 86 86 85 85 85 128 126 125 84 84 44 83 80 79 79 45
33 32 31 94 30 82 79 79 25 33 88 87 86 100 96 158 96 94 31

Using classical and p-value approaches, test the hypothesis that the mean career passing touchdowns is 70 against that it is not, at the 2% level of significance. Assume that NFL touchdowns are normally distributed.

Solution   Here variance is unknown and touchdowns are normally distributed. We carry out the required test using the t-distribution.

$$H_0:\mu = 70 \text{ vs. } H_1:\mu \neq 70;$$

$$\overline{x} = \frac{2980}{38} = 78.42; s^2 = \frac{\sum\limits_{i=1}^{38} x_i^2 - n\overline{x}^2}{n - 1} = \frac{267630 - 38(78.42)^2}{38 - 1} = 917.1693;$$

$$s = \sqrt{917.1693} = 30.2848$$

$$Z_c = \frac{\sqrt{38}(78.42 - 70)}{30.2848} = 1.71$$

With $\alpha = 0.02$, $\Rightarrow \alpha/2 = 0.01$. From the table of standard normal distribution (Appendix II), we obtain the critical values, based as $Z_{0.01} = 2.33$ so that is $t_{(37, 0.01)} = 2.33$ and -2.33.

Classical Approach

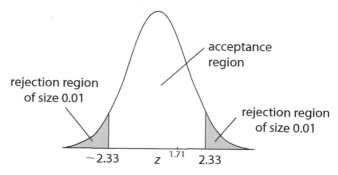

■ **FIGURE 9.18**   **Critical Values, Rejection and Acceptance Regions for Example 9.11**

We accept the null hypothesis and conclude that the mean number of NFL career touchdowns is 70.

p-Value Approach

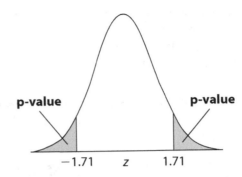

p-value   p-value

−1.71   z   1.71

■ **FIGURE 9.19**   *p*-Value for Example 9.11

To obtain the *p*-value, we read the table of the normal distribution in Appendix II.

$$p\text{-value} = P(Z < -1.71) + P(Z > 1.71) = 0.0436 + (1 - 0.9564) = 0.0972$$

The *p*-value is greater than $\alpha = 0.02$, so we accept the null hypothesis and conclude that mean career NFL touchdowns is 70.

■ **EXAMPLE 9.12**   **(Using Central Limit Theorem for Example 9.7)**
The amount people spend on eating out affects businesses such as restaurants and related establishments. The following is a sample of how much people spend on eating out every week (in dollars):

10 10   8 19   20   18 15 21 25 23 25 38 33 35 40 38 40 45 59
65 58 98 96 102 103 87 87 68 56 43 65 80 84 33 26 28 30

It is believed that an individual spends on average $50 a week eating out, but a restaurant owner disputes it, thinking that it is less. Test this claim at 5% level of significance. Assume that people's weekly expenses for eating out are normally distributed.

Solution   Again, here variance is unknown and no assumption of normality was made for the data. We use central limit theorem, and carry out a test assuming normality because of the large sample of size thirty-seven.

$$H_0{:}\mu = 50 \text{ vs. } H_1{:}\mu < 50;$$

$$\bar{x} = \frac{1731}{37} = 46.784; \quad s^2 = \frac{\sum_{i=1}^{37} x_i^2 - n\bar{x}^2}{n-1} = \frac{111001 - 37(46.784)^2}{37-1} = 833.8408;$$

$$s = \sqrt{833.8408} = 28.8763$$

$$Z_c = \frac{\sqrt{37}(46.784 - 50)}{28.8763} = -0.68$$

With $\alpha = 0.05$. From the table of normal distribution (Appendix II), we obtain the critical value, based on $Z_{0.05}$, that is $-Z_{0.05} = -1.645$.

Classical Approach

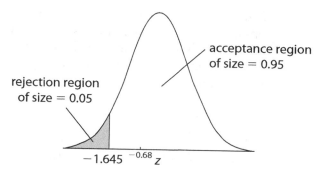

■ **FIGURE 9.20   Rejection Region for Example 9.12**

We accept the null hypothesis and conclude that the mean weekly amount a person spends eating out is $50.00 at a 5% level of significance.

***p*-Value Approach**

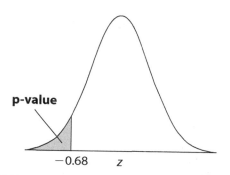

■ **FIGURE 9.21   *p*-Value for Example 9.12**

In the normal distribution table (Appendix II), we read the value for $P(Z \le -Z_c)$. The *p*-value $= P(Z_c < -0.68) = 0.2483$

The *p*-value is greater than $\alpha = 0.05$, so we do not reject the null hypothesis and conclude that mean weekly expenditure for eating out is $50.00 at a 5% level of significance.

9.5.2 Hypotheses Tests About a Population Proportion

In chapter seven, we used $\hat{p}$ as a point estimate of the proportion of a population p which has an attribute of interest to us where:

$$\hat{p} = \frac{x}{n}$$

is the sample proportion; and x is the number out of n which have the attribute of interest. It can also represent the number of times out of n trials we observed "success" as opposed to "failure" which was observed $n - x$ times. Here, x represents the number of "successes," where "success" is defined as the aspects of the outcomes of the trial we defined to be of interest to us. We also found the sampling distribution of $\hat{p}$ as $\hat{p} \sim N\left[p, \dfrac{p(1-p)}{n}\right]$ provided that $np(1-p) \geq 10$. On the basis of this sampling distribution, we found the $100(1-\alpha)\%$ confidence interval for the population proportion, p. In this section, we shall test hypothesis about p, and the $\hat{p}$ and its sampling distribution will play major roles in the test.

The hypotheses we test about a single population proportion take the following three possible forms:

Case 1: $H_0{:}p = p_0$ vs. $H_1{:}p > p_0$

Case 2: $H_0{:}p = p_0$ vs. $H_1{:}p < p_0$

Case 3: $H_0{:}p = p_0$ vs. $H_1{:}p \neq p_0$,

where p is the population proportion. If we want to carry out a test of any of the above hypotheses, we take a sample of size n from the target population and obtain the sample proportion as:

$$\hat{p} = \frac{x}{n} = \frac{\text{number with desired attribute}}{\text{sample size}}.$$

Thereafter, we can obtain the test statistic as follows:

$$Z_c = \frac{\hat{p} - p_0}{\sqrt{\dfrac{p_0(1 - p_0)}{n}}}$$

$Z_c \sim N(0, 1)$, so we can read the values of Z_c from the tables of the standard normal distribution table (Appendix II).

The same figures which depicted the decision regions for classical tests of hypotheses for single population mean, Figures 9.1–9.3 can be used to represent the rejection/acceptance regions corresponding to Cases 1–3 of the classical tests of hypothesis for the single population proportion. We do not intend to reproduce them here. We can, however, state the decision rules in terms of the critical values.

Decision Rules

For Case 1: If $Z_{cal} > Z_\alpha$, we reject H_0 and accept H_1

For Case 2: If $Z_{cal} < -Z_\alpha$, we reject H_0 and accept H_1

For Case 3: If $Z_{cal} < -Z_{\alpha/2}$, or $Z_{cal} > Z_{\alpha/2}$, we reject H_0 and accept H_1

■ **EXAMPLE 9.13** A survey of freshmen on Hamilton College Campus showed that out of 1024, there were 377 smokers. Test the hypotheses at the (a) 10%, and (b) 2% levels of significance that the proportion of freshmen smokers is not equal to 34%.

Solution We are testing the following hypothesis:

$$H_0{:}p = 0.34 \text{ vs. } H_1{:}p \neq 0.34$$

From the sample, we see that:

$$\hat{p} = \frac{377}{1024} = 0.36816406$$

$$\sqrt{\frac{p_0(1 - p_0)}{n}} = \sqrt{\frac{0.34(1 - 0.34)}{1024}} = 0.014803399.$$

Thus,

$$Z_{cal} = \frac{0.36816406 - 0.34}{0.014803399} = 1.902.$$

(a) Testing at 10% level of significance, we see that the critical values are $Z_{0.10/2} = Z_{0.05} = +1.645$ and $-Z_{0.05} = 1.645$. Since $Z_c > Z_{0.05}$, we reject the null hypothesis and conclude that the population proportion is not equal to 0.34.

(b) Testing at 2% level of significance, we see that the critical values are $Z_{0.02/2} = Z_{0.01} = 2.33$. In this case, since $Z_c < \alpha/2$, we cannot reject the null hypothesis and conclude that the population proportion is probably 0.34 at 2% level of significance.

Use of p-Value

Just as was previously described we can use the p-value approach to carry out the tests of hypothesis in Example 9.13(a) and 9.13(b). The test statistic is $Z_c = 1.902$. From the standard normal distribution, we see that for two-tailed test, the p-value is depicted in Figure 9.22.

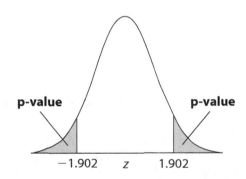

■ **FIGURE 9.22**   *P*-Value for Example 9.13

$$p\text{-value} = P(Z < -1.902) + P(Z > 1.902) = 0.0274 + 0.0274 = 0.0548.$$

For Example 9.13(a) we compare the p-value with 10%; since p-value < 0.10, so we reject the null hypothesis. For example 9.13(b), p-value > 0.02, so we accept H_0, the null hypothesis.

■ **EXAMPLE 9.14**   In a sample of 450 students from SCSU, 197 were found to have been working for more than ten hours a week during an academic session. Carry out a test of a hypothesis that the proportion of students who worked more than ten hours a week during the session is greater than 36%. Use a level of significance of 3%.

Solution   We are testing the following hypothesis:

$$H_0{:}p = 0.36 \text{ vs. } H_1{:}p > 0.36.$$

From the sample, we see that:

$$\hat{p} = \frac{197}{450} = 0.4377777$$

$$\sqrt{\frac{p_0(1 - p_0)}{n}} = \sqrt{\frac{0.36(1 - 0.36)}{450}} = 0.02262741.$$

Thus,

$$Z_{cal} = \frac{0.4377777 - 0.36}{0.02262741} = 3.437.$$

Classical Approach

Testing at 3% level of significance, we see that the critical values are $Z_{0.03} = +1.881$. Since $Z_c > Z_{0.03}$, we reject the null hypothesis and conclude that the proportion of students who worked more than 10 hours a week was greater than 36%.

Using p-Value

This is a one-sided test to the right. Extreme values are to the right.

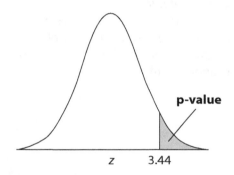

■ **FIGURE 9.23** *p*-Value for Example 9.14

From the tables $P(Z > 3.44) = 0.0003$. Thus, *p*-value $< 3\%$, the level of significance, we reject the null hypothesis and conclude that the population proportion who worked more than ten hours a week is probably greater than 0.36 at 3% level of significance.

■ **EXAMPLE 9.15** In a similar survey the number of female freshmen interviewed at SCSU was 420, out of which 133 were found to be smokers. Test the hypothesis that the proportion of female freshmen smokers is less than 36%. Test at a 4% level of significance.

$$\hat{p} = \frac{133}{420} = 0.316667$$

$$\sqrt{\frac{p_0(1 - p_0)}{n}} = \sqrt{\frac{0.36(1 - 0.36)}{450}} = 0.02262741.$$

Thus,

$$Z_{cal} = \frac{0.316667 - 0.36}{0.02262741} = -1.85014.$$

Classical Approach

Testing at 4% level of significance, from the table we see that the critical value are $-Z_{0.04} = -1.75$. Since $Z_c < -Z_{0.04}$, we reject the null hypothesis and conclude that the proportion of female freshmen smokers is probably less than 0.36 at the 4% level of significance.

Using *p*-Value

This is a one-sided test to the left, extreme values are to the left.

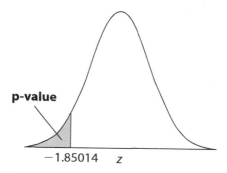

■ FIGURE 9.24 *p*-Value for Example 9.15

$$p\text{-value} = P(Z < -1.85) = 0.0322.$$

Since *p*-value is less than the level of significance (4%), we reject the null hypothesis. We conclude that the proportion of female freshmen smokers is probably less than 36% at a 4% level of significance.

9.6 Use of MINITAB for Hypotheses Tests Involving a Single Mean or Proportion

In this section, we use the MINITAB software to perform the same tests that we performed manually in previous sections treated in this chapter. We discuss both data entry and the required MINITAB commands that would lead to the tests of the hypotheses.

9.6.1 Use of MINITAB to Perform Tests for a Single Population Mean When Variance is Known

Recall Example 9.3. We now analyze the data with MINITAB.

■ EXAMPLE 9.16   We do a third example to illustrate Case 2:

A Biotech company claims that the time it takes for a biomass to degrade is normally distributed with mean of 1500 hrs. and variance of 157.25 hrs². This is disputed by a rival company which believes that it is not equal to 1500 hours. The lifetimes of a sample of twenty of these biomasses were obtained as follows:

1493 1494 1496 1508 1504 1483 1502 1486 1493 1484
1506 1508 1478 1496 1501 1486 1485 1491 1488 1498

Can we support the claim of the producer at 5% level of significance?

Solution

```
MTB > name c3 'biomass'
MTB > set c3;
DATA> 1493 1494 1496 1508 1504 1483 1502 1486 1493 1484 1506
      1508 1478 1496 1501 1486 1485 1491 1488 1498
DATA> end
MTB > OneZ c3;
SUBC> alternative -1;
SUBC> confidence 95;
SUBC> sigma 12.53994;
SUBC> test 1500.
```

One-Sample Z: Biomass

```
Test of mu = 1500 vs < 1500
The assumed standard deviation = 12.5399
```

Variable	N	Mean	StDev	SE Mean	95% Upper Bound	Z	P
biomass	20	1494.00	8.90	2.80	1498.61	-2.14	0.016

The output p-value < 0.05, so we reject H_0.
Next we do another example:

■ **EXAMPLE 9.17** As part of an effort of residents of a city who live near the airport to persuade the city to take some action to contain levels of noise from aircraft that were taking off they collected a sample of measurements for noise made by twenty aircraft and obtained the following values in decibels:

79 80 82 79 74 91 87 74 88 86 84 84 81 77 77 89 76 85 86 78

The city engineer measurements show that the noise has a variance of 43, and contends that the mean is 78 decibels. Protesters believe it is not exactly 78. Test the hypothesis that the mean is 78 against that it is not, at 3% level of significance.

Solution

```
MTB > name c2 'NOISE'
MTB > set c2
DATA> 79 80 82 79 74 91 87 74 88 86 84 84 81 77 77 89 76 85 86 78
DATA> end
MTB > ONEZ C2;
SUBC> SIGMA 6.5574;
SUBC> CONFIDENCE 97;
SUBC> TEST 78.
```

One-Sample Z: Noise

```
Test of mu = 78 vs not = 78
The assumed standard deviation = 6.5574
```

Variable	N	Mean	StDev	SE Mean	97% CI	Z	P
NOISE	20	81.85	5.09	1.47	(78.67, 85.03)	2.63	0.009

The test is equivalent to Case 3, a two-sided test. The output states some descriptive statistics of the variable noise, and displays the confidence interval at 97% confidence level. The confidence interval does not contain 78, which means the test is significant. The postulated mean is 78, the value of the test statistic obtained is ($Z = 2.63$) while the computed p-value as ($p = 0.009$). On the basis of the p-value which is less than $\alpha = 0.03$, we reject the null hypothesis, $H_0{:}\mu = 78$ and accept the alternative $H_1{:}\mu \neq 78$. So we conclude that the mean is not equal to 78 decibels.

EXAMPLE 9.18 Y follows the normal distribution, whose variance is 66. The following is a sample taken from Y: 70 76 86 96 88 100 74 75 79 77 81 85 86 79 87

Using MINITAB and the data, test the hypothesis that the mean of Y is 78 against that it is more at a 2% level of significance.

Solution In MINITAB, pull down the MENU, Editor and select Enable commands. Thereafter appears the prompt:

MTB>

Enter the name of the variable whose values will be inserted in c1 as follows:

MTB> name c1 'Y' (press enter)
MTB>

Enter the data in column one of MINITAB as follows:

MTB> set c1 (press enter)
DATA> 70 76 86 96 88 100 74 75 79 77 81 85 86 79 87 (press enter)
DATA> end (press enter)

The data has now been entered into MINITAB. Next we perform the test required by entering these commands:

```
MTB > SET C1
DATA> 70 76 86 96 88 100 74 75 79 77 81 85 86 79 87
DATA> END
MTB > oneZ c1;
SUBC> alternative +1;
SUBC> confidence 99;
SUBC> test 78;
SUBC> sigma 8.12404.
```

One-Sample Z: C1
```
Test of mu = 78 vs > 78
The assumed standard deviation = 8.12404
```

Variable	N	Mean	StDev	SE Mean	99% Upper Bound	Z	P
Y	15	82.60	8.25	2.10	77.72	2.19	0.014

In the above MINITAB analysis, the command of oneZ c1 was used with the subcommand of alternative +1 to show that we are carrying out test of the Case 1 type, one-sided to the right. The hypothesized mean is included in the subcommand as test 78 while the standard deviation is entered as 8.12404. The subcommand confidence 99 shows that we want the 99% confidence interval for the mean to be computed. Since it is a one-sided test to the right, only the lower bound is output. The output of the test is presented following the command inputs.

The output states some descriptive statistics of the variable Y: the lower bound for its mean, 77.019, which is less than 78 the postulated mean; the value of the test statistic is obtained ($T = 2.19$); and MINITAB computes the p-value as ($p = 0.014$). On the basis of the p-value which is less than $\alpha = 0.02$, we reject the null hypothesis $H_0{:}\mu = 78$, and accept the alternative $H_1{:}\mu > 78$.

9.6.2 Use of MINITAB to Perform Tests for a Single Population Mean When Variance is Unknown

In this case, we do not know the variance, so MINITAB will compute the variance and use it to compute the value of the test statistic, and output the attendant p-value with which we will make decisions to reject or not reject the null hypothesis.

Recall that the sets of hypotheses were described as follows:

Case 1: $H_0{:}\mu = \mu_0$ vs. $H_0{:}\mu > \mu_0$
Case 2: $H_0{:}\mu = \mu_0$ vs. $H_0{:}\mu < \mu_0$
Case 3: $H_0{:}\mu = \mu_0$ vs. $H_0{:}\mu \neq \mu_0$

In MINITAB data are entered in columns. All the data for a particular data set to be used for a test of a hypothesis must be entered into one column. When the variance of the variable is unknown, we can test the hypotheses represented by Cases 1–3 above using the t-test.

Let us use some examples to illustrate the test of hypotheses using this software.

■ **EXAMPLE 9.19**   Recall Example 9.8; we now carry out the tests required using MINITAB.
A manager of a market garden chain was planning for inventory purposes and believed that a particular shop in the chain sold, on average, sixty plants a day. He took a random sample of plant sales for twenty-one days and obtained the following figures:

71 50 76 62 43 22 63 66 93 68 63 10 81 25 61 61 66 32 59 20 67

Assuming that daily sale of plants is normally distributed, test the claim of the manager using a level of significance of 5% against that it is less.

Solution
```
MTB > Onet 'plants';
SUBC> Test 60;
SUBC> Alternative -1.
```

One-Sample T: Plants
```
Test of mu = 60 vs < 60
```

Variable	N	Mean	StDev	SE Mean	95% Upper Bound	T	P
plants	21	55.19	21.88	4.77	63.42	-1.01	0.163

The hypothesis to be tested fits Case 2, a one-sided test to the left. The test statistic is -1.01. The upper bound is higher than sixty (60 is included within its bound) indicating that the test is not significant. The *p*-value of 0.163, which is larger than 0.05, shows that the test is not significant at the 5% level of significance.

■ **EXAMPLE 9.20** The scores in a mathematics placement test at Saint Cloud State University used to determine the level of mathematics courses a student could comfortably cope with is believed to be normally distributed. A sample of twenty-two scores shows the following:

23 25 28 24 27 26 24 26 25 24 17 18 27 26 25 25 26 27 20 17 26 27

Test the hypothesis that the mean is 23 against that it is not at a 5% level of significance.

Solution
```
MTB > name c1 'placement'
MTB > onet c1;
SUBC> confidence 95;
SUBC> test 23.
```

One-Sample T: Placement
```
Test of mu = 23 vs not = 23

Variable   N   Mean   StDev  SE Mean      95% CI        T     P
placement  22  24.227  3.280   0.699  (22.773, 25.681)  1.76  0.094
```

The hypothesis to be tested fits Case 3, a two-sided test. The test statistic is 1.76. The confidence interval includes 23, at 95% confidence interval, indicating that the test is not significant. The *p*-value of 0.094, which is larger than 0.05, shows that the test is not significant at the 5% level of significance, so we failed to reject H_0.

■ **EXAMPLE 9.21** The following are the ages of a sample of statistics textbooks that are still in print, albeit in an Edition that may be different from the original Edition:

6 20 36 28 8 19 8 23 30 25 10 19 14 23 12

Assume that the ages are normally distributed. Test the hypothesis that the mean age of a statistics text book in print is fifteen years against that it is greater at a 5% level of **significance.**

Solution
```
MTB > set c1
DATA > 6 20 36 28 8 19 8 23 30 25 10 19 14 23 12
DATA > end
MTB > desc c1
MTB > name c1 'text-age'
MTB > onet c1;
SUBC> confidence 97;
SUBC> alternative 1;
SUBC> test 15.
```

One-Sample T: Text-Age
```
Test of μ = 15 vs > 15

                                    97% Lower
Variable   N   Mean   StDev  SE Mean   Bound    T     P
text-age   15  18.73   8.96    2.31    14.00   1.61  0.065
```

The hypothesis to be tested fits Case 2, a one-sided test to the right. The test statistic is 1.61. The 97% lower bound is 14 below the postulated mean of 15, indicating that the test is not significant. The p-value of 0.065 which is larger than 0.03 shows that the test is not significant at a 3% level of significance, so we failed to reject H_0.

■ **EXAMPLE 9.22** The following data is a sample of NFL career passing touchdowns for NFL professionals:

88 67 87 86 86 85 85 85 128 126 125 84 84 44 83 80 79 79 45
33 32 31 94 30 82 79 79 25 33 88 87 86 100 96 158 96 94 31

Using MINITAB, test the hypothesis that the mean career passing touchdowns is 70 against that it is not, at a 2% level of significance.

Solution Data were entered into c1 which was named "touchdown."

```
MTB > name c2 'touchdown'
MTB > onet c2;
SUBC> confidence 98;
SUBC> test 70.
```

One-Sample T: Touchdown

```
Test of μ = 70 vs ≠ 70
```

Variable	N	Mean	StDev	SE Mean	98% CI	T	P
touchdown	38	78.68	30.28	4.93	(66.70, 90.67)	1.76	0.086

```
MTB >
```

This is a two-sided test. The descriptive statistics show that the standard deviation (STDEV) is 30.28. Using that in the Onet command, the test statistic has a value of 1.71; with the p-value = 0.095 and a level of significance of 2%, we failed to reject H_0. This is the same result obtained by the classical analysis carried out previously.

We repeat the test for Example 9.7

■ **EXAMPLE 9.23** The amount people spend on eating out affects businesses such as restaurants and related establishments. The following is a sample of how much people spend on eating out every week (in dollars):

10 10 8 19 20 18 15 21 25 23 25 38 33 35 40 38 40 45 59
65 58 98 96 102 103 87 87 68 56 43 65 80 84 33 26 28 30

It is believed that an individual spends on average $50 a week eating out, but a restaurant owner disputes it, thinking that it is less. Using MINITAB, test this claim at a 5% level of significance.

Solution
```
MTB > name c2 'EatOutCost'
MTB > onet c2;
SUBC> alternative -1;
SUBC> test 50.
```

One-Sample T: EatOutCost
```
Test of mu = 50 vs < 50


                                           95% Upper
    Variable    N    Mean   StDev  SE Mean    Bound       T     P
    EatOutCost  37  46.78   28.88     4.75    54.80   -0.68  0.251
```

This is a one-sided test to the left. The descriptive statistics show that the standard deviation (STDEV) is 28.88. Using that in the Onet command, the test statistic has a value of -0.68; with the p-value $= 0.251$ and level of significance of 5%, we failed to reject H_0. This is again the same result obtained by the classical analysis carried out previously.

9.6.3 Use of MINITAB to Perform Tests for a Single Population Proportion

This test is carried out differently in MINITAB. One can use the *Onep* command to do it, in which case only confidence limits would be displayed along with the p-value, which enables a decision to be reached about the population proportion. The sets of hypothesis that could be tested are:

Case 1: $H_0{:}p = p_0$ vs. $H_1{:}p > p_0$
Case 2: $H_0{:}p = p_0$ vs. $H_1{:}p < p_0$
Case 3: $H_0{:}p = p_0$ vs. $H_1{:}p \neq p_0$

We illustrate with examples.

■ **EXAMPLE 9.24** An apple grower working with the Agricultural Extension service of a University is studying the proportion of initial apples on the trees that are cast off early and fall without maturing for harvest. Out of 5296 apples in a cluster sample of ten apple trees, she found that that of the apples that were initially on the trees, only 3975 were left. It is believed that for that species of apple, 25% of the apples are lost by being cast off early. Does the finding of the grower agree with the norm? Use MINITAB to carry out the test at 3% level of significance.

Solution   **Method A:**

```
MTB > POne 5296 1321;
SUBC> Test 0.25;
SUBC> Confidence 97.0.
```

Test and CI for One Proportion
```
Test of p = 0.25 vs p not = 0.25


                                                    Exact
    Sample    X     N   Sample p        97% CI      P-Value
         1  1321  5296  0.249434  (0.236605, 0.262593)  0.924
```

Method B:

```
MTB > POne 5296 1321;
SUBC> Test 0.25;
SUBC> Confidence 97.0;
SUBC> UseZ.
```

Test and CI for One Proportion

```
Test of p = 0.25 vs p not = 0.25

Sample    X      N   Sample p        97% CI        Z-Value  P-Value
   1    1321   5296  0.249434  (0.236531, 0.262336)  -0.10    0.924
```

Using the normal approximation.

Both methods capture the essence of the test, and reach the same conclusion. Method B however is the method we used previously in manual analysis and is aptly described as the normal approximation, as it has something to do with the normal approximation to the binomial distribution. Here, a two-sided test was performed. With a low Z test statistic and an attendant very high p-value, we fail to reject H_0 for this test. The finding of the grower agrees with the norm. Note that 1321 represents the number of apples that were cast by the trees, which had to be calculated and used in the program.

■ **EXAMPLE 9.25**   The National Survey of Child and Adolescent Well-Being reports that the percentage of children fourteen years and younger in the welfare system who have serious health care needs has reached 80%. Suppose that a survey of 10,000 such children show that 7915 had such issues. Test the claim against that it is less at a 1% level of significance.

Solution

```
MTB > POne 10000 7915;
SUBC> Test 0.8;
SUBC> Confidence 99.0;
SUBC> Alternative -1;
SUBC> UseZ.
```

Test and CI for One Proportion

```
Test of p = 0.8 vs p < 0.8

Sample    X      N    Sample p  99% Upper Bound  Z-Value  P-Value
   1    7915   10000  0.791500     0.800950       -2.13    0.017
```

Using the normal approximation.

The test is not significant at a 1% level of significance, a stricter restriction than a 5% level of significance for which it could have been significant.

Exercises

9.1 These days, people spend a lot of time on popular websites for both business and pleasure. The time (in minutes) spent by a sample of sixteen individuals on one of such websites are given below:

60 78 105 123 140 127 90 117 106 189 76 167 88 145 135 154

Assuming that time people spend on the website is normally distributed, test the hypothesis that the mean time spent on this website is one hundred minutes against that it is not. Test at the 1% level of significance.

9.2 These days, many people spend their leisure time playing video games; others spend their leisure time playing games online. A small sample of times people spent, per week, playing online-games are given below (in hours).

8.9 10.7 11.9 12.8 7.8 12 8 10 9.7 9.8
7.5 8.8 9.3 8.2 8.5 11 7.4 9.5 10.3

Assuming that time people spend online playing games is normally distributed, test the hypothesis that the mean time spent on this activity is 8.5 hours against that it greater; test at a 5% level of significance.

9.3 A car is marketed by its manufacturer by claiming that it gets on average thirty-two miles per gallon. Under test conditions, the car was put through its paces, and the number of miles it attained per gallon are presented below:

29 28 25 33 34 33 31 28 30 31 34 32 29 27 32 33 29

Assuming that miles/gallon achieved by this car is normally distributed, test the claim of the manufacturer against the hypothesis that the mean is less than thirty-two miles per gallon, at 5% level of significance.

9.4 Recall Exercise 8.12 about young resident doctors working hours. A sample of weekly working hours of a sample of young resident doctors showed that they had worked:

70 65 76 78 80 60 67 70 65 65 70 72 75 79 71 68 80 74

If the weekly hours they worked are normally distributed, test the hypothesis that the mean weekly working hours is sixty-eight against that it is greater at a 5% level of significance.

9.5 Recall exercise 8.14 about the students' determinations of concentrations of a compound by titration. A sample of determinations of the concentrations of a compound made by students during titration, gave the following values:

4.4 3.4 4.0 3.8 4.1 4.4 3.7 3.9 3.5 4.5 5.0
5.1 3.8 4.4 5.1 5.5 5.6 4.5 5.2 4.9 5.2

Assume that the determinations are normally distributed; test the hypothesis that the mean concentration is 4.2 against that it is not at a 5% level of significance.

9.6 Guests give gifts to couples when they get married. The following are the costs of gifts (in $) given by a sample of guests to a couple that got married:

156 78 112 120 89 134 78 103 140 106 97 88 159 109

If the cost of the gifts is normally distributed, test the hypothesis that the mean cost of the wedding gift is $100.00 against that it is not. Test at a 2% level of significance.

9.7 In Minnesota, Local Property taxes pay for many local projects especially education and related projects. Here is a sample of local property taxes paid in Stearns County (rounded to the nearest $10).

2240 2150 2280 2050 2290 2240 2100
2210 2140 2270 2350 2170 2350

If local property tax is normally distributed, test the hypothesis that the mean is $2150.00 on houses whose market prices are in the vicinity of $200,000 against that it is greater. Test at a 2.5% level of significance.

9.8 Coconut palm is a multiple use tree, and is considered as one of the ten most useful trees in the world. It plays and important part in the nutrition and economy of some nations. The heights of a sample of fifteen full-grown coconut trees are given below in meters:

20.5 30.1 26.7 22.7 27.9 30 32.4 20.7
28.4 25.9 28.5 23.8 28 29.4 26.9

If the heights are normally distributed, test that the mean height of a coconut tree is twenty-five meters against that it is greater. Test at a 5% level of significance.

9.9 We wait in lines in order to get some services that are essential to life. We particularly wait in lines to check out at a supermarket. The waiting times are even longer for week-ends. A sample of week-end waiting times (in minutes) of a random sample of customers at a local popular supermarket is given below:

15.3 10.4 5.9 13.7 18.5 20.5 13.2 10.6
17.8 19.7 21.0 13.0 15.4 16.8 25.6 24.1

Assume that the waiting times are normally distributed, and test that the mean week-end waiting time for a customer is thirteen minutes against that it is not. Test at 2.5% level of significance.

9.10 Hundreds of pounds of apple fruit can be harvested from a single apple tree in a season. Apple is a very important fruit for health and weight loss. Here is a sample of weights of apples harvested from single trees from an orchard:

204 256 351 224 303 289 247 233
178 199 187 227 247 311 229 314

Assume the weight of apples harvested from single trees from the orchard is normally distributed. Test that the mean weight of apples harvested from a single tree is 220 lbs. against that it is not at a 5% level of significance.

Tests of Hypothesis for Single Population Mean (Large Samples)

9.11 The following is a sample of high school GPAs for students accepted into the Medicine program of a prominent university.

3.33 3.97 3.50 3.70 4.00 3.70 3.40 3.60 3.87 3.78 3.65 4.00
3.85 3.98 3.82 3.60 3.70 4.00 3.80 3.52 3.50 3.96 4.00 3.71
3.89 3.89 3.30 3.97 3.76 3.97 3.75 3.56 3.91 3.65

Test the hypothesis that the mean GPA for those admitted to the program is 3.65 against that it is not at a 1% level of significance. Use both classical and p-value approaches.

9.12 The following is a sample of times (in days) from planting to first maturation of green beans (bush variety):

69 60 87 67 89 90 61 69 72 75 77 75 84 90 84 69 74
78 88 69 78 72 74 88 84 82 84 65 77 79 85 88 90 65

Test the hypothesis that the mean period to maturation is seventy-four days against that it is greater at a 2% level of significance.

9.13 **Recall the nutrition problem of Exercise 8.6.** Nutritionists often need to know how much of a nutrient is in a quantity of food in order to determine what a reasonable daily intake amount is. Here is a sample of determinations of the percentage of nutrient content of a fruit made by a research food scientist.

3.56 6.40 5.60 4.56 4.55 5.33 5.56 6.12 3.45 6.34 4.33 5.40 5.70
4.89 6.20 5.70 4.56 5.33 5.34 5.10 6.00 5.77 5.45 6.10 5.41 6.30
6.40 5.40 5.33 4.98 4.88 4.12 5.14 4.13 6.00 4.87 4.56 4.10 5.20

Test the hypothesis that the mean percentage nutrient content of the fruit is 4.9 against that it is not, at a 5% level of significance.

9.14   **Recall exercise 8.8** The following are the times (in minutes) used by a sample of students at a high school to complete a titration experiment in a chemistry class.

    16.70 19.70 22.50 26.90 18.12 19.70 23.50 23.10 24.70
    34.00 23.50 23.60 25.00 23.10 24.70 34.00 23.50 23.60
    25.80 25.70 29.70 32.50 36.50 28.20 29.70 23.50 20.80
    24.00 34.00 28.90 27.90 28.90 27.90 31.30 32.10 34.50

Test the hypothesis that the mean time for a student to complete the titration is twenty-five minutes against that it is greater. Test at 3% level of significance.

9.15   Quite a few people have their wisdom teeth extracted sometime in life. The following is a sample of the costs (in $) of extracting a wisdom tooth (also called a third molar).

    256 249 206 187 210 198 250 234 215 212 198 246 189
    234 238 156 165 191 225 227 182 197 212 194 184 182
    223 247 178 188 239 159 167 213 223 236 214

Test the hypothesis that the mean cost of extraction of a wisdom tooth is $200 against that it is greater. Test at a 5% level of significance.

9.16   The following is a sample of final scores in different games for a basketball team:

    100 100 112 109 123 120 108 127 95 99 107 89 106 87 96 116
    109 97 99 133 138 88 107 98 98 92 96 97 109 125 98 99 133

Test the hypothesis that the mean score of the team is 112 against that it is less. Test at the 1% level of significance.

9.17   It is advisable that when we exercise, we should drink more water to replenish the loss due to perspiration. A survey of regular exercisers found that during exercise, they consume roughly the following amount of water (in milliliters):

    400 410 560 600 450 560 580 480 500 560 470 550
    490 560 590 480 490 450 620 580 610 570 630 560
    580 470 490 530 580 600 450 400 420 470

Test the hypothesis that the mean water consumption is 500 millimeters against that it is not at a 2% level of significance.

9.18   The average number of miles a car is driven in a year affects how long it lasts. A sample of number of miles a car is driven in a year is given below (in thousands of miles):

    7.65 8.96 10.78 12.56 14.67 11.93 10.45 9.67 8.77 12.36 11.34 9.68
    8.86 7.56 7.78 10.23 13.45 15.12 6.56 5.77 8.45 10.22 11.67 12.88
    13.65 11.66 12.97 14.34 14.72 10.66 11.52 9.78 10.44 11.34 14.26

Test the hypothesis that the mean number of miles (in thousands) a car is driven per year is ten against that it is not at a 4% level of significance.

9.19 The life-span of a dog depends on the breed. Some breeds usually live longer than other breeds because of factors such as breed health, body structure, and genetics. It is reputed that the lifespan of the Labrador Retriever (LR) is 12.5 years. The following is a sample of life spans of LRs:

10.5 13.5 11.6 13.6 9.6 8.6 10.6 11.7 9.4 10.6 14 13.5 12.7 13.2 14 10.6 12.7 13.5 13.4 12.8 13.4 11.5 13 12.5 12.8 10.6 9.6 13.6 12.7 10.4 11.6 13 12.5 13.7 12.2 12 12.6 13.0 12.5 12.4 12.5 12.8

Test that the above reputation is correct against that it is less at a 5% level of significance.

9.20 Greyhounds are reputed to be very fast, and are specially raced against each other. It is claimed that the mean speed of a greyhound is 18.7 m/s. Here is a sample of speeds recorded for greyhounds:

16.6 19.6 17.7 19.7 15.7 14.7 16.7 17.8 15.5 16.7 20.1 19.6 18.8 19.3 20.1 16.7 18.8 19.6 19.5 18.9 19.5 17.6 19.1 18.6 18.9 16.7 15.7 19.7 18.8 16.5 17.7 19.1 18.6 19.8 18.3 18.1 18.7 19.1

Test the hypothesis that the mean speed of greyhounds is 18.7 m/s against that it is less, at a 5% level of significance.

Exercise Problems on Testing Hypotheses About a Single Population Proportion

Use table below for questions 9.21–9.22. Samples of patients treated by two doctors, Lenny and Blaine, yield the following data about patients they treated for sports-related injuries:

Doctors	Did Not Fully Recover	Fully Recovered
Lenny	66	90
Blaine	65	129

9.21 Test the hypothesis that the proportion of Dr. Lenny's patients who did not fully recover is 39% against that it is greater. Test at a 2.5% level of significance. Use both classical and p-value approaches.

9.22 Test the hypothesis that the proportion of Dr. Blaine's patients who fully recovered is 70% against that it is not. Test at a 4% level of significance. Use both classical and p-value approaches.

9.23 The proportion of eggs of a rare bird species which hatch when incubated is believed to be 27%. Of the 520 eggs that were incubated, it was found that 178 hatched. Can we say that the proportion of eggs hatched is more than claimed? Use a level of significance of 2%. Test by classical and p-value approaches.

9.24 It is reported by Minnesota DNR that 33% of licensed hunters harvested at least one antlerless deer in a season. A survey of 500 licensed hunters showed that 141 harvested at least one antlerless deer in that season. Is the proportion of antlerless deer harvested 33% or less? Test at a 5% level of significance.

9.25 A production process is set up produce only 5% defectives. In a pilot operation, out of 1500 items produced, ninety items were found to be defective. Is the process operating as it was set up to do? Test at a 1% level of significance.

9.26 Two formulations of a treatment for a disease were applied to two groups of patients.

	Allergic	Not Allergic
D1	276	471
D2	195	239

Test the hypothesis that the proportion of patients who are allergic to both D_1 and D_2 is 42% against that it is not. Test (i) by classical approach, and (ii) by p-value approach. Test at a 2% level of significance.

9.27 Test the hypothesis that the proportion of allergic patients who received D_1 is 54% against that it is greater. Test (i) by classical approach, and (ii) by p-value approach. Test at 3% level of significance.

9.28 Thirty-six percent of state public high school graduates who enrolled in a state public college or university in Minnesota took at least one remedial course in their first two years of college. A survey of 1040 freshmen/sophomores showed that 401 needed remedial courses. Does this represent an increase in the proportion which needed remedial courses in the first two years? Test at 5% level of significance.

9.29 The proportion of part-time students nationwide in the USA is 40%. A survey of 1235 students, nationwide, it was found that 470 were part time students. Does this indicate a decrease in the proportion of part-time students? Test at a 3% level of significance.

9.30 The proportion of adult Americans who want to retire at sixty-five has dropped dramatically. It is believed that only 29% of adult Americans now plan to retire at sixty-five. A poll of 1020 found that 350 want to retire at sixty-five. Does this disagree with the belief? Test at a 5% level of significance.

CHAPTER **10**

Simple Linear Regression Analysis

10.0   Introduction

There are two ways in which we can generate data. We can carry out an experiment or we can observe a process. In experiments, the researcher intervenes and sets the values of the variables governing the conditions that impinge on the system and leads to the observed responses. In observational studies, the researcher simply observes what is happening and records the outcomes of interest. In both experimental and pure observational studies, it is possible to find that a relationship exists between two or more variables.

Statisticians often explore these relationships with the aim of predicting one of the variables based on the knowledge of other variables in the relationship. One of the variables may depend on the values of other variables, with its value changing with changes in the values of the other variables. The **dependent** variable is called the **response variable.** It is called a response variable because it apparently changes with changes in the settings of other variables in the system. It is also called the uncontrolled variable, especially in the experimental context, because the researcher does not control its value, but observes or measures it subsequent to setting or controlling the values of other variables in the relationship. Each of the variables upon which the response variable depends is called a **regressor, predictor,** or **explanatory** variable. The explanatory variables are also called the **independent** or **controlled** variables. We may be interested in predicting a value or values of the response variable which is within the ranges of observation or experimentation for the other variables, which were not particularly studied. Such prediction using the functional relationship between the variables is called **interpolation.** Other times, we may be interested in predicting the values of the response variable for values of the other variables which are outside the ranges that were studied, but nevertheless in their vicinity. Predicting in such vicinity outside the range is called **extrapolation.**

10.1   The Simple Linear Regression

Consider for instance, an ice-cream salesman who discovers that there is a relationship between the temperature in a summer's day and ice cream sales. This has a huge implication for the salesman as it relates to inventory, cost, and potential profits. If the salesman can find a functional relationship between sales and predicted daily temperature, he could predict sales, and can then easily modify the inventory as needed. This will reduce transportation and storage costs. Finding this functional relationship between temperature and sales would be a smart move on the part of the salesman. There are several such relationships between variables which when exploited are beneficial in business, government, and industry. This is why regression analysis is important.

In general if a variable y is known to depend on k independent variables, $x_1, x_2, \ldots, x_k$ the functional relationship between y and $x_1, x_2, \ldots, x_k$ is called the **regression equation.** Here, y is the dependent, response, or uncontrolled variable, while $x_1, x_2, \ldots, x_k$ are the independent, or controlled variables, or regressors. The model which represents the relationship between y and $x_1, x_2, \ldots, x_k$, is a function:

$$y = f(x_1, x_2, \ldots, x_k) \tag{10.1}$$

Normally this functional relationship between y and $x_1, x_2, \ldots, x_k$ is unknown. Regression analysis is the statistical method for fitting this model, using data gathered for y that arise for different settings of $x_1, x_2, \ldots, x_k$. In regression analysis, sample measurements of y for the different values set for $x_1, x_2, \ldots, x_k$ are used to estimate the model parameters and therefore fit the functional relationship y and the predictors, $x_1, x_2, \ldots, x_k$. Sometimes, in a relationship between two variables only, it is not easy to decide which one is the independent variable and which one is the response variable. This is more so if the variables arise from observational rather than an experimental situation. For instance, a researcher on the relationship between the diameter of a tree and its height would only observe all the information needed. He/she is not in a position to control anything that influence heights or diameters. Specifically, he/she cannot, for instance, fix the height of the tree to see what diameter would result, or vice versa. The question the researcher may have is: does height of the tree give rise to the diameter of the tree, or does the diameter of the tree influence the height of the tree?

10.1.1 Exploration of Relationships Using Scatterplots

Since the true functional relationship in Equation 10.1 is unknown to the investigator, it is the practice to choose a polynomial of some order to approximate that functional relationship. The choice of the polynomial equation often depends on the initial exploratory analysis of the data. In particular, when the number of predictors $k = 1$ the exploratory method which is used involves the plotting of the response variable y against the independent variable x called a scatterplot. That initial plot could indicate that a linear, quadratic, cubic, or other relationship may be adequate.

EXAMPLE 10.1   The following data show the yield of a chemical process Y, for coded temperature values X.

X	1	2	3	4	5	6	7	8	5	7	10	10	10	9	4	3	10	11	12	12
Y	3	3	4	5	6	7	8	10	7	10	10	8	12	11	6	5	11	13	10	14

Obtain the scatterplot for the data and check what type of relationship exists between X and Y. Check the data with a software program that provides a scatterplot.

Solution   We produce the same scatterplot as in Figure 10.1 (a). The purpose of the scatterplot is to examine relationships between X and Y. It therefore does no harm if, as in Figure 10.1 (b), we try to visualize the relationship by imposing a suggested relationship. The imposed relationship is only a perception that may or may not eventually stand up under further analysis and scrutiny, but it is useful as an exploratory action.

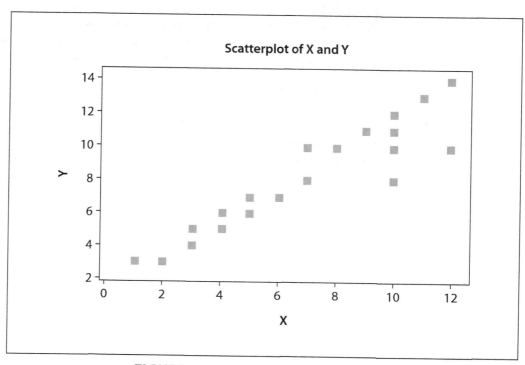

FIGURE 10.1 (A):   Scatterplot of X and Y

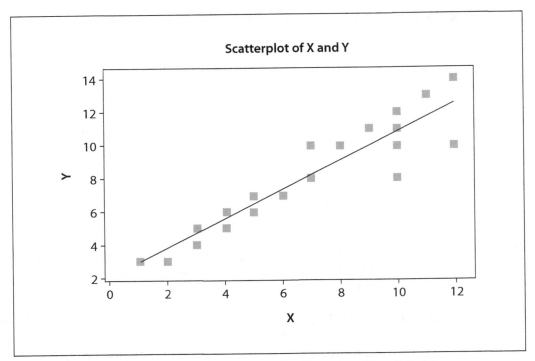

FIGURE 10.1 (B)   Scatterplot of X and Y Showing Possible Linear Relationship

The relationship between X and Y above seems linear.

As we pointed out earlier, the line we imposed may or may not stand scrutiny as a good fit for the line that represents the relationship between X and Y. The line is a guess. What are the ways that we can guess the equation of the line in this situation? We use what we know to make a guess. One of the things we know about straight lines is from our coordinate geometry we learned in high school. In high school coordinate geometry, we learned that the equation of a straight line is usually stated as:

$$y = mx + b \tag{10.2}$$

We also know that we can always find the equation of a straight line if we know two points on the line whose coordinates are (x_1, y_1) and (x_2, y_2). We can then use the coordinates of the two points to obtain the slope of the line as:

$$m = \frac{y_1 - y_2}{x_1 - x_2} \tag{10.3}$$

The equation is found by using either of these relationships:

$$\frac{y - y_2}{x - x_2} = m \tag{10.4}$$

or

$$\frac{y - y_1}{x - x_1} = m \tag{10.5}$$

Both Equation 10.4 and 10.5 yield the desired equation for the line. Let us explore the suggested line that will fit the relationship between X and Y in Example 10.1 using this idea. Looking at Figure 10.1 (b), we see two points which appear to be on the suggested line, they are $(1, 3)$ and $(10, 11)$. Using them we find m as:

$$m = \frac{11 - 3}{10 - 1} = \frac{8}{9} \tag{10.6}$$

Also, by using Equation 10.4, we see that:

$$\frac{y - 3}{x - 1} = \frac{8}{9} \Rightarrow y - 3 = \frac{8}{9}(x - 1) \Rightarrow y = \frac{19}{9} + \frac{8}{9}x$$
$$\Rightarrow y = 2.1111 + 0.88889x$$

Since the points lie on the line suggested, we can use this equation to represent the linear relationship between X and Y. Let us see how well the equation predicts the values that were observed during this experiment. Let $\hat{y}$ represent the estimated values of y for corresponding x. We obtain the $\hat{y}$ and the corresponding residual or deviation from y for each observation.

x	y	$\hat{y}$	Residuals
1	3	3	0
2	3	3.888889	−0.88889
3	4	4.777778	−0.77778
3	5	4.777778	0.222222
4	5	5.666667	−0.66667
4	6	5.666667	0.333333
5	6	6.555556	−0.55556
5	7	6.555556	0.444444
6	7	7.444444	−0.44444
7	8	8.333333	−0.33333
7	10	8.333333	1.666667
8	10	9.222222	0.777778
9	11	10.11111	0.888889
10	8	11	−3
10	10	11	−1
10	11	11	0
10	12	11	1
11	13	11.88889	1.111111
12	10	12.77778	−2.77778
12	14	12.77778	1.222222

The line we fitted seems to be very sensitive to the values of y used in building the model, that is 3 and 11 as found in the two points (1, 3) and (10, 11) and does a reasonable job as a predictor of the relationship between X and Y for other values. The fitted line may be a good predictor; but it is more sensitive to the two points (1, 3) and (10, 11) than other points in the sample. In fact, even if the fit was good, and sensitive to all points, it probably would not be the best line that can be fitted to the data. We need a different strategy for finding the line of best fit than what we learned in high school. The new strategy has to take into account all the n sample points in our data, that is $(x_1, y_1), (x_2, y_2), \ldots, (x_n, y_n)$. It also has to give the best fit.

10.1.2 Regression Using Least Squares Approach

The new strategy which takes into account all the n sample points in the data set, and produces the line of best fit, is called the least squares linear regression. Inferential statistics has shown that when we estimate the line by least squares, the line so fitted is BLUE, which means that it is the BEST, LINEAR, UNBIASED ESTIMATE. There are other implications of BLUE but the most important thing to us for this text, is that it is the BEST. Thus, to obtain a line of best fit, we need to use this method. To obtain the line of best fit by this method, we have to think about the scatter plot again, and use it to illustrate the process. The scatterplot is now presented in Figure 10.2, showing the differences between the points and the estimated line.

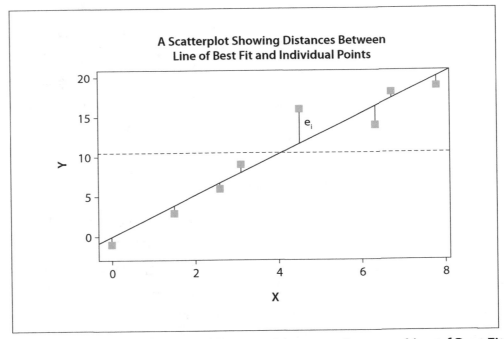

■ **FIGURE 10.2**   **A Scatterplot Showing Distances Between Line of Best Fit and Individual Points**

Figure 10.2 shows the distances between the points and the proposed line of best fit, denoted as ε_i for the point (x_i, y_i). The goal is to fit a line to the data that fits the data so well that each of the distances ε_i is as small as possible. One of the qualities of the least squares line is that it minimizes the distances between the points and the lines, or tries as much as possible to make every one of the n points lie on the fitted line.

Figures 10.3 (a), (b), (c), and (d) show the scatterplots, which suggest positive linear, negative linear, quadratic, or cubic relationships between y and x.

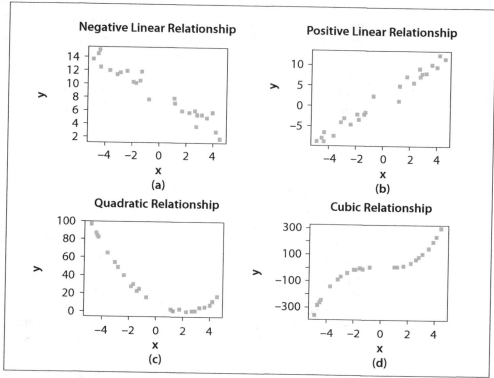

■ **FIGURE 10.3** **Sundry Scatterplots for Linear and Non-Linear Relationships Between X and Y**

Figure 10.4 shows the case when there is no linear relationship between X and Y.

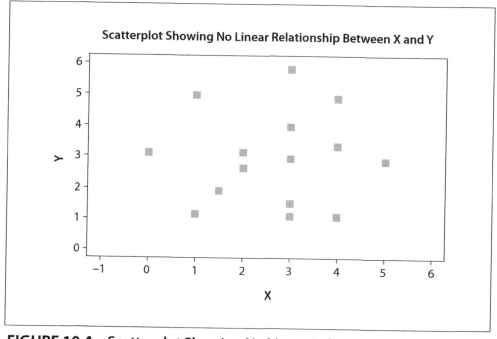

■ **FIGURE 10.4** **Scatterplot Showing No Linear Relationship Between X and Y**

10.1.3 The Least Squares Simple Linear Regression

Suppose that there is only one variable, x (the independent or predictor variable), which is related to the response variable, y. We may wish to explore the relationship by using regression to obtain a model that describes it. Then experiments are carried out, and for different settings of x, we observe y. If data arises from a pure observational study, we simply observe changes in y as the values of x change. For n pairs of observations, (x, y), we would use a the scatterplot to get an idea of the functional relationship between them. If the scatterplot (of y vs. x) indicates that it could adequately be modeled by a straight line, then the model to be fitted is of the form:

$$y = \beta_0 + \beta_1 x + \varepsilon \tag{10.7}$$

which is similar to $y = mx + b$, and the parameters, β_o and β_1 are unknown quantities, which we shall estimate using the responses of our experiments or pure outcomes of observational studies. In the model of (10.7) β_0 is the equivalent of b, while β_1 is the equivalent of m in a typical equation of a straight line. ε is the difference between the observed value of y and the line of best fit and random error $\varepsilon_{ij} \sim N(0, \sigma^2)$. The model (Equation 10.7) assumes that the errors are normally and independently distributed, which means they are uncorrelated.

We can fit a straight line of best fit to the n points, $(x_1, y_1), (x_2, y_2), \ldots, (x_n, y_n)$ obtained from the experiments or pure observational studies, using the Least Squares method. This is a mathematical technique (see Figure 10.2) which seeks to minimize the sum of the squares of the errors, ε_i, which occur between the actual observations, y_i, and the fitted straight line; we note that if we rewrite (10.7) in terms of the ith observation, then:

$$y_i = \beta_0 + \beta_1 x_i + \varepsilon_i; i = 1, 2, \ldots, n \tag{10.8}$$

From Equation 10.8, it is easy to make ε_i the subject of the equation, so that:

$$\varepsilon_i = y_i - \beta_0 - \beta_1 x_i; i = 1, 2, \ldots, n \tag{10.9}$$

The function to be minimized is the sum of the squares of the errors, ε_i, denoted by $g(\varepsilon)$

$$g(\varepsilon) = \sum_{i=1}^{n} \varepsilon_i^2 = \sum_{i=1}^{n} (y_i - \beta_0 - \beta_1 x_i)^2 \tag{10.10}$$

Minimization requires calculus which is not emphasized in this text. However, after minimization, the estimates for the parameters are:

$$\hat{\beta}_0 = \bar{y} - \hat{\beta}_1 \bar{x} \tag{10.11}$$

$$\hat{\beta}_1 = \frac{\sum_{i=1}^{n} y_i(x_i - \bar{x})}{\sum_{i=1}^{n} (x_i - \bar{x})^2} \tag{10.12}$$

Equations 10.11 and 10.12 lead to the fitted model of the form:

$$y_i = \hat{\beta}_0 + \hat{\beta}_1 x_i \tag{10.13}$$

for the ith observation, y_i. Using the assumptions of the model, statisticians using theory of statistical inference have determined that the values of β_0 and β_1 estimated this way and used to estimate the line, results in a best linear unbiased estimate which captures the linear relationship contained in the n sample data points:

$$\hat{y}_i = \hat{\beta}_o + \hat{\beta}_1 x_i \tag{10.14}$$

$\hat{\beta}_0$ is an estimate of the intercept of the line on the Y axis, while $\hat{\beta}_1$ is the estimate of the slope of the line.

From (10.11 & 10.12), we note that if we define:

$$S_{xy} = \sum_{i=1}^{n} (y_i - \bar{y})(x_i - \bar{x}) = \sum_{i=1}^{n} x_i y_i - n\bar{x}\bar{y} \tag{10.15}$$

$$S_{xx} = \sum_{i=1}^{n} (x_i - \bar{x})^2 = \sum_{i=1}^{n} (x_i - \bar{x})^2 = \sum_{i=1}^{n} x_i^2 - n\bar{x}^2$$

$$S_{yy} = \sum_{i=1}^{n} (y_i - \bar{y})^2 = \sum_{i=1}^{n} y_i^2 - n\bar{y}^2 \tag{10.16}$$

To obtain the intercept of the model, we see that

$$\hat{\beta}_0 = \bar{y} - \hat{\beta}\bar{x} \tag{10.17}$$

The slope is obtained for the model as:

$$\hat{\beta}_1 = \frac{S_{xy}}{S_{xx}} \tag{10.18}$$

S_{xy} is the corrected sum of the cross products of x and y, while
S_{xx} is the corrected sum of squares of x, and
S_{yy} is the corrected sum of squares of y.

10.1.4 Assumptions of the Least Squares Linear Model

Before we can use the results of least squares analysis in further investigations, particularly in making inference about the parameters of the fitted line and related attributes, it is important that we note that there are some assumptions which are necessary for the theory of the analyses. These assumptions can be encapsulated in a single assumption, which relates to the errors ε_i of (10.9). It is necessary to assume that $\varepsilon_i \sim NID(0, \sigma^2)$. This means that the errors ε_i are normal and independent with constant variance across all observations y_i.

We have to assume that each of y_i we observe is obtained independent of any other. Another way to put this is to say that the $y_1, y_2, \ldots, y_n$ which is associated with $x_1, x_2, \ldots, x_n$ is a random sample. Offshoots of this assumption are:

1. There are two parameters β_0 and β_1 which need to be estimated in order to fit the straight line, and the mean of y given x is $\beta_0 + \beta_1 x$.

2. For a given value of x the distribution of the response variable y, is normal.
3. Each y_i is independent of another, or the $y_1, y_2, \ldots, y_n$ constitutes a random sample.
4. For a given value of x, the variance of the corresponding y is the same for all values of x, and that common variance is σ^2.

Standard Error of Estimate

The idea of offshoot statement 4 indicates that there is a common variance for all y for any given x. This means that we can estimate this variance because it is usually not known in advance of our sampling for regression. The best estimate of this error is obtained by using the SSE, the sum of squares error, which we had denoted as $g(\varepsilon)$ when we previously encountered it, as follows:

$$\hat{\sigma}^2 = \frac{SSE}{n-2} = \sum_{i=1}^{n} \frac{\varepsilon_i^2}{n-2} = \sum_{i=1}^{n} \frac{(y_i - \hat{y}_i)^2}{n-2} \tag{10.19}$$

The **standard error of the estimate** is used for gauging how much, on average, the predicted responses $\hat{y}_i$ differ from their corresponding observed values y_i. This can be obtained from (10.19) as:

$$\hat{\sigma} = \sqrt{\frac{SSE}{n-2}} = \sqrt{\sum_{i=1}^{n} \frac{\varepsilon_i^2}{n-2}} = \sqrt{\sum_{i=1}^{n} \frac{(y_i - \hat{y}_i)^2}{n-2}} \tag{10.20}$$

■ **EXAMPLE 10.2**   Now we refit the line to the data of Example 10.1, the yield of a chemical process for coded temperature values, and see whether we get a better fit than we did using techniques learned in high school. We will also obtain the SSE and the corresponding standard error of estimate.

■ **TABLE 10.1**

x	y	XY	x^2	y^2
1	3	3	1	9
2	3	6	4	9
3	4	12	9	16
3	5	15	9	25
4	5	20	16	25
4	6	24	16	36
5	6	30	25	36
5	7	35	25	49
6	7	42	36	49
7	8	56	49	64
7	10	70	49	100
8	10	80	64	100
9	11	99	81	121

Continued.

■ **TABLE 10.1—cont'd**

x	y	XY	x^2	y^2
10	8	80	100	64
10	10	100	100	100
10	11	110	100	121
10	12	120	100	144
11	13	143	121	169
12	10	120	144	100
12	14	168	144	196
139	**163**	**1333**	**1193**	**1533**

$$n = 20; \overline{X} = \frac{139}{20} = 6.95; \overline{Y} = \frac{163}{20} = 8.15;$$

$$\sum_{i=1}^{20} x_i y_i = 1333; \sum_{i=1}^{20} x_i^2 = 1193; \sum_{i=1}^{20} y_i^2 = 1533;$$

$$S_{xx} = \sum_{i=1}^{20} x_i^2 - n\overline{x}^2 = 1193 - 20(6.95)^2 = 226.95$$

$$S_{yy} = \sum_{i=1}^{20} y_i^2 - n\overline{y}^2 = 1553 - 20(8.15)^2 = 204.55$$

$$S_{xy} = \sum_{i=1}^{n} x_i y_i - n\overline{xy} = 1333 - 20(6.95)(8.15) = 200.15$$

$$\hat{\beta}_1 = \frac{S_{xy}}{S_{xx}} = \frac{200.15}{226.95} = 0.881912$$

$$\hat{\beta}_0 = \overline{y} - \hat{\beta}_1\overline{x} = 8.15 - 0.881912(6.95) = 2.020709$$

The fitted line is:

$$\hat{y} = 2.020709 + 0.881912x \tag{10.21}$$

10.2 Fitted Values and Residuals in Least Squares Simple Linear Regression

With the above fitted least squares (10.21), we can find the estimated value of the response y for each value of x_i that we have used in the models. They are denoted as $\hat{y}_i$ and are referred to as fitted values. They are also referred to as estimated values. Thus, when $x = 3$, the fitted value is:

$$\hat{y} = 2.020709 + 0.881912(3) = 4.666446.$$

Similarly, when $x = 5$, then the fitted value is:

$$\hat{y} = 2.020709 + 0.881912(5) = 6.430271.$$

We can also find the difference between each observed value y_i and fitted value, $\hat{y}_i$. These differences are called residuals or errors. Thus the residual for a typical value y_i is:

$$\text{residual for } y_i = y_i - \hat{y}_i \qquad \qquad \textbf{(10.22)}$$

Thus, for the two values of y when $x = 3$, there are two residuals obtained as follows:

1st residual for y_i when $x = 3$, is: $y_i - \hat{y}_i = 4 - 4.666446 = -0.666446$
2nd residual for y_i when $x = 3$, is: $y_i - \hat{y}_i = 5 - 4.666446 = 0.333554$

Similarly, when $x = 5$ there are two values of y leading to the following two residuals:

1st residual for y_i when $x = 5$, is: $y_i - \hat{y}_i = 6 - 6.430271 = -0.430271$
2nd residual for y_i when $x = 5$, is: $y_i - \hat{y}_i = 7 - 6.430271 = 0.569729$

Thus, we can find the residuals for all the values of y in the data set. The residuals for all the responses have been calculated and shown in Table 10.2.

10.2.1 Residuals, Sum of Squares Error, and Sum of Squares Regression

Each residual shows the difference between the value of y that was observed and its value that was estimated via least squares regression. This deviation from y_i represents the part of the deviation from the mean of y, $\bar{y}$, which could not be explained by our fitting the least squares straight line to the data. The typical deviation that allows us to calculate the variance or measure variation in y is $y_i - \bar{y}$, which features in the calculation of the sum of squares of y, a measure of the total variation in y. This is understood better when the sum of squares, y, is expressed as:

$$S_{yy} = \sum_{i=1}^{n} (y_i - \bar{y})^2 \qquad \qquad \textbf{(10.23)}$$

It means that this deviation can be subdivided as follows:

$$y_i - \bar{y} = (\hat{y}_i - \bar{y}) + (y_i - \hat{y}_i) \qquad \qquad \textbf{(10.24)}$$

This subdivision of the deviation is captured in Figure 10.5.

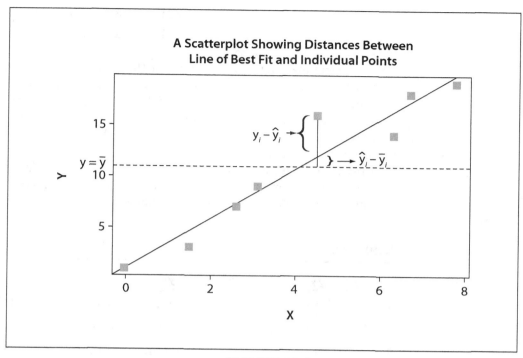

A Scatterplot Showing Distances Between Line of Best Fit and Individual Points

■ **FIGURE 10.5**

From Equation 10.24, it can be shown (but the proof is outside the scope of this text) that:

$$\sum_{i=1}^{n} (y_i - \overline{y})^2 = \sum_{i=1}^{n} (\hat{y}_i - \overline{y})^2 + \sum_{i=1}^{n} (y_i - \hat{y}_i)^2 \tag{10.25}$$

The implication of Equation 10.25 is that:

$$S_{yy} = SSR + SSE \tag{10.26}$$

The SSR in Equation 10.26 is the sum of squares due to regression, or fitting the straight line. Therefore the total variation in y, can be subdivided into variation that is explained by the regression (the fitting of the straight line) and variation which is due to error (residual or unexplained regression).

10.2.2 Interpolation and Extrapolation

We can also use the same equation to find the estimated values of y, when the value of y is within the range of values investigated, but was not originally investigated or observed in the data set used for building the model. When the estimated value for y is found for such a value of x using our estimated model, then we are engaged in **interpolation.** For instance values of y for x values of 5.5, 6.8, 8.2 were not originally investigated in the data set but are within the range of x that was used in the model. We can obtain the estimated values of y for each of them as follows:

$$x = 5.5; \hat{y} = 2.020709 + 0.881912(5.5) = 6.871225$$

$$x = 6.8; \hat{y} = 2.020709 + 0.881912(6.8) = 8.017711$$
$$x = 8.2; \hat{y} = 2.020709 + 0.881912(8.2) = 9.252387$$

Extrapolation involves the use of the same model to obtain values for y for values of x that were not originally investigated in the data set but are outside the range of x that was used in the model. We are careful to keep such values of x in the vicinity of the ends of the x values, and not to stray too far. For instance, the values of x in our current model lie between 1 and 12, inclusive. It would not make sense to try and find the value of y when $x = 17$, $x = 20$, or $x = -3$ as that would be straying too far, and our estimates would be unreliable. But we can use the estimated model for values of $x = 13$, or $x = 0$, for instance, for which the corresponding y's from our model are 2.0202709 and 13.48557.

■ **TABLE 10.2**

x	y	xy	x^2	y^2	$\hat{y}$	$y - \hat{y}$	$(y - \hat{y})^2$	$(\hat{y} - \bar{y})$	$(\hat{y} - \bar{y})^2$
1	3	3	1	9	2.902622	0.097378	0.009483	−5.15	26.5225
2	3	6	4	9	3.784534	−0.78453	0.615494	−5.15	26.5225
3	4	12	9	16	4.666446	−0.66645	0.444151	−4.15	17.2225
3	5	15	9	25	4.666446	0.333554	0.111258	−3.15	9.9225
4	5	20	16	25	5.548359	−0.54836	0.300697	−3.15	9.9225
4	6	24	16	36	5.548359	0.451641	0.20398	−2.15	4.6225
5	6	30	25	36	6.430271	−0.43027	0.185133	−2.15	4.6225
5	7	35	25	49	6.430271	0.569729	0.324591	−1.15	1.3225
6	7	42	36	49	7.312183	−0.31218	0.097458	−1.15	1.3225
7	8	56	49	64	8.194096	−0.1941	0.037673	−0.15	0.0225
7	10	70	49	100	8.194096	1.805904	3.261291	1.85	3.4225
8	10	80	64	100	9.076008	0.923992	0.853761	1.85	3.4225
9	11	99	81	121	9.95792	1.04208	1.08593	2.85	8.1225
10	8	80	100	64	10.83983	−2.83983	8.064649	−0.15	0.0225
10	10	100	100	100	10.83983	−0.83983	0.705319	1.85	3.4225
10	11	110	100	121	10.83983	0.160167	0.025654	2.85	8.1225
10	12	120	100	144	10.83983	1.160167	1.345988	3.85	14.8225
11	13	143	121	169	11.72174	1.278255	1.633936	4.85	23.5225
12	10	120	144	100	12.60366	−2.60366	6.779031	1.85	3.4225
12	14	168	144	196	12.60366	1.396343	1.949773	5.85	34.2225
139	**163**	**1333**	**1193**	**1533**	**163**	**0**	**28.03525**	**1.42E-14**	**204.55**

From Table 10.2, we see that $S_{yy} = 204.55$ and $SSE = 28.03525$;
$\therefore SSR = S_{yy} - SSE = 204.55 - 28.03535 = 176.5148.$

We estimate the common variance, σ^2 as follows:

$$\hat{\sigma}^2 = \frac{SSE}{n-2} = \frac{28.03525}{20-2} = 1.5575 \Rightarrow \hat{\sigma} = \sqrt{1.5575} = 1.248004$$

Thus, the standard error of estimate is $\hat{\sigma} = 1.248004$.

10.3   Testing Hypothesis About the Slope of the Line, β_1

Hypotheses can be tested about the slope of the line, β_1. If the assumptions of the linear least squares model which was described previously are satisfied, it can be shown that:

$$t_{cal} = \frac{\hat{\beta}_1 - \beta_1}{\hat{\sigma}/\sqrt{S_{xx}}} \sim t_{(n-2)} \tag{10.27}$$

The quantity t_{cal} is distributed as t, with $n-2$ degrees of freedom. This means that we can test hypothesis that the slope is zero, against that it is not zero. In that case, t_{cal} is reduced to the form:

$$t_{cal} = \frac{\hat{\beta}_1 - \beta_1}{\hat{\sigma}/\sqrt{S_{xx}}} = \frac{\hat{\beta}_1}{\hat{\sigma}/\sqrt{S_{xx}}} \sim t_{(n-2)} \tag{10.28}$$

The form in Equation 10.28 is very important and is used to test the hypothesis:

$$H_0:\beta_1 = 0 \text{ vs. } H_1:\beta_1 \neq 0$$

It is important to test this hypothesis because if indeed $\beta_1 = 0$, then the fitted line is not significant and therefore not valid.

■ EXAMPLE 10.3   For health reasons, as well as beauty, people (especially women) often watch their weight. In this process, body fat measurements are often made. The following data show the ages of fourteen females and their percentages of body fat.

Age	23	41	40	47	48	47	47	51	54	53	56	57	58	58
Body Fat	28.8	33.4	28.2	26.3	33.1	37.4	38.2	40.1	43.5	45.2	41.3	40.6	41.8	41.4

Is there a linear relationship between ages of the women and their percentages body fat? Fit the least squares linear relationship to the data. Test that the slope is zero at a 5% level of significance.

Solution   Here is the scatterplot. It suggests that a linear relationship exists.

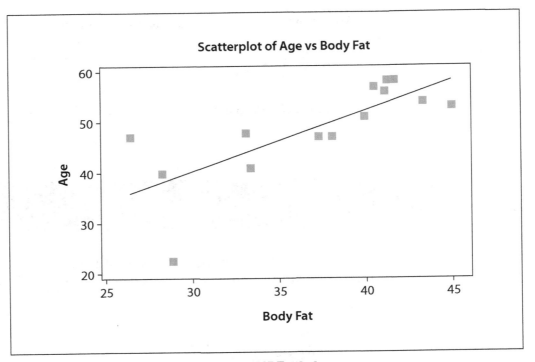

■ **FIGURE 10.6**

x	y	xy	x^2	y^2
23	28.8	662.4	529	829.44
41	33.4	1369.4	1681	1115.56
40	28.2	1128	1600	795.24
47	26.3	1236.1	2209	691.69
48	33.1	1588.8	2304	1095.61
47	37.4	1757.8	2209	1398.76
47	38.2	1795.4	2209	1459.24
51	40.1	2045.1	2601	1608.01
53	45.2	2395.6	2809	2043.04
54	43.5	2349	2916	1892.25
56	41.3	2312.8	3136	1705.69
57	40.6	2314.2	3249	1648.36
58	41.8	2424.4	3364	1747.24
58	41.4	2401.2	3364	1713.96
680	**519.3**	**25780.2**	**34180**	**19744.09**

$$n = 14; \overline{X} = \frac{680}{14} = 48.57143; \overline{Y} = \frac{163}{14} = 37.09286;$$

$$\sum_{i=1}^{14} x_i y_i = 25780.2; \sum_{i=1}^{14} x_i^2 = 34180; \sum_{i=1}^{14} y_i^2 = 19744.09;$$

$$S_{xx} = \sum_{i=1}^{14} x_i^2 - n\overline{x}^2 = 34180 - 14(48.57143)^2 = 1151.429$$

$$S_{yy} = \sum_{i=1}^{14} y_i^2 - n\overline{y}^2 = 19744.09 - 14(37.09286)^2 = 481.7693$$

$$S_{xy} = \sum_{i=1}^{n} x_i y_i - n\overline{xy} = 25780.2 - 14(48.57143)(37.09286) = 557.0571$$

$$\hat{\beta}_1 = \frac{S_{xy}}{S_{xx}} = \frac{557.0571}{1151.429} = 0.483797$$

$$\hat{\beta}_0 = \overline{y} - \hat{\beta}_1 \overline{x} = 37.09286 - 0.483797(48.57143) = 13.59417$$

The fitted line is:

$$\hat{y} = 13.59417 + 0.483797x \tag{10.27}$$

Next, we test that $H_0: \beta_1 = 0$ vs. $H_1: \beta_1 \neq 0$.
First we find SSE, $\hat{\sigma}$ and the value of the test statistic, t_{cal}.

x	y	xy	x^2	y^2	$\hat{y}_i$	$y_i - \hat{y}_i$	$(y_i - \hat{y}_i)^2$	$\hat{y}_i - \overline{y}$	$(\hat{y}_i - \overline{y})^2$
23	28.8	662.4	529	829.44	24.7215	4.078499	16.63415	−12.3714	153.0505
41	33.4	1369.4	1681	1115.56	33.42985	−0.02985	0.000891	−3.66301	13.41764
40	28.2	1128	1600	795.24	32.94605	−4.74605	22.52499	−4.14681	17.19601
47	26.3	1236.1	2209	691.69	36.33263	−10.0326	100.653600	−0.76023	0.577947
48	33.1	1588.8	2304	1095.61	36.81643	−3.71643	13.81182	−0.27643	0.076414
47	37.4	1757.8	2209	1398.76	36.33263	1.067371	1.139281	−0.76023	0.577947
47	38.2	1795.4	2209	1459.24	36.33263	1.867371	3.487074	−0.76023	0.577947
51	40.1	2045.1	2601	1608.01	38.26782	1.832183	3.356895	1.17496	1.380531
53	45.2	2395.6	2809	2043.04	39.23541	5.964589	35.57632	2.142554	4.590537
54	43.5	2349	2916	1892.25	39.71921	3.780792	14.29439	2.626351	6.897719
56	41.3	2312.8	3136	1705.69	40.6868	0.613198	0.376012	3.593945	12.91644
57	40.6	2314.2	3249	1648.36	41.1706	−0.5706	0.325583	4.077742	16.62798
58	41.8	2424.4	3364	1747.24	41.6544	0.145604	0.021201	4.561539	20.80764
58	41.4	2401.2	3364	1713.96	41.6544	−0.2544	0.064717	4.561539	20.80764
680	**519.3**	**25780.2**	**34180**	**19744.09**	**519.3003**	**−0.00034**	**212.267**	**0.00034**	**269.5028**

$$SSE = 212.267 \Rightarrow \hat{\sigma}^2 = \frac{212.267}{14 - 2} = 17.6889 \Rightarrow \hat{\sigma} = 4.20582$$

$$t_{cal} = \frac{\hat{\beta}_1}{\hat{\sigma}/\sqrt{S_{xx}}} = \frac{0.483797}{4.20582/\sqrt{1151.429}} = 3.9033$$

Since $n = 14$, the level of significance is $\alpha = 0.05$, and we are carrying out a two-sided test, we read the tables in Appendix III for $t_{(14-2,\,0.025)} = t_{(12,\,0.025)}$ and the value is 2.179. The picture below shows the t-distribution with the acceptance and rejection regions:

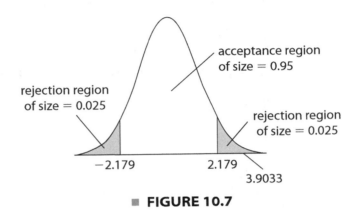

■ FIGURE 10.7

We reject the null hypothesis, and conclude that $\beta_1 \neq 0$. The fitted line is significant.

10.4   Another Form for the Regression Equations

There are other ways of writing the regression equations which lead to the estimation of the parameters, β_0 and β_1, for ease of computation—especially when using a calculator. By doing a little algebra on Equation 10.18 it can be shown that:

$$\hat{\beta}_0 = \bar{y} - \hat{\beta}\bar{x} \tag{10.28}$$

$$\hat{\beta}_1 = \frac{n\sum_{i=1}^{n} x_i y_i - \sum_{i=1}^{n} x_i \sum_{i=1}^{n} y_i}{n\sum_{i=1}^{n} x_i^2 - \left(\sum_{i=1}^{n} X_i\right)^2} \tag{10.29}$$

In this form, we do not need to find S_{xx} and S_{xy} before calculating the estimated values of β_0 and β_1.

■ EXAMPLE 10.4   The following data shows the weights of some individuals, and their corresponding heights.

Weight	Heights	Weight	Heights	Weight	Heights
170	75	172	70	177	74
177	75	185	74	168	68
166	66	163	65	167	67
175	75	177	74	158	65
160	65	165	66	157	65
185	74	182	72	160	63
165	65	172	71		

Obtain the linear relationship between weights (x) and heights (y) using the alternative form of the formulae for estimating the parameters β_0 and β_1. Extract residuals and find the total sum of square, sum of squares regression, and sum of squares error. Find the estimate for the common variance, and hence find the standard error of estimate, and test the hypothesis that the slope is zero against that it is not.

Solution

Weight (x)	Heights (y)	xy	x^2	y^2	$\hat{y}$	$y - \hat{y}$	$(y - \hat{y})^2$	$\hat{y}_i - \bar{y}$	$(\hat{y} - \bar{y})^2$
170	75	12750	28900	5625	69.42821	5.571787	31.04481	0.021787	0.000475
177	75	13275	31329	5625	72.47839	2.521609	6.35851	−3.02839	9.171155
166	66	10956	27556	4356	67.68525	−1.68525	2.840081	1.764746	3.114329
175	75	13125	30625	5625	71.60691	3.393088	11.51305	−2.15691	4.652269
160	65	10400	25600	4225	65.07082	−0.07082	0.005015	4.379185	19.17726
185	74	13690	34225	5476	75.96431	−1.96431	3.858513	−6.51431	42.43623
165	65	10725	27225	4225	67.24951	−2.24951	5.060314	2.200486	4.842138
172	70	12040	29584	4900	70.29969	−0.29969	0.089816	−0.84969	0.721977
185	74	13690	34225	5476	75.96431	−1.96431	3.858513	−6.51431	42.43623
163	65	10595	26569	4225	66.37803	−1.37803	1.898979	3.071965	9.436972
177	74	13098	31329	5476	72.47839	1.521609	2.315293	−3.02839	9.171155
165	66	10890	27225	4356	67.24951	−1.24951	1.561286	2.200486	4.842138
182	72	13104	33124	5184	74.65709	−2.65709	7.060129	−5.20709	27.11379
172	71	12212	29584	5041	70.29969	0.700307	0.49043	−0.84969	0.721977
177	74	13098	31329	5476	72.47839	1.521609	2.315293	−3.02839	9.171155
168	68	11424	28224	4624	68.55673	−0.55673	0.309952	0.893267	0.797925
167	67	11189	27889	4489	68.12099	−1.12099	1.256627	1.329006	1.766258
158	65	10270	24964	4225	64.19934	0.800664	0.641063	5.250664	27.56948
157	65	10205	24649	4225	63.7636	1.236404	1.528695	5.686404	32.33519
160	63	10080	25600	3969	65.07082	−2.07082	4.288276	4.379185	19.17726
3401	**1389**	**236816**	**579755**	**96823**	**1389**	**−4.3E-14**	**88.29464**	**1.42E-14**	**268.6554**

$$\hat{\beta}_1 = \frac{20(236816) - 3401(1389)}{20(579755) - 3401^2} = 0.43574$$

$$\hat{\beta}_0 = \frac{1389}{20} - 0.435745\left(\frac{3401}{20}\right) = -4.64755$$

Fitted line is:

$$\hat{y} = -4.64755 + 0.43574x$$

$$\overline{Y} = \frac{1389}{20} = 69.45$$

$$S_{yy} = \sum_{i=1}^{n} y_i^2 - n\overline{y}^2 = 96823 - 20(69.45)^2 = 356.95$$

$$SSE = 88.29464;$$

$$SSR = S_{yy} - SSE = 268.6554$$

$$\sigma^2 = \frac{SSE}{n-2} = \frac{88.29464}{20-2} = 4.90526 \Rightarrow \sigma = \sqrt{4.90526} = 2.214782$$

Standard error of estimate = 2.214782
Next, we test that $H_0:\beta_1 = 0$ vs. $H_1:\beta_1 \neq 0$
We have already found SSE, $\hat{\sigma}$ and the value of the test statistic; we need to find S_{xx} and the test statistic, t_{cal}

$$S_{xx} = 579755 - 20(3401/20)^2 = 1414.95$$

$$t_{cal} = \frac{\hat{\beta}_1}{\hat{\sigma}/\sqrt{S_{xx}}} = \frac{0.43574}{2.214782/\sqrt{1414.95}} = 7.4006$$

Since $n = 20$, the level of significance is $\alpha = 0.05$, and we are carrying out a two-sided test, we read the tables Appendix III for $t_{(20-2, 0.025)} = t_{(18, 0.025)}$ and the value is 2.101. The picture below shows the t-distribution with the acceptance and rejection regions:

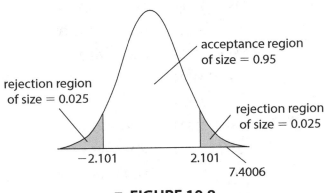

■ FIGURE 10.8

We reject the null hypothesis, and conclude that $\beta_1 \neq 0$. The fitted line is significant.

10.5 Use of MINITAB for Simple Linear Regression

EXAMPLE 10.5 Carry out the analysis of data of Example 10.1 and 10.2 using MINITAB.

Solution Here we use MINITAB to carry out the regression analysis of Example 10.1. First we enter the data using the SET to read X values into c1 and Y values into c2.

```
MTB > set c1
DATA> 1 2 3 3 4 4 5 5 6 7 7 8 9 10 10
DATA> 10 10 11 12 12
DATA> end
MTB > set c2
DATA> 3 3 4 5 5 6 6 7 7 8 10 10 11 8 10
DATA> 11 12 13 10 14
DATA> end
```

Next, we name them accordingly as presented below:

```
MTB > name c1 'x' c2 'y'
```

The command required for the analysis is:

```
MTB > regress c2 1 c1
```

The output of MINITAB includes what we present below and some other things we have not discussed because they are beyond the scope of this book, which we chose not to present:

Regression Analysis: *y* versus *x*

```
The regression equation is
y = 2.02 + 0.882 x
```

Predictor	Coef	SE Coef	T	P
Constant	2.0207	0.6398	3.16	0.005
x	0.88191	0.08284	10.65	0.000

```
S = 1.24800  R-Sq = 86.3%  R-Sq(adj) = 85.5%
```

The MINITAB output shows that the fitted line is $\hat{y} = 2.02 + 0.882x$.

It also shows that two sets of hypotheses: (i) that the intercept (represented by Constant) is zero against that it is not, and (ii) that the slope (represented by x) is zero against that it is not; were carried out.

From the output, if the level of significance that we would allow for both sets of hypotheses is $\alpha = 1\%$, then the p-values (0.005, and less than 0.0001 in the output) are both less than 1%, and we should reject the null hypotheses and conclude that neither the intercept nor the slope is zero at a 1% level of significance.

The rest of the output shows $\hat{\sigma} = 1.24800$, $R^2 = 86.3\%$ or 0.863, and adjusted R^2 which is beyond the scope of our discussion in this text. The MINITAB output agrees with the manual analysis we previously carried out.

■ **EXAMPLE 10.6   Analyze the Data of Example 10.3 Using MINITAB**

Solution   Once again, we use MINITAB to carry out the regression analysis, this time of Example 10.3. First we enter the data using the SET to read X values into c1 and Y values into c2.

```
MTB > set c1
DATA> 23 41 40 47 48 47 47 51 54 53 56 57 58 58
DATA> end
MTB > set c2
DATA> 28.8 33.4 28.2 26.3 33.1 37.4 38.2 40.1 43.5 45.2 41.3
DATA> 40.6 41.8 41.4
DATA> end
```

We again name them accordingly as presented below:

```
MTB > name c1 'age' c2 'bodyfat'
```

Using the same command we used previously:

```
MTB > regress c2 1 c1
```

The output of MINITAB includes what we present below and some other things we have not discussed because they are beyond the scope of this book, which we chose not to present:

Regression Analysis: y versus x

Regression Analysis: c2 versus c1
```
    The regression equation is
    bodyfat = 13.6 + 0.484 age


    Predictor    Coef    SE Coef    T      P
    Constant    13.594    6.124    2.22   0.046
     Bodyfat    0.4838   0.1239    3.90   0.002


    S = 4.20582 R-Sq = 55.9% R-Sq(adj) = 52.3%
```

The MINITAB output shows that the fitted line is:

$$\hat{y} = 13.6 + 0.484x$$

It also tests two sets of hypotheses: (i) that the intercept (represented by Constant) is zero against that it is not, and (ii) that the slope (represented by x) is zero against that it is not. If the level of significance that we would allow in both cases is $\alpha = 5\%$, then the p-values (0.046, and 0.002 in the output) are both less than 5%, and we should reject the null hypotheses and conclude that neither the intercept nor the slope is zero at a 1% level of significance.

The rest of the output shows $\hat{\sigma} = 4.20582$, $R^2 = 55.9\%$ or 0.559, and adjusted R^2 which is beyond the scope of our discussion in this text. The MINITAB output agrees with the manual analysis we previously carried out.

Exercises

10.1

x	y
12	8
14	9
11	6
11	4
10	3
9	8
8	3
6	2
7	4
8	3

For data of X and Y in the table above:
(a) Find S_{xx}, S_{xy}, S_{yy}.
(b) Obtain the equation of the linear relationship between X and Y.
(c) Obtain the SSE and test the hypothesis that the slope is zero against that it is not at a 5% level of significance.

10.2

x	y
0	2.22
1	1.33
2	1.18
4	0.72
5	0.63
7	0.57
8	0.49
9	0.45

For data of X and Y in the table:
(a) Find S_{xx}, S_{xy}, S_{yy}.
(b) Obtain the equation of the linear relationship between X and Y.
(c) Obtain the SSE and test the hypothesis that the slope is zero against that it is not at a 5% level of significance.

10.3

x	y
3.7	650
2	490
3.3	580
2.8	530
3.5	640
3.15	615
3.7	650
2.2	498
3.3	590
2.8	550
3.3	630
3.2	625

For data of X and Y in the table:
(a) Find S_{xx}, S_{xy}, S_{yy}.
(b) Obtain the equation of the linear relationship between X and Y.
(c) Obtain the SSE and test the hypothesis that the slope is zero against that it is not at a 1% level of significance.

10.4

x	y
0	1.3
2	3
3	7
5	8
7	17
8	19
10	22
11	30

For data of X and Y in the table:
(a) Find S_{xx}, S_{xy}, S_{yy}.
(b) Obtain the equation of the linear relationship between X and Y.
(c) Obtain the SSE and test the hypothesis that the slope is zero against that it is not at a 1% level of significance.

10.5   The following are the heights (X) and the diameters (Y) of a tree:

Tree	1	2	3	4	5	6	7	8	9	10
X	1.8	1.9	1.8	2.4	5.1	3.1	5.5	5.1	8.3	13.7
Y	22.0	32.5	23.6	40.8	63.2	27.9	41.4	44.3	63.5	92.7

(a) Obtain a scatterplot of the data. Does it suggest a linear relationship?
(b) Obtain the sums of squares S_{xx}, S_{yy}, and S_{xy}.
(c) Obtain least squares estimated linear equation relating Y and X.
(d) For each value of X, obtain the estimated Y.
(e) Find the residuals for the analysis.
(f) Find SSE.
(g) Test that the population slope is zero against that it is not, at a 5% level of significance.

10.6

Height (x)	Size (y)
66	5.5
63.96	5.5
63	3.5
66	5.5
62.04	3.5
64.56	5.5
63.6	8
64.92	5
64.92	5
60	3
67.68	7
66	4
66.96	6
66	6
66.96	6.5

The above data show the heights of some juniors in a European university and their shoe sizes. Fit a least square regression line for the relationship between heights and shoe sizes. Obtain S_{xx}, S_{xy}, S_{yy} and SSE. Obtain the standard error of the estimate and hence test the hypothesis that the slope is zero against that it is not at a 2% level of significance.

10.7

Height (ins.)	Weight (lbs.)
63	154
65	145
70	166
66	150
58	145
69	164
67	156
64	158
65	154
68	155

The above data show the heights of some adults and their weights. Fit a least square regression line for the relationship between heights and weights. Obtain S_{xx}, S_{xy}, S_{yy} and SSE. Obtain the standard error of the estimate and hence test the hypothesis that the slope is zero against that it is not at a 2% level of significance.

10.8

Weekly Working Hours	Annual Salary
19	23.4
17	24.1
18	20.45
15	19.55
25	28.67
30	33.75
24	27.88
17	25.6
32	36.9
28	33.26
19	22.34
24	26.9
30	36.67

The above data show the average weekly working hours of First Flight officers of a regional Airline and their annual salaries. Obtain a scatterplot of the data. Fit a least square regression line for the relationship between their weekly working hours and their annual salaries. Obtain S_{xx}, S_{xy}, S_{yy} and SSE. Obtain the standard error of the estimate and hence test the hypothesis that the slope is zero against that it is not at a 2% level of significance.

10.9   A sample of instructional costs at twelve institutions, and the average wages of faculty were collected. The aim was to find out whether they are linearly related. All values were recorded in thousands of dollars as below:

Cost	Wage
5.66	11.96
15.6	12
10.51	11.44
6.24	11.44
1.76	9.61
3.94	11.31
4.28	9.99
2.98	9.97
15.3	11.29
12.28	11.67
2.34	8.84
2.74	8.85

Obtain the least squares equation of the linear relationship between the cost of instruction and average wages for faculty in the institutions. Obtain S_{xx}, S_{xy}, S_{yy} and SSE. Obtain the standard error of the estimate and hence test the hypothesis that the slope is zero against that it is not at a 2% level of significance.

10.10   In the same study as described earlier, a sample of instructional costs at fifteen institutions and the number of students were collected and presented in thousands. The aim was to find out whether they are linearly related. All cost values were recorded in thousands of dollars as below:

Cost	Number of Students
1.08	1.58
5.66	3.8
15.7	11.4
6.95	5.36
10.32	8.04
4.88	2.86
15.35	12.24
15.91	4.38
5.56	13.45
9.29	7.52
12.09	9.78
15.85	13.95
19.54	16.65
1.83	1.15
2.65	4.06

Obtain the least squares equation of the linear relationship between the cost of instruction and the number of students in the institutions. Obtain S_{xx}, S_{xy}, S_{yy} and SSE. Obtain the standard error of the estimate and hence test the hypothesis that the slope is zero against that it is not at a 1% level of significance.

10.11   In the same study as described earlier, a sample of instructional costs at fifteen institutions and the number of faculty were collected and presented. The aim was to find out whether they are linearly related. All cost values were recorded in thousands of dollars as below:

Cost	Number of Faculty
1.08	81
5.66	290
15.7	466
6.95	435
10.32	475
4.88	225
15.35	959
15.91	848
5.56	340
9.29	485
12.09	643
15.85	875
19.54	1071
1.83	99
2.65	220

Obtain the least squares equation of the linear relationship between the cost of instruction and number of faculty in the institutions. Obtain S_{xx}, S_{xy}, S_{yy} and SSE. Obtain the standard error of the estimate and hence test the hypothesis that the slope is zero against that it is not at a 1% level of significance.

10.12 The following are the ages of a sample of people and the number of hours of television they watched per week:

X (age)	Y (weekly TV watching hours)
10	20
10	21
13	22
13	26
15	22
17	20
18	25
19	27
19	23
23	30
25	28
28	30

Obtain a scatterplot of the data. Obtain, by the least squares regression method, the equation of the linear relationship between age and weekly TV watching. Obtain S_{xx}, S_{xy}, S_{yy} and SSE. Obtain the standard error of the estimate and hence test the hypothesis that the slope is zero against that it is not at a 2% level of significance.

10.13 Is there a linear relationship between scores in STAT 193 class and average number of hours per week of study for the course? A study matched students' recollections of number of hours of study per week with their final scores in STAT 193.

X (Weekly Hours of Study)	Y (Final Score in STAT 193)
2	70
2	75
2.5	72
2.5	78
2.75	80
3	81
3	70
3.5	85
3	80
3.25	70
3.5	87
3.5	85
3.75	90
4	90

Obtain a scatterplot of the data. Obtain, by the least squares regression method, the equation of the linear relationship between weekly study hours and final score in STAT 193. Obtain S_{xx}, S_{xy}, S_{yy} and SSE. Obtain the standard error of the estimate and hence test the hypothesis that the slope is zero against that it is not by a 1% level of significance.

10.14

x	y
0	6
3	10
4	12
5	12
7	15
3	9
10	19
13	22
11	17
15	27
18	29
24	27
12	19
11	13

The two variables X and Y are believed to be linearly related. Obtain S_{xx}, S_{xy}, S_{yy} and SSE. Find the least squares estimate of the linear relationship between X and Y. Obtain the standard error of the estimate and hence test the hypothesis that the population slope is zero against that it is not 1% level of significance.

10.15

X	Y
3.2	7.4
3	6.5
7.2	5.5
6	6.2
7	6.4
5.4	4.3
8	4
12.2	3.7
12.5	3.3
14.7	3.1
13.4	3.8

Obtain, by the least squares regression method, the equation of the linear relationship between X and Y. Obtain S_{xx}, S_{xy}, S_{yy} and SSE. Find the standard error of the estimate of the slope and hence test the hypothesis that the population slope is zero against that it is not, using a 5% level of significance.

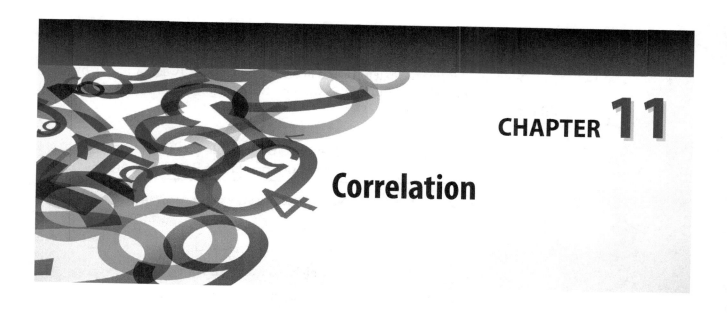

CHAPTER 11

Correlation

11.0 Introduction

In linear least squares regression analysis, we learned how to fit a linear relationship between two variables X (an independent or controllable variable) and Y (a response, outcome, or dependent variable). We considered linear regression and showed how, by least squares method, we can fit a line to a sample of size n, using pairs of values (x, y) in which the Y values were determined or observed for the corresponding values of X. We tested hypotheses for the slope and determined whether the slope is zero. If the slope is zero, then the line is parallel to the X-axis and the values of Y do not change with changes in X. It is often of interest to the investigator to find out how linearly related the two variables are. This is the subject of discussion in this section.

In general, when we talk about correlation between two quantitative variables X and Y, we are referring to the relationship between the variables. The relationship may be linear; in that case, we are speaking about linear correlation. The relationship may also be nonlinear—quadratic, cubic, exponential, etc. Correlation examines the co-relationship between the two variables, X and Y.

11.1 Linear Correlation Coefficient

The linear correlation coefficient, denoted by r, measures the strength of the linear relationship between the variables X and Y. The Linear correlation coefficient is actually called the Pearson product-moment correlation coefficient, named after Karl Pearson[1] who derived the measure. The value of r lies between -1 and $+1$, so that mathematically we can write:

$$-1 \le r \le +1.$$

We can interpret the extent of the linear relationship between X and Y by looking at the values of r. In Table 11.1, we present some possible ways of interpreting the different values of r.

■ **TABLE 11.1**   **Some Examples of Possible Interpretations of the Extent of Correlation Between X and Y**

$-0.99, -0.98, -0.97$	Near perfect negative linear correlation
$0.99, 0.98, 0.97$	Near perfect positive linear correlation
$-0.9, -0.89$	Very high negative linear correlation
$0.9, 0.89$	Very high positive linear correlation
$-0.76, -0.71$	Moderately high negative linear correlation
$0.76, 0.71$	Moderately high negative linear correlation
$-0.45, -0.48$	Low negative linear correlation
$0.45, 0.48$	Low positive linear correlation

Pictorially, when we carry out a scatterplot of the n points in the sample used for regression, we can visualize the different levels of the linear relationship between the variables X and Y which can occur. Here in Figure 11.1, we present a few:

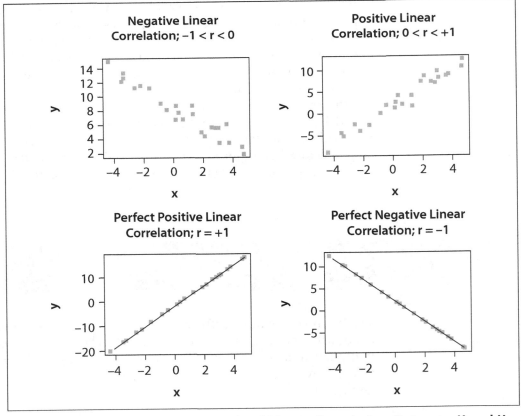

■ **FIGURE 11.1**   **Different Levels of Linear Correlation Between X and Y**

The correlation coefficient can be calculated by using the following formula:

$$r = \frac{n \sum_{i=1}^{n} x_i y_i - \sum_{i=1}^{n} x_i \sum_{i=1}^{n} y_i}{\sqrt{\left[n \sum_{i=1}^{n} x_i^2 - \left(\sum_{i=1}^{n} x_i \right)^2 \right]\left[n \sum_{i=1}^{n} y_i^2 - \left(\sum_{i=1}^{n} y_i \right)^2 \right]}} \tag{11.1}$$

The formula has been rendered in different ways. The following formula approaches the calculation by using the sums of squares S_{xx}, S_{xy}, and S_{yy} which we had used in regression analysis in Chapter 10.

$$r = \frac{S_{xy}}{\sqrt{S_{yy} S_{xx}}} \tag{11.2}$$

where

$$S_{yy} = \sum_{i=1}^{n} y_i^2 - \frac{\left(\sum_{i=1}^{n} y_i \right)^2}{n}; \tag{11.3a}$$

$$S_{xy} = \sum_{i=1}^{n} x_i y_i - \frac{\left(\sum_{i=1}^{n} y_i \right)\left(\sum_{i=1}^{n} x_i \right)}{n}; \tag{11.3b}$$

$$S_{xx} = \sum_{i=1}^{n} x_i^2 - \frac{\left(\sum_{i=1}^{n} x_i \right)^2}{n}. \tag{11.3c}$$

Yet another form of the formula for the correlation coefficient requires the calculation of standard deviations for X and Y and using them in the formula as follows:

$$r = \frac{\frac{1}{n-1} \sum_{i=1}^{n} (x_i - \bar{x})(y_i - \bar{y})}{s_x s_y} \tag{11.4}$$

where

$$s_y = \sqrt{\frac{\sum_{i=1}^{n} y_i^2 - n\bar{y}^2}{n-1}}; \, s_x = \sqrt{\frac{\sum_{i=1}^{n} x_i^2 - n\bar{x}^2}{n-1}} \tag{11.5}$$

The three formulas stated in 11.1–11.5 are the same. In manual calculation, these are optional forms that one can use. One can use the one that he/she finds most convenient and easiest to use. They all yield the same value for r. In the solutions of our examples, we will refer to the use of the formula in Equation 11.1 as method 1, the formula in Equations 11.2 & 11.3 as method 2, while the use of the formulae in Equations 11.4 & 11.5 will be referred to as method 3. We consider some examples.

■ **EXAMPLE 11.1**   The following data give the values of X and Y as follows:

X	y
74	5
66	6
81	8
52	11
73	12
62	15
52	16
45	17
62	18
46	18
60	19
46	20
36	20

Obtain the correlation coefficient for the linear relationship between X and Y and interpret it.

Solution   Here we use all three methods to show that they yield the same value for r:

	X	Y	XY	X²	Y²
	74	5	370	5476	25
	66	6	396	4356	36
	81	8	648	6561	64
	52	11	572	2704	121
	73	12	876	5329	144
	62	15	930	3844	225
	52	16	832	2704	256
	45	17	765	2025	289
	62	18	1116	3844	324
	46	18	828	2116	324
	60	19	1140	3600	361
	46	20	920	2116	400
	36	20	720	1296	400
Totals→	755	185	10113	45971	2969

Using method 1:

$$n = 13; \sum_{i=1}^{n} x_i = 755; \sum_{i=1}^{n} x_i^2 = 45971; \sum_{i=1}^{n} y_i = 185; \sum_{i=1}^{n} y_i^2 = 2969; \sum_{i=1}^{n} x_i y_i = 10113;$$

$$r = \frac{n \sum_{i=1}^{n} x_i y_i - \sum_{i=1}^{n} x_i \sum_{i=1}^{n} y_i}{\sqrt{\left[n \sum_{i=1}^{n} x_i^2 - \left(\sum_{i=1}^{n} x_i \right)^2 \right]\left[n \sum_{i=1}^{n} y_i^2 - \left(\sum_{i=1}^{n} y_i \right)^2 \right]}}$$

$$= \frac{13(10113) - 755(185)}{\sqrt{[13(45971) - 755^2][13(2969) - 185^2]}} = -0.74706$$

Using method 2:

$$S_{yy} = \sum_{i=1}^{13} y_i^2 - \frac{\left(\sum_{i=1}^{13} y_i \right)^2}{13} = 2969 - \frac{185^2}{13} = 336.3077$$

$$S_{xy} = \sum_{i=1}^{13} x_i y_i - \frac{\left(\sum_{i=1}^{13} y_i \right)\left(\sum_{i=1}^{13} x_i \right)}{13} = 10113 - \frac{185(755)}{13} = -631.231$$

$$S_{xx} = \sum_{i=1}^{13} x_i^2 - \frac{\left(\sum_{i=1}^{13} x_i \right)^2}{13} = 45971 - \frac{755^2}{13} = 2122.923$$

$$r = \frac{S_{xy}}{\sqrt{S_{yy}S_{xx}}} = \frac{-631.231}{\sqrt{2122.923 \times 336.3077}} = -0.74706$$

We can find the same correlation using method 3 as follows:

x	y	$x_i - \bar{x}$	$y_i - \bar{y}$	$\dfrac{x_i - \bar{x}}{s_x}$	$\dfrac{y_i - \bar{y}}{s_y}$	$\dfrac{x_i - \bar{x}}{s_x} \cdot \dfrac{y_i - \bar{y}}{s_y}$
74	5	15.92308	−9.23077	1.197155285	−1.74365	−2.08742
66	6	7.923077	−8.23077	0.595685963	−1.55476	−0.92615
81	8	22.92308	−6.23077	1.723440942	−1.17697	−2.02843
52	11	−6.07692	−3.23077	−0.45688535	−0.61028	0.278827
73	12	14.92308	−2.23077	1.12197162	−0.42138	−0.47278
62	15	3.923077	0.769231	0.294951302	0.145304	0.042858
52	16	−6.07692	1.769231	−0.45688535	0.3342	−0.15269
45	17	−13.0769	2.769231	−0.9831710	0.523096	−0.51429
62	18	3.923077	3.769231	0.294951302	0.711992	0.210003
46	18	−12.0769	3.769231	−0.9079873	0.711992	−0.64648
60	19	1.923077	4.769231	0.144583972	0.900887	0.130254

Continued.

x	y	$x_i - \bar{x}$	$y_i - \bar{y}$	$\dfrac{x_i - \bar{x}}{s_x}$	$\dfrac{y_i - \bar{y}}{s_y}$	$\dfrac{x_i - \bar{x}}{s_x} \cdot \dfrac{y_i - \bar{y}}{s_y}$
46	20	−12.0769	5.769231	−0.9079873	1.089783	−0.98951
36	20	−22.0769	5.769231	−1.6598239	1.089783	−1.80885
Total→						−8.96466
Total/(n-1)						−0.74706

We see that $\dfrac{\sum_{i=1}^{13}(x_i - \bar{x})(y_i - \bar{y})}{s_x s_y} = -8.96466$ and $\dfrac{1}{(13-1)} \dfrac{\sum_{i=1}^{13}(x_i - \bar{x})(y_i - \bar{y})}{s_x s_y} = -0.74706$.

■ **EXAMPLE 11.2**   For X and Y below, obtain both the linear regression equation and correlation coefficient:

x	y
1	3
2	3
3	4
4	5
5	6
6	7
7	8
8	10

Solution   We proceed as follows, using methods 1 and 2, but first, let us fit the least squares equation:

x	y	xy	x^2	y^2	$\hat{y}$	$y - \hat{y}$	$(y - \hat{y})^2$
1	3	3	1	9	2.25	0.75	0.5625
2	3	6	4	9	3.25	−0.25	0.0625
3	4	12	9	16	4.25	−0.25	0.0625
4	5	20	16	25	5.25	−0.25	0.0625
5	6	30	25	36	6.25	−0.25	0.0625
6	7	42	36	49	7.25	−0.25	0.0625
7	8	56	49	64	8.25	−0.25	0.0625
8	10	80	64	100	9.25	0.75	0.5625
36	**46**	**249**	**204**	**308**	**46**	**0**	**1.5**

$$\sum_{i=1}^{8} x_i = 36; \quad \sum_{i=1}^{8} y_i = 46; \quad \sum_{i=1}^{8} x_i y_i = 249; \quad \sum_{i=1}^{8} x_i^2 = 204; \quad \sum_{i=1}^{8} y_i^2 = 308$$

Finding the equation, we need to find $\hat{\beta}_0$ and $\hat{\beta}_1$:

$$\hat{\beta}_1 = \frac{n \sum_{i=1}^{n} x_i y_i - \sum_{i=1}^{n} x_i \sum_{i=1}^{n} y_i}{\sum_{i=1}^{n} x_i^2 - \left(\sum_{i=1}^{n} x_i\right)^2} = \frac{8(249) - 36(46)}{8(204) - 36^2} = 1$$

$$\hat{\beta}_0 = \bar{y} - \hat{\beta}_1 \bar{x} = \frac{46}{8} - 1 \times \frac{36}{8} = 1.25$$

The fitted equation is: $\hat{y} = \hat{\beta}_0 + \hat{\beta}_1 x = 1.25 + 1x$
Next, we obtain the S_{yy}, S_{xx} and S_{xy} as follows:

$$S_{yy} = \sum_{i=1}^{8} y_i^2 - \frac{\left(\sum_{i=1}^{8} y_i\right)}{8} = 308 - \frac{(46)^2}{8} = 43.5$$

$$S_{xx} = \sum_{i=1}^{8} x_i^2 - \frac{\left(\sum_{i=1}^{8} x_i\right)}{8} = 204 - \frac{(36)^2}{8} = 42$$

$$S_{xy} = \sum_{i=1}^{8} x_i y_i - \frac{\sum_{i=1}^{8} x_i \sum_{i=1}^{8} y_i}{8} = 249 - \frac{36(46)}{8} = 42$$

To obtain the correlation coefficient, r, we use method 2:

$$r = \frac{S_{xy}}{\sqrt{S_{xx} S_{yy}}} = \frac{42}{\sqrt{43.5 \times 42}} = 0.982607$$

We find r again using method 1:

$$n = 8; \quad \sum_{i=1}^{8} x_i y_i = 249; \quad \sum_{i=1}^{8} y_i = 46; \quad \sum_{i=1}^{8} x_i = 36; \quad \sum_{i=1}^{8} x_i^2 = 249; \quad \sum_{i=1}^{8} y_i^2 = 308;$$

$$r = \frac{n \sum_{i=1}^{n} x_i y_i - \sum_{i=1}^{n} x_i \sum_{i=1}^{n} y_i}{\sqrt{\left[n \sum_{i=1}^{n} x_i^2 - \left(\sum_{i=1}^{n} x_i\right)^2\right]\left[n \sum_{i=1}^{n} y_i^2 - \left(\sum_{i=1}^{n} y_i\right)^2\right]}}$$

$$= \frac{8(249) - 36(46)}{\sqrt{[8(204) - 36^2][8(308) - 46^2]}} = 0.982607$$

■ **EXAMPLE 11.3**   The data on age of women and their percentage of body fat, from Example 10.3, are presented below:

Age	23	41	40	47	48	47	47	51	54	53	56	57	58	58
Body Fat	28.8	33.4	28.2	26.3	33.1	37.4	38.2	40.1	43.5	45.2	41.3	40.6	41.8	41.4

Obtain the correlation coefficient for the linear relationship between age and body fat of the women.

x	y	xy	x^2	y^2
23	28.8	662.4	529	829.44
41	33.4	1369.4	1681	1115.56
40	28.2	1128	1600	795.24
47	26.3	1236.1	2209	691.69
48	33.1	1588.8	2304	1095.61
47	37.4	1757.8	2209	1398.76
47	38.2	1795.4	2209	1459.24
51	40.1	2045.1	2601	1608.01
53	45.2	2395.6	2809	2043.04
54	43.5	2349	2916	1892.25
56	41.3	2312.8	3136	1705.69
57	40.6	2314.2	3249	1648.36
58	41.8	2424.4	3364	1747.24
58	41.4	2401.2	3364	1713.96
680	**519.3**	**25780.2**	**34180**	**19744.09**

Solution Using method 1, we determine the value of r as follows:

$$n = 14; \quad \sum_{i=1}^{14} x_i y_i = 25780.2; \quad \sum_{i=1}^{14} y_i = 519.3; \quad \sum_{i=1}^{14} x_i = 680; \quad \sum_{i=1}^{14} x_i^2 = 34180; \quad \sum_{i=1}^{14} y_i^2 = 19744.09;$$

$$r = \frac{n \sum_{i=1}^{n} x_i y_i - \sum_{i=1}^{n} x_i \sum_{i=1}^{n} y_i}{\sqrt{\left[n \sum_{i=1}^{n} x_i^2 - \left(\sum_{i=1}^{n} x_i \right)^2 \right]\left[n \sum_{i=1}^{n} y_i^2 - \left(\sum_{i=1}^{n} y_i \right)^2 \right]}}$$

$$= \frac{14(25780.2) - 680(519.3)}{\sqrt{[14(34180) - 680^2][14(19744.09) - 519.3^2]}} = 0.747931$$

■ **EXAMPLE 11.4** Consider the data of Example 10.4 on the heights and weights of people. Find r, the correlation coefficient, for the linear relationship between heights and weights.

Weight (x)	Heights (y)	Weight (x)	Heights (y)
170	75	177	74
177	75	165	66
166	66	182	72
175	75	172	71
160	65	177	74
185	74	168	68
165	65	167	67
172	70	158	65
185	74	157	65
163	65	160	63

Solution

Weight (x)	Heights (y)	xy	x^2	y^2
170	75	12750	28900	5625
177	75	13275	31329	5625
166	66	10956	27556	4356
175	75	13125	30625	5625
160	65	10400	25600	4225
185	74	13690	34225	5476
165	65	10725	27225	4225
172	70	12040	29584	4900
185	74	13690	34225	5476
163	65	10595	26569	4225
177	74	13098	31329	5476
165	66	10890	27225	4356
182	72	13104	33124	5184
172	71	12212	29584	5041
177	74	13098	31329	5476
168	68	11424	28224	4624
167	67	11189	27889	4489
158	65	10270	24964	4225
157	65	10205	24649	4225
160	63	10080	25600	3969
3401	**1389**	**236816**	**579755**	**96823**

$$n = 20; \quad \sum_{i=1}^{20} x_i y_i = 236816; \quad \sum_{i=1}^{20} y_i = 1389; \quad \sum_{i=1}^{20} x_i = 3401; \quad \sum_{i=1}^{20} x_i^2 = 579755; \quad \sum_{i=1}^{20} y_i^2 = 96823;$$

$$r = \frac{n \sum_{i=1}^{n} x_i y_i - \sum_{i=1}^{n} x_i \sum_{i=1}^{n} y_i}{\sqrt{\left[n \sum_{i=1}^{n} x_i^2 - \left(\sum_{i=1}^{n} x_i \right)^2 \right]\left[n \sum_{i=1}^{n} y_i^2 - \left(\sum_{i=1}^{n} y_i \right)^2 \right]}}$$

$$= \frac{20(236816) - 3401(1389)}{\sqrt{[20(579755) - 3401^2][20(96823) - 1389^2]}} = 0.867549$$

11.2 Testing Whether the Population Correlation Coefficient is Zero

Each time we used the observations $(x_1, y_1), (x_2, y_2), \ldots, (x_n, y_n)$ to determine the linear relationship between X and Y, we were using a sample from the population of points (x, y) to estimate the linear relationship between X and Y. The correlation coefficient, r, derived on that basis is a statistic. The true population linear correlation coefficient, ρ, is therefore being estimated by r. Once we have the extent of the linear relationship determined by finding a value r, we may wish to test the hypothesis that the true population linear correlation coefficient, ρ, is zero against some alternatives. Just like we asserted that $-1 \leq r \leq +1$, we can also state that the population linear correlation coefficient is also $-1 \leq \rho \leq +1$.

We can test the following sets of hypotheses:

Case 1: $H_0{:}\rho = 0$ vs. $H_1{:}\rho \neq 0$
Case 2: $H_0{:}\rho = 0$ vs. $H_1{:}\rho > 0$
Case 3: $H_0{:}\rho = 0$ vs. $H_1{:}\rho < 0$

In Case 1 we test to determine that X and Y are correlated. Case 2 is used for testing for positive linear correlation, while Case 3 is seeking to establish that X and Y are negatively linearly correlated. The test statistic used for testing all these hypotheses has a t-distribution provided that the following assumptions which we mentioned in the context of linear regression are satisfied.

We have to assume that each of y_i we observe is obtained independent of any other. Another way to put this is to say that the $y_1, y_2, \ldots, y_n$ which is associated with $x_1, x_2, \ldots, x_n$ is a random sample. Offshoots of this assumption are:

1. There are two parameters β_0 and β_1 which need to be estimated in order to fit the straight line and the mean of y given x is $\beta_0 + \beta_1 x$.
2. For a given value of x the distribution of the response variable y is normal.
3. Each y_i is independent of another, or the $y_1, y_2, \ldots, y_n$ constitute a random sample.
4. For a given value of x, the variance of the corresponding y is the same for all values of x, and that common variance is σ^2.

Under these assumptions, the test statistic is:

$$t_{cal} = \frac{r}{\sqrt{\dfrac{1 - r^2}{n - 2}}}.$$

t_{cal} follows t-distribution with $n - 2$ degrees of freedom. We will do some examples.

■ EXAMPLE 11.5 Test the hypothesis that the population correlation coefficient, ρ, for the data of Example 11.1 is zero against that it is less than zero at a 5% level of significance.

Solution

$$S_{yy} = \sum_{i=1}^{13} y_i^2 - \frac{\left(\sum_{i=1}^{13} y_i\right)^2}{13} = 2969 - \frac{185^2}{13} = 336.3077$$

$$S_{xy} = \sum_{i=1}^{13} x_i y_i - \frac{\left(\sum_{i=1}^{13} y_i\right)\left(\sum_{i=1}^{13} x_i\right)}{13} = 10113 - \frac{185(755)}{13} = -631.231$$

$$S_{xx} = \sum_{i=1}^{13} x_i^2 - \frac{\left(\sum_{i=1}^{13} x_i\right)^2}{13} = 45971 - \frac{755^2}{13} = 2122.923$$

$$r = \frac{S_{xy}}{\sqrt{S_{yy} S_{xx}}} = \frac{-631.231}{\sqrt{2122.923 \times 336.3077}} = -0.74706$$

We test the hypothesis $H_0{:}\rho = 0$ vs. $H_1{:}\rho < 0$

$$t_{cal} = \frac{r}{\sqrt{\dfrac{1 - r^2}{n - 2}}} = \frac{-0.74706}{\sqrt{\dfrac{1 - (-0.74706)^2}{13 - 2}}} = -3.7273$$

From the tables, we see that $t(13 - 2, 0.05) = 1.796$. Thus at a 5% level of significance, we fail to reject H_0 and conclude that ρ is negative; X and Y are negatively linearly correlated.

■ EXAMPLE 11.6 Test the hypothesis that the population correlation coefficient, ρ, for the data of Example 11.2 is zero against that it is greater than zero at a 1% level of significance.

Solution $n = 8; \sum_{i=1}^{8} x_i y_i = 249; \sum_{i=1}^{8} y_i = 46; \sum_{i=1}^{8} x_i = 36; \sum_{i=1}^{8} x_i^2 = 249; \sum_{i=1}^{8} y_i^2 = 308;$

$$r = \frac{n \sum_{i=1}^{n} x_i y_i - \sum_{i=1}^{n} x_i \sum_{i=1}^{n} y_i}{\sqrt{\left[n \sum_{i=1}^{n} x_i^2 - \left(\sum_{i=1}^{n} x_i\right)^2\right]\left[n \sum_{i=1}^{n} y_i^2 - \left(\sum_{i=1}^{n} y_i\right)^2\right]}}$$

$$= \frac{8(249) - 36(46)}{\sqrt{[8(204) - 36^2][8(308) - 46^2]}} = 0.982607$$

We test the hypothesis $H_0:\rho = 0$ vs. $H_1:\rho > 0$

$$t_{cal} = \frac{r}{\sqrt{\dfrac{1-r^2}{n-2}}} = \frac{0.982607}{\sqrt{\dfrac{1-(0.98607)^2}{8-2}}} = 12.9613$$

From the tables, we see that $t(8-2, 0.01) = 3.143$. Thus at a 1% level of significance, we reject H_0 and conclude that ρ is positive; X and Y are positively correlated.

■ **EXAMPLE 11.7**   Test the hypothesis that the population correlation coefficient, ρ, for the data of Example 11.3 is zero against that it is not equal to zero at a 1% level of significance.

Solution

$$n = 14; \sum_{i=1}^{14} x_i y_i = 25780.2; \sum_{i=1}^{14} y_i = 519.3; \sum_{i=1}^{14} x_i = 680; \sum_{i=1}^{14} x_i^2 = 34180; \sum_{i=1}^{14} y_i^2 = 19744.09;$$

$$r = \frac{n\sum_{i=1}^{n} x_i y_i - \sum_{i=1}^{n} x_i \sum_{i=1}^{n} y_i}{\sqrt{\left[n\sum_{i=1}^{n} x_i^2 - \left(\sum_{i=1}^{n} x_i\right)^2\right]\left[n\sum_{i=1}^{n} y_i^2 - \left(\sum_{i=1}^{n} y_i\right)^2\right]}}$$

$$= \frac{14(25780.2) - 680(519.3)}{\sqrt{[14(34180) - 680^2][14(19744.09) - 519.3^2]}} = 0.747931$$

We test the hypothesis $H_0:\rho = 0$ vs. $H_1:\rho \neq 0$

$$t_{cal} = \frac{r}{\sqrt{\dfrac{1-r^2}{n-2}}} = \frac{0.747931}{\sqrt{\dfrac{1-(0.747931)^2}{14-2}}} = 3.9033$$

From the tables, we see that $t(14-2, 0.005) = 3.055$. Thus at a 1% level of significance, we reject H_0 and conclude that ρ is not zero. Body fat and age are correlated.

■ **EXAMPLE 11.8**   Test the hypothesis that the population correlation coefficient, ρ, for the data of Example 11.4 is zero against that it is not equal to zero at a 1% level of significance.

Solution

$$n = 20; \sum_{i=1}^{20} x_i y_i = 236816; \sum_{i=1}^{20} y_i = 1389; \sum_{i=1}^{20} x_i = 3401; \sum_{i=1}^{20} x_i^2 = 579755; \sum_{i=1}^{20} y_i^2 = 96823;$$

$$r = \frac{n\sum_{i=1}^{n} x_i y_i - \sum_{i=1}^{n} x_i \sum_{i=1}^{n} y_i}{\sqrt{\left[n\sum_{i=1}^{n} x_i^2 - \left(\sum_{i=1}^{n} x_i\right)^2\right]\left[n\sum_{i=1}^{n} y_i^2 - \left(\sum_{i=1}^{n} y_i\right)^2\right]}}$$

$$= \frac{20(236816) - 3401(1389)}{\sqrt{[20(579755) - 3401^2][20(96823) - 1389^2]}} = 0.867549$$

Here we test the hypothesis $H_0: \rho = 0$ vs. $H_1: \rho \neq 0$

$$t_{cal} = \frac{r}{\sqrt{\dfrac{1 - r^2}{n - 2}}} = \frac{0.867549}{\sqrt{\dfrac{1 - (0.867549)^2}{20 - 2}}} = 7.4006$$

From the tables, we see that $t(20 - 2, 0.005) = 2.878$. Thus at a 1% level of significance, we reject H_0 and conclude that ρ is not zero. Weight and age are correlated.

11.3 Coefficient of Determination

In Chapter 10, we showed in Equation 10.24 that the typical deviation from the mean $\bar{y}$ of the ith observation can be subdivided into two parts as illustrated in Figure 10.2. The first part is a deviation of the fitted value from the mean, and the second a deviation of the ith observation from the fitted value. We reproduce Equation 10.24 here:

$$y_i - \bar{y} = (\hat{y}_i - \bar{y}) + (y_i - \hat{y}_i) \tag{10.24}$$

As stated in chapter 10, from 10.24, it can be shown (but the proof is outside the scope of this text) that:

$$\sum_{i=1}^{n} (y_i - \bar{y})^2 = \sum_{i=1}^{n} (\hat{y}_i - \bar{y})^2 + \sum_{i=1}^{n} (y_i - \hat{y}_i)^2 \tag{10.25}$$

The implication of Equation 10.25 is that:

$$S_{yy} = SSR + SSE \tag{10.26}$$

The SSR in Equation 10.26 is the sum of squares due to regression or fitting the straight line. Therefore the total variation in y, can be subdivided into variation that is explained by the regression (the fitting of the straight line) and variation which is due to error (residual variation).

If we divide both sides of Equation 10.26 by S_{yy}, we have:

$$\frac{S_{yy}}{S_{yy}} = \frac{SS_R}{S_{yy}} + \frac{SSE}{S_{yy}} \Rightarrow 1 = \frac{SS_R}{S_{yy}} + \frac{SSE}{S_{yy}} \tag{11.1}$$

$$1 - \frac{SSE}{S_{yy}} = \frac{SS_R}{S_{yy}} \text{ or } \frac{SS_R}{S_{yy}} = 1 - \frac{SSE}{S_{yy}} \tag{11.2}$$

$\dfrac{SS_R}{S_{yy}}$ is called the coefficient of determination and is conventionally denoted by R^2. From Equations 11.1 and 11.2, we see that it is the ratio of the sum of squares due to regression and the total sum of squares. It represents the proportion of variation in the responses Y which is explained or accounted for by the fitting of the regression line. $\frac{SSE}{S_{yy}}$ is the proportion of variation in Y due to residual variation or due to error.

$R^2 = \frac{SS_R}{S_{yy}} = 1 - \frac{SSE}{S_{yy}}$ is the coefficient of determination. It measures the proportion of variation in Y which is explained by the fitted regression line. It is often used as a measure of how well the model of Y (in this case, a straight line that describes the values of Y in terms of X) fits the data.

Actually, quantitatively, R^2 is equal to the square of the linear correlation coefficient r. It is not recommended that we find r as the square root of R^2 unless we have already fitted the line and know whether the slope is positive or negative, which will help us determine whether r is positive or negative.

EXAMPLE 11.9 Fit a straight line to the data of Example 11.1. Obtain the coefficient of determination and interpret it.

Solution

X	Y	XY	X²	Y²	$\hat{y}$	$y - \hat{y}$	$(y - \hat{y})^2$
74	5	370	5476	25	9.496195	−4.4962	20.21577
66	6	396	4356	36	11.87492	−5.87492	34.51467
81	8	648	6561	64	7.414813	0.585187	0.342444
52	11	572	2704	121	16.03768	−5.03768	25.37826
73	12	876	5329	144	9.793536	2.206464	4.868484
62	15	930	3844	225	13.06428	1.93572	3.747012
52	16	832	2704	256	16.03768	−0.03768	0.00142
45	17	765	2025	289	18.11907	−1.11907	1.25231
62	18	1116	3844	324	13.06428	4.93572	24.36133
46	18	828	2116	324	17.82173	0.178274	0.031782
60	19	1140	3600	361	13.65896	5.341039	28.5267
46	20	920	2116	400	17.82173	2.178274	4.744877
36	20	720	1296	400	20.79513	−0.79513	0.632232
755	185	10113	45971	2969	185	3.55E-15	148.6173

$$S_{yy} = \sum_{i=1}^{13} y_i^2 - \frac{\left(\sum_{i=1}^{13} y_i\right)^2}{13} = 2969 - \frac{185^2}{13} = 336.3077$$

$$S_{xy} = \sum_{i=1}^{13} x_i y_i - \frac{\left(\sum_{i=1}^{13} y_i\right)\left(\sum_{i=1}^{13} x_i\right)}{13} = 10113 - \frac{185(755)}{13} = -631.231$$

$$S_{xx} = \sum_{i=1}^{13} x_i^2 - \frac{\left(\sum_{i=1}^{13} x_i\right)^2}{13} = 45971 - \frac{755^2}{13} = 2122.923$$

$$\hat{\beta}_1 = \frac{S_{xy}}{S_{xx}} = \frac{-631.231}{2122.923} = -0.29734;$$

$$\hat{\beta}_0 = \overline{Y} - \hat{\beta}_1\overline{X} = \frac{185}{13} - (-0.29734)\left(\frac{755}{13}\right) = 31.49938$$

The fitted line is:

$$\hat{y} = 31.49938 - 0.29734x \tag{11.3}$$

With Equation 11.3, we obtain the $\hat{y}_i$ in the above table, and the residuals, $y_i - \hat{y}_i$ and subsequently, $(y_i - \hat{y}_i)^2$.

But $SSE = \sum_{i=1}^{n} (y_i - \hat{y}_i)^2$ so that we can obtain $R^2 = 1 - \frac{SSE}{S_{yy}}$; from the table above, $SSE = 148.6173$.

In this case, $R^2 = 1 - \frac{SSE}{S_{yy}} = 1 - \frac{148.6173}{336.3077} = 0.558091$.

11.4 Use of MINITAB to Find the Correlation Coefficient and Test its Population Value is Not Zero

The use of MINITAB for finding the correlation coefficient, r, and testing that $\rho = 0$ is relatively easy and straight forward. We provide two simple illustrations using the data of Examples 11.4 and 11.9.

Reanalyzing Data of Example 11.4 Using MINITAB

```
MTB > set c1
DATA> 170 177 166 175 160 185 165 172 185 163 177 165 182
DATA> 172 177 168 167 158 157 160
DATA> end
MTB > set c2
DATA> 75 75 66 75 65 74 65 70 74 65 74 66 72 71 74
DATA> 68 67 65 65 63
DATA> end
MTB > name c1 'X' c2 'Y'

MTB > corr c1 c2
```

Correlations: X, Y

```
Pearson correlation of X and Y = 0.868
p-Value = 0.000
```

The p-value (< 0.0001) indicates that data are highly significant, which means that $\rho \neq 0$.

Reanalyzing Data of Example 11.9 Using MINITAB

Solution

```
MTB > set c1
DATA> 74 66 81 52 73 62 52 45 62 46 60 46 36
DATA> end
MTB > set c2
DATA> 74 66 81 52 73 62 52 45 62 46 60 46 36
DATA> end
MTB > name c1 'X' c2 'Y'
MTB > corr c1 c2
```

Output:

Correlations: X, Y

```
Pearson correlation of X and C4 = -0.747
p-Value = 0.003
```

The p-value (< 0.003) indicates that data are highly significant, which means that $\rho \neq 0$.

Reference

Pearson, Karl (1896). Mathematical Contributions to the Theory of Evolution III. Regression, Heredity, and Panmixia. In *Karl Pearson's Early Statistical Papers.* London: Cambridge University Press, 1956.

Exercises

11.1

x	y
12	8
14	9
11	6
11	4
10	3
9	8
8	3
6	2
7	4
8	3

For data of X and Y in the table above:
(a) Obtain r, the correlation coefficient, for the linear relationship between X and Y.
(b) Test the hypothesis that the population correlation coefficient, ρ, is zero against that it is not at a 5% level of significance.
(c) Find the coefficient of determination for the fitted linear relationship.

11.2

x	y
0	2.22
1	1.33
2	1.18
4	0.72
5	0.63
7	0.57
8	0.49
9	0.45

For data of X and Y in the table above:
(a) Obtain r, the correlation coefficient, for the linear relationship between X and Y.
(b) Test the hypothesis that the population correlation coefficient, ρ, is zero against that it is not at a 1% level of significance.
(c) Find the coefficient of determination for the fitted linear relationship.

11.3

x	y
3.7	650
2	490
3.3	580
2.8	530
3.5	640
3.15	615
3.7	650
2.2	498
3.3	590
2.8	550
3.3	630
3.2	625

(a) Obtain r, the correlation coefficient, for the linear relationship between X and Y.
(b) Test the hypothesis that the population correlation coefficient, ρ, is zero against that it is greater at a 1% level of significance.
(c) Find the coefficient of determination for the fitted linear relationship.

11.4

x	y
0	1.3
2	3
3	7
5	8
7	17
8	19
10	22
11	30

(a) Obtain r, the correlation coefficient, for the linear relationship between X and Y.
(b) Test the hypothesis that the population correlation coefficient, ρ, is zero against that it is not at a 1% level of significance.
(c) Find the coefficient of determination for the fitted linear relationship.

11.5   The following are the heights (X) and the diameters (Y) of trees:

Tree	1	2	3	4	5	6	7	8	9	10
X	1.8	1.9	1.8	2.4	5.1	3.1	5.5	5.1	8.3	13.7
Y	22.0	32.5	23.6	40.8	63.2	27.9	41.4	44.3	63.5	92.7

(a) Obtain the linear correlation coefficient for X and Y and interpret it.
(b) Test that the population linear correlation coefficient is zero against that it is not, at a 5% level of significance.
(c) Obtain the coefficient of determination for the fitted linear relationship between X and Y.

11.6

Height (x)	Shoe Size (y)
66	5.5
63.96	5.5
63	3.5
66	5.5
62.04	3.5
64.56	5.5
63.6	8
64.92	5
64.92	5
60	3
67.68	7
66	4
66.96	6
66	6
66.96	6.5

Obtain the correlation coefficient for the linear relationship between height and shoe size. Test the hypothesis that the population correlation coefficient is zero against that it is greater at a 2% level of significance. Obtain the coefficient of determination for the fitted linear relationship between X and Y.

11.7

Height (ins.)	Weight (lbs.)
63	154
65	145
70	166
66	150
58	145
69	164
67	156
64	158
65	154
68	155

Obtain the correlation coefficient for the linear relationship between height and weight. Test the hypothesis that the population correlation coefficient is zero against that it is greater at a 2% level of significance. Obtain the coefficient of determination for the fitted linear relationship between X and Y.

11.8

Weekly Working Hours	Annual Salary
19	23.4
17	24.1
18	20.45
15	19.55
25	28.67
30	33.75
24	27.88
17	25.6
32	36.9
28	33.26
19	22.34
24	26.9
30	36.67

Obtain the correlation coefficient for the linear relationship between weekly working hours and annual salary. Test the hypothesis that the population correlation coefficient is zero against that it is not at a 1% level of significance. Obtain the coefficient of determination for the fitted linear relationship between X and Y.

11.9   A sample of instructional costs at twelve institutions and the average wages of faculty were collected. The aim was to find out whether they are linearly related. All values were recorded in thousands of dollars as below:

Cost	Wage
5.66	11.96
15.6	12
10.51	11.44
6.24	11.44
1.76	9.61
3.94	11.31
4.28	9.99
2.98	9.97
15.3	11.29
12.28	11.67
2.34	8.84
2.74	8.85

Obtain the correlation coefficient for the linear relationship between cost and annual wage. Test the hypothesis that the population correlation coefficient is zero against that it is not at a 1% level of significance. Obtain the coefficient of determination for the fitted linear relationship between X and Y.

11.10   In the same study as described earlier, a sample of instructional costs at fifteen institutions and the number of students were collected and presented. The aim was to find out whether they are linearly related. All cost values were recorded in thousands of dollars, and number of students is in thousands:

Cost	Number of Students	Cost	Number of Students
1.08	1.58	5.56	13.45
5.66	3.8	9.29	7.52
15.7	11.4	12.09	9.78
6.95	5.36	15.85	13.95
10.32	8.04	19.54	16.65
4.88	2.86	1.83	1.15
15.35	12.24	2.65	4.06
15.91	4.38		

Obtain the correlation coefficient for the linear relationship between cost and number of students. Test the hypothesis that the population correlation coefficient is zero against that it is not at a 5% level of significance.

11.11   In the same study as described earlier, a sample of instructional costs at fifteen institutions and the number of faculty were collected and presented. The aim was to find out whether they are linearly related. All cost values were recorded in thousands of dollars as below:

Cost	Number of Faculty	Cost	Number of Faculty
1.08	81	5.56	340
5.66	290	9.29	485
15.7	466	12.09	643
6.95	435	15.85	875
10.32	475	19.54	1071
4.88	225	1.83	99
15.35	959	2.65	220
15.91	848		

Obtain the correlation coefficient for the linear relationship between cost and number faculty. Test the hypothesis that the population correlation coefficient is zero against that it is greater at a 1% level of significance. Obtain the coefficient of determination for the fitted linear relationship between X and Y.

11.12   The following are the ages of a sample of people and the number of hours of television they watched per week.

X (age)	Y (weekly TV watching hours)
10	20
10	21
13	22
13	26
15	22
17	20
18	25
19	27
19	23
23	30
25	28
28	30

Obtain the correlation coefficient for the linear relationship between age and number of hours of TV watching per week. Test the hypothesis that the population correlation coefficient is zero against that it is greater at a 1% level of significance. Obtain the coefficient of determination for the fitted linear relationship between X and Y.

11.13 Is there a linear relationship between scores in STAT 193 class and the average number of hours per week of study for the course? A study matched students' recollections of number of hours of study per week with their final scores in STAT 193.

X (Weekly Hours of Study)	Y (Final Score in STAT 193)
2	70
2	75
2.5	72
2.5	78
2.75	80
3	81
3	70
3.5	85
3	80
3.25	70
3.5	87
3.5	85
3.75	90
4	90

Obtain the correlation coefficient for the linear relationship between weekly hours of study and final score in STAT 193. Test the hypothesis that the population correlation coefficient is zero against that it is greater at a 1% level of significance. Obtain the coefficient of determination for the fitted linear relationship between X and Y.

11.14

x	y	x	y
0	6	13	22
3	10	11	17
4	12	15	27
5	12	18	29
7	15	24	27
3	9	12	19
10	19	11	13

Obtain the correlation coefficient for the linear relationship between X and Y. Test the hypothesis that the population correlation coefficient is zero against that it is not at a 1% level of significance. Obtain the coefficient of determination for the fitted linear relationship between X and Y.

11.15

X	Y
3.2	7.4
3	6.5
7.2	5.5
6	6.2
7	6.4
5.4	4.3
8	4
12.2	3.7
12.5	3.3
14.7	3.1
13.4	3.8

Obtain the correlation coefficient for the linear relationship between X and Y. Test the hypothesis that the population correlation coefficient is zero against that it is less at a 1% level of significance. Obtain the coefficient of determination for the fitted linear relationship between X and Y.

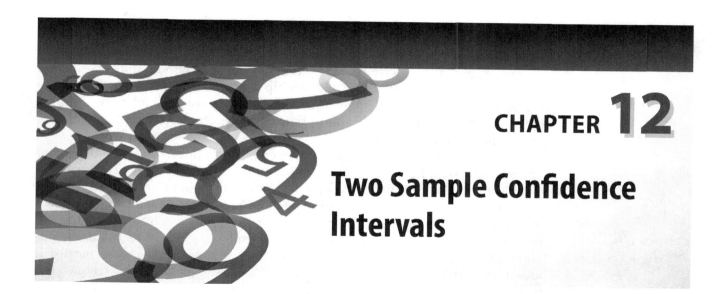

CHAPTER 12

Two Sample Confidence Intervals

Introduction

The sampling distribution of the difference between two sample means and proportions form part of the core knowledge that lead to understanding of statistical inference when comparing two population parameters. In this chapter, we consider confidence intervals related to differences between two means and two proportions. A knowledge of the distributions of these differences will be useful in understanding the theory of confidence intervals for differences between two means or two proportions which will be treated later. We briefly discuss the sampling distributions of these differences as an introductory material for this chapter.

12.1 **Sampling Distribution of Difference Between Two Sample Means**

In Chapter 7, we discussed the sampling distribution of the sample mean. The results of that section were important in preparing us for the construction of the confidence interval for the population mean, μ which was discussed in Chapter 8. By that theory which applied to the single sample mean, we found that if

$X \sim N(\mu, \sigma^2)$, then for a sample of size n from X, $\overline{x} \sim N\left(\mu, \dfrac{\sigma^2}{n}\right)$.

An extension of that theory shows that if $X_1 \sim N(\mu_1, \sigma_1^2)$ and $X_2 \sim N(\mu_2, \sigma_2^2)$ then if two independent samples of sizes n_1 and n_2 are taken from X_1 and X_2, respectively, the difference between their sample means:

$$\overline{x}_1 - \overline{x}_2 \sim N\left[\mu_1 - \mu_2, \ \frac{\sigma_1^2}{n_1} + \frac{\sigma_2^2}{n_2}\right] \tag{12.1}$$

The results above have already been part of established statistics theory. Without attempting to prove the relationship in (12.1), which is beyond the scope of this text, we can easily demonstrate its veracity by using simulations. Using a Statistics software, and writing appropriate code, we obtained the compendium of pictures in Figures 12.1 and 12.2.

In Figure 12.1, the differences between two sample means were simulated and plotted. Here, samples of sizes 20, 30, and 50 were taken from normally distributed variables, $X_1 \sim N(50, 49)$ and $X_2 \sim N(30, 33)$. The sample means were calculated each time for each variable, and the differences between their sample means were determined in 5,000 simulations. As we can see, the differences were approximately normally distributed. This goes to buttress the result stated in (12.1). In those simulation results, the pictures show that the higher the sample size, the better the distribution approximates the normal. In Figure 12.2, the sample sizes were retained but the number of simulations was raised to 50,000. Figure 12.2 does not only support the result stated in (12.1), but also emphasizes that the approximation gets better for larger sample sizes, and as the number of simulations increases.

The results as depicted in Figures 12.1 and 12.2 give insight into the established theory encapsulated in equation (12.1). This means that we can obtain the probabilities of the differences between the sample means via conversion to the standard normal variable, Z as shown in equation (12.2):

$$Z = \frac{\bar{x}_1 - \bar{x}_2 - (\mu_1 - \mu_2)}{\sqrt{\dfrac{\sigma_1^2}{n_1} + \dfrac{\sigma_2^2}{n_2}}} \sim N(0, 1) \qquad (12.2)$$

We apply this theory in the next example.

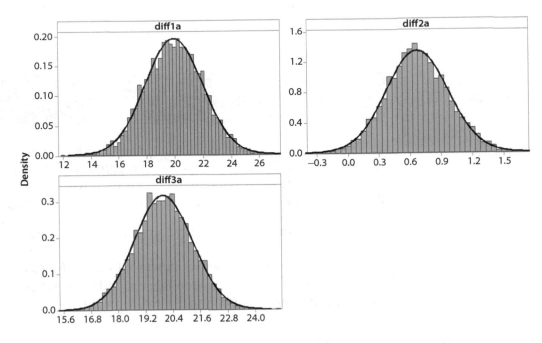

FIGURE 12.1 Histogram of 5,000 Simulations of Differences Between xbar1 and xbar2 for Samples $n_1 = n_2 = 20$, $n_1 = n_2 = 30$, and $n_1 = n_2 = 50$.

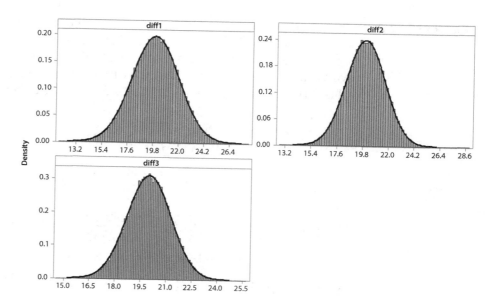

FIGURE 12.2   Histogram of 50,000 Simulations of Differences Between xbar1 and xbar2 for Samples $n_1 = n_2 = 20$, $n_1 = n_2 = 30$, and $n_1 = n_2 = 50$.

EXAMPLE 12.1   Suppose that X_1 is normally distributed with mean 90 and variance 78

Suppose that X_2 is normally distributed with mean 80 and variance 88

Take samples of sizes 15 and 20 from X_1 and X_2, respectively. What is the probability that the difference in sample means is: (a) greater than 5, (b) less than 15, and (c) lies between 6 and 13.

(a) We use the following formula to convert the samples means to standard normal variates:

$$Z = \frac{\overline{X}_1 - \overline{X}_2 - (\mu_1 - \mu_2)}{\sigma_{\overline{x}_1 - \overline{x}_2}}$$

$$\sigma_{\overline{x}_1 - \overline{x}_2} = \sqrt{\frac{\sigma_1^2}{n_1} + \frac{\sigma_2^2}{n_2}} = \sqrt{\frac{78}{15} + \frac{88}{20}} = 3.098387$$

$$P[(\overline{X}_1 - \overline{X}_2) > 5] = P\left(Z > \frac{5 - (90 - 80)}{3.098387}\right) = p(Z > -1.61) = 1 - 0.0537 = 0.9463$$

(b) $$\sigma_{\overline{x}_1 - \overline{x}_2} = \sqrt{\frac{\sigma_1^2}{n_1} + \frac{\sigma_2^2}{n_2}} = \sqrt{\frac{78}{15} + \frac{88}{20}} = 3.098387$$

$$P[(\overline{X}_1 - \overline{X}_2) < 15] = P\left(Z < \frac{15 - (90 - 80)}{3.098387}\right) = p(Z < 1.61) = 0.9463$$

(c) $$\sigma_{\overline{x}_1 - \overline{x}_2} = \sqrt{\frac{\sigma_1^2}{n_1} + \frac{\sigma_2^2}{n_2}} = \sqrt{\frac{78}{15} + \frac{88}{20}} = 3.098387$$

$$P[6 < (\overline{X}_1 - \overline{X}_2) < 13] = P\left(\frac{6 - (90 - 80)}{3.098387} < Z < \frac{13 - (90 - 80)}{3.098387}\right)$$

$$= p(-1.29 < Z < 0.97) = 0.8365 - 0.0985 = 0.738$$

EXAMPLE 12.2   Suppose population one has a mean of 40 and a variance of 140 and population two has a mean of 35 and a variance of 164. If a sample of 30 observations is taken from population one and a sample of 26 observations is taken from population two. What is the probability that the difference between the means of the samples will be (a) less than 9, (b) greater than 1.5, and (c) lies between 4 and 8?

$$\sigma_{\bar{x}_1 - \bar{x}_2} = \sqrt{\frac{\sigma_1^2}{n_1} + \frac{\sigma_2^2}{n_2}} = \sqrt{\frac{140}{30} + \frac{164}{26}} = 3.312757$$

(a) $P[(\bar{X}_1 - \bar{X}_2) < 9] = P\left(Z < \dfrac{9 - (40 - 35)}{3.312757}\right) = P(Z < 1.21) = 0.8869$

(b) $P[(\bar{X}_1 - \bar{X}_2) > 1.5] = P\left(Z > \dfrac{1.5 - (40 - 35)}{3.312757}\right) = P(Z > -1.06)$

$$= 1 - 0.1446 = 0.8554$$

(c) $P[4 < (\bar{X}_1 - \bar{X}_2) < 8] = P\left(\dfrac{4 - (40 - 35)}{3.312757} < Z < \dfrac{8 - (40 - 35)}{3.312757}\right)$

$$= P(-0.30 < Z < 0.91)$$

$$= 0.8186 - 0.3821 = 0.4365$$

12.1.1 Distribution of the Difference Between Two Sample Means When the Populations Are Not Normal (Central Limit Theorem)

In the earlier examples, the samples were drawn from two normal distributions and we can see that the differences follow the normal distribution: what happens if we sample from two nonnormal distributions? The law of large numbers and the central limit theorem have proven that the differences in this case, for sufficiently large samples, will also be approximately normally distributed. Again, proof is being the scope of this text.

12.2 Sampling Distribution of the Difference Between Two Sample Proportions

As we learned in Chapter 7, each time we take a sample of size n from a population which has a true proportion p with an attribute of interest, and determine the $\hat{p}$, we get a different value. The $\hat{p}$ is a variable, which varies from one sample to another. By the same token, if we take an independent sample of size n_1 from a population with true population value p_1, and an independent sample of size n_2 from a population with true population value p_2, we can calculate the values of $\hat{p}_1 - \hat{p}_2$. We can explore its distribution by plotting the histogram of the values of $\hat{p}_1 - \hat{p}_2$ and examining the underlying shape which the histogram suggests. We want to determine whether it is symmetric, bell-shaped, skewed to the left or to the right? Is it humped in more than one place? We can therefore simulate the behavior of $\hat{p}_1 - \hat{p}_2$ in the long run. We carried out such a simulation, and the result is shown in Figures 12.3 and 12.4.

Figures 12.3 and 12.4 show the results of 10,000 and 50,000 simulations of the differences between two proportions from populations where the true proportions that have a desired attribute are 0.40 and 0.60, respectively. The pictures show that the differences are normally distributed with a mean of 0.20. The approximation to normality was better when we simulated for 50,000 times rather than 10,000 times.

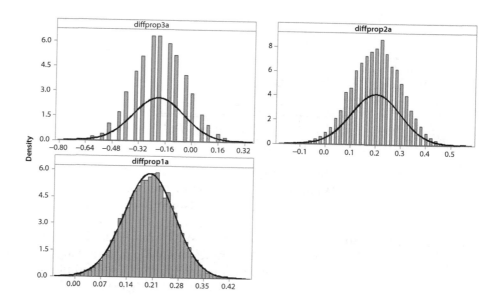

■ FIGURE 12.3   10,000 Simulations of Differences of Proportions for Populations With $p_2 = 0.6$ and $p_1 = 0.4$, and Sample Sizes $n_1 = n_2 = 20, 50, 100$.

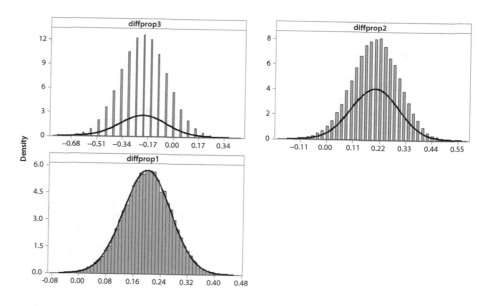

■ FIGURE 12.4   50,000 Simulations of Differences of Proportions for Populations With $p_2 = 0.6$ and $p_1 = 0.4$, and Sample Sizes $n_1 = n_2 = 20, 50, 100$.

Statisticians have long established that in the long run (i.e., for large number of trials), the distribution of $\hat{p}_1 - \hat{p}_2$ is approximately normal with mean of $p_1 - p_2$ and a variance of $\dfrac{p_1(1-p_1)}{n_1} + \dfrac{p_2(1-p_2)}{n_2}$.

Therefore, if samples of sizes n_1 and n_2 are drawn from populations 1 and 2 which are known to have respective proportions of p_1 and p_2 of individuals with an attribute of interest, then

$$\hat{p}_1 - \hat{p}_2 \sim N\left[p_1 - p_2, \frac{p_1(1-p_1)}{n_1} + \frac{p_2(1-p_2)}{n_2}\right] \tag{12.3}$$

That leads to:

$$Z = \frac{(\hat{p}_1 - \hat{p}_2) - (p_1 - p_2)}{\sqrt{\dfrac{p_1(1-p_1)}{n_1} + \dfrac{p_2(1-p_2)}{n_2}}} \tag{12.4}$$

The above simulations serve to demonstrate without proof the results stated in (12.3) and (12.4). With the results we can determine the probabilities of different values of the differences between two sample proportions. Here is an example:

EXAMPLE 12.3 Suppose that it is known that 28% of men and 21% of women aged 60 and above have lower back pain. If samples of 1,090 and 1,200 are taken from men and women aged 60 and above, what is the probability that the difference in proportions found of men and women who have lower back pain (a) is less than 12%, (b) between 4% and 11%.

Solution (A) We need to find $P(\hat{p}_m - \hat{p}_f > 0.12)$. First we note that $p_m - p_f = 0.28 - 0.21 = 0.07$
We also find that:

$$\sqrt{\frac{p_m(1-p_m)}{n_m} + \frac{p_f(1-p_f)}{n_f}} = \sqrt{\frac{0.28(1-0.28)}{1090} + \frac{0.21(1-0.21)}{1200}} = 0.017978$$

Using the formula that $Z = \dfrac{(\hat{p}_m - \hat{p}_f) - (p_m - p_f)}{\sqrt{\dfrac{p_m(1-p_m)}{n_1} + \dfrac{p_f(1-p_f)}{n_2}}}$, then

$$P(\hat{p}_m - \hat{p}_f > 0.12) = P\left(Z > \frac{0.12 - 0.07}{0.017978}\right)$$

$$= P(Z < 2.78) = 0.9973.$$

(B) We need to find $P(0.04 < \hat{p}_m - \hat{p}_f < 0.11)$.

Then $P(0.04 < \hat{p}_m - \hat{p}_f < 0.11) = P\left(\dfrac{0.04 - 0.07}{0.017978} < Z < \dfrac{0.11 - 0.07}{0.017978}\right)$

$$= P(-1.67 < Z < 2.22)$$

$$= 0.9868 - 0.0475 = 0.9393.$$

EXAMPLE 12.4 Suppose it is known that 40% of male juniors in a college have binged on drinks at least once, while the equivalent percentage for females is 32%. Suppose that samples of 120 juniors were taken from each of female juniors and male juniors. What is the probability that $\hat{p}_m - \hat{p}_f$:

(a) Is less than 10%
(b) Lie between 6% and 13%

Solution (a) We need to find $P(\hat{p}_m - \hat{p}_f > 0.10)$. First, we find that $p_m - p_f = 0.40 - 0.32 = 0.08$
We also find that:

$$\sqrt{\frac{p_m(1-p_m)}{n_m} + \frac{p_f(1-p_f)}{n_f}} = \sqrt{\frac{0.32(1-0.32)}{120} + \frac{0.40(1-0.40)}{120}} = 0.061752$$

Using the formula that $Z = \dfrac{(\hat{p}_m - \hat{p}_f) - (p_m - p_f)}{\sqrt{\dfrac{p_m(1-p_m)}{n_1} + \dfrac{p_f(1-p_f)}{n_2}}}$, then

$$P(\hat{p}_m - \hat{p}_f > 0.10) = P\left(Z > \frac{0.10 - 0.08}{0.061752}\right)$$

$$= P(Z < 0.32) = 0.6255.$$

(b) We need to find $P(0.06 < \hat{p}_m - \hat{p}_f < 0.13)$.
Then

$$P(0.06 < \hat{p}_m - \hat{p}_f < 0.13) = P\left(\frac{0.06 - 0.08}{0.061752} < Z < \frac{0.13 - 0.08}{0.061752}\right) = P(-0.32 < Z < 0.81)$$

$$= 0.79103 - 0.3745 = 0.4165.$$

12.3 The Two Sample Confidence Intervals

As we discussed in Chapter 8, we are often interested in attaching some probability to the chance that a parameter of interest to us would lie in an interval. This probability is a measure of the confidence we can attach to that interval being an interval estimate of the parameter or a measure of the confidence we have that the parameter we are estimating would lie within the interval. This is why we call these intervals, confidence interval estimates. These confidence intervals are obtained on the basis of the sampling distribution of the statistic being used to estimate the parameter. From sampling theory, and large sample theory, we know that most of the parameters would be normally or approximately normally distributed.

In this chapter, we extend the ideas of confidence interval for the estimation of a parameter of a population, to the estimation of the difference between the values of the same parameter for two different populations.

12.4 Confidence Intervals for the Difference Between Two Parameters

The general principle described earlier for a single parameter would apply to the difference between the parameter for two different populations. As usual, we need to know the sampling distribution of the difference between the two parameters. On the basis of this distribution, we can construct the $100(1 - \alpha)\%$ confidence interval for the difference between the two parameters.

Typically, the confidence interval for the difference between the parameters $(\theta_1 - \theta_2)$ would be written as the $100(1 - \alpha)\%$ confidence interval, converting the probability attached to the confidence interval into a percentage. Provided that the sampling distribution of the estimate of the difference between the parameters $(\theta_1 - \theta_2)$ is normally distributed, then, $100(1 - \alpha)\%$ for difference between θ_1 and θ_2, $(\hat{\theta}_1 - \hat{\theta}_2)$ is:

$$(\hat{\theta}_1 - \hat{\theta}_2) \pm Z_{\alpha/2}\sigma_{(\hat{\theta}_1 - \hat{\theta}_2)}$$

Although other confidence intervals can be constructed in this section, we will only consider the differences between two population proportions, and the differences between two populations' means, under different possible sampling distributions of the estimates of their differences.

12.5 Confidence Interval for the Difference Between Two Population Proportions

In the single population situation, we determined the $100(1 - \alpha)\%$ confidence interval by sampling from a binomial process in which p is the probability of success and $q = 1 - p$ is the probability of failure. We found that the confidence interval estimate for the proportion of successes, p, by using the proportion of successes in our sample, denoted by $\hat{p}$ as follows:

$$\hat{p} \pm Z_{\alpha/2}\sigma_{\hat{p}}$$

Provided that we were sampling from a very large or an infinite population, or from a finite population with replacement, then the confidence interval could be written as:

$$100(1 - \alpha)\% \text{ CI} = \hat{p} \pm Z_{\alpha/2}\sqrt{\frac{\hat{p}(1 - \hat{p})}{n}}$$

Of more interest to us in this section is the confidence interval for the difference between two population proportions. It is like the confidence interval for a single population proportion if we replace the estimated single proportion $\hat{p}$ by the estimated difference between the two proportions $\hat{p}_1 - \hat{p}_2$ and replace the standard error of $\hat{p}$ by the standard error of the difference between the two proportions, $\sigma_{(\hat{p}_1 - \hat{p}_2)}$. If n_1 and n_2 respectively are the sample sizes from which $\hat{p}_1$ and $\hat{p}_2$ were obtained and both sample sizes are large $(n_1, n_2 \geq 30)$, we can establish the $100(1 - \alpha)\%$ confidence interval for the difference between the two proportions as:

$$100(1 - \alpha)\% \text{ CI} = (\hat{p}_1 - \hat{p}_2) \pm Z_{\alpha/2}\hat{\sigma}_{(\hat{p}_1 - \hat{p}_2)}.$$

Where the standard error for the difference between the proportions, $\hat{\sigma}_{(\hat{p}_1 - \hat{p}_2)}$ is:

$$\hat{\sigma}_{(\hat{p}_1 - \hat{p}_2)} = \sqrt{\frac{\hat{p}_1(1 - \hat{p}_1)}{n_2} + \frac{\hat{p}_2(1 - \hat{p}_2)}{n_2}}.$$

We do a few examples.

■ **EXAMPLE 12.5**   Samples of migraine patients treated by Drs. Lilian and Beth show the recoveries presented in the table below:

Doctors	Recovered	Did not recover
Lilian	67	91
Beth	66	130

Obtain a 89% confidence interval for the difference between the proportions of patients who recovered for Lilian and Beth.

Solution

$$\hat{p}_L = \frac{67}{67 + 91} = 0.42405063; \hat{p}_B = \frac{66}{66 + 130} = 0.3367347 \Rightarrow n_1 = 158, n_2 = 196$$

$$(\hat{p}_L - \hat{p}_B) \pm Z_{\frac{\alpha}{2}} \hat{\sigma}_{\hat{p}_L - \hat{p}_B} \text{ where } \hat{\sigma}_{\hat{p}_L - \hat{p}_B} = \sqrt{\frac{\hat{p}_L(1 - \hat{p}_L)}{n_L} + \frac{\hat{p}_B(1 - \hat{p}_B)}{n_B}}$$

$$(0.42405063 - 0.3367347) \pm Z_{\frac{0.11}{2}} \hat{\sigma}_{\hat{p}_L - \hat{p}_B}$$

$$\hat{\sigma}_{\hat{p}_L - \hat{p}_B} = \sqrt{\frac{0.42405063(1 - 0.42405063)}{158} + \frac{0.3367347(1 - 0.3367347)}{196}}$$

$$= 0.051819713$$

The 89% confidence interval means that $\alpha = 0.11$ and $\alpha/2 = 0.055 \Rightarrow Z_{0.055} = 1.60$. Thus, $100(1 - 0.11)\%$ c. i. is:

$$0.08731593 \pm 1.60 \times 0.051819713 = (0.004404, 0.170227)$$

■ **EXAMPLE 12.6**   A survey of freshmen in Hamilton College Campus showed, the number of female-freshmen interviewed were 422 while the male-freshmen were 602. If they found 137 female-smokers and 242 male-smokers among the interviewees, find the 94% confidence interval for the difference in the proportions of male-freshmen and female-freshmen who smoked.

Solution

$$\hat{p}_m = \frac{242}{602} = 0.4019933; \hat{p}_f = \frac{137}{422} = 0.3246445 \Rightarrow n_m = 602, n_f = 420$$

$$(\hat{p}_m - \hat{p}_f) \pm Z_{\frac{\alpha}{2}} \sigma_{\hat{p}_m - \hat{p}_f} \text{ where } \hat{\sigma}_{\hat{p}_m - \hat{p}_f} = \sqrt{\frac{\hat{p}_m(1 - \hat{p}_m)}{n_m} + \frac{\hat{p}_f(1 - \hat{p}_f)}{n_f}}$$

$$\hat{\sigma}_{\hat{p}_m - \hat{p}_f} = \sqrt{\frac{0.401993(1 - 0.401993)}{602} + \frac{0.324645(1 - 0.324645)}{422}} = 0.030313$$

94% confidence interval $= 100(1 - 0.06)\%$ confidence interval $\Rightarrow \alpha = 0.06$ and $\alpha/2 = 0.03$ $Z_{0.03} = 1.88$:

$$(0.4019933 - 0.3246445) \pm Z_{0.03} \times 0.030313$$
$$= 0.077349 \pm 1.88 \times 0.030313$$
$$= (0.02036, 0.134337)$$

■ **EXAMPLE 12.7** A sample of 450 of ex-prisoners who were not treated for behavioral aberrations showed that 197 re-offended. A sample of 600 who were treated for behavioral aberrations showed that 102 re-offended.

Obtain the (a) 96% (b) 99% confidence interval for the difference in proportion of ex-prisoners who were not treated for behavioral aberrations who re-offended and the proportion of ex-prisoners who were treated for behavioral aberrations who re-offended.

Solution

$$\text{let treated} = t; \text{not treaded} = nt; p_t = \frac{197}{450} = 0.437778; p_{nt} = \frac{102}{600} = 0.17;$$

$$\hat{\sigma}_{t-nt} = \sqrt{\frac{\hat{p}_t(1-\hat{p}_1)}{n_t} + \frac{\hat{p}_{nt}(1-\hat{p}_{nt})}{n_{nt}}} = \sqrt{\frac{0.437778(1-0.437778)}{450} + \frac{0.17(1-0.17)}{600}}$$

$$= 0.027966$$

(a) $96\% \Rightarrow 100(1 - 0.04)\% \Rightarrow \alpha = 0.04 \Rightarrow \alpha/2 = 0.02$

96% CI for $p_1 - p_2$: $\hat{p}_1 - \hat{p}_2 \pm Z_{0.02} \times \hat{\sigma}_{t-nt}$

$= (0.437778 - 0.17) \pm 2.05 \times 0.027966 = (0.210447, 0.325109)$

(b) $99\% \Rightarrow 100(1 - 0.01)\% \Rightarrow \alpha = 0.01 \Rightarrow \alpha/2 = 0.005$

99% CI for $p_1 - p_2$: $\hat{p}_1 - \hat{p}_2 \pm Z_{0.005} \times \hat{\sigma}_{t-nt}$

$= (0.437778 - 0.17) \pm 2.575 \times 0.027966 = (0.195764, 0.339791)$

■ **EXAMPLE 12.8** Independent samples of 747 U.S. men and 434 U.S. women were taken. Of those sampled, 276 men and 195 women said that they would pay extra tax towards preserving wildlife habitats.

Obtain the (a) 96% and (b) 99% confidence interval for the difference between proportion of males and females who would pay extra tax.

Solution

$$\text{let female} = f; \text{male} = m; \hat{p}_f = \frac{195}{434} = 0.449309; \hat{p}_m = \frac{276}{747} = 0.369478;$$

$$\hat{\sigma}_{m-f} = \sqrt{\frac{\hat{p}_f(1-\hat{p}_f)}{n_f} + \frac{\hat{p}_m(1-\hat{p}_m)}{n_m}} = \sqrt{\frac{0.449309(1-0.449309)}{434} + \frac{0.369478(1-0.369478)}{747}}$$

$$= 0.029698$$

(a) $96\% \Rightarrow 100(1 - 0.04)\% \Rightarrow \alpha = 0.04 \Rightarrow \alpha/2 = 0.02$

96% CI for $p_f - p_m$: $\hat{p}_f - \hat{p}_m \pm Z_{0.02} \times \hat{\sigma}_{m-f}$

$= (0.449309 - 0.369478) \pm 2.05 \times 0.029698 = (0.01895, 0.140712)$

(b) $99\% \Rightarrow 100(1 - 0.01)\% \Rightarrow \alpha = 0.01 \Rightarrow \alpha/2 = 0.005;$

99% CI for $p_f - p_m$: $\hat{p}_f - \hat{p}_m \pm Z_{0.005} \times \hat{\sigma}_{m-f}$

$= (0.449309 - 0.369478) \pm 2.575 \times 0.029698 = (0.003358, 0.156304)$

■ **EXAMPLE 12.9** A survey was carried out to study regular use of sunscreen by teenagers and young women. Of 330 teenage girls surveyed, 167 were found to use sunscreen regularly, while 149 of 400 young women in their twenties surveyed used sunscreen regularly.

Obtain the (a) 94%, and (b) 99% confidence interval for the difference between the proportion of teenage girls and the proportion of young women who used sunscreen regularly.

Solution

$$x_G = 167; n_G = 330, \hat{p}_G = \frac{167}{330} = 0.506061; x_y = 149; n_y = 400, \hat{p}_y = \frac{149}{400} = 0.3725$$

(a) $(\hat{p}_G - \hat{p}_y) \pm Z_{\alpha/2}\sigma_{\hat{p}_G - \hat{p}_y}$

$\hat{p}_G - \hat{p}_y = 0.506061 - 0.3725 = 0.133561$

$$\sigma_{\hat{p}_G - \hat{p}_y} = \sqrt{\frac{(0.506061)(1 - 0.506061)}{330} + \frac{(0.3725)(1 - 0.3725)}{400}}$$

$= 0.036631$

94% CI for $p_G - p_y$; 94% $= 100(1 - 0.06)\% \Rightarrow \alpha = 0.06; \alpha/2 = 0.03$;

$0.133561 \pm 1.88 \times 0.036631 = [0.064694, 0.202427]$

(b) 99% CI for $p_G - p_y$; 99% $= 100(1 - 0.01)\% \Rightarrow \alpha = 0.01; \alpha/2 = 0.005$;

$0.135661 \pm 2.575 \times 0.036631 = [0.039236, 0.227885]$

■ **EXAMPLE 12.10**   A survey found that 159 out of the 500 female students, and 270 of the 620 male students, have binged on alcohol at least once. Obtain (i) the 96%, and (ii) the 93% confidence interval for the difference between the proportion males and female who have binged on alcohol.

Solution

$$\hat{p}_m = \frac{270}{620} = 0.435484; \hat{p}_f = \frac{159}{500} = 0.318 \Rightarrow n_m = 620, n_f = 500$$

$[(\hat{p}_m - \hat{p}_f) - Z_{\frac{\alpha}{2}}\hat{\sigma}_{\hat{p}_m - \hat{p}_f}, (\hat{p}_m - \hat{p}_f) + Z_{\frac{\alpha}{2}}\hat{\sigma}_{\hat{p}_m - \hat{p}_f}]$

where $\hat{\sigma}_{\hat{p}_m - \hat{p}_f} = \sqrt{\dfrac{\hat{p}_m(1 - \hat{p}_m)}{n_m} + \dfrac{\hat{p}_f(1 - \hat{p}_f)}{n_f}}$

$(0.435484 - 0.318) \pm Z_{\frac{0.07}{2}}\hat{\sigma}_{\hat{p}_m - \hat{p}_f}$

where $\hat{\sigma}_{\hat{p}_m - \hat{p}_f} = \sqrt{\dfrac{0.435484(1 - 0.435484)}{620} + \dfrac{0.318(1 - 0.318)}{500}}$

$= 0.028814$

The 96% confidence interval means that $\alpha = 0.04$ and $\alpha/2 = 0.02 \Rightarrow Z_{0.02} = 2.05$. Thus, $100(1 - 0.07)\%$ CI $= (0.117484 - 2.05 \times 0.028814,\ 0.0.117484 + 2.05 \times 0.028814) = (0.058415, 0.176553)$.

The 93% confidence interval means that $\alpha = 0.07$ and $\alpha/2 = 0.035 \Rightarrow Z_{0.035} = 1.81$. Thus, $100(1 - 0.07)\%$ CI $= (0.117484 - 1.81 \times 0.028814,\ 0.117484 + 1.81 \times 0.028814) = (0.0653313, 0.169638)$.

12.6   Confidence Intervals for Differences Between Two Means

Sometimes we need to establish the confidence interval as an estimate for the difference between the means of two independent populations represented by X_1 and X_2. In this case, we are looking for the confidence interval for the difference in their means, $\mu_1 - \mu_2$. If X_1 and X_2 are normally distributed and the two samples are truly independent, then the difference between the means of the samples $\overline{X}_1 - \overline{X}_2$ would be normally distributed, so that by comparison with the previous section on the confidence interval for a single mean in Chapter 8, the $100(1 - \alpha)\%$ confidence interval for the difference of means would be:

$$(\overline{X}_1 - \overline{X}_2) \pm Z_{\alpha/2}\hat{\sigma}_{\overline{x}_1 - \overline{x}_2};$$

where

$$\hat{\sigma}_{\overline{x}_1 - \overline{x}_2} = \sqrt{\frac{\sigma_1^2}{n_1} + \frac{\sigma_2^2}{n_2}}.$$

The problem with the above formula is that we often do not know the values of the population variances σ_1^2 and σ_2^2. Because of this, two sample z-tests are rarely used when σ_1^2 and σ_2^2 are unknown. Instead, two-sample t-tests are used. We treat two-sample t- tests later.

Meanwhile we will find confidence intervals when the variances are known in the next examples.

■ **EXAMPLE 12.11**   The following data represent degrading times in days for two types of biomasses, X_1 and X_2:

X_1: 65 82 81 77 67 99 66 75 82 80 79 84 86 76
X_2: 64 76 71 69 83 74 59 82 65 67 79

If both degrading times are believed to be normally distributed, and it was known that the variances of X_1 and X_2 are 95 and 71 respectively, obtain the (a) 97%, and (b) 99% confidence interval for the difference between the mean of X_1 and the mean of X_2.

Solution   It is rare for the variances to be known as stated above. Nevertheless, in this case the confidence interval has the following formula:

$$100(1 - \alpha)\% \text{ CI} = (\overline{X}_1 - \overline{X}_2) \pm Z_{\alpha/2} \times \hat{\sigma}_{\overline{x}_1 - \overline{x}_2}; \hat{\sigma}_{\overline{x}_1 - \overline{x}_2} = \sqrt{\frac{\sigma_{\overline{x}_1}^2}{n_1} + \frac{\sigma_{\overline{x}_2}^2}{n_2}}$$

$$\overline{X}_1 = \frac{1099}{14} = 78.5; \overline{X}_2 = \frac{789}{11} = 71.727272$$

(a) $(\overline{X}_1 - \overline{X}_2) \pm Z_{\frac{\alpha}{2}}\sqrt{\frac{\sigma_1^2}{n_1} + \frac{\sigma_2^2}{n_2}}$

97% CI for difference between means μ_1 and μ_2:

$$(\overline{X}_1 - \overline{X}_2) \pm Z_{0.015}\sqrt{\frac{95}{14} + \frac{71}{11}}$$

$$(78.5 - 71.72727) \pm 2.17\sqrt{\frac{95}{14} + \frac{71}{11}}$$

$$= 6.72727 \pm 7.89602$$
$$= (-1.12329, 14.66874)$$

(b) $(\overline{X}_1 - \overline{X}_2) \pm Z_{\frac{\alpha}{2}}\sqrt{\dfrac{\sigma_1^2}{n_1} + \dfrac{\sigma_2^2}{n_2}}$

99% CI for difference between the means μ_1 and μ_2:

$(\overline{X}_1 - \overline{X}_2) \pm Z_{0.005}\sqrt{\dfrac{95}{14} + \dfrac{71}{11}}$

$(78.5 - 71.72727) \pm 2.575\sqrt{\dfrac{95}{14} + \dfrac{71}{11}}$

$= 6.72727 \pm 9.369696$

$= (-2.59697, 16.14243)$

■ **EXAMPLE 12.12** A survey of the heights of eighteen year olds in North Jade County—264 males, and 202 females—was taken. It is known that the variances for heights of eighteen-year-old males and females are $0.077\,\text{m}^2$ and $0.066\,\text{m}^2$ respectively. The mean height for the males from the samples was $1.67\,\text{m}$ and the mean height for the females was $1.56\,\text{m}$. Obtain the (a) 94%, and (b) 98% confidence limits for the difference in the mean heights of eighteen-year-old males and females from this county.

Solution Since the variances are known this means that the applicable confidence interval is obtained by the formula:

(a) 94% CI $= 100(1 - 0.06)\% \Rightarrow \alpha = 0.06; \alpha/2 = 0.03;$

$Z_{0.03} = 1.88; \hat{\sigma}_{\overline{x}_1 - \overline{x}_2} = \sqrt{\dfrac{0.077}{264} + \dfrac{0.066}{202}} = 0.024068;$

94% CI $= (\overline{X}_1 - \overline{X}_2) \pm Z_{0.03} \times \hat{\sigma}_{\overline{x}_1 - \overline{x}_2} = (1.67 - 1.56) \pm 1.88(0.024068)$
$= (0.063249, 0.156751)$

(b) 98% CI $= 100(1 - 0.02)\% \Rightarrow \alpha = 0.02; \alpha/2 = 0.01;$

$Z_{0.01} = 2.33; \hat{\sigma}_{\overline{x}_1 - \overline{x}_2} = \sqrt{\dfrac{0.077}{264} + \dfrac{0.066}{202}} = 0.024068;$

98% CI $= (\overline{X}_1 - \overline{X}_2) \pm Z_{0.01} \times \hat{\sigma}_{\overline{x}_1 - \overline{x}_2} = (1.67 - 1.56) \pm 2.33(0.024068)$
$= (0.052058, 0.167942)$

12.7 Confidence Intervals for the Difference Between the Means for Two Independent Samples

In this section we consider confidence intervals for the difference between the means of two populations when independent samples are drawn from each of the populations and the variances are unknown. When the variances are unknown, we have to resort to the use of formula different from those used earlier. This is because the **difference** between the means **has a different distribution which is not normal.** Suppose that sample X_1 and X_2 contain n_1, and n_2 observations. There are two possibilities here for obtaining confidence interval estimates. If the two populations are normal, one possibility is that the variances of the two populations are equal. The other possibility is that the variances are not equal. We consider both situations; first, when the variances are equal.

12.7.1 Confidence Interval for the Difference Between Means for Two Independent Samples When Variances are Equal

If we have reason to believe that the two samples might have come **from normal populations with equal variance,** then if we know this common variance we can establish a confidence interval for the difference between the two means by using the formula:

$$100(1 - \alpha)\% \text{ CI} = (\overline{X}_1 - \overline{X}_2) \pm Z_{\alpha/2}\sqrt{\sigma^2\left(\frac{1}{n_1} + \frac{1}{n_2}\right)}.$$

However, if the variances are unknown, and **often the common variance is unknown,** we can then **pool the two sample variances** in order to estimate the **common variance.** Then the estimated common variance is obtained as s_p^2, where

$$s_p^2 = \frac{(n_1 - 1)s_1^2 + (n_2 - 1)s_2^2}{n_1 + n_2 - 2}.$$

The standard error for the difference between the two means is:

$$\hat{\sigma}_{\overline{x}_1 - \overline{x}_2} = s_p\sqrt{\left(\frac{1}{n_1} + \frac{1}{n_2}\right)}.$$

When the variances are equal, the difference in the sample means is t-distributed, so that:

$$(\overline{X}_1 - \overline{X}_2) \sim t_{(n_1+n_2-2)}.$$

The confidence interval this time is obtained by using the t-distribution with $n_1 + n_2 - 2$ degrees of freedom, so that the $100(1 - \alpha)\%$ confidence interval for $(\mu_1 - \mu_2)$ is:

$$\overline{X}_1 - \overline{X}_2 \pm t_{(n_1+n_2-2,\ \alpha/2)} \cdot s_p\sqrt{\frac{1}{n_1} + \frac{1}{n_2}}.$$

Here are examples for cases when the variances of the two samples are believed to be equal. This confidence interval is referred to as **the pooled variance t confidence interval:**

■ **EXAMPLE 12.13** The following samples were taken from two variables A and B, which are believed to be normally distributed and of equal variance.

A: 82 64 62 66 68 43 38 26 68 56 54 66
B: 70 26 34 41 53 49 40 46 34 66 38 46 44 35 33

Obtain the (a) 90%, and (b) 95% confidence intervals for the difference between their means.

Solution $\overline{X}_A = \dfrac{693}{12} = 57.75$; $\overline{X}_B = \dfrac{655}{15} = 43.6667$;

$$s_A^2 = \frac{\displaystyle\sum_{i=1}^{n_1} X_A^2 - n(\overline{X}_A)^2}{n_A - 1} = \frac{42645 - 12(57.75)^2}{12 - 1} = 238.5682$$

$$s_B^2 = \frac{\displaystyle\sum_{i=1}^{n_1} X_B^2 - n(\overline{X}_B)^2}{n_B - 1} = \frac{30661 - 15(43.6667)^2}{15 - 1} = 147.0952$$

$$S_P^2 = \frac{(n_A - 1)s_A^2 + (n_B - 1)s_B^2}{n_A + n_B - 2} = \frac{(12 - 1)(238.5682) + (15 - 1)(147.0952)}{12 + 15 - 2}$$

$$= 187.3433 \Rightarrow s_p = \sqrt{187.3433} = 13.68734$$

(a) 90% CI $\Rightarrow 100(1 - 0.10)\%$ CI $\Rightarrow \alpha = 0.10$; $\alpha/2 = 0.05$
$t_{(12+15-2,\,0.05)} = 1.708$

$$100(1 - \alpha)\% \text{ CI for } \mu_1 = \overline{X}_A - \overline{X}_B \pm t_{(n_A + n_B - 2,\, \alpha/2)} \cdot s_p \sqrt{\frac{1}{n_A} + \frac{1}{n_B}}$$

90% CI for $\mu_A - \mu_B = 57.75 - 43.6667 \pm 1.708 \times 13.68734 \times \sqrt{\dfrac{1}{12} + \dfrac{1}{15}}$

$$= (5.029081,\ 23.13759)$$

(b) 95% CI $\Rightarrow 100(1 - 0.05)\%$ CI $\Rightarrow \alpha = 0.05$; $\alpha/2 = 0.025$
$t_{(12+15-2,\,0.025)} = 2.06$

95% CI for $\mu_A - \mu_B = 57.75 - 43.6667 \pm 2.06 \times 13.68734 \times \sqrt{\dfrac{1}{12} + \dfrac{1}{15}}$

$$= (3.163099,\ 25.00357)$$

■ **EXAMPLE 12.14** The times to metastasis of two different types of cancer were recorded for a sample of eighteen patients for cancer type X and a sample of fifteen patients from cancer type Y:

X: 22.3 26.7 23.8 25 24.9 26.1 24.4 25.2 23.9 24.8 24.6 25.7 25.3 24.7
24.2 25.7 24.8 24.7
Y: 26.7 27.3 24.4 28.3 26.2 24.3 25.5 23.2 28.4 27.3 25.5 24.8 24.9
26.7 27.5

Assume that variances of X and Y are equal, obtain the (a) 90%, and (b) 99% confidence intervals for the difference between the mean of X and the mean of Y.

Solution $\overline{X} = \dfrac{446.8}{18} = 24.8222$; $\overline{Y} = \dfrac{391}{15} = 26.0667$;

$$s_x^2 = \frac{\displaystyle\sum_{i=1}^{n_1} X_i^2 - n_1(\overline{X})^2}{n_1 - 1} = \frac{11106.58 - 18(24.8222)^2}{18 - 1} = 0.94183$$

$$s_y^2 = \frac{\displaystyle\sum_{i=1}^{n_2} Y_i^2 - n_2(\overline{Y})^2}{n_2 - 1} = \frac{10226.14 - 15(26.06667)^2}{15 - 1} = 2.43381$$

$$s_P^2 = \frac{(n_1 - 1)s_x^2 + (n_2 - 1)s_y^2}{n_1 + n_2 - 2} = \frac{(18 - 1)(0.94183) + (15 - 1)(2.43381)}{18 + 15 - 2}$$

$$= 1.615627 \Rightarrow s_p = \sqrt{1.615627} = 1.271073$$

(a) 90% CI $\Rightarrow 10(1 - 0.10)\%$ CI $\Rightarrow \alpha = 0.10; \alpha/2 = 0.05$

$t_{(18+15-2, \, 0.05)} = 1.696$

$100(1 - \alpha)\%$ CI for $\mu_1 - \mu_2 = \overline{X} - \overline{Y} \pm t_{(n_1+n_2-2, \, \alpha/2)} \cdot s_P \sqrt{\frac{1}{n_1} + \frac{1}{n_2}}$

90% CI for $\mu_A - \mu_B = (24.8222 - 26.06667) \pm 1.696 \times 1.271073$

$\times \sqrt{\frac{1}{18} + \frac{1}{15}} = (-1.99812, -0.49082)$

(b) 99% CI for $\mu_1 - \mu_2$:

99% CI $\Rightarrow 100(1 - 0.01)\%$ CI $\Rightarrow \alpha = 0.01; \alpha/2 = 0.005;$

$t_{(18+15-2, \, 0.005)} = 2.744$

$100(1 - \alpha)\%$ CI for $\mu_1 - \mu_2 = \overline{X} - \overline{Y} \pm t_{(n_1+n_2-2, \, \alpha/2)} \cdot s_P \sqrt{\frac{1}{n_1} + \frac{1}{n_2}}$

99% CI for $\mu_1 - \mu_2 = (24.8222 - 26.06667) \pm 2.744 \times 1.271073$

$\times \sqrt{\frac{1}{18} + \frac{1}{15}} = (-2.46382, -0.02512)$

■ **EXAMPLE 12.15**   In a pilot project in the textile industry, two similar machines were being compared for speed of production. The time it took to produce the same piece of cloth was recorded fifteen times for each of the machines. Obtain the (a) 95%, and (b) 98% confidence intervals for the difference between the means for both machines.

Machine1 (X_1)	Machine2 (X_2)
1603	1602
1604	1597
1605	1596
1605	1601
1602	1599
1601	1603
1596	1604
1598	1602
1599	1601
1602	1607
1614	1600
1612	1596
1607	1595
1593	1606
1604	1597

Solution $\overline{X}_1 = \dfrac{24006}{15} = 1600.4; \ \overline{X}_2 = \dfrac{24045}{15} = 1603;$

$$s_1^2 = \dfrac{\displaystyle\sum_{i=1}^{n_1} X_{1i}^2 - n_1(\overline{X}_1)^2}{n_1 - 1} = \dfrac{38419396 - 15(1600.4)^2}{15 - 1} = 13.82857$$

$$s_2^2 = \dfrac{\displaystyle\sum_{i=1}^{n_2} X_{2i}^2 - n_2(\overline{X}_2)^2}{n_2 - 1} = \dfrac{38544559 - 15(1603)^2}{15 - 1} = 30.28571$$

$$s_P^2 = \dfrac{(n_1 - 1)s_x^2 + (n_2 - 1)s_y^2}{n_1 + n_2 - 2} = \dfrac{(15 - 1)(13.82857) + (15 - 1)(30.28571)}{15 + 15 - 2}$$

$$= 22.05714 \Rightarrow s_p = \sqrt{22.05714} = 4.696503$$

(a) $95\% \ \text{CI} \Rightarrow 10(1 - 0.05)\% \ \text{CI} \Rightarrow \alpha = 0.05; \ \alpha/2 = 0.025$

$t_{(15+15-2,\,0.025)} = 2.048$

$$100(1 - \alpha)\% \ \text{CI for } \mu_1 - \mu_2 = \overline{X}_1 - \overline{X}_2 \pm t_{(n_1+n_2-2,\,\alpha/2)} \cdot s_P\sqrt{\dfrac{1}{n_1} + \dfrac{1}{n_2}}$$

$$90\% \ \text{CI for } \mu_2 - \mu_1 = (1603 - 1600.4) \pm 2.048 \times 4.696503 \times \sqrt{\dfrac{1}{15} + \dfrac{1}{15}}$$

$$= (-0.91216,\ 6.11216)$$

(b) $98\% \ \text{CI for } \mu_1 - \mu_2$:

$98\% \ \text{CI} \Rightarrow 100(1 - 0.02)\% \ \text{CI} \Rightarrow \alpha = 0.02; \ \alpha/2 = 0.01;$

$t_{(15+15-2,\,0.01)} = 2.467$

$$100(1 - \alpha)\% \ \text{CI for } \mu_1 - \mu_2 = \overline{X}_1 - \overline{X}_2 \pm t_{(n_1+n_2-2,\,\alpha/2)} \cdot s_P\sqrt{\dfrac{1}{n_1} + \dfrac{1}{n_2}}$$

$$99\% \ \text{CI for } \mu_1 - \mu_2 = (1603 - 1600.4) \pm 2.467 \times 4.696503 \times \sqrt{\dfrac{1}{15} + \dfrac{1}{15}}$$

$$= (-1.63071,\ 6.830709)$$

■ **EXAMPLE 12.16** The following are samples of the red blood corpuscles measurements for males and females:

male blood corpuscles
5.44 5.07 4.80 5.14 5.99 4.82 4.57 4.59 4.77 5.12 5.37
5.17 4.97 4.80 5.05 5.11 5.13

female blood corpuscles
3.58 4.22 4.36 4.59 3.63 4.54 3.92 4.24 4.21 5.06 4.71
4.23 4.68 3.83 3.95 3.87 4.20 4.07 3.74

Assume that the red blood corpuscles measurements are normally distributed for both males and females, and that their variances are equal. Obtain the (a) 90%, and (b) 98% confidence interval for difference between the means of the red blood corpuscles for males and females.

Solution $\overline{X}_1 = \dfrac{85.91}{17} = 5.053529; \ \overline{X}_2 = \dfrac{79.63}{19} = 4.191053;$

$$s_1^2 = \frac{\sum\limits_{i=1}^{n_1} X_{1i}^2 - n_1(\overline{X}_1)^2}{n_1 - 1} = \frac{436.0291 - 17(5.053529)^2}{17 - 1} = 0.117524$$

$$s_2^2 = \frac{\sum\limits_{i=1}^{n_2} X_{2i}^2 - n_2(\overline{X}_2)^2}{n_2 - 1} = \frac{336.5829 - 19(4.191053)^2}{19 - 1} = 0.158299$$

$$s_P^2 = \frac{(n_1 - 1)s_x^2 + (n_2 - 1)s_y^2}{n_1 + n_2 - 2} = \frac{(17 - 1)(0.117524) + (19 - 1)(0.158299)}{17 + 19 - 2}$$

$$= 0.139111 \Rightarrow s_p = \sqrt{0.139111} = 0.372976$$

(a) 90% CI $\Rightarrow 10(1 - 0.1)\%$ CI $\Rightarrow \alpha = 0.1; \alpha/2 = 0.05$

$t_{(17+19-2,\, 0.05)} = 1.691$

$$100(1 - \alpha)\% \text{ CI for } \mu_1 - \mu_2 = \overline{X}_1 - \overline{X}_2 \pm t_{(n_1+n_2-2,\, \alpha/2)} \cdot s_P\sqrt{\frac{1}{n_1} + \frac{1}{n_2}}$$

90% CI for $\mu_2 - \mu_1 = (5.053529 - 4.191053) \pm 1.691 \times 0.372976$

$$\times \sqrt{\frac{1}{17} + \frac{1}{19}} = (0.651918, 1.073036)$$

(b) 98% CI for $\mu_1 - \mu_2$:

98% CI $\Rightarrow 100(1 - 0.02)\%$ CI $\Rightarrow \alpha = 0.02; \alpha/2 = 0.01;$

$t_{(17+19-2,\, 0.01)} = 2.441$

$$100(1 - \alpha)\% \text{ CI for} \mu_1 - \mu_2 = \overline{X}_1 - \overline{X}_2 \pm t_{(n_1+n_2-2,\, \alpha/2)} \cdot s_P\sqrt{\frac{1}{n_1} + \frac{1}{n_2}}$$

98% CI for $\mu_1 - \mu_2 = (5.053529 - 4.191053) \pm 2.441 \times 0.372976$

$$\times \sqrt{\frac{1}{17} + \frac{1}{19}} = (0.55853, 1.166424)$$

12.7.2 Confidence Interval for the Difference Between Means When Variances Are Unknown and Are Not Equal

If we have reason to believe that the two small samples might have come **from normal populations with unequal variances**, then if we know the variances we can establish a confidence interval for the difference between the two means by using the formula:

$$100(1 - \alpha)\% \text{ CI} = (\overline{X}_1 - \overline{X}_2) \pm Z_{\alpha/2}\sqrt{\left(\frac{\sigma_1^2}{n_1} + \frac{\sigma_2^2}{n_2}\right)}.$$

The variances σ_1^2 and σ_2^2 are more often than not unknown, and therefore we can rarely, if ever, obtain the $100(1 - \alpha)\%$ confidence interval on the basis of the normal distribution as stated above. It is known, however, that under the conditions stated, $\overline{X}_1 - \overline{X}_2$ is approximately t-distributed with a standard error of $\sqrt{\left(\frac{s_1^2}{n_1} + \frac{s_2^2}{n_2}\right)}$ and the following degrees of freedom:

$$\nu = \frac{[(s_1^2/n_1) + (s_2^2/n_2)]^2}{\dfrac{(s_1^2/n_1)^2}{n_1 - 1} + \dfrac{(s_2^2/n_2)^2}{n_2 - 1}}.$$

Because the degrees of freedom obtained this way could be fractional, it is rounded down to the nearest integer. With this we can establish a $100(1 - \alpha)\%$ confidence interval for the difference between the means as follows:

$$100(1 - \alpha)\% \text{ CI} = (\overline{X}_1 - \overline{X}_2) \pm t_{(\nu, \alpha/2)}\sqrt{\left(\frac{s_1^2}{n_1} + \frac{s_2^2}{n_2}\right)}.$$

Here, we provide some examples.

■ **EXAMPLE 12.17** The times used by a sample of eighteen chemistry students from School A and a sample of fifteen chemistry students from School B to complete a titration experiment are as follows:

A: 2.3 6.7 3.8 5 4.9 6.1 4.4 5.2 3.9 4.8 4.6 5.7 5.3 4.7 4.2 5.7 4.8 4.7
B: 6.7 7.3 4.4 8.3 6.2 4.3 5.5 3.2 8.4 7.3 5.5 4.8 4.9 6.7 7.5

Obtain a 95% confidence interval for the difference between the means of School A and School B.

Solution

$$\overline{X}_1 = \frac{86.8}{18} = 4.8222; \overline{X}_2 = \frac{91}{15} = 6.0667;$$

$$s_1^2 = \frac{434.58 - 18(4.8222)^2}{18 - 1} = 0.9418$$

$$s_2^2 = \frac{586.14 - 15(6.0667)^2}{15 - 1} = 2.43381$$

$$\nu = \frac{\left[\dfrac{0.9418}{18} + \dfrac{2.43381}{15}\right]^2}{\left(\dfrac{1}{18 - 1}\right)\left(\dfrac{0.9418}{18}\right)^2 + \left(\dfrac{1}{15 - 1}\right)\left(\dfrac{2.43381}{15}\right)^2} = 22.5383 \approx 22$$

$$100(1 - \alpha)\% \text{ CI} = (\overline{X}_1 - \overline{X}_2) \pm t_{(\nu, \alpha/2)}\sqrt{\left(\frac{s_1^2}{n_1} + \frac{s_2^2}{n_2}\right)}$$

$$95\% \text{ CI} = 100(1 - 0.05)\% \text{ CI} \Rightarrow \alpha = 0.05; \alpha/2 = 0.025;$$

$$(4.8222 - 6.0667) \pm t_{(22, 0.025)}\sqrt{\left(\frac{0.9418}{18} + \frac{2.43381}{15}\right)}$$

$$= 1.2445 \pm 2.074 \times 0.463224 = (-2.205226, -0.283774)$$

■ **EXAMPLE 12.18** The times it took for samples of eighth grade students from two different schools to complete a computer-based task are presented below:

X_1: 48.2 54.6 58.3 47.8 51.4 52.0 55.2 49.1 49.9 52.6 49.6 47.5
X_2: 52.3 57.4 55.6 53.2 61.3 58.0 59.8 54.8 58.9 57.7 58.6

Assume that times used for the task for both groups are normally distributed. Obtain the 99% confidence interval for the difference between their means.

Solution

$$\overline{X}_1 = \frac{616.2}{12} = 51.35; \ \overline{X}_2 = \frac{627.6}{11} = 57.05;$$

$$s_1^2 = \frac{31765 - 12(51.35)^2}{12 - 1} = 11.2$$

$$s_2^2 = \frac{35885 - 11(57.05)^2}{11 - 1} = 7.745$$

$$\nu = \frac{\left[\dfrac{11.2}{12} + \dfrac{7.745}{11}\right]^2}{\left(\dfrac{1}{12-1}\right)\left(\dfrac{11.2}{12}\right)^2 + \left(\dfrac{1}{11-1}\right)\left(\dfrac{7.745}{11}\right)^2} = 20.82 \approx 20$$

$$100(1-\alpha)\% \ \mathrm{CI} = (\overline{X}_1 - \overline{X}_2) \pm t_{(\nu, \alpha/2)}\sqrt{\left(\frac{s_1^2}{n_1} + \frac{s_2^2}{n_2}\right)}$$

$$99\% \ \mathrm{CI} = 100(1 - 0.01)\% \ \mathrm{CI} \Rightarrow \alpha = 0.01; \ \alpha/2 = 0.005;$$

$$(51.35 - 57.05) \pm t_{(20, 0.005)}\sqrt{\left(\frac{11.2}{12} + \frac{7.745}{11}\right)} = -5.7 \pm 2.845 \times 1.28$$

$$= (-9.35, -2.06)$$

■ **EXAMPLE 12.19**   Independent samples of equal sizes are chosen from the populations Y_1 and Y_2, yielding the following values for a variable of interest:

Y_1: 10 11 12 12 9 8 3 5 6 9 21 19 14 0 11
Y_2: 9 4 3 8 6 10 11 9 8 14 7 3 2 13 4

Assuming that the values are normally distributed, obtain the 98% confidence interval for the difference between the means of Y_1 and Y_2.

Solution

$$\overline{Y}_1 = \frac{150}{15} = 10; \ \overline{Y}_2 = \frac{111}{15} = 7.4;$$

$$s_1^2 = \frac{1924 - 15(10)^2}{15 - 1} = 30.28571$$

$$s_2^2 = \frac{1015 - 15(7.4)^2}{15 - 1} = 13.82857$$

$$\nu = \frac{\left[\dfrac{30.28571}{15} + \dfrac{13.82857}{15}\right]^2}{\left(\dfrac{1}{15-1}\right)\left(\dfrac{30.28571}{15}\right)^2 + \left(\dfrac{1}{15-1}\right)\left(\dfrac{13.82857}{15}\right)^2} = 24.57927 \approx 24.$$

$$100(1-\alpha)\% \ \mathrm{CI} = (\overline{X}_1 - \overline{X}_2) \pm t_{(\nu, \alpha/2)}\sqrt{\left(\frac{s_1^2}{n_1} + \frac{s_2^2}{n_2}\right)}$$

$$98\% \ \mathrm{CI} = 100(1 - 0.02)\% \ \mathrm{CI} \Rightarrow \alpha = 0.02; \ \alpha/2 = 0.01;$$

$$(10 - 7.4) \pm t_{(24, 0.01)}\sqrt{\left(\frac{30.28571}{15} + \frac{13.82857}{15}\right)} = 2.6 \pm 2.492 \times 1.714921$$

$$= (-1.67358, 6.873582)$$

EXAMPLE 12.20   The following data show the levels of a disease found in independent samples taken from two populations.

$$Y_1: 7.7 \ 4.8 \ 34.4 \ 20.1 \ 11.2 \ 26.4 \ 5.0 \ 40.2 \ 2.2 \ 73.5 \ 55.9 \ 65.8 \ 45.6 \ 15.5$$
$$Y_2: 65.1 \ 0.0 \ 40.0 \ 48.8 \ 15.0 \ 8.8 \ 13.5 \ 6.1 \ 29.4 \ 20.5 \ 48.4 \ 28.7$$

Obtain the 95% confidence interval for the difference between the mean disease levels for the two populations.

Solution

$$\overline{Y}_1 = \frac{408.3}{14} = 29.16429; \ \overline{Y}_2 = \frac{324.3}{12} = 27.025;$$

$$s_1^2 = \frac{19314.29 - 14(29.16429)^2}{14 - 1} = 569.7317$$

$$s_2^2 = \frac{13192.31 - 12(27.025)^2}{12 - 1} = 402.5457$$

$$\nu = \frac{\left[\dfrac{569.7317}{14} + \dfrac{402.5457}{12} \right]^2}{\left(\dfrac{1}{14 - 1} \right)\left(\dfrac{569.7317}{14} \right)^2 + \left(\dfrac{1}{12 - 1} \right)\left(\dfrac{402.5457}{12} \right)^2} = 23.99593 \approx 23.$$

$$100(1 - \alpha)\% \ \text{CI} = (\overline{X}_1 - \overline{X}_2) \pm t_{(\nu, \alpha/2)}\sqrt{\left(\frac{s_1^2}{n_1} + \frac{s_2^2}{n_2} \right)}$$

$$95\% \ \text{CI} = 100(1 - 0.05)\% \ \text{CI} \Rightarrow \alpha = 0.05; \ \alpha/2 = 0.025;$$

$$(29.16429 - 27.025) \pm t_{(23, \, 0.025)}\sqrt{\left(\frac{569.7317}{14} + \frac{402.5457}{12} \right)}$$

$$= 2.139286 \pm 2.069 \times 8.616298 = (-15.6878, 19.9664)$$

EXAMPLE 12.21   The vital capacity, the volume of gas that can be expelled from the lungs from a position of full inspiration, with no limit to duration of inspiration, was measured for adults in two different occupations, and the data obtained are as follows:

Vital capacity measurements for adults in occupation I:
4.44 4.81 4.30 5.88 5.25 5.02 6.31 3.90 5.83 6.47
4.18 4.61 5.33 5.66 6.17 4.72 6.05 4.57 5.65 6.23
4.57 4.36 5.41 6.10 6.57 4.26 3.54 4.02 4.71 5.33

Vital capacity measurements for adults in occupation II:
3.68 2.96 3.44 4.13 4.75 4.16 3.19 3.19 4.76 4.77 3.46 2.87 3.73 5.05
4.55 4.07 3.06 4.03 4.73 4.88 3.98 3.30 3.82 3.70 4.65 3.34 4.18 5.36
6.44 4.21 3.72 4.14 4.75 5.73 4.33 4.02 4.45 4.89 5.76 4.45

Obtain the (a) 90%, and (b) 95% confidence interval for the difference in the mean vital capacity for adults in occupation I and those in occupation II.

Solution

$$\overline{X}_1 = \frac{154.25}{30} = 5.14167; \ \overline{X}_2 = \frac{168.68}{40} = 4.217;$$

$$s_1^2 = \frac{\displaystyle\sum_{i=1}^{n} X_{1i}^2 - n_1 \overline{X}_1^2}{n_1 - 1} = \frac{814.2621 - 30 \times 5.14167^2}{30 - 1} = 0.729656;$$

$$s_2^2 = \frac{\sum\limits_{i=1}^{n} X_{2i}^2 - n_2\overline{X}_2^2}{n_2 - 1} = \frac{736.6764 - 40 \times 4.217^2}{40 - 1} = 0.650073$$

$$\hat{\sigma}_{\overline{x}_1 - \overline{x}_2} = \sqrt{\frac{0.729656}{30} + \frac{0.650073}{40}} = 0.201429$$

$$\nu = \frac{\left[\dfrac{0.729656}{30} + \dfrac{0.650073}{40}\right]^2}{\left(\dfrac{1}{30 - 1}\right)\left(\dfrac{0.729656}{30}\right)^2 + \left(\dfrac{1}{40 - 1}\right)\left(\dfrac{0.650073}{40}\right)^2} = 60.58813 \approx 60.$$

(a) 90% CI $= 100(1 - 0.10)\% \Rightarrow \alpha = 0.10; \alpha/2 = 0.05;$
   $t_{(60,\,0.05)} = 1.671;$
   90% CI $= (\overline{X}_1 - \overline{X}_2) \pm t_{(60,\,0.05)} \times \hat{\sigma}_{\overline{x}_1 - \overline{x}_2} = (5.14167 - 4.217)$
   $\pm 1.671(0.201428) = (0.588082, 1.261258)$

(b) 95% CI $= 100(1 - 0.05)\% \Rightarrow \alpha = 0.05; \alpha/2 = 0.025;$
   $t_{(62,\,0.025)} = 2.00;$
   95% CI $= (\overline{X}_1 - \overline{X}_2) \pm t_{(60,\,0.025)} \times \hat{\sigma}_{\overline{x}_1 - \overline{x}_2} = (5.14167 - 4.217)$
   $\pm 2.00(0.201428) = (0.521812, 1.327528)$

■ **EXAMPLE 12.22**   Samples of times it took for patients hospitalized for the same disease at hospitals A and B to fully recover are recorded below:

Hospital A:
249.10 202.50 222.20 214.40 205.90 196.70 188.6
214.30 195.10 213.30 225.50 191.40 198.50 189.99
201.10 239.80 245.70 213.00 238.80 176.77 192.45
171.10 222.00 212.50 201.70 184.90 189.90 196.66
248.30 209.70 233.90 229.80 217.90 199.56 199.00

Hospital B:
199.50 184.00 173.20 186.00 214.10 198.60 190.6
125.50 143.50 190.40 152.00 165.70 165.5   167.8
154.70 145.30 154.60 190.30 135.40 155.80 153.78
167.70 203.40 186.70 155.30 195.90 198.60 178.42
168.90 166.70 178.60 150.20 212.40 166.4   166.78

Obtain the (a) 95%, and (b) 99% confidence interval for the difference in the means of the recovery times for hospitals A and B.

Solution $\overline{X}_1 = \dfrac{7332.03}{35} = 209.487; \overline{X}_2 = \dfrac{6042.28}{35} = 172.637;$

$$s_1^2 = \frac{\sum\limits_{i=1}^{n} X_{1i}^2 - n_1\overline{X}_1^2}{n_1 - 1} = \frac{1549887 - 35 \times 209.487^2}{35 - 1} = 409.5639;$$

$$s_2^2 = \frac{\sum\limits_{i=1}^{n} X_{2i}^2 - n_2\overline{X}_2^2}{n_2 - 1} = \frac{1059491 - 35 \times 172.637^2}{35 - 1} = 481.3917;$$

$$\hat{\sigma}_{\overline{x}_1 - \overline{x}_2} = \sqrt{\frac{409.5639}{35} + \frac{481.3917}{35}} = 5.054133$$

$$\nu = \frac{\left[\dfrac{409.5639}{35} + \dfrac{481.3917}{35}\right]^2}{\left(\dfrac{1}{35-1}\right)\left(\dfrac{409.5639}{35}\right)^2 + \left(\dfrac{1}{35-1}\right)\left(\dfrac{481.3917}{35}\right)^2} = 67.56089 \approx 67$$

(a) 95% CI = $100(1 - 0.05)\% \Rightarrow \alpha = 0.05$; $\alpha/2 = 0.025$;
$t_{(67, 0.025)} = 1.996$;
95% CI = $(\overline{X}_1 - \overline{X}_2) \pm t_{(67, 0.025)} \times \hat{\sigma}_{\overline{x}_1 - \overline{x}_2} = (209.487 - 172.637)$
$\pm 1.966(5.045381) = (26.76195, 46.93805)$

(b) 99% CI = $100(1 - 0.01)\% \Rightarrow \alpha = 0.01$; $\alpha/2 = 0.005$; $t_{(67, 0.005)} = 2.651$;
99% CI = $(\overline{X}_1 - \overline{X}_2) \pm t_{(67, 0.025)} \times \hat{\sigma}_{\overline{x}_1 - \overline{x}_2} = (209.487 - 172.637)$
$\pm 2.651(5.045381) = (23.45149, 50.24851)$

■ **EXAMPLE 12.23**   Samples of heights of adult women and men were taken and the data obtained are as follows:

Heights of women:
66 58 66 68 64 62 66 63 64 65 64 64 62 69 62
67 63 66 65 71 63 67 64 60 65 67 59 60 58 72
63 67 62 69 63 64 63 64 60 65

Heights of men:
70 69 64 71 68 66 74 73 62 69 67 68 72 66 72
68 71 67 71 75 64 65 67 72 72 72 67 71 66 75
69 70 69 62 66 76 69 68 66 68

Obtain the (a) 90% and (b) 98% confidence intervals for the difference between means of the heights of men and women.

Solution $\overline{y}_1 = \dfrac{2570}{40} = 64.25$; $\overline{y}_2 = \dfrac{2757}{40} = 68.925$

$s_1^2 = \dfrac{165522 - 40(64.25)^2}{40 - 1} = 10.24359$; $s_2^2 = \dfrac{190481 - 40(68.925)^2}{40 - 1} = 11.6609$

$$\nu = \frac{\left[\dfrac{10.24359}{40} + \dfrac{11.6609}{40}\right]^2}{\left(\dfrac{1}{40-1}\right)\left(\dfrac{10.24359}{40}\right)^2 + \left(\dfrac{1}{40-1}\right)\left(\dfrac{11.6609}{40}\right)^2} = 77.6748 \approx 77$$

(a) $100(1 - \alpha)\%$ CI = $(\overline{y}_1 - \overline{y}_2) \pm t_{(\nu, \alpha/2)}\sqrt{\dfrac{s_1^2}{n_1} + \dfrac{s_2^2}{n_2}}$;
90% CI = $100(1 - 0.1)\%$ CI $\Rightarrow \alpha = 0.1$; $\alpha/2 = 0.05$; $t_{(77, 0.05)} = 1.665$;
90% CI = $(68.925 - 64.25) \pm 1.665\sqrt{\left(\dfrac{10.24359}{40} + \dfrac{11.6609}{40}\right)}$
$= 4.675 \pm 1.665 \times 0.740008 = (3.442887, 5.907113)$

(b) $100(1 - \alpha)\%$ CI = $(\overline{y}_1 - \overline{y}_2) \pm t_{(\nu, \alpha/2)}\sqrt{\dfrac{s_1^2}{n_1} + \dfrac{s_2^2}{n_2}}$;
98% CI = $100(1 - 0.02)\%$ CI $\Rightarrow \alpha = 0.02$; $\alpha/2 = 0.01$; $t_{(77, 0.01)} = 2.376$;

$$98\% \text{ CI} = (68.925 - 64.25) \pm 2.376\sqrt{\left(\frac{10.24359}{40} + \frac{11.6609}{40}\right)}$$

$$= 4.675 \pm 2.376 \times 0.740008 = (2.916741, 6.433259)$$

12.8 Use of Central Limit Theorem for Constructing Confidence Intervals for the Difference Between Two Population Means for Large Samples

We have emphasized that if we draw samples from normal populations, the best way to find the confidence interval is by using the t-distribution as in the preceding sections. When we take large samples, n_1 and n_2, from two nonnormal populations for which the variances are known, we can base our $100(1 - \alpha)\%$ confidence intervals on central limit theorem (CLT), so that:

$$100(1 - \alpha)\% \text{ CI for } \mu_1 - \mu_2 = (\overline{X}_1 - \overline{X}_2) \pm Z_{\alpha/2}\sqrt{\left(\frac{\sigma_1^2}{n_1} + \frac{\sigma_2^2}{n_2}\right)};$$

where σ_1^2 and σ_2^2 are the known population variances of the nonnormal variables X_1 and X_2 respectively. However, if we do not know the population variances we replace them with their sample equivalents, so that the $100(1 - \alpha)\%$ confidence interval becomes:

$$100(1 - \alpha)\% \text{ CI for } \mu_1 - \mu_2 = (\overline{X}_1 - \overline{X}_2) \pm Z_{\alpha/2}\sqrt{\left(\frac{s_1^2}{n_1} + \frac{s_2^2}{n_2}\right)}.$$

where s_1^2 and s_2^2 are the sample variances of the nonnormal variables X_1 and X_2, respectively, obtained from the samples. These same methods were endorsed by Daniel (2009), p. 243; and Devore and Farnum (2005), p. 355; Sullivan III, p. 466. We intend to present this method for comparison. In this section, we will find the confidence intervals for the differences between the means considered in Examples 12.17–12.19 using this method, because the sample sizes are large in each case.

■ **EXAMPLE 12.24** **Using Z Due to Large Samples**
The vital capacity, the volume of gas that can be expelled from the lungs from a position of full inspiration, with no limit to duration of inspiration, was measured for adults in two different occupations, and the data obtained are as follows:

Vital capacity measurements for adults in occupation I:
4.44 4.81 4.30 5.88 5.25 5.02 6.31 3.90 5.83 6.47
4.18 4.61 5.33 5.66 6.17 4.72 6.05 4.57 5.65 6.23
4.57 4.36 5.41 6.10 6.57 4.26 3.54 4.02 4.71 5.33

Vital capacity measurements for adults in occupation II:
3.68 2.96 3.44 4.13 4.75 4.16 3.19 3.19 4.76 4.77 3.46 2.87 3.73 5.05
4.55 4.07 3.06 4.03 4.73 4.88 3.98 3.30 3.82 3.70 4.65 3.34 4.18 5.36
6.44 4.21 3.72 4.14 4.75 5.73 4.33 4.02 4.45 4.89 5.76 4.45

Obtain the (a) 90%, and (b) 95% confidence interval for the difference in the mean vital capacity for adults in occupation I and those in occupation II.

Solution

$$\overline{X}_1 = \frac{154.25}{30} = 5.14167; \quad \overline{X}_2 = \frac{168.68}{40} = 4.217;$$

$$s_1^2 = \frac{\sum_{i=1}^{n} X_{1i}^2 - n_1 \overline{X}_1^2}{n_1 - 1} = \frac{814.2621 - 30 \times 5.14167^2}{30 - 1} = 0.729656;$$

$$s_2^2 = \frac{\sum_{i=1}^{n} X_{2i}^2 - n_2 \overline{X}_2^2}{n_2 - 1} = \frac{736.6764 - 40 \times 4.217^2}{40 - 1} = 0.650073;$$

$$\hat{\sigma}_{\overline{x}_1 - \overline{x}_2} = \sqrt{\frac{0.729656}{30} + \frac{0.650063}{40}} = 0.201429$$

(a) 90% CI = $100(1 - 0.10)\% \Rightarrow \alpha = 0.10; \alpha/2 = 0.05;$
$Z_{0.05} = 1.645;$
90% CI = $(\overline{X}_1 - \overline{X}_2) \pm Z_{0.05} \times \hat{\sigma}_{\overline{x}_1 - \overline{x}_2} = (5.14167 - 4.217)$
$\pm 1.645(0.201429) = (0.593332, 1.25602)$

(b) 95% CI = $100(1 - 0.05)\% \Rightarrow \alpha = 0.05; \alpha/2 = 0.025;$
$Z_{0.025} = 1.96;$
95% CI = $(\overline{X}_1 - \overline{X}_2) \pm Z_{0.025} \times \hat{\sigma}_{\overline{x}_1 - \overline{x}_2} = (5.14167 - 4.217)$
$\pm 1.96(0.201429) = (0.52987, 1.31947)$

■ **EXAMPLE 12.25** **Using z Due to Large Samples**

Samples of times it took for patients hospitalized for the same disease at hospitals A and B to fully recover are recorded below:

Hospital A:
249.10 202.50 222.20 214.40 205.90 196.70 188.60 214.30 195.10 213.30
225.50 191.40 198.50 189.99 201.10 239.80 245.70 213.00 238.80 176.77
192.45 171.10 222.00 212.50 201.70 184.90 189.90 196.66 248.30 209.70
233.90 229.80 217.90 199.56 199.00

Hospital B:
199.50 184.00 173.20 186.00 214.10 198.60 190.60 125.50 143.50 190.40
152.00 165.70 165.50 167.80 154.70 145.30 154.60 190.30 135.40 155.80
153.78 167.70 203.40 186.70 155.30 195.90 198.60 178.42 168.90 166.70
178.60 150.20 212.40 166.40 166.78

Obtain the (a) 95%, and (b) 99% confidence interval for the difference in the means of the recovery times for hospitals A and B.

Solution

$$\overline{X}_1 = \frac{7332.03}{35} = 209.487; \quad \overline{X}_2 = \frac{6042.28}{35} = 172.637;$$

$$s_1^2 = \frac{\sum_{i=1}^{n} X_{1i}^2 - n_1 \overline{X}_1^2}{n_1 - 1} = \frac{1549887 - 35 \times 209.487^2}{35 - 1} = 409.5639;$$

$$s_2^2 = \frac{\sum\limits_{i=1}^{n} X_{2i}^2 - n_2 \overline{X}_2^2}{n_2 - 1} = \frac{1049591 - 35 \times 4.217^2}{35 - 1} = 484.4852;$$

$$\hat{\sigma}_{\overline{x}_1 - \overline{x}_2} = \sqrt{\frac{409.5639}{35} + \frac{484.4852}{35}} = 5.054133$$

(a) 95% CI $= 100(1 - 0.05)\% \Rightarrow \alpha = 0.05$; $\alpha/2 = 0.025$;
   $Z_{0.025} = 1.96$;
   95% CI $= (\overline{X}_1 - \overline{X}_2) \pm Z_{0.025} \times \hat{\sigma}_{\overline{x}_1 - \overline{x}_2} = (209.487 - 172.637)$
   $\pm 1.96(5.054133) = (26.9439, 46.7561)$

(b) 99% CI $= 100(1 - 0.01)\% \Rightarrow \alpha = 0.01$; $\alpha/2 = 0.005$;
   $Z_{0.005} = 2.575$;
   99% CI $= (\overline{X}_1 - \overline{X}_2) \pm Z_{0.005} \times \hat{\sigma}_{\overline{x}_1 - \overline{x}_2} = (209.487 - 172.637)$
   $\pm 2.575(5.054133) = (23.8356, 49.86439)$

■ **EXAMPLE 12.26** **Using Z Due to Large Samples**
Samples of heights of adult women and men were taken and the data obtained are as follows:

Heights of women:
66 58 66 68 64 62 66 63 64 65 64 64 62 69 62 67 63 66 65 71 63 67
64 60 65 67 59 60 58 72 63 67 62 69 63 64 63 64 60 65

Heights of men:
70 69 64 71 68 66 74 73 62 69 67 68 72 66 72 68 71 67 71 75 64 65
67 72 72 72 67 71 66 75 69 70 69 62 66 76 69 68 66 68

Obtain the (a) 90%, and (b) 98% confidence intervals for the difference between means of the heights of men and women.

Solution

$$\overline{y}_1 = \frac{2570}{40} = 64.25; \quad \overline{y}_2 = \frac{2757}{40} = 68.925;$$

$$s_1^2 = \frac{165522 - 40(64.25)^2}{40 - 1} = 10.24359;$$

$$s_2^2 = \frac{190481 - 40(68.925)^2}{40 - 1} = 11.6609;$$

$$\sigma_{\overline{x}_1 - \overline{x}_2} = \sqrt{\left(\frac{10.24359}{40} + \frac{11.6609}{40}\right)} = 0.740008;$$

(a) $100(1 - \alpha)\%$ CI $= (\overline{y}_1 - \overline{y}_2) \pm Z_{\alpha/2}\sqrt{\frac{s_1^2}{n_1} + \frac{s_2^2}{n_2}};$
   90% CI $= 100(1 - 0.1)\%$ CI $\Rightarrow \alpha = 0.1$; $\alpha/2 = 0.05$; $Z_{0.05} = 1.645$;
   90% CI $= (68.925 - 64.25) \pm 1.645(0.740008)$
   $= 4.675 \pm 1.645 \times 0.740008 = (3.44577, 5.89231)$

(b) $100(1 - \alpha)\%$ CI $= (\overline{y}_1 - \overline{y}_2) \pm Z_{\alpha/2}\sqrt{\frac{s_1^2}{n_1} + \frac{s_2^2}{n_2}};$
   98% CI $= 100(1 - 0.02)\%$ CI $\Rightarrow \alpha = 0.02$; $\alpha/2 = 0.01$; $Z_{0.01} = 2.33$;

$$98\% \text{ CI} = (68.925 - 64.25) \pm 2.33(0.740008)$$
$$= 4.675 \pm 2.33 \times 0.740008 = (2.95078, 6.39922)$$

12.9 Use of MINITAB for Confidence Intervals for Means and Proportions

Minitab easily provides confidence intervals for the difference between two means and proportions in the context of testing hypotheses about these differences which are discussed in Chapter 13. We therefore provide MINITAB approaches in Chapter 13.

Reading List

Bartlett, M. S. (1939). Complete Simultaneous Fiducial Distribution, *Ann Math Stat*, 10:129.

Bartlett, M. S. (1936). The Information Available in Small Samples, *Proc Cambridge Philosophical Soc*, 32:560.

Cochran W. G. and G. M. Cox (1950). *Experimental Designs*, John Wiley and Sons, New York, New York

Daniel, W. W. (2009). Biostatistics, a Foundation for Analysis in the Health Sciences, 9th Ed., New York, New York: Wiley.

Devore, J. L. and N. R. Farnum (2005). *Applied Statistics for Engineers and Scientists*, Belmont, CA: Brooks/Cole.

Dixon, W. J. and F. J. Massey (1951). *Introduction to Statistical Analysis*, New York: McGraw-Hill Book Co. Inc.

Larson, R. J. and M. L. Marx (2006). *An Introduction to Mathematical Statistics and Its Applications, 4th Ed.*, New Jersey: Pearson.

Satterthwaite, F. E. (1946). An Approximate Distribution of Estimates of Variance Components, *Biometrics*, 2:110.

Smith, H. F. (1936). The Problem of Comparing the Result of Two Experiments with Unequal Errors, *J Sci Ind Research, India*, 9:211.

Sukhatme, P. V. (1938). On the Fisher-Behrens Test of Significance for the Difference in Means of Two Normal Samples, *Sankhyā*, 4:39.

Triola M. M. and F. T. Triola (2006). *Biostatistics for the Biological and Health Sciences*, Boston: Pearson.

Weis, N. A. (2012). *Elementary Statistics, 8th Ed.*, Addison-Wesley.

Welch B. L. (1937). Significance of the Difference between Two Means when the Population Variances are Unequal, *Biometrika*, 29:350.

References

Daniel, W. W. (2009). Biostatistics, a Foundation for Analysis in the Health Sciences, 9th Ed., Wiley.

Devore, J. L. and N. R. Farnum (2005). *Applied Statistics for Engineers and Scientists*, Belmont, CA: Brooks/Cole.

Sullivan III, M (2011) Fundamentals of Statistics, 3rd Ed., Prentice Hall, Boston

Name _____ Date _____

Exercises

Exercises on sampling distribution of differences between two means, two proportions

12.1 70% of Republicans and 60% of Democrats in Boston are believed to favor mandatory sentencing for violent crimes. If samples of 400 and 300 Republicans and Democrats were taken from the city, what is the probability that the difference in percentage of support between Republicans and Democrats found would:

(a) exceed 5.9%, (b) lie between 4% and 15%, and (c) be less than 8%?

12.2 Two similar machines A and B produce the same item for a manufacturing concern. Machine A is known to produce 12% defectives while Machine B produces 7% defectives. If samples of 650 and 420 items made by machines A and B, respectively, are chosen, what is the probability that the difference in percentage of defectives found would:

(a) exceed 6.8%, (b) lie between 7.2% and 9%, and (c) be less than 2.7%?

12.3 In City A, mean cost for plumbers to tune-up a typical home heating system for winter is normally distributed with mean of 114 dollars, and variance of 560 dollars2. In City B area, mean cost for plumbers to tune-up a typical home heating system for winter is normally distributed with mean 102 dollars, and variance of 481 dollars2. If samples of costs of tune-ups for 50 households in City A are taken, while a sample of 42 households is taken from city B, what is the probability that the difference in the mean cost of tune-ups is (a) less than 18, (b) between 3 and 15, and (c) greater than 20 dollars?

12.4 Two populations A and B are believed to be of equal means with variances, 368 and 567, respectively. If samples of sizes 50 and 62 are taken from A and B, respectively, what is the probability that the difference in their sample means (a) is greater than 10, (b) between 6 and 9, and (c) less than 12.

12.5 Suppose that samples of sizes 200 were taken from populations that contain an attribute of interest in the proportions 0.6 and 0.4, respectively, what is the probability that difference between the proportions would (a) exceed 0.27, (b) lie between 0.14 and 0.28, and (c) be less than 0.25?

12.6 Find the 99% confidence interval for $\hat{p}_1 - \hat{p}_2$: $n_1 = 750$; $x_1 = 257$; $n_2 = 560$; $x_2 = 149$.

12.7 Find the 98% confidence interval for $\hat{p}_1 - \hat{p}_2$: $n_1 = 1050$; $x_1 = 194$; $n_2 = 1260$; $x_2 = 114$.

12.8 Find the 95% confidence interval for $\hat{p}_1 - \hat{p}_2$: $n_1 = 609$; $x_1 = 211$; $n_2 = 806$; $x_2 = 217$.

12.9 Find the 96% confidence interval for $\hat{p}_1 - \hat{p}_2$: $n_1 = 975$; $x_1 = 102$; $n_2 = 789$; $x_2 = 93$.

12.10 Find the 95% confidence interval for $\hat{p}_1 - \hat{p}_2$: $n_1 = 1035$; $x_1 = 149$; $n_2 = 1106$; $x_2 = 103$.

12.11 Find the 98% confidence interval for $\hat{p}_1 - \hat{p}_2$: $n_1 = 534$; $x_1 = 146$; $n_2 = 442$; $x_2 = 89$.

12.12 A survey of 1356 customers by a company showed that 546 preferred their brand of a product. After two months of the advertisement campaign pushing their product, 1645 customers were surveyed out of which 844 preferred their brand of the product. Obtain the 90% confidence interval for difference between the proportions that preferred their brand after and before.

12.13 A survey of 1230 college age females showed that 234 of them preferred a popular brand of beverage above all others. Similar survey of 1678 male students showed that 567 preferred the popular beverage over all others. Obtain a 95% confidence interval for the difference between proportions of males and females who preferred this beverage over all others.

12.14 To manage the treatment of a disease, a study was carried out to determine the proportions of readmissions for patients who were discharged early and for those who were discharged late. The following data was obtained:

	Readmitted	Were Not Readmitted
Early Discharged	167	1091
Late Discharge	76	1304

Obtain the 99% confidence interval for the difference between the proportion of readmissions for patients who were discharged early and the proportion of readmissions for those who were discharged late.

12.15 In a study of a new medication for preventing flu, samples of 120 men and 140 women were given the medication. At the end of the study 100 men and 96 women did not catch the flu. Obtain the 99% confidence interval for the difference between the proportion of men and women who do not catch the flu due to the medication.

12.16 A sample of 300 students of SCSU who entered in 2000 showed that 211 were still enrolled after six semesters. A sample of 400 students who entered SCSU in 2005 showed that 302 were still enrolled after six semesters. Obtain a 98% confidence interval for the difference in proportions of students retained among 2005 entrants and 2000 entrants.

12.17 In a poll during the health care reform debate of 2010, 900 Republicans and 1020 Democrats were asked "Would you be willing to pay a little more for health insurance so that coverage may be extended to all uninsured people?" 191 Republicans and 702 Democrats responded "yes." Obtain the 95% confidence interval for the difference between proportions of Democrats who responded "yes" and the proportion of Republicans who responded "yes."

12.18 In a survey of enrollment of eighth graders for Algebra I it was found that of 1002 students from Massachusetts, 466 enrolled for Algebra I, while out of 1010 students from Minnesota 313 enrolled. Obtain the 96% confidence interval for difference between proportions of eighth graders from Massachusetts and eighth graders from Minnesota who enrolled for Algebra I.

12.19 In a study, 566 patients were treated with a medication for a disease of which 202 recovered fully. In the same study, 365 patients were a given placebo (an inert substance that looks like the medication), 96 recovered fully. Obtain a 98% confidence interval for the difference between proportions of patients who recovered fully for both groups.

12.20 In a study of the marketing of credit-cards, of 1100 freshmen students credit cardholders surveyed, 112 opened their accounts while in high school; another independent survey of freshmen students showed that of 2306 credit card holding students, 471 got their credit cards after high school, but before attending college. Obtain the 96% confidence interval for the difference between the proportions of students who obtained a card after high school and before college and those who obtained credit cards while in high school.

12.21 A prenatal intervention to increase birth weights of babies was being studied. Data set A is the sample of birth weights for babies whose mothers participated in the program, while the data set B is a sample of birth weights for babies whose mothers did not participate.

A: 6.1 5.5 5.6 7.0 6.9 7.4 8.0 9.0 7.8 6.9 8.2 7.4 8.3 6.8 7.4
B: 6.0 6.1 5.5 5.8 5.3 5.9 5.9 6.3 5.7 6.6 7.0 4.5 5.5 5.2

Obtain a 95% confidence interval for difference between mean weights for babies whose mothers participated and for babies whose mothers did not participate. Assume birth weights for both groups are normally distributed with equal variance.

12.22 The following show the number of hours that samples of men and women spend on news and related items per week:

Male: 10 15 16 12 8 13 17 13 15 20 23 14 15 6 12
Female: 7 9 6 9 0 5 10 14 16 7 8 10 5 12 4 0

Obtain a 98% confidence interval for the difference between the weekly mean number of hours spent by men and women on news and related items. Assume weekly hours spent on news and related items are normally distributed with equal variance.

12.23 A chemical production process was piloted using two different formulations of the catalyst. The yields of the process per unit time were recorded for independent samples for both catalysts:

Formulation I: 52 72 91 72 83 54 72 60 89 97 96 68 90 82 80
                88 77 60 70 85 77 64
Formulation II: 70 60 80 51 81 53 73 72 59 73 75 60 59 60 64
                72

Assume yields for both formulations are normally distributed with equal variance. Obtain the 95% confidence interval for the difference between the means of Formulations I and II.

12.24 The following are samples of heights of adult males and females.

Males: 71.9 66.1 72.4 73.0 68.0 68.7 70.3 63.7 71.1 65.6 68.3 66.3
Females: 63.4 64.1 62.7 61.3 58.2 63.2 60.5 65.0 61.8 68.0 67.0 57.0

Assume both are normally distributed with equal variance. Obtain the 99% confidence interval for difference between their mean heights.

12.25   The following are the weights of samples of college males and females:

Males: 146.0 154.0 150.0 142.0 124.0 141.0 150.0 170.0 173.0
            186.0 176.0 167.0 187.0 120.0 176.0 142.0 140.0 180.0
Females: 133.0 134.0 126.0 156.0 132.0 126.0 144.0 160.0 156.0
            136.0 147.0 137.0

Assume weights are both normally distributed with equal variance. Obtain the 95% confidence interval for difference between their mean heights.

12.26   The following are samples of math scores for students who are taking the SAT for the first time, and for students taking the SAT for the second time.

Scores of candidates sitting first time for the SAT:
500 515 525 530 500 522 530 509 523 506 521 512 507
498 505 507
Scores for candidates sitting second time for the SAT:
501 506 509 520 522 513 595 550 568 563 570 565 538

Assume scores are normally distributed for both groups. Obtain the 95% confidence interval for the difference between the second time takers of the SAT and first time takers of the SAT.

12.27   The same production process is being utilized by two plants for the production of an item. Samples of the percentage yields of the active ingredients resulting from the process for two plants are presented below:

Plant I: 8.7 9.3 8.5 7.8 6.3 7.1 5.8 6.1 3.9 5.3 4.7 4.9
Plant II: 11.2 9.7 10.5 9.4 7.5 8.5 6.8 7.7 6.0 4.1 4.3 5.1

Obtain a 98% confidence interval for the difference between the mean percentage of active ingredients for plant II and plant 1. Assume observations are normally distributed for both plants.

12.28   Measurements of an attribute of interest was made for two separate groups, choosing samples of sizes fourteen from each group.

Group 1: 5.5 5.5 3.5 8.0 5.0 3.0 7.0 4.0 8.0 9.0 6.0 9.5 7.0 7.5
Group 2: 4.5 4.5 7.0 4.5 4.5 5.0 4.0 6.5 6.0 6.0 5.0 5.5 5.5 5.0

Assume both measurements to be normally distributed. Obtain the 90% confidence interval for the difference between the means.

12.29   Another attribute was measured for the same sample taken from the groups in Exercise 12.23. The values obtained were:

Group 1: 21.7 21.4 21.1 20.0 20.1 20.2 20.6 21.1 20.9 20.3 20.5 21.5
            20.7 21.1
Group 2: 18.0 21.3 20.4 18.2 20.8 21.1 20.9 16.5 20.6 20.0 19.1 20.9
            20.3 20.2

Assume the measurements are normally distributed, and find the 95% confidence intervals for the difference between their means.

CHAPTER 13

Tests of Hypotheses Involving Two Samples

13.0 Introduction

In Chapter 8, we studied the confidence intervals for single population means and proportions. We also tested hypothesis about a single population mean and proportion in Chapter 9. In each of those cases, our focus was always on a single population from which we took a single sample and either constructed a confidence interval or tested hypotheses on the single population parameter of interest (mean and proportion). In this chapter, we seek to compare two means and two proportions. We will always be drawing independent samples, one from each of the two populations of interest, and making inference by comparing the same parameter for both populations.

13.1 Testing Hypotheses about Two Population Proportions

We also studied the confidence interval for the difference between two population proportions in Chapter 12. In this chapter, we shall learn how to test hypotheses to compare two population proportions. The typical hypotheses that are tested are usually of the general forms:

Case 1: $H_0:p_1 = p_2$ vs. $H_1:p_1 < p_2$
Case 2: $H_0:p_1 = p_2$ vs. $H_1:p_1 > p_2$
Case 3: $H_0:p_1 = p_2$ vs. $H_1:p_1 \neq p_2$

We can also recast the hypotheses so that they are stated in terms of differences between proportions.

Case 1: $H_0:p_1 - p_2 = 0$ vs. $H_1:p_1 - p_2 < 0$
Case 2: $H_0:p_1 - p_2 = 0$ vs. $H_1:p_1 - p_2 > 0$
Case 3: $H_0:p_1 - p_2 = 0$ vs. $H_1:p_1 - p_2 \neq 0$

In this form the theme of the hypotheses becomes the difference between the proportions as in construction of confidence intervals which we encountered earlier. Later we will show how testing these types of hypotheses relates to confidence interval estimation.

The testing of the above set of hypotheses depends on the distribution of $\hat{p}_1 - \hat{p}_2$ where $\hat{p}_1 = \dfrac{x_1}{n_1}$ and $\hat{p}_1 = \dfrac{x_2}{n_2}$. Here, a sample of size n_1 is taken from population 1, and x_1 of the items are found to have an attribute of interest. Similarly, a sample of size n_2 is taken from population 2, and x_2 of the items are seen to have the attribute of interest.

It is known that if we assume that the null hypothesis is true, then:

$$(\hat{p}_1 - \hat{p}_2) \sim N\left[p_1 - p_2, \bar{p}(1 - \bar{p})\left(\frac{1}{n_1} + \frac{1}{n_2}\right)\right],$$

where

$$\bar{p} = \frac{x_1 + x_2}{n_1 + n_2}$$

is an estimate of the common proportion which is required to satisfy the condition under H_0 that $p_1 = p_2$.

On the basis of the above distribution for $\hat{p}_1 - \hat{p}_2$, a test statistic for the above set of hypotheses is obtained as follows:

$$Z_{cal} = \frac{\hat{p}_1 - \hat{p}_2}{\sqrt{\bar{p}(1 - \bar{p})\left[\dfrac{1}{n_1} + \dfrac{1}{n_2}\right]}},$$

where $Z_{cal} \sim N(0, 1)$. With it we can test all the hypotheses stated in Case 1 to Case 3. We present an example:

■ **EXAMPLE 13.1**   Samples of migraine patients treated by Drs. Lilian and Beth show the recoveries presented in the table below:

Doctors	Recovered	Did Not Recover
Lilian	67	91
Beth	66	130

Test the hypothesis that the proportion of their patients who recovered are the same for Lilian and Beth against that they are not equal at 5% level of significance. Test by classical and p-value approaches.

Solution

Physician	Recovered	Did Not Recover	Total
Lilian	67	91	158
Beth	66	130	196

$$\hat{p}_L = \frac{67}{67 + 91} = 0.42405063;$$

$$\hat{p}_B = \frac{66}{66 + 130} = 0.3367347 \Rightarrow n_1 = 158, n_2 = 196$$

$$\bar{p} = \frac{67 + 66}{158 + 196} = 0.375706;$$

$$\sqrt{\bar{p}(1 - \bar{p})\left(\frac{1}{n_1} + \frac{1}{n_2}\right)} = \sqrt{0.375706(1 - 0.375706)\left(\frac{1}{158} + \frac{1}{196}\right)} = 0.05178$$

The hypothesis being tested in this case is Case 3:

Case 3: $H_0: p_1 = p_2$ vs. $H_1: p_1 \neq p_2$

$$Z_{cal} = \frac{\hat{p}_1 - \hat{p}_2}{\sqrt{\bar{p}(1 - \bar{p})\left[\frac{1}{n_1} + \frac{1}{n_2}\right]}} = \frac{0.424051 - 0.336735}{0.05178} = 1.68626$$

Classical Approach

For the classical test, we establish the critical values and associated rejection regions. Since we are testing at a 5% level of significance, the critical values are $\pm Z_{0.025} = \pm 1.96$.

For this particular test, we present the regions:

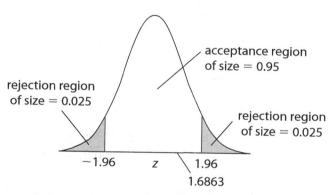

acceptance region
of size = 0.95

rejection region
of size = 0.025

rejection region
of size = 0.025

−1.96   z   1.96

1.6863

■ **FIGURE 13.1   The Critical Values and Rejection Regions for Example 13.1**

Clearly, the value of the test statistic lies in the acceptance region. We therefore do not reject H_0 and subsequently conclude that the proportion of patients of Drs. Lilian and Beth who fully recovered are the same at a 5% level of significance.

p-Value Approach

For this approach we use the value of the test statistic to show the areas of the distribution, under the assumption that H_0 is true, that represent the p-value. Figure 13.2 is a depiction of the p-value for a two-sided test.

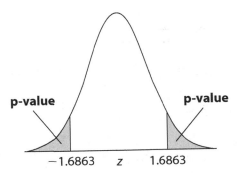

■ **FIGURE 13.2** The p-Value for Example 13.1

The p-value $= P(Z_{cal} \leq -1.6863) + P(Z \geq 1.6863) = 0.0455 + 0.0455 = 0.0910$. Here we see that p-value is greater than the level of significance of 5%, consequently, we conclude that we could not reject H_0, meaning that the proportion of patients who recovered for both physicians are the same at a 5% level of significance.

■ **EXAMPLE 13.2**   Independent samples of 747 U.S. men and 434 U.S. women were taken. Of those sampled, 276 men and 195 women said that they would pay extra tax towards preserving wildlife habitats.

Test the hypothesis that the proportion of females who would pay extra tax towards preserving wildlife habitats is equal to the proportion of males who would do the same against that they are not equal at a 2.5% level of significance using both classical and p-value approaches.

Solution $n_m = 747; x_m = 276; n_{ff} = 434; x_f = 195;$

$$\hat{p}_m = \frac{276}{747} = 0.369478; \hat{p}_f = \frac{195}{434} = 0.449309$$

$$H_0:p_m = p_f \text{ vs. } H_1:p_m \neq p_f$$

$$\bar{p} = \frac{276 + 195}{747 + 434} = 0.398815$$

$$\sigma_{\hat{p}_m - \hat{p}_f} = \sqrt{0.398815(1 - 0.398815)\left(\frac{1}{747} + \frac{1}{434}\right)} = 0.0295535$$

$$Z_{cal} = \frac{\hat{p}_m - \hat{p}_f}{\sigma_{\hat{p}_m - \hat{p}_f}} = \frac{0.449309 - 0.369478}{0.0295535} = 2.701223 \simeq 2.70$$

Classical approach: At 2.5% l.o.s, we see from the tables that the critical value, $Z_{0.0125} = 2.24$

Since $Z_{cal} > Z_{0.0125}$, we reject H_0 and conclude that probably, $p_m \neq p_f$

P-value approach: The p-value $= P[Z \leq -Z_{cal}] + P[Z \geq Z_{cal}] = P[Z \leq -2.7] + P[Z \geq 2.70] = 0.0035 + 0.0035 = 0.0070$

Since p-value $< <0.025$, the level of significance, we reject H_0 and conclude that probably $p_m \neq p_f$.

■ **EXAMPLE 13.3**   A survey was carried out to study regular use of sunscreen by teenage girls and women in their twenties. Of 340 teenage girls surveyed, 167 were found to use sunscreen regularly, while 179 of 400 women in their twenties surveyed used sunscreen regularly. Using both classical and p-value approaches, test at a 4% l.o.s. that the proportion of teenage girls who used sunscreen regularly is equal to the proportion of women in their twenties who used sunscreen regularly against that they are not equal.

Solution

$$x_{teen} = 167; n_{teen} = 340, \hat{p}_{\text{teen}} = \frac{167}{340} = 0.491; x_{\text{twenty}} = 179;$$

$$n_{\text{twenty}} = 400, \hat{p}_{\text{twenty}} = \frac{179}{400} = 0.448$$

$$\bar{p} = \frac{167 + 179}{400 + 340} = 0.468;$$

$$\hat{\sigma}_{\bar{p}} = \sqrt{\frac{0.468(1 - 0.468)}{400} + \frac{0.468(1 - 0.468)}{340}} = 0.0368$$

$$H_0: p_{\text{teen}} = p_{\text{twenty}} \text{ vs. } H_1: p_{\text{teen}} \neq p_{\text{twenty}}$$

$$Z_{cal} = \frac{0.491 - 0.448}{\sqrt{\dfrac{0.468(1 - 0.468)}{400} + \dfrac{0.468(1 - 0.468)}{340}}} = \frac{0.491 - 0.448}{0.0368} = 1.18672$$

Classical Approach

With $\alpha = 0.04$, then the critical value $Z_{0.02} = 2.05$.

We accept H_0 and conclude that proportion of teenage girls and girls who are in their twenties who regularly use sunscreen are equal, that is $p_{\text{teen}} = p_{\text{twenty}}$.

p-Value Approach

p-value $= P(Z \leq -1.19) + P(Z \geq 1.19) = 2(0.1170) = 0.2340 > 0.04$, so we accept H_0 and conclude that proportion of teenage girls and girls who are in their twenties who regularly use sunscreen are equal, that is $p_{\text{teen}} = p_{\text{twenty}}$.

■ **EXAMPLE 13.4**   In a study, a sample of laboratory fruit-flies which is made up of 1020 black flies and 1200 gray flies which were kept under the same conditions showed that 163 black flies had died, while 240 gray flies had died, in three months. Can we conclude that gray flies die in a larger proportion than black flies? Use a level of significance of 1%.

Solution

$$n_g = 1200; x_g = 240; n_b = 1020; x_b = 163;$$

$$\hat{p}_g = \frac{240}{1200} = 0.2; \hat{p}_f = \frac{163}{1020} = 0.159804$$

$$H_0: p_g = p_b \text{ vs. } H_1: p_g > p_b$$

$$\bar{p} = \frac{163 + 240}{1020 + 1200} = 0.181532$$

$$\sigma_{\bar{p}} = \sqrt{0.181532(1 - 0.181532)\left(\frac{1}{1020} + \frac{1}{1200}\right)} = 0.016416$$

$$Z_{cal} = \frac{\hat{p}_g - \hat{p}_b}{\sigma_{\bar{p}}} = \frac{0.2 - 0.159804}{0.016416} = 2.448623 \simeq 2.45$$

Classical Approach

At 1% level of significance, we see from the tables that the critical value is $Z_{0.01} = 2.33$

Since $Z_{cal} > Z_{0.01}$, we reject H_0 and conclude that probably, gray flies die in greater proportions than black flies; that is $p_g > p_b$.

p-Value Approach
The *p*-value $= P[Z \geq Z_{cal}] = P[Z \geq 2.45] = 0.0075$

Since *p*-value $< <0.01$, the level of significance, we reject H_0 and conclude that probably gray flies die in greater proportions than black flies; that is $p_g > p_b$.

13.2 Making Inferences about the Means of Two Populations

In the previous section we learned how to make inferences about two population proportions. In each case, we took samples from the two populations and constructed a relevant test statistic for our test. Because we knew the distribution of the test statistic under the assumption that the null hypothesis was true, we used it to test the sets of hypotheses that were tested in order to make inference about the proportions. In making inferences about population means, the sets of hypotheses tested are of the form:

Case 1: $H_0{:}\mu_1 = \mu_2$ vs. $H_1{:}\mu_1 < \mu_2$

Case 2: $H_0{:}\mu_1 = \mu_2$ vs. $H_1{:}\mu_1 > \mu_2$

Case 3: $H_0{:}\mu_1 = \mu_2$ vs. $H_1{:}\mu_1 \neq \mu_2$

Just as we have done for comparing two population proportions, we can construct relevant test statistics for these tests regarding the means of two populations. In general, if we take independent samples of sizes n_1 and n_2 from two variables, X_1 and X_2, whose values are measured or obtained from two populations, we can make inferences about the means of the populations by using the statistic $\overline{X}_1 - \overline{X}_2$ to construct a test statistic. If X_1 and X_2 are normally distributed and their variances, σ_1^2 and σ_2^2, are known, then:

$$\overline{X}_1 - \overline{X}_2 \sim N\left[\mu_1 - \mu_1, \frac{\sigma_1^2}{n_1} + \frac{\sigma_2^2}{n_2}\right]$$

The appropriate test statistic would be:

$$Z_c = \frac{\overline{X}_1 - \overline{X}_2}{\sqrt{\dfrac{\sigma_1^2}{n_1} + \dfrac{\sigma_2^2}{n_2}}}$$

Z_c has the standard normal distribution. Often, or almost always, the variances σ_1^2 and σ_2^2 are unknown, and therefore the above test statistic, although relevant to the inferences we wish to make, is not useful.

We will not pursue tests carried out under the assumptions that lead to Z_c above as they are not of practical importance. However if σ_1^2 and σ_2^2 are unknown, under some assumptions we can carry out inferences about two populations which depend on the *t*-distribution. The two cases that depend on the *t*-distribution are:

(a) If X_1 and X_2 are normally distributed and their variances, σ_1^2 and σ_2^2, are unknown but assumed equal, the **pooled variance *t*-test** can be used for inferences regarding their means.

(b) If X_1 and X_2 are normally distributed and their variances, σ_1^2 and σ_2^2, are unknown but assumed not to be equal, **an approximate *t*-test** can be used to make inferences about their means.

13.3 Making Inferences about Two Population Means, the Pooled Variance *t*-Test

For the pooled variance *t*-test, we need to take independent samples of sizes n_1 and n_2 for a variable measured as X_1 and X_2 on two different populations. Under the pooled variance *t*-test, we assume that the variances of X_1 and X_2 are equal for the populations. In this case, the statistic $\overline{X}_1 - \overline{X}_2$ is used to construct a test statistic. If, in addition to the forgoing, X_1 and X_2 are normally distributed then it is known that:

$$\overline{X}_1 - \overline{X}_2 \sim t[n_1 + n_2 - 2].$$

The set of hypothesis being tested are the same as Case 1 to Case 3 stated earlier. The test statistic is of the form:

$$t_c = \frac{\overline{X}_1 - \overline{X}_2}{s_p\sqrt{\dfrac{1}{n_1} + \dfrac{1}{n_2}}}$$

Here,

$$s_p^2 = \frac{(n_1 - 1)s_1^2 + (n_2 - 1)s_2^2}{n_1 + n_2 - 2}$$

and

$$s_p = \sqrt{\frac{(n_1 - 1)s_1^2 + (n_2 - 1)s_2^2}{n_1 + n_2 - 2}}$$

where s_1^2 and s_2^2 are the sample variances of X_1 and X_2. s_p^2 is an estimate of the common variance for X_1 and X_2. The test statistic, t_c is *t*-distributed with $n_1 + n_2 - 2$ degrees of freedom.

The pooled variance *t*-test is only valid in situations where it makes sense to assume equal variance. We cannot use this test if, by the nature of the two populations being compared, we could not assume that the variances are equal.

We consider an example:

■ **EXAMPLE 13.5**   You are a test engineer, and you wish to test the mileages attained by the same make of car run on two new formulations of biofuels. The following are mileages recorded for the two types of fuel, each having been studied with fifteen cars. In the study, cars were chosen at random from a large pool and assigned to the fuels at random.

Fuel A	Fuel B
29	38
30	33
31	32
31	37
28	35
27	39
22	40
24	38
25	37
28	43
40	36
38	32
33	31
19	42
30	33

Assuming that the miles per gallon are drawn from two normally distributed populations with equal variance, test the hypothesis that the mean miles/gallon of fuel B is greater than the mean miles/per gallon for fuel A at a 1% level of significance.

Solution

$$\overline{X}_A = \frac{435}{15} = 29; \overline{X}_B = \frac{546}{15} = 36.4;$$

$$s_A^2 = \frac{\sum_{i=1}^{n_A} X_{Ai}^2 - n_A(\overline{X}_A)^2}{n_A - 1} = \frac{13039 - 15(29)^2}{15 - 1} = 30.2857$$

$$s_B^2 = \frac{\sum_{i=1}^{n_B} X_{Bi}^2 - n_B(\overline{X}_B)^2}{n_B - 1} = \frac{20068 - 15(36.4)^2}{15 - 1} = 13.8286$$

$$s_P^2 = \frac{(n_A - 1)s_A^2 + (n_B - 1)s_B^2}{n_A + n_B - 2} = \frac{(15 - 1)(30.2857) + (15 - 1)(13.8286)}{15 + 15 - 2}$$

$$= 22.0571 \Rightarrow s_p = \sqrt{22.0571} = 4.6965$$

The hypothesis to be tested is the type represented by Case 2:

$$H_0: \mu_B = \mu_A \text{ vs. } H_1: \mu_B > \mu_A$$

The test statistic is:

$$t_c = \frac{\overline{X}_B - \overline{X}_A}{s_P\sqrt{\frac{1}{n_A} + \frac{1}{n_B}}} = \frac{36.4 - 29}{4.6965\sqrt{\frac{1}{15} + \frac{1}{15}}} = 4.3151$$

The test statistic is distributed as t with $n_A + n_B - 2$ degrees of freedom. Since we are testing at a 1% level of significance, then in the t-table (Appendix III), we read $t_{(15+15-2, 0.01)} = t_{(28, 0.01)}$ whose value is 2.467.

Decision: Since $t_c > t_{(28, 0.01)}$ we reject the null hypothesis, and conclude that the mean mileage/gallon for fuel B is greater than mean mileage/gallon for fuel A at a 1% level of significance.

■ **EXAMPLE 13.6**   In a particular year, in a state, there was an interest in comparing ACT science scores for two groups of students:

(a) students who took only biology in high school, and
(b) students who took biology, physics, and chemistry in high school.

A sample of ACT science scores of twenty-one students were randomly taken from those who took only biology in high school, while ACT science scores of twenty-three students from those who took biology, physics, and chemistry in high school were randomly chosen. The ACT science scores are presented below:

Biology (A)	Biology, Physics, and Chemistry (B)
19.4	24.7
21.4	27.7
17.4	19.9
24.6	25.9
18.4	22.1
16.4	22.5
18.2	23.1
20.4	23.9
23.4	26.9
15.6	19.1
21.6	25.1
17.8	21.3
18.2	21.7
18.8	22.3
21.4	26.5
22.4	27.5
16.2	21.3
21	26.1
17.1	22.2
18.9	24
18.8	23.9
	24
	23.4

Assuming that scores are drawn from two normally distributed populations with equal variance, test the hypothesis that the mean score is greater for those who took biology, physics, and chemistry at a 1% level of significance.

Solution

$$\overline{X}_A = \frac{407.4}{21} = 19.4; \overline{X}_B = \frac{545.1}{23} = 23.7;$$

$$s_A^2 = \frac{\sum\limits_{i=1}^{n_A} X_{Ai}^2 - n_A(\overline{X}_A)^2}{n_1 - 1} = \frac{8022.38 - 21(19.4)^2}{21 - 1} = 5.941$$

$$s_B^2 = \frac{\sum\limits_{i=1}^{n_B} X_{Bi}^2 - n_B(\overline{X}_B)^2}{n_2 - 1} = \frac{13040.99 - 23(23.7)^2}{23 - 1} = 5.55091$$

$$s_P^2 = \frac{(n_A - 1)s_A^2 + (n_B - 1)s_B^2}{n_A + n_B - 2} = \frac{(21 - 1)(5.941) + (23 - 1)(5.55091)}{21 + 23 - 2}$$

$$= 5.73667 \Rightarrow s_p = \sqrt{5.73667} = 2.39513$$

The hypothesis to be tested is the type represented by Case 2:

$$H_0: \mu_B = \mu_A \text{ vs. } H_1: \mu_B > \mu_A.$$

The test statistic is:

$$t_c = \frac{\overline{X}_B - \overline{X}_A}{s_P\sqrt{\dfrac{1}{n_A} + \dfrac{1}{n_B}}} = \frac{23.7 - 19.4}{2.39513\sqrt{\dfrac{1}{21} + \dfrac{1}{23}}} = 5.9482.$$

The test statistic is distributed as t with $n_A + n_B - 2$ degrees of freedom. Since we are testing at a 1% level of significance, then in the t-table (Appendix III), we read $t_{(23+21-2, 0.01)} = t_{(42, 0.01)}$ whose value is 2.418.

Decision: Since $t_c > t_{(42, 0.01)}$ we reject the null hypothesis, and conclude that the mean ACT science score for students who took biology, physics, and chemistry is greater than the mean ACT science score for those students who took only biology at a 1% level of significance.

■ **EXAMPLE 13.7** Independently, twenty males and twenty females were randomly chosen from residents of a city in order to compare their heights. The following heights (in inches) were obtained:

Male	Female	Male	Female	Male	Female
62.9	58.4	71.4	64.8	66.9	62.0
65.0	64.5	67.8	59.8	70.7	65.1
69.5	65.7	69.7	70.9	67.9	61.0
74.3	60.8	65.3	62.0	68.9	62.8
67.5	65.2	68.5	67.4	72.3	63.8
64.8	67.8	66.9	63.2	68.2	65.2
71.5	60.7	67.9	68.8		

Assume that heights are normally distributed, and the variances are equal. Test the hypothesis that the mean heights for males is not equal to the mean height for females at a 5% level of significance.

Solution

$$\overline{X}_m = \frac{1367.9}{20} = 68.4; \ \overline{X}_f = \frac{1279.9}{20} = 64;$$

$$s_m^2 = \frac{\sum_{i=1}^{n_A} X_{mi}^2 - n_m(\overline{X}_m)^2}{n_m - 1} = \frac{93705.2 - 20(68.4)^2}{20 - 1} = 7.7721$$

$$s_f^2 = \frac{\sum_{i=1}^{n_B} X_{fi}^2 - n_f(\overline{X}_f)^2}{n_f - 1} = \frac{82101.0 - 20(64)^2}{20 - 1} = 10.19839$$

$$s_P^2 = \frac{(n_m - 1)s_m^2 + (n_f - 1)s_f^2}{n_m + n_f - 2} = \frac{(20 - 1)(7.7721) + (20 - 1)(10.19839)}{20 + 20 - 2}$$

$$= 8.985237 \Rightarrow s_p = \sqrt{8.985237} = 2.99754$$

The hypothesis to be tested is the type represented by Case 3:

$$H_0 : \mu_m = \mu_f \text{ vs. } H_1 : \mu_m \neq \mu_f.$$

The test statistic is:

$$t_c = \frac{\overline{X}_m - \overline{X}_f}{s_P \sqrt{\dfrac{1}{n_m} + \dfrac{1}{n_f}}} = \frac{68.4 - 64}{2.99754 \sqrt{\dfrac{1}{20} + \dfrac{1}{20}}} = 4.6.$$

The test statistic is distributed as t with $n_m + n_f - 2$ degrees of freedom. Since we are testing at a 5% level of significance, then in the t-table (Appendix III), we read $t_{(20+20-2, \ 0.025)} = t_{(38, \ 0.025)}$ whose value is 2.024.

Decision: Since $t_c > t_{(38, \ 0.025)}$ we reject the null hypothesis, and conclude that the mean height of males is not equal to mean height of females at a 5% level of significance.

■ **EXAMPLE 13.8** Two different formulations of the same fertilizer were being piloted on small plots of land. Twenty-two similar plots which are contiguous were randomized for application of the fertilizers. The following are the yields in bushels/plot under the different fertilizer treatments.

Plot	Fertilizer A	Fertilizer B	Plot	Fertilizer A	Fertilizer B
1	45	39	7	59	50
2	88	71	8	55	60
3	40	42	9	62	51
4	32	27	10	50	48
5	29	28	11	66	49
6	34	30			

Assume that yields for different fertilizers are normally distributed and their variances are equal, test whether mean yield per plot for fertilizer formulation A is significantly different from mean yield per plot for fertilizer B at a 5% level of significance.

Solution

$$\overline{X}_A = \frac{560}{11} = 50.9091; \overline{X}_B = \frac{495.0}{11} = 45;$$

$$s_A^2 = \frac{\sum_{i=1}^{n_A} X_{Ai}^2 - n_A(\overline{X}_A)^2}{n_1 - 1} = \frac{31596 - 11(50.9091)^2}{11 - 1} = 308.7$$

$$s_B^2 = \frac{\sum_{i=1}^{n_B} X_{Bi}^2 - n_B(\overline{X}_B)^2}{n_2 - 1} = \frac{24145 - 11(45)^2}{11 - 1} = 187.0$$

$$s_P^2 = \frac{(n_A - 1)s_A^2 + (n_B - 1)s_B^2}{n_A + n_B - 2} = \frac{(11 - 1)(308.7) + (11 - 1)(187.0)}{11 + 11 - 2}$$

$$= 247.8455 \Rightarrow s_p = \sqrt{247.8455} = 15.74311$$

The hypothesis to be tested is the type represented by Case 3:

$$H_0: \mu_A = \mu_B \text{ vs. } H_1: \mu_A > \mu_B$$

The test statistic is:

$$t_c = \frac{\overline{X}_B - \overline{X}_A}{s_P\sqrt{\frac{1}{n_A} + \frac{1}{n_B}}} = \frac{50.9091 - 45}{15.74311\sqrt{\frac{1}{11} + \frac{1}{11}}} = 0.8803$$

The test statistic is distributed as t with $n_A + n_B - 2$ degrees of freedom. Since we are testing at a 5% level of significance, then in the t-table (Appendix III), we read $t_{(11+11-2, 0.05)} = t_{(20, 0.05)}$ whose value is 1.725.

Decision: Since $t_c < t_{(20, 0.05)}$ we do not reject the null hypothesis, and conclude that the mean yield for different formulations of the fertilizer are the same at a 5% level of significance.

13.4 Comparing Two Means when Populations are Normal and Variances are Unknown

When the two samples are drawn from two populations which are normally distributed with unknown variances, we can test hypotheses about the difference between the means of the populations by using a test statistic which is approximately t-distributed with degrees of freedom which may be fractional. The set of hypotheses tested are still those stated in cases 1 to 3. The test statistic in this case is:

$$t_c = \frac{\overline{X}_1 - \overline{X}_2}{\sqrt{\frac{s_1^2}{n_1} + \frac{s_2^2}{n_2}}}$$

The test statistic $t_c \sim t_{(\nu)}$, where:

$$\nu = \frac{[(s_1^2/n_1) + (s_2^2/n_2)]^2}{\dfrac{(s_1^2/n_1)^2}{n_1 - 1} + \dfrac{(s_2^2/n_2)^2}{n_2 - 1}}$$

We will use this statistic to test the hypotheses involving means from populations who satisfy conditions we described above. Here are some examples.

■ **EXAMPLE 13.9** In a pilot project to study the effects of a particular protein on baby weight gains, sixteen and eleven baby rabbits were respectively fed for a period with high and low protein versions of formula, and their weight changes recorded as follows:

> High protein: 103 144 133 123 127 90 139 85 100 127
>   168 127 87 99 123 134
> Low protein: 80 128 110 96 104 137 98 76 81 75 94

Assume the weight gains are normally distributed, test the hypothesis that the means are equal against that the mean weight gained is greater at a 1% level of significance.

Solution

$H_0{:}\mu_h = \mu_L$ vs. $H_1{:}\mu_h > \mu_L$

$$\overline{X}_h = \frac{1909}{16} = 119.3125; \quad \overline{X}_L = \frac{1079}{11} = 98.09091;$$

$$s_h^2 = \frac{\sum\limits_{i=1}^{n_A} X_{hi}^2 - n_h(\overline{X}_h)^2}{n_h - 1} = \frac{235875 - 16(119.3125)^2}{16 - 1} = 540.4958$$

$$s_L^2 = \frac{\sum\limits_{i=1}^{n_B} X_{Li}^2 - n_L(\overline{X}_L)^2}{n_L - 1} = \frac{110087 - 11(98.09091)^2}{11 - 1} = 424.6909$$

$$\nu = \frac{\left[\left(\dfrac{540.4958}{16}\right) + \dfrac{424.6909}{11}\right]^2}{\dfrac{\left(\dfrac{540.4958}{16}\right)^2}{16 - 1} + \dfrac{\left(\dfrac{424.6909}{11}\right)_2}{11 - 1}} = 23.27564 \approx 23$$

$$t_c = \frac{\overline{X}_1 - \overline{X}_2}{\sqrt{\dfrac{s_L^2}{n_1} + \dfrac{s_h^2}{n_h}}} = \frac{119.3125 - 98.09091}{\sqrt{\dfrac{540.4958}{16} + \dfrac{424.6909}{11}}} = 2.4943$$

From the t tables $t_{(23, 0.01)} = 2.5$. We conclude that there is no significant difference between the mean weight gains for the rabbits on high or low protein baby formula.

■ **EXAMPLE 13.10** Two methods of instruction are being used for a workplace training offered by a company. Method A uses a human instructor, while method B is an online instruction with minimal face to face instruction. The students are subjected to same test at the end of instruction. Samples of twelve workers were taken from both instruction groups and their scores were obtained as follows:

A: 73 77 75 67 71 70 71 68 76 73 70 69
B: 86 74 79 80 71 72 67 77 72 75 69 69

Assume the scores of workers who were subjected to each teaching method are normally distributed. Test at a 5% level of significance that mean score for method A is less than mean score for method B.

Solution

$H_0 : \mu_A = \mu_B$ vs. $H_1 : \mu_A < \mu_B$

$$\overline{X}_1 = \frac{860}{12} = 71.6667; \overline{X}_2 = \frac{891}{12} = 74.25;$$

$$s_1^2 = \frac{61744 - 12(71.6667)^2}{12 - 1} = 10.06061$$

$$s_2^2 = \frac{66487 - 12(74.25)^2}{12 - 1} = 30.02273$$

$$\nu = \frac{\left[\dfrac{10.06061}{12} + \dfrac{30.02273}{12}\right]^2}{\left(\dfrac{1}{12-1}\right)\left(\dfrac{10.06061}{12}\right)^2 + \left(\dfrac{1}{12-1}\right)\left(\dfrac{30.02273}{12}\right)^2} = 17.6279 \approx 17$$

$$t_c = \frac{\overline{X}_1 - \overline{X}_2}{\sqrt{\dfrac{s_L^2}{n_1} + \dfrac{s_h^2}{n_h}}} = \frac{71.6667 - 74.25}{\sqrt{\dfrac{10.06061}{12} + \dfrac{30.02273}{12}}} = -1.41348$$

From the tables, $t_{(17,\,0.05)} = 1.74$. Hence, we could not reject the null hypothesis. We therefore conclude that mean scores are the same for both methods of instruction.

13.5 Linking Hypotheses Testing and Confidence Intervals

We can demonstrate that we can test hypotheses for the difference between the means by using confidence intervals, especially for the two-sided test represented by Case 3. For the other cases, confidence bounds can also be found, but that is beyond the scope of this book. Let us use some cases for which we constructed confidence intervals in Chapter 12 to illustrate this link between confidence intervals and tests of hypotheses. By this we wish to emphasize the link between decisions we make in hypotheses testing and the confidence intervals we construct. Basically, if the interval encloses the postulated value of zero for the difference between the means then there is no significant difference between the means. However if the difference between the means is postulated to be equal to a constant, then if the interval includes the constant the postulated value of the difference is supported.

Recall Example 12.13, if we rewrite the problem in terms of a test of hypothesis, we have:

■ **EXAMPLE 13.11**   The times used by a sample of eighteen chemistry students from School A and a sample of fifteen chemistry students from school B to complete a titration experiment are as follows:

A: 2.3 6.7 3.8 5.0 4.9 6.1 4.4 5.2 3.9 4.8 4.6 5.7 5.3 4.7 4.2 5.7 4.8 4.7
B: 6.7 7.3 4.4 8.3 6.2 4.3 5.5 3.2 8.4 7.3 5.5 4.8 4.9 6.7 7.5

Test at a 5% level of significance that the means for both schools are equal, against that they are not.

Solution $H_0 : \mu_A = \mu_B$ vs. $H_1 : \mu_A \neq \mu_B$ or $H_0 : \mu_A - \mu_B = 0$ vs. $H_1 : \mu_A - \mu_B \neq 0$

$$\overline{X}_1 = \frac{86.8}{18} = 4.8222; \ \overline{X}_2 = \frac{91}{15} = 6.0667;$$

$$s_1^2 = \frac{434.58 - 18(4.8222)^2}{18 - 1} = 0.9418$$

$$s_2^2 = \frac{586.14 - 15(6.0667)^2}{15 - 1} = 2.43381$$

$$\nu = \frac{\left[\dfrac{0.9418}{18} + \dfrac{2.43381}{15}\right]^2}{\left(\dfrac{1}{18 - 1}\right)\left(\dfrac{0.9418}{18}\right)^2 + \left(\dfrac{1}{15 - 1}\right)\left(\dfrac{2.43381}{15}\right)^2} = 22.5383 \approx 22$$

$$t_c = \frac{\overline{X}_1 - \overline{X}_2}{\sqrt{\dfrac{s_L^2}{n_1} + \dfrac{s_h^2}{n_h}}} = \frac{4.8222 - 6.0667}{\sqrt{\dfrac{0.9418}{18} + \dfrac{2.43381}{15}}} = -2.6845$$

From the tables, $t_{(22, 0.025)} = 2.074$. Hence, we reject the null hypothesis, and conclude that mean titration times are not the same for schools A and B.

Let us consider the above test of the hypothesis via a 95% confidence interval for the difference between the means of A and B. Recall that the 95% confidence interval we obtained in Chapter 12 for the data is as shown below:

$$100(1 - \alpha)\% \ \text{CI} = (\overline{X}_1 - \overline{X}_2) \pm t_{(\nu, \alpha/2)} \sqrt{\left(\frac{s_1^2}{n_1} + \frac{s_2^2}{n_2}\right)}$$

$$95\% \ \text{CI} = 100(1 - 0.05)\% \ \text{CI} \Rightarrow \alpha = 0.05; \ \alpha/2 = 0.025;$$

$$(4.8222 - 6.0667) \pm t_{(22,0.025)} \sqrt{\left(\frac{0.9418}{18} + \frac{2.43381}{15}\right)}$$

$$= -1.2445 \pm 2.074 \times 0.463224 = (-2.205226, -0.283774)$$

Clearly, the interval $(-2.205226, -0.283774)$ does not contain the value zero. We can conclude that there is significant difference between the means of A and B at the 5% level of significance based on the idea that the postulated difference between the means of A and B (to wit, zero) is not included in the interval at a 95% confidence level.

13.6 Paired Comparison or Matched Pair *t*-Test

Frequently, we need to compare two samples taken from what appears to be two different populations in order to ascertain whether the populations are actually different from each other. For instance, we may have taken observations from individuals before applying some treatment to them then taken another set of measurements from the same individuals after treatment, and we may wish to compare the measurements **before** and **after** treatment in order to find out whether there has been a treatment effect.

This type of comparison of pairs of observations is called a **paired comparison test, or the matched pair *t*-test.** If the two sets of observations are basically the same, then if a matched pair *t*-test is applied to the two data sets, it will find no significant change.

If a sample of *n* people had their heart rates measured before they were started on a tread mill set at the same speed and level of resistance, and their heart rates were measured again at the end of thirty minutes on the tread mill, then under normal circumstances we should expect their heart rates to rise. But the question we need to answer is whether the rise in heart rates is significant or just due to normal fluctuations in heart rates. The **paired comparison test can be used to test for a significant increase in heart rate in this situation.** Here we test for increase, while in other situations we may seek to decrease the level of a variable we observed after some treatment(s) has been applied. This principle could be applied to medical treatments in which the level of disease could be measured before and after treatment(s) to ascertain whether there is an improvement in the patient's condition after treatment, which in this case would be signified by a decrease in the level of the disease. There are several other situations which give rise to matched pairs of observations. The test procedure we discuss below applies.

Consider *n* matched pairs of observations, if we take the differences between pair i $(i = 1, 2, \ldots, n)$, we obtain the difference d_i (see the table). We now use the differences d_i in our inference as follows:

Pairs (i)	Before Treatment	After Treatment	Differences (d_i)
1	X_{11}	X_{12}	$d_1 = X_{12} - X_{11}$
2	X_{21}	X_{22}	$d_2 = X_{22} - X_{21}$
3	X_{31}	X_{32}	$d_3 = X_{32} - X_{32}$
4	X_{41}	X_{42}	$d_4 = X_{42} - X_{41}$
...	...	...	...
...	...	...	...
$n-1$	X_{n-11}	X_{n-11}	$d_{n-1} = X_{n-12} - X_{n-11}$
n	X_{n1}	X_{n1}	$d_n = X_{n2} - X_{n1}$

We set up the null and alternative hypotheses:

(1) $H_o: \mu_d = 0$ vs. $H_1: \mu_d > 0$ (one-sided test)

Other forms of the test are:

(2) $H_o : \mu_d = 0$ vs. $H_1 : \mu_d < 0$ (one-sided test)

or

(3) $H_o : \mu_d = 0$ vs. $H_1 : \mu_d \neq 0$ (two-sided test)

The test statistic for testing all the hypotheses set out above is:

$$t_c = \frac{\bar{d}}{s_d / \sqrt{n}}$$

where $\bar{d} = \dfrac{1}{n} \displaystyle\sum_{i=1}^{n} d_i;\ s_d^2 = \dfrac{\displaystyle\sum_{i=1}^{n} d_i^2 - n\bar{d}^2}{n-1}$

The test statistics, t_c is t-distributed with $n - 1$ degrees of freedom.

Decision Rule: The decision rule for (1), requires that t_c be compared with the tabulated value $t(\alpha, n - 1)$ and if $t_c > t(\alpha, n - 1)$, reject the null hypothesis and accept the alternative. For (2) if $t_c < -t(\alpha, n - 1)$, we reject the null hypothesis and accept the alternative. For (3), if $t_c < -t(\alpha/2, n - 1)$ or $t_c > t(\alpha/2, n - 1)$ we reject the null hypothesis and accept the alternative.

EXAMPLE 13.12   A chocolatier who was experimenting on two possible formulations of a chocolate candy, chose a sample of twelve customers to taste the formulations and score them on a scale of 1 to 10 (1 = poor and 10 = excellent). The chocolatier made sure that hours elapsed between tastes, and the order in which each candidate tasted the chocolates was randomized. The following were the scores:

Individual	Chocolate A	Chocolate B
1	9	7
2	7	6
3	9	3
4	8	5
5	7	7
6	9	6
7	8	8
8	9	5
9	7	9
10	7	8
11	8	6
12	9	7

Test at the 5% level of significance that the formulations are equally preferred against that chocolate A is more preferred.

Individual	Chocolate A	Chocolate B	d_i	d_i^2
1	9	7	2	4
2	7	6	1	1
3	9	3	6	36
4	8	5	3	9
5	7	7	0	0
6	9	6	3	9
7	8	8	0	0
8	9	5	4	16
9	7	9	−2	4
10	7	8	−1	1
11	8	6	2	4
12	9	7	2	4
			20	**88**

$H_0 : \mu_d = 0$ vs. $H_1 : \mu_d > 0$

$$\bar{d} = \frac{20}{12} = 1.6667; \quad s_d^2 = \frac{\sum_{i=1}^{n} d_i^2 - n\bar{d}^2}{n-1} = \frac{88 - 12(1.6667)^2}{12 - 1} = 4.969697$$

$$t_{cal} = \frac{1.6667}{\sqrt{(4.969697/12)}} = 2.5898$$

$$t_{(11, \, 0.05)} = 1.796$$

The data are significant. Therefore chocolate A is significantly more preferred to chocolate B at the 5% level of significance.

■ **EXAMPLE 13.13**   Resting membrane potentials (RMPs) of eight guinea-pig heart cells were measured before and fifteen minutes after the cells were infused with an ischemia inducing drug. The values in RMPs are:

Cell	Control	15 min.
1	−94.10	−91.42
2	−84.93	−83.90
3	−75.99	−69.95
4	−80.01	−64.10
5	−85.47	−83.96
6	−78.99	−72.63
7	−83.86	−75.27
8	−79.36	−63.95

Is there any significant increase in RMP due to the inducement? Test at a 5% level of significance.

Solution

Cell	Control	15min.	d_i	d_i^2
1	−94.10	−91.42	−2.68	7.1824
2	−84.93	−83.90	−1.03	1.0609
3	−75.99	−69.95	−6.04	36.4816
4	−80.01	−64.10	−15.91	253.1281
5	−85.47	−83.96	−1.51	2.2801
6	−78.99	−72.63	−6.36	40.4496
7	−83.86	−75.27	−8.59	73.7881
8	−79.36	−63.95	−15.41	237.4681

Since the values are negative, a negative significant value for the test statistics would signify an increase in the RMP.

$$H_0{:}\mu_d = 0 \text{ vs. } H_1{:}\mu_d > 0$$

$$\bar{d} = \frac{-57.53}{8} = -7.19125; \quad s_d^2 = \frac{\sum_{i=1}^{n} d_i^2 - n\bar{d}^2}{n-1} = \frac{651.8389 - 8(7.19125)^2}{8-1}$$

$$= 34.01084$$

$$t_{cal} = \frac{-7.19125}{\sqrt{(34.01084/8)}} = -3.48734$$

The data are significant. We reject H_0 and conclude that there is significant increase in RMP at the 5% level of significance.

13.7 Use of MINITAB for Testing Hypotheses and Constructing Confidence Intervals

As discussed earlier, MINITAB often obtains confidence intervals in the context of testing hypotheses. We have also, in the text, demonstrated the relationship between tests of hypothesis and confidence intervals. Here we demonstrate the use of MINITAB to achieve both aims.

13.7.1 Use of MINITAB for Testing for Difference Between Two Proportions, and Constructing $100(1 - \alpha)\%$ Confidence Interval for the Difference

■ **EXAMPLE 13.14** **Using MINITAB for Two Proportions**

Two populations were surveyed for their views on a political proposal. Out of 200 chosen from population I, seventy-six were in favor, while out of 224 chosen from population II, fifty-six were in favor. Using MINITAB, Test at a 1% level of significance that the proportions in favor for both populations are the same against that they are not. Obtain the 99% confidence interval of the difference between the proportions.

Solution
```
MTB > PTwo 200 76 224 56;
SUBC> Confidence 99.0;
SUBC> Pooled.
```

Output

Test and CI for Two Proportions
```
Sample   X      N     Sample p
  1     56    224.0   0.250000
  2     76    200.0   0.380000
```

```
Difference = p (1) - p (2)
Estimate for difference: 0.13
99% CI for difference: (0.0143727, 0.245627)
Test for difference = 0 (vs not = 0): Z = 2.89 P-Value = 0.004
```

```
Fisher's exact test: P-Value = 0.005
```

In this analysis, the pooled estimate of the p was used for testing the hypotheses. The p-value shows that there is highly significant difference between the proportions. This is also reflected by the 99% confidence interval whose limits are all positive (the same sign, or does not include zero).

■ **EXAMPLE 13.15** **Using MINITAB for Data of Example 12.2**
Consider the data for Example 12.2:

A survey of freshmen in Hamilton College Campus showed, the number of female freshmen interviewed were 422 while the male freshmen were 602. If they found 137 female smokers and 242 male smokers among the interviewees, find the 94% confidence interval for the difference in the proportions of male freshmen and female freshmen who smoked.

To obtain the 94% confidence interval in MINITAB, we need to test the hypothesis that the proportions are equal at a 6% level of significance. Let us do that using MINITAB:

Solution
```
MTB > PTwo 602 242 422 137;
SUBC> Confidence 94.0;
SUBC> Pooled.
```

Output

Test and CI for Two Proportions
```
Sample    X      N     Sample p
 male    242   602.0   0.324645
female   137   422.0   0.401993
```

```
Difference = p (1) - p (2)
Estimate for difference: 0.0773488
94% CI for difference: (0.134361, 0.0203363)
Test for difference = 0 (vs not = 0): Z = 2.52 P-Value = 0.012

Fisher's exact test: P-Value = 0.012
```

In this analysis, the pooled estimate of the p was used for testing the hypothesis. The p-value shows that there is a highly significant difference between the proportions at 6% since p-value = 0.012. This is also reflected by the 94% confidence interval whose limits are all positive (the same sign, or does not include zero).

13.7.2 Using MINITAB for Confidence Interval and Hypotheses Tests for the Difference Between Means When Variances are not Known but are Assumed Equal

■ **EXAMPLE 13.16** **Using MINITAB for Data of Example 12.9**

The following samples were taken from two variables A and B which are believed to be normally distributed and of equal variance.

A: 82 64 62 66 68 43 38 26 68 56 54 66
B: 70 26 34 41 53 49 40 46 34 66 38 46 44 35 33

Obtain the (a) 90%, and (b) 95% confidence intervals for the difference between their means.

Use MINITAB to test that the means are equal at a 5% level of significance and construct a 95% confidence interval for the difference between the two means.

Solution
```
MTB > set c1
DATA> 82 64 62 66 68 43 38 26 68 56 54 66
DATA> end
MTB > set c2
DATA> 70 26 34 41 53 49 40 46 34 66 38 46 44 35 33
DATA> end
MTB > name c1 'A' c2 'B'
MTB > TwoSample c1 c2;
SUBC> Pooled.
```

Two-Sample T-Test and CI: *A, B*
```
Two-sample T for A vs B

      N    Mean   StDev   SE Mean
A    12    57.8    15.4     4.5
B    15    43.7    12.1     3.1

Difference = mu (A) - mu (B)
Estimate for difference: 14.08
95% CI for difference: (3.17, 25.00)
T-Test of difference = 0 (vs not =): T-Value = 2.66 P-Value =
   0.014 DF = 25
Both use Pooled StDev = 13.6873
```

The data are significant at a 5% level of significance because p-value is less than 0.05, also the limits of the 95% confidence interval are of the same sign, which means that the interval does not contain zero.

■ **EXAMPLE 13.17** **Using MINITAB for Data of Example 12.10**

The times to metastasis of two different types of cancer were recorded for samples of eighteen patients for cancer type X and a sample of fifteen patients from cancer type Y:

X: 22.3 26.7 23.8 25 24.9 26.1 24.4 25.2 23.9 24.8 24.6 25.7 25.3 24.7 24.2 25.7 24.8 24.7

Y: 26.7 27.3 24.4 28.3 26.2 24.3 25.5 23.2 28.4 27.3 25.5 24.8 24.9 26.7 27.5

Assume that variances of X and Y are equal. **Use MINITAB to test that the means are equal at a 1% level of significance and construct a 99% confidence interval for the difference between the two means.**

Solution

```
MTB > set c2
DATA> 26.7, 27.3, 24.4, 28.3, 26.2, 24.3, 25.5, 23.2, 28.4, 27.3,
      25.5, 24.8, 24.9, 26.7, 27.5
DATA> end
MTB > set c1
DATA> 22.3, 26.7, 23.8, 25, 24.9, 26.1, 24.4, 25.2, 23.9, 24.8,
      24.6, 25.7, 25.3, 24.7, 24.2, 25.7, 24.8, 24.7
DATA> end
MTB > name c1 'x' c2 'y'
MTB > TwoSample c1 c2;
SUBC> Pooled;
SUBC> Confidence 99.0.
```

Two-Sample *T*-Test and CI: *x, y*

```
Two-sample T for x vs y

       N    Mean    StDev   SE Mean
  x   18   24.822   0.970    0.23
  y   15   26.07    1.56     0.40


Difference = mu (x) - mu (y)
Estimate for difference: -1.244
99% CI for difference: (-2.464, -0.025)
T-Test of difference = 0 (vs not =): T-Value = -2.80 P-Value =
   0.009 DF = 31
Both use Pooled StDev = 1.2711
```

The data are significant at a 1% level of significance because p-value is less than 0.01, also the limits of the 99% confidence interval are of the same sign, which means that the interval does not contain zero.

13.7.3 Using MINITAB for Confidence Interval and Hypotheses Tests for the Difference Between Means when Variances Are Unknown and are not Equal

■ **EXAMPLE 13.18**   **Using MINITAB for Data of Example 12.13**
The times used by a sample of eighteen chemistry students from School A and a sample of fifteen chemistry students from School B to complete a titration experiment are as follows:

A: 2.3 6.7 3.8 5 4.9 6.1 4.4 5.2 3.9 4.8 4.6 5.7 5.3 4.7 4.2 5.7 4.8 4.7
B: 6.7 7.3 4.4 8.3 6.2 4.3 5.5 3.2 8.4 7.3 5.5 4.8 4.9 6.7 7.5

Use MINITAB to test for difference between the means School A and School B at 5% level of significance and obtain a 95% confidence interval for the difference between the means.

Solution

```
MTB > set c1
DATA> 2.3, 6.7, 3.8, 5, 4.9, 6.1, 4.4, 5.2, 3.9, 4.8, 4.6, 5.7,
      5.3, 4.7, 4.2, 5.7, 4.8, 4.7
DATA> end
MTB > set c2
DATA> 6.7, 7.3, 4.4, 8.3, 6.2, 4.3, 5.5, 3.2, 8.4, 7.3, 5.5, 4.8,
      4.9, 6.7, 7.5
DATA> end
MTB > name c1 'A' c2 'B'
MTB > TwoSample c1 c2.
```

Two-Sample *T*-Test and CI: *A*, *B*

```
Two-sample T for A vs B

        N   Mean   StDev   SE Mean
  x    18   4.822  0.970    0.23
  y    15   6.07   1.56     0.40


Difference = mu (A) - mu (B)
Estimate for difference: -1.244
95% CI for difference: (-2.205, -0.284)
T-Test of difference = 0 (vs not =): T-Value = -2.69 P-Value =
   0.013 DF = 22
```

Since p-value is less than 0.05, and the limits of the 95% confidence interval are of the same sign, which means that the interval does not contain zero, then the data are significant at the 5% level of significance.

■ **EXAMPLE 13.19**   **Using MINITAB for Data of Example 12.14**
The times it took for samples of eighth grade students from two different schools to complete a computer-based task are presented below:

X_1: 48.2 54.6 58.3 47.8 51.4 52.0 55.2 49.1 49.9 52.6 49.6 47.5
X_2: 52.3 57.4 55.6 53.2 61.3 58.0 59.8 54.8 58.9 57.7 58.6

Assume that times used for the task for both groups are normally distributed. Using MINITAB, test the hypothesis that the means are equal at a 1% level of significance and obtain the 99% confidence interval for the difference between their means.

Solution

```
MTB > set c1
DATA> 48.2 54.6 58.3 47.8 51.4 52.0 55.2 49.1 49.9 52.6 49.6
DATA> 47.5
DATA> end
MTB > set c2
DATA> 52.3 57.4 55.6 53.2 61.3 58.0 59.8 54.8 58.9
DATA> 57.7 58.6
DATA> end
MTB > name c1 'X1' c2 'X2'
MTB > TwoSample c1 c2;
SUBC> Confidence 99.0.
```

Two-Sample *T*-Test and CI: *X*1, *X*2

```
Two-sample T for X1 vs X2

       N    Mean   StDev   SE Mean
X1    12   51.35    3.35     0.97
X2    11   57.05    2.78     0.84


Difference = mu (X1) - mu (X2)
Estimate for difference: -5.70
99% CI for difference: (-9.35, -2.06)
T-Test of difference = 0 (vs not =): T-Value = -4.46 P-Value =
0.000 DF = 20
```

Again, since *p*-value is less than 0.01, and the limits of the 99% confidence interval are of the same sign, which means that the interval does not contain zero, then the data are significant at a 1% level of significance.

13.7.4 Using MINITAB to Carry out Paired Comparison or Matched Pair *t*-Test and Obtaining the Confidence Interval for the Mean Difference

■ **EXAMPLE 13.20**   Using MINITAB, carry out the test of Example 13.13 at a 5% level of significance and obtain the 95% confidence interval for the mean of the difference.

Solution

```
MTB > set c5
DATA> 9 7 9 8 7 9 8 9 7 7 8 9
DATA> end
MTB > set c6
DATA> 7 6 3 5 7 6 8 5 9 8 6 7
DATA> end
MTB > name c5 'chocA' c6 'chocB'
MTB > Paired 'chocA' 'chocB'.
```

Paired *T*-Test and CI: *chocA, chocB*

```
Paired T for chocA - chocB

                 N    Mean   StDev   SE Mean
      chocA     12   8.083   0.900    0.260
      chocB     12   6.417   1.621    0.468
 Difference     12   1.667   2.229    0.644

95% CI for mean difference: (0.250, 3.083)
T-Test of mean difference = 0 (vs not = 0): T-Value = 2.59
  P-Value = 0.025
```

Exercises

13.1 Two varieties of palm seeds were planted under the same conditions and the heights (in inches) of samples of the palm trees after six months are given below:

Variety I	Variety II
24.6	24.9
23.7	23.9
30.1	28.4
28.0	20.3
31.0	18.5
27.5	28.5
23.2	17.3
26.1	16.5
20.4	23.1
24.7	25.4
26.2	21.0
18.2	24.4
	16.5
	14.8

Test whether the mean heights for the two varieties are the same against that they are not at a 5% level of significance. Assume data are normal with equal variance.

13.2 The results of the standard tests for two independent groups of students who were taught by two different methods are presented below:

Student	Method 1	Method 2	Student	Method 1	Method 2
1	34	52	6	23	27
2	38	35	7	35	45
3	26	52	8	32	41
4	33	35	9	15	17
5	32	46	10	33	41

Test whether the mean scores are the same against that they are not at a 5% level of significance. Assume data are normal with equal variance.

13.3 A company claims to have found a way of making three-strand sisal rope lines that are better than three-strand manila rope lines. The breaking strengths of equal size $\frac{1}{2}$ inch diameter rope samples were measured by stretching them until they broke. Compare the breaking mean strengths for both materials.

Manila	Sisal
2218	2224
2230	2236
2209	2219
2206	2215
2227	2237
2211	2230
2234	2250
2245	2251
2244	2253
2224	2221
2241	2247
2231	2237
	2218
	2219
	2239

Assuming breaking strengths to be normally distributed with equal variance, is the claim of the company true at a 5% level of significance?

13.4 The following data are samples of heights of males and females:

Female heights:
58.6 64.7 65.3 61.0 65.4 67.4 60.9 63.1 60.0 71.1
62.2 67.2 63.4 68.4 62.2 64.7 59.6 61.0 64.0 65.4

Male heights
62.5 64.6 69.1 73.9 67.1 64.4 71.1 71.0 67.4 69.3
64.9 68.1 66.5 67.5 66.5 70.3 67.5 68.5 71.9 67.8

Assume the variances of the male and female heights are equal. Test the hypothesis that the means for male and female heights are not equal at a 5% level of significance.

13.5 A1C level is related to blood sugar and is measured for patients who have Type 2 diabetes or a tendency to have it. It should normally be between 6.5–7.0%. The following patients had their A1C levels measured before they were put on a diet/exercise regimen. They had their A1C measured after they had subjected themselves to diet/exercise regimen for six months.

A1C Before	A1C After
7.9	6.9
7.8	7.1
8.3	7.5
7.4	7.3
7.3	7.3
7.4	7.5
7.8	6.8
7.6	6.7
7.3	6.8
7.4	6.7
7.3	7.1
7.4	7
7.2	7.4

Has the diet/exercise regimen lowered the A1C levels of the patients at the 5% level of significance?

13.6 High LDL type of cholesterol is bad for our health. The LDL levels (mg/dl) for some patients were obtained before and after they had been put on a cholesterol lowering treatment for three months. The observed values of LDL for each individual in the study are presented in the table:

LDL Before	LDL After	LDL Before	LDL After
121	106	142	131
136	138	138	129
149	147	127	120
127	107	180	144
164	133	176	136
170	141	164	144
169	137	153	132

Has the treatment significantly lowered the level of LDL in the patients? Test at a 1% level of significance.

13.7 A company claims that their training improves keyboard skills for secretaries. The following sample of secretaries had enrolled in their program and their skills were tested before and after the first two weeks of training and the results are presented here:

Keyboard Skills Before	Keyboard Skills After	Keyboard Skills Before	Keyboard Skills After
50	55	58	67
55	60	60	65
59	56	62	61
60	61	45	49
45	49	58	59
48	50	61	67
45	46	55	70
55	61	44	52

Test that the skills had improved significantly at the 1% level of significance.

13.8 Two methods are used to make determinations of the percent nutrient contents of a fruit. A sample of sixteen fruits was used. Each fruit was split into two equal parts and assigned randomly to the two methods for determination of nutrient content. The results are given below:

Method I	Method II
13.2	16.3
6	7.8
16.3	16.3
14.6	16
12.7	13.9
11.9	11.2
14.9	15.7
10.5	11.9
4.6	4.1
6.9	6.1
12.7	13.5
5.3	5.7
15.1	15.8
10.78	11.6
13.9	15.7
9.7	10.9

Are the two methods different from each other? Test at the 2% level of significance.

13.9. The following data represent samples of growths achieved by plant types X and Y in a single month in *mm*:

Y: 27 27 36 31 30 26 35 30 37 32 38 38 39 33 22
X: 56 38 55 37 44 43 57 35 39 63 42 60 44 28 55

Assume that monthly growth for plants types X and Y are normally distributed; test the hypothesis that their mean growths are equal, against that they are not at a 1% level of significance.
Recall Example 12.17–12.19 on confidence intervals; we now convert the problems to problems on tests of hypotheses:

13.10 The vital capacity, the volume of gas that can be expelled from the lungs from a position of full inspiration with no limit to duration of inspiration, was measured for adults in two different occupations, and the data obtained are as follows:

Vital capacity measurements for adults in occupation I:
4.44 4.81 4.30 5.88 5.25 5.02 6.31 3.90 5.83 6.47
4.18 4.61 5.33 5.66 6.17 4.72 6.05 4.57 5.65 6.23
4.57 4.36 5.41 6.10 6.57 4.26 3.54 4.02 4.71 5.33

Vital capacity measurements for adults in occupation II:
3.68 2.96 3.44 4.13 4.75 4.16 3.19 3.19 4.76 4.77
3.46 2.87 3.73 5.05 4.55 4.07 3.06 4.03 4.73 4.88
3.98 3.30 3.82 3.70 4.65 3.34 4.18 5.36 6.44 4.21
3.72 4.14 4.75 5.73 4.33 4.02 4.45 4.89 5.76 4.45

Test the hypothesis that the means of the vital capacities are equal against that they are not at a 2% level of significance.

13.11 Samples of times it took for patients hospitalized for the same disease at hospitals A and B to fully recover are recorded below:

Hospital A:
249.10 202.50 222.20 214.40 205.90 196.70 188.60
214.30 195.10 213.30 225.50 191.40 198.50 189.99
201.10 239.80 245.70 213.00 238.80 176.77 192.45
171.10 222.00 212.50 201.70 184.90 189.90 196.66
248.30 209.70 233.90 229.80 217.90 199.56 199.00

Hospital B:
199.50 184.00 173.20 186.00 214.10 198.60 190.60
125.50 143.50 190.40 152.00 165.70 165.50 167.80
154.70 145.30 154.60 190.30 135.40 155.80 153.78
167.70 203.40 186.70 155.30 195.90 198.60 178.42
168.90 166.70 178.60 150.20 212.40 166.40 166.78

Test at a 5% level of significance that the means of the recovery times are equal for hospitals A and B against that it is longer for Hospital A.

13.12   Samples of heights of adult women and men were taken and the data obtained are as follows:

Heights of women:
66 58 66 68 64 62 66 63 64 65 64 64 62 69 62 67 63 66 65 71
63 67 64 60 65 67 59 60 58 72 63 67 62 69 63 64 63 64 60 65

Heights of men:
70 69 64 71 68 66 74 73 62 69 67 68 72 66 72 68 71 67 71 75
64 65 67 72 72 72 67 71 66 75 69 70 69 62 66 76 69 68 66 68

Test the hypothesis that the mean height of women is less than the mean height of men at a 5% level of significance.

13.13   Samples of students of STAT 319:01 and STAT 319:02 of SCSU were taken to determine how many hours they studied weekly. The following data were obtained:

319:01 students:
14 17 10 17 20 15 16 18 10 8 12 17 16 14

319:02 students:
12 13 9 14 17 13 13 14 10 18 14 15 10 11 9 10

Test the hypothesis that the mean weekly study time of STAT 319:01 is greater than the mean weekly study time of students of STAT 319:02 at a 5% level of significance.

13.14   The following data show samples of weekly expenditure on snacks by male and female students:

Male
22 31 15 21 33 24 20 22 24 17 25 21 17 33 17 17

Female
10 15 20 7 25 13 24 15 10 14 9 12 14 14 12 24

Test the hypothesis that the mean weekly expenditure of male students on snacks is greater than mean weekly expenditure for snacks by female students at the 5% level of significance.

13.15   To manage the treatment of a disease, a study was carried out to determine the proportions of readmissions for patients by gender. The following data was obtained:

	Readmitted	Were Not Readmitted
Men	160	1340
Women	83	1055

Test the hypothesis that the proportion of readmissions for men and women is the same against that the proportion of male readmissions is more than the proportion of women readmissions at the 1% level of significance.

13.16   In a study of a new medication for preventing flu, samples of 120 men and 140 women, were given the medication. At the end of the study 100 men and 96 women did not catch the flu. Test at the 1% level of significance that there is no difference between the proportion of men and women who do not catch the flu due to the medication.

13.17   A sample of 300 students of SCSU who entered in 2000 showed that 211 were still enrolled after six semesters. A sample of 400 students who entered SCSU in 2005 showed that 302 were still enrolled after six semesters. Test at the 5% level of significance that the proportion of students retained by SCSU did not increase between 2000 and 2005.

13.18   In a poll during the health care reform debate of 2010, 900 Republicans and 1020 democrats were asked "Would you be willing to pay a little more for health insurance so that coverage may be extended to all uninsured people?" 191 Republicans and 702 Democrats responded "yes." Test the hypothesis that there is no difference between proportions of Democrats who responded "yes" and the proportion of Republicans who responded "yes" against that there is a difference at a 1% level of significance.

13.19   In a survey of enrollment of eighth graders in Algebra I it was found that of 1002 students from Massachusetts, 466 enrolled for Algebra I, while out of 1010 students from Minnesota 313 enrolled. Test the hypothesis that there is no difference between proportions of eighth graders from Massachusetts and eighth graders from Minnesota who enrolled for Algebra I, against that there is a difference at a 5% level of significance.

13.20   In a study, 566 patients were treated with a medication for a disease of which 202 recovered fully; in the same study, 365 patients were given placebo (an inert substance that looks like the medication), of which 96 recovered fully. Test at the 2% level of significance that the proportion of patients who recovered fully for both groups are the same against the proportion that took the actual medication that fully recovered is greater.

CHAPTER 14

Chi-Square Goodness of Fit and Analyses of Categorized Data

14.1 Data Analyses Based on Chi-Square

In many studies, observations are recorded in categories with associated counts. One class of such data are said to be arranged in a contingency table (meaning a two-way classification involving two categorical variables). The chi-square distribution plays a central role in the analyses of collected data which are categorized.

There are three types of chi-square-based analyses that we intend to study in this section. One of them is related to multinomial experiments, leading to investigation of the goodness of fit. Here observed data are analyzed to see whether they fit a theoretically postulated distribution for the data. The other analysis is related to experiments in which the outcomes are assigned into a contingency table but one of the variables is controlled by the experimenter so that the row totals are fixed before the study. For this class of studies, the chi-square is used as a tool in the test for homogeneity across populations from which the row samples were drawn.

The third class of analyses involving the chi-square distribution investigates the independence of two categorical variables arranged in a contingency table. In this case, inference about association between the two variables by which the contingency table was classified is being made using a process which involves the chi-square distribution. Usually a single sample is taken and the outcomes of the study are classified according to the categories of the two categorical variables. The researcher does not control the counts that should belong to any of the categories of the classifying variables.

Thus, one of the major roles the chi-square distribution plays is in the making of inferences about data which are arranged in contingency tables. The purpose of this chapter is to explore these methods of analyses of categorized data which are based on the chi-square distribution.

We discuss in the ensuing sections how to use the chi-square distribution to carry out the analyses described previously.

14.2   The Chi-Square Distribution

Let X be normally distributed so that $X \sim N(\mu, \sigma^2)$, then the standard normal variate is obtained by converting any independent value, x, selected as a sample from X, to $z = \dfrac{x - \mu}{\sigma}$. z is normally distributed with mean of zero and variance of one. We can also define another variable, based on z, that is $z^2 = \left(\dfrac{x - \mu}{\sigma}\right)^2$. It can be shown that z^2 follows the chi-square distribution with one degree of freedom. If we select a sample of size k, from X, that is $x_1, \ldots, x_k$, then it can be shown that $z_1^2 + z_2^2 + \ldots + z_k^2$ which is derived from that sample has chi-square distribution with k degrees of freedom.

14.2.1 Multinomial Experiments and Goodness of Fit Tests

There is a class of experiments for which the number of independent trials is fixed; the outcome of each trial must be assigned to one and only one of the k categories, and the probability of each category is fixed, and remains the same throughout the study. These are called multinomial experiments.

Such studies are motivated by a desire to prove or disprove a postulate one would have made about the distribution of the data in the k categories to which the outcomes of the experiment are assignable. Such a process in the end involves a hypothesis test by which one can confirm or otherwise that the postulated distribution is a good fit for the data. These types of tests utilize the chi-square distribution in the process of deciding whether the distribution is or is not a good fit to what was expected. Thus, such chi-square-based tests are called *goodness-of-fit tests*.

For a typical multinomial experiment, the outcomes (called the observed counts) of the experiment, along with the expected counts, would be classed in a table as shown in the following:

	Categories						
	1	2	3	4	...	k	Row Total
Observed counts	o_1	o_2	o_3	o_4	...	o_k	n
Expected counts	e_1	e_2	e_3	e_4		e_k	n

The expected counts are obtained via the postulated distribution for the categories. For each of the observed, o_i ($i = 1, \ldots, k$), the associated expected e_i ($i = 1, \ldots, k$) is obtained as the product of the fixed number of trials, n and the fixed probability of the category i, so that:

$$e_i = np_i$$

where p_i is the fixed probability of the ith category. The null hypothesis associated with the chi-square test that would ensue, would be specific to each study, but will invariably state that the observations of the experiments confirm that the postulated distribution of the data is a good fit. The alternative hypothesis would state the opposite.

The test statistic for the hypothesis is:

$$\chi^2_{cal} = \sum_{i=1}^{k} \frac{(o_i - e_i)^2}{e_i}$$

To make decision, the value of χ^2_{cal} is compared with the upper α-percentile value of the chi-square distribution with $k - 1$ degrees of freedom, α being the level of significance for the test. If the χ^2_{cal} is greater than the upper α-percentile value of the chi-square distribution, the null hypothesis would be rejected in favor of the alternative.

If some parameters are to be estimated in order to perform the analysis, then the number of degrees of freedom will be reduced by the number of parameters estimated. Suppose p parameters are estimated for data arranged in k categories, then, the degree of freedom for the chi-square test would be:

$$df = k - p - 1$$

It is also suggested that this test should not be used unless the following criteria are met:

(a) The number of categories k should be greater than or equal to 3.
(b) The smallest expectation for the categories e_i is at least equal to $5\left(\frac{r}{k}\right)$ where r is the number of categories with e_i less than 5.

If condition (b) above is not met, then categories could be combined.
In the next example, we illustrate the analysis of the outcome of such an experiment.

■ EXAMPLE 14.1 It is postulated that in human births, a boy is equally likely as a girl. 350 families with five children were surveyed for the distribution of girls. The following were the outcomes of the experiment.

Number of Girls	0	1	2	3	4	5
Observed	4	61	125	115	42	3

Test that the stated distribution is a good fit of the outcomes of this experiment at 1% level of significance.

Solution The probability of having a girl is the same as the probability of having a boy that is 0.5. By this idea, the postulated distribution for girls within the five-child families is binomial (5, 0.5), because the number of children is 5, representing $n = 5$ trials, in a binomial situation with outcomes of boy and girl.

Thus, the hypothesis to be tested is stated as follows:

H_0: The distribution for girls within the five-child families is binomial (5, 0.5).

Vs.

H_1: The distribution of girls within the five-child families is not binomial (5, 0.5).

The null hypothesis determines the probabilities to be assigned to the categories of girls on the basis of which the expected frequencies would be calculated as follows:

Number of Girls	0	1	2	3	4	5
Probabilities	0.03125	0.15625	0.3125	0.3125	0.15625	0.03125

Since the number of performances of the full designed experiment is 350 (the number of families), then we can obtain the expected values as $350p_i$, where p_i is the probability of the i_{th} category ($i = 0, 1, 2, \ldots , 5$).

Number of Girls	0	1	2	3	4	5	Total
Observed Frequency	4	61	125	115	42	3	350
Expected Frequency	10.9375	54.6875	109.375	109.375	54.6875	10.9375	350

$$\chi^2_{cal} = \frac{(4-10.9375)^2}{10.9375} + \frac{(61-54.6875)^2}{54.6875} + \frac{(125-109.375)^2}{109.375} +$$
$$\frac{(115-109.375)^2}{109.375} + \frac{(42-54.6875)^2}{54.6875} + \frac{(3-10.9375)^2}{10.9375}$$
$$= 16.3543$$

The number of categories based on number of girls, k is 6 for this study. From the table of the chi-square distribution (Appendix IV, page 521), $\chi^2_{[5,0.01]} = 15.1$. Clearly, $\chi^2_{cal} > \chi^2_{[5,0.01]}$. We, therefore, reject the null hypothesis that the fit is good. The distribution of girls in the five-child families is not binomial [5, 0.5] at 1% level of significance.

■ EXAMPLE 14.2 A study was conducted in a conservation area regarding the health of bird nests in the forest. Bird nests in 81-square-mile plots were investigated to find if they have been ravaged by snakes. The number of nests ravaged in this way was collected and categorized as stated in the following frequency table:

Number of Nests	0	1	2	3	4	5
Counts	25	31	10	8	4	2

Test the hypothesis that the distribution of ravaged nests follows the Poisson distribution at 5% level of significance.

Solution The hypothesis being tested here is:

H_0: The distribution of ravaged nests per square mile follows the Poisson distribution.

Vs.

H_1: The distribution of ravaged nests per square mile does not follow the Poisson distribution.

Since the parameters for the postulated Poisson distribution is not given, we have to estimate it by statistical theory. The Poisson distribution is given as:

$$p(x) = \frac{\lambda^x e^{-\lambda}}{x!}; x = 0,1,2,...$$

The parameter to be estimated is λ. Statistical theory (see Wackerly, Mendenhall, and Sheaffer 1996, 403, 625) states that the desired estimate, usually referred to as the minimum variance unbiased estimate (MVUE) of this parameter λ is obtained via maximum-likelihood estimation. This process indicates that the MVUE for λ is $\bar{x}$. To obtain the $\bar{x}$ for the given data, we present the following table, and the attendant calculations:

i	x_i	f_i	$f_i x_i$
1	0	25	0
2	1	31	31
3	2	10	20
4	3	8	24
5	4	4	16
6	5	2	10
7	≥ 6	0	0
Total→		**80**	**101**

$$\bar{x} = \frac{\sum_{i=1}^{7} f_i x_i}{\sum_{i=1}^{7} f_i} = \frac{101}{80} = 1.2625 = \hat{\lambda}$$

Thus, the underlying distribution that we intend to fit on the basis of the null hypothesis that the distribution of ravaged nests is Poisson:

$$p(x) = \frac{1.2625^x e^{-1.2625}}{x!}; x = 0,1,2,...$$

If the observations are independent, the values in the data would be distributed as multinomial; with the associated probability calculated from the above distribution, we will find the expected value for each category as in the following table, with $e_i = np_i$:

i	x_i	o_i	Probability	e_i
1	0	25	0.282946	22.6357
2	1	31	0.357219	28.5775
3	2	10	0.225495	18.0396
4	3	8	0.094896	7.5916
5	4	4	0.029951	2.3961

Continued.

i	x_i	o_i	Probability	e_i
6	5	2	0.007563	0.6050
7	6	0	0.001931	0.1273
Total→		80		80

It is suggested that for those categories for which their individual expected values are less than 5, they should be combined to produce a category with expected value greater than 5, and that the number of categories be reduced accordingly before the final calculation of the test statistic (see Cochran 1952, 1954) and reaching conclusion about goodness of the fit.

Clearly, the expected frequencies for some of the categories in the previous table are less than 5, so we need to combine some categories to make the combined expected frequency to be above 5.0. We therefore combine the categories, for $i = 4$, 5, 6, and 7 into one, as category 5, before we calculate the value of the test statistic:

i	x_i	o_i		e_i
1	0	25	0.282946	22.6357
2	1	31	0.357219	28.5775
3	2	10	0.225495	18.0396
4	3	8	0.094896	7.5916
5	≥4	6	0.039444	3.15552
Total→		80		80

Again, category 5 has expectation of less than 5, and it is the only one with expectation less than 5. We need to consider whether we should proceed with the test. Criterion (b) in page 2 says we cannot proceed with the test unless e_5 is at least $5(r/k)$, where r = number with expectation less than 5, and k = number of categories. Here $5(r/k) = 5(1/5) = 1$, since there are $k = 5$ categories, and $r = 1$ category with expectation less than 5. $e_5 = 3.15552$, way above $5(r/k) = 1$. We therefore proceed with the test without any further combining of categories.

The value of the test statistic is:

$$\chi^2_{cal} = \frac{(25-22.6357)^2}{22.6357} + \frac{(31-28.5775)^2}{28.5775} + \frac{(10-18.0396)^2}{18.0396}$$
$$+ \frac{(8-7.5916)^2}{7.5916} + \frac{(6-3.15552)^2}{3.15552}$$
$$= 6.621$$

The degree of freedom of the chi-square statistic is $df = k - p - 1 = 5 - 1 - 1 = 3$. The degree of freedom was reduced by 1 because $\bar{x}$ was estimated for the purpose of the fit. From the tables, $\chi^2_{[k-p-1,0.05]} = \chi^2_{[3,0.05]} = 7.81$, the calculated chi-square, χ^2_{cal} is less than the tabulated at 5% level of significance; the data does not provide sufficient evidence for us to conclude that we should reject the null hypothesis. Our analysis confirms that Poisson distribution with $\lambda = 1.2625$ is probably a good fit for the data at 5% level of significance.

14.3 Chi-Square Test of Homogeneity

This test applies when there is a single categorical variable with c levels, and simple random samples are taken from r different populations, and the outcomes (counts) are subsequently arranged in a contingency table. In the setup, the row categories are the different r populations. Because samples have to be taken from each of the populations, their sizes (and therefore totals for the rows) are determined by the researcher. By this token, the rows are controlled by the researcher.

The data from the r samples taken from the r populations are put into a two-way table of r rows with c columns, reflecting the categories of the column variable. The column variable is the one being studied.

Populations	Column Variable (Categories)						Row Total
	1	2	3	4	...	c	
1	O_{11}	O_{12}	O_{13}	O_{14}	...	O_{1c}	R_1
2	O_{21}	O_{22}	O_{23}	O_{24}	...	O_{2c}	R_2
3	O_{31}	O_{32}	O_{33}	O_{34}	...	O_{3c}	R_3
.	.	.	.	.	.	.	.
.	.	.	.	.	.	.	.
.	.	.	.	.	.	.	.
r	O_{r1}	O_{r2}	O_{r3}	O_{r4}	...	O_{rc}	R_c
Column Total	C_1	C_2	C_3	C_4	...	C_c	N

The O_{ij}'s refer to the number of counts out of the sample of the population i which identify with or fall into category j of the research variable in the column.

The total for row i is represented by R_i while the total for column j is C_j and N is the total for the entire study. The row totals are usually fixed by the researchers. These totals are used for calculating the expected values for the cells. To obtain the expected values for cell (i,j), the formula used is:

$$e_{ij} = \frac{R_i \times C_j}{N}$$

The use of the above formula results in the expected frequency table, containing the expected values for each cell as follows:

Populations	Column Variable (Categories)						Row Total
	1	2	3	4	...	c	
1	e_{11}	e_{12}	e_{13}	e_{14}	...	e_{1c}	R_1
2	e_{21}	e_{22}	e_{23}	e_{24}	...	e_{2c}	R_2
3	e_{31}	e_{32}	e_{33}	e_{34}	...	e_{3c}	R_3
.	.	.	.	.	.	.	.
.	.	.	.	.	.	.	.
.	.	.	.	.	.	.	.
r	e_{r1}	e_{r2}	e_{r3}	e_{r4}	...	e_{rc}	R_c
Column Total	C_1	C_2	C_3	C_4	...	C_c	N

The totals for the expected frequency table should be the same as those of the observed frequency table, that is N.

The test itself is called the test of homogeneity because its purpose is to determine that the proportions for the different levels of the categorical variable (the column variable) are the same across the populations (the rows).

The test is appropriate if for each row (population), the method of the sampling is simple random sampling, and the single variable under study is categorical. When the sampling results are presented in a contingency table as stated earlier, then for the test to be valid, the expected counts per cell must be greater than or equal to five. Based on the forgoing, the set of null hypotheses to be tested must be tied to the different levels of the categorical variable displayed in c columns for each of the rows. Therefore, the hypotheses to be tested will be stated for the r populations and c levels of the categorical variable:

$$H_0 : p_{11} = p_{21} = \ldots = p_{r1}$$
$$H_0 : p_{12} = p_{22} = \ldots = p_{r2}$$
$$H_0 : p_{13} = p_{23} = \ldots = p_{r3}$$
$$\vdots$$
$$H_0 : p_{1c} = p_{2c} = \ldots = p_{rc}$$

Vs.

H_1: At least one of the set of null hypotheses is not true.

The test statistic is obtained by using the formula:

$$\chi^2_{cal} = \sum_{i=1}^{r} \sum_{j=1}^{c} \frac{(o_{ij} - e_{ij})^2}{e_{ij}}$$

This test statistic is chi-square distributed with $(r - 1)(c - 1)$ degrees of freedom, if row variable has r rows and column variable has c columns. To decide to reject the null hypotheses, the value of the test statistic must exceed the tabulated $\chi^2_{[(r-1)(c-1),\alpha]}$ which is based on the level of significance, α. In the next example, we consider an application of the chi-square test for homogeneity.

■ **EXAMPLE 14.3**   From registered voters who identify themselves as Republican, Democratic, and Independent, samples of 200 were drawn using simple random sampling for each group. The results (counts) are presented in a contingency table as follows:

	Support for Female Presidency		
Political Leaning	**Yes**	**No**	**Undecided**
Republican	80	60	60
Democratic	95	55	50
Independent	65	60	75

Analyze the data and test whether the proportions are the same for all categories of the responses to the "Support for Female Presidency?" variable.

Solution

Observed

	Support for Female Presidency			
Political Leaning	Yes	No	Undecided	Total
Republican	80	60	60	200
Democratic	95	55	50	200
Independent	65	60	75	200
Total	240	175	185	600

$$H_0 : p_{RY} = p_{DY} = p_{IY}$$
$$H_0 : p_{RN} = p_{DN} = p_{IN}$$
$$H_0 : p_{RU} = p_{DU} = p_{IU}$$

Vs.

H_1: At least one of the above hypotheses in H_0 is untrue.

Expected

We calculate the expected values for each cell using the formula $e_{ij} = \dfrac{R_i \times C_j}{N}$:

$$e_{11} = \frac{200 \times 240}{600} = 80; e_{12} = \frac{200 \times 175}{600} = 58.3333; e_{13} = \frac{200 \times 185}{600} = 61.6667;$$
$$...e_{31} = \frac{200 \times 240}{600} = 80; e_{32} = \frac{200 \times 175}{600} = 58.3333; e_{33} = \frac{200 \times 185}{600} = 61.6667.$$

	Support for Female Presidency			
Political Leaning	Yes	No	Undecided	Total
Republican	80	58.33333333	61.66666667	200
Democratic	80	58.33333333	61.66666667	200
Independent	80	58.33333333	61.66666667	200
Total	240	175	185	600

Next, we obtain the value of the test statistic:

$$\chi^2_{cal} = \frac{(80-80)^2}{80} + \frac{(60-58.3333)^2}{58.3333} + \frac{(60-61.6667)^2}{61.6667} + ... +$$
$$\frac{(65-80)^2}{80} + \frac{(60-58.3333)^2}{58.3333} + \frac{(75-61.6667)^2}{61.6667}$$
$$= 11.0458$$

From the tables of the chi-square distribution, $\chi^2_{[(r-1)(c-1),0.05]} = \chi^2_{[4,0.05]} = 9.4877$, the calculated chi-square, χ^2_{cal} is greater than the tabulated at 5% level of significance; the data **does not** provide sufficient evidence for homogeneity that is the proportions for the different populations **are not** the same across levels of **"Support for female presidency"** variable. We conclude that at least one of the set of null hypotheses is not true.

■ **EXAMPLE 14.4**   A city plans to reserve an area that cuts across five of its wards as a recreation area. They decide to sample the five wards for support of the proposal. Simple random sampling was used to select 150, 170, 200, 200, and 180 from wards 1–5 to participate in the survey. The results were presented in a contingency table as follows:

Wards	Attitude to Proposal for Recreation Area		
	Favor	Against	Indifferent
1	79	44	27
2	95	43	32
3	106	54	40
4	94	65	41
5	101	47	32

Use the chi-square to test for homogeneity of proportions across wards at 5% level of significance.

Solution   By using F = Favor, A = Against, I = Indifferent, the following are the hypotheses to be tested:

$$H_0 : p_{1F} = p_{2F} = p_{3F} = p_{4F} = p_{5F}$$
$$H_0 : p_{1A} = p_{2A} = p_{3A} = p_{4A} = p_{5A}$$
$$H_0 : p_{1I} = p_{2I} = p_{3I} = p_{4I} = p_{5I}$$

Vs.

H_1: At least one of the above hypotheses in H_0 is untrue.

Observed

Wards	Attitude to Proposal for Recreation Area			Total
	Favor	Against	Indifferent	
1	79	44	27	150
2	95	43	32	170
3	106	54	40	200
4	94	65	41	200
5	101	47	32	180
Total	475	253	172	900

Expected   The expected values are calculated as follows:

$$e_{11} = \frac{150 \times 475}{900} = 79.1667; e_{12} = \frac{150 \times 253}{900} = 42.1667; e_{13} = \frac{150 \times 172}{900} = 28.1667; \ldots$$

$$e_{51} = \frac{180 \times 475}{900} = 95; e_{52} = \frac{180 \times 253}{900} = 50.6; e_{53} = \frac{180 \times 172}{900} = 34.4.$$

Wards	Attitude to Proposal for Recreation Area			Total
	Favor	Against	Indifferent	
1	79.1667	42.1667	28.6667	150
2	89.7222	47.7889	32.4889	170
3	105.5556	56.2222	38.2222	200
4	105.5556	56.2222	38.2222	200
5	95	50.6	34.4	180
Total	475	253	172	900

The value of the test statistic is calculated as follows:

$$\chi^2_{cal} = \frac{(79-79.1667)^2}{79.1667} + \frac{(44-42.1667)^2}{42.1667} + \frac{(27-28.6667)^2}{28.6667} + \dots$$
$$+ \frac{(101-95)^2}{95} + \frac{(47-50.6)^2}{50.6} + \frac{(32-34.4)^2}{34.4}$$
$$= 4.7869$$

From the tables, $(r-1)(c-1) = (5-1)(3-1) = 8$; $\chi^2_{[(r-1)(c-1),0.05]} = \chi^2_{[8,0.05]} = 15.51$. Thus, the calculated chi-square of 4.7869 is less than the tabulated 15.51 at 5% level of significance; the data does not provide sufficient evidence for us to conclude that we should reject the null hypothesis. We therefore accept the claim at 5% level of significance of homogeneity of proportions of the responses for each ward to the proposal for recreation area.

14.4 Chi-Square Test for Independence or Association Between Two Categorical Variables

In the research that leads to test for association between, or independence of, two categorical variables, a single random sample is taken from a single population and the individuals are assigned to a two-way table depending on which cells they fall into within the levels of the categorical variables that were used for the classification.

The data from the single sample are set into a two-way table with r rows and c columns, reflecting r categories of the row variable, and c columns of the column variable, as shown in the following:

Row Variable(Categories)	Column Variable (Categories)						Row Total
	1	2	3	4	...	c	
1	O_{11}	O_{12}	O_{13}	O_{14}	...	O_{1c}	R_1
2	O_{21}	O_{22}	O_{23}	O_{24}	...	O_{2c}	R_2
3	O_{31}	O_{32}	O_{33}	O_{34}	...	O_{3c}	R_3
.	.	.		.		.	.
.	.	.		.		.	.
.	.	.		.		.	.
r	C_1	C_2	C_3	C_4		C_c	N

The cells are filled with O_{ij}, the observed frequencies for the cell (i,j). This is the number observed out of the sample that belong to categories i of the row variable, and category j of the column variable. The total for row i is R_i, while the column total for column j is the C_j.

Calculating Expected Frequencies

Again, to obtain the expected values for each cell, (i,j), we use the corresponding row and column totals along with the overall total, as previously described, in the formula:

$$e_{ij} = \frac{R_i \times C_j}{N}$$

Use of the above formula results in expected frequencies in the same way as was described under the test for homogeneity.

Stating the Hypotheses

The hypotheses stated under test for association or independence are different from those stated for homogeneity. Under the test for association, the hypotheses are stated specific to the two categorical variables being studied. Typically, the null hypothesis would state as follows:

H_0: There is no association between the row variable and the column variable.
OR H_0: The row and the column variables are independent.
The alternative hypothesis statements corresponding to the above are:
H_1: There is an association between the row variable and the column variable.
OR H_1: The row and the column variables are not independent.

The Test Statistic, Its Distribution and Decision Making

Again, the formula for the test statistic is the same as was stated earlier:

$$\chi^2_{cal} = \sum_{i=1}^{r} \sum_{j=1}^{c} \frac{(o_{ij} - e_{ij})^2}{e_{ij}}$$

The test statistic follows the chi-square distribution with $(r-1)(c-1)$ degrees of freedom, where there are r rows for the row variable, and c columns for the column variable. Decision is made to reject the null hypothesis when the value of the calculated test statistic exceeds the critical value read from the table, that is, if $\chi^2_{cal} > \chi^2_{[(r-1)(c-1),\alpha]}$. Again, α here, refers to the level of significance for the test.

The next example illustrates application of the chi-square in testing for association between two categorical variables.

EXAMPLE 14.5   Four possible education policies were being proposed and to see whether preferences would be independent of the level of education attained by respondents, a single sample of size 310 was taken and the result of the survey was assigned to a two-way table as follows:

Highest Educ Level	Policy Favored			
	I	II	III	IV
College Grad	8	44	23	10
High School Grad	15	100	30	5
Grade-School Grad	15	40	10	10

Test the hypothesis that the policy favored is independent of the highest education level attained at 5% level of significance.

Solution

Highest Educ Level	Policy Favored				Total
	I	II	III	IV	
College Grad	8	44	23	10	85
High School Grad	15	100	30	5	150
Grade-School Grad	15	40	10	10	75
Total	38	184	63	25	310

Hypothesis to Be Tested

H_0: There is no association between policy favored and the highest education attained.
Vs.
H_1: There is an association between policy favored and the highest education attained.

The expected values are calculated as follows:

$$e_{11} = \frac{85 \times 38}{310} = 10.419; e_{12} = \frac{85 \times 184}{310} = 50.452; e_{13} = \frac{85 \times 63}{310} = 17.274; e_{14} = \frac{85 \times 25}{310} = 6.85;$$

$$...e_{31} = \frac{75 \times 38}{310} = 9.1935; e_{32} = \frac{75 \times 184}{310} = 44.516; e_{33} = \frac{75 \times 63}{310} = 15.242; e_{34} = \frac{75 \times 25}{310} = 6.0484$$

The following table shows the expected frequencies for each cell. Note that the sum of frequencies is the same for both the observed and the expected.

Highest Educ Level	Policy Favored				Total
	I	II	III	IV	
College Grad	10.419	50.452	17.274	6.8548	85
High School Grad	18.387	89.032	30.484	12.097	150
Grade-School Grad	9.1935	44.516	15.242	6.0484	75
Total	38	184	63	25	310

Next, we obtain the test statistic:

$$\chi^2_{cal} = \frac{(8-10.419)^2}{10.419} + \frac{(44-50.452)^2}{50.452} + \frac{(23-17.274)^2}{17.274} + \frac{(10-6.85)^2}{6.85} \cdots$$

$$+ \frac{(15-9.1935)^2}{9.1935} + \frac{(40-44.516)^2}{44.516} + \frac{(10-15.242)^2}{15.242} + \frac{(10-6.0484)^2}{6.0484}$$

$$= 19.3838$$

From the tables, $df = (r-1)(c-1) = (3-1)(4-1) = 6; \chi^2_{[(r-1)(c-1),0.05]} = \chi^2_{[6,0.05]}$ $= 12.6$. Thus, the calculated chi-square is greater than the tabulated chi-square, $\chi^2_{[6,0.05]} = 12.6$, at 5% level of significance; we reject the claim at 5% level of significance that there is no association between policy favored and the levels of education attained.

EXAMPLE 14.6   The following data show a sample of 224 patients taken from hospital records who were suffering from same disease; each of them was found to have been treated with one of four different treatment regimens; the frequencies observed for levels of recovery after 6 months were as recorded in the following. Do these data indicate that there is an association between the treatment regimens and the levels of recovery? Use a 2.5% level of significance:

	Levels of Recovery			
Treatment Regimens	**Fully Recovered**	**Mildly Recovered**	**No Change**	**Worsened**
I	17	19	6	5
II	23	35	6	5
III	10	14	15	11
IV	19	9	20	10

Analysis   The following are the proper ways of stating the hypotheses to be tested:

H_0: There is no association between treatment regimens and the levels of recovery.

H_1: There is an association between treatment regimens and the levels of recovery.

	Levels of Recovery				
Treatment Regimen	**Fully Recovered**	**Mildly Recovered**	**No Change**	**Worsened**	**Total R_i**
I	17	19	6	5	47
II	23	35	6	5	69
III	10	14	15	11	50
IV	19	9	20	10	58
Total C_j	69	77	47	31	$224 = N$

We find the expected values for each cell, e_{ij} as follows:

$$e_{11} = \frac{47 \times 69}{224} = 14.48, \; e_{12} = \frac{47 \times 77}{224} = 16.16, \; \dots \; e_{43} = \frac{58 \times 47}{224} = 12.17, e_{44} = \frac{58 \times 31}{224} = 8.03$$

Treatment Regimen	Levels of Recovery				
	Fully Recovered	Mildly Recovered	No Change	Worsened	Total R_i
I	14.48	16.16	9.86	6.50	47
II	21.25	23.72	14.48	9.55	69
III	15.40	17.19	10.49	6.92	50
IV	17.87	17.19	12.17	8.03	58
Total C_j	69	77	47	31	224 = N

$$\chi^2_{cal} = \sum_{i=1}^{r} \sum_{j=1}^{c} \frac{(o_{ij} - e_{ij})^2}{e_{ij}} = \frac{(17 - 14.48)^2}{14.48} + \frac{(19 - 16.16)^2}{16.16} + \frac{(6 - 9.86)^2}{9.86} + \frac{(5 - 6.5)^2}{6.5}$$

$$+ \frac{(23 - 21.25)^2}{21.25} + \frac{(35 - 23.72)^2}{23.72} + \frac{(6 - 14.48)^2}{14.48} + \frac{(5 - 9.55)^2}{9.55} + \frac{(10 - 15.40)^2}{15.40}$$

$$+ \frac{(14 - 17.19)^2}{17.19} + \frac{(15 - 10.49)^2}{10.49} + \frac{(11 - 6.92)^2}{6.92} + \frac{(19 - 17.87)^2}{17.87} + \frac{(9 - 17.19)^2}{17.19}$$

$$+ \frac{(20 - 12.17)^2}{12.17} + \frac{(10 - 8.03)^2}{8.03}$$

$$= 0.439 + 0.501 + 1.512 + 0.348 + 0.143 + 5.366 + 4.964 + 2.167 + 1.895 + 0.591$$
$$+ 1.938 + 2.406 + 0.072 + 6.00 + 5.038 + 0.485 = 33.866$$

There are $c = 4$ columns, and $r = 4$ rows in this table so that the degrees of freedom, $df = (r-1)(c-1) = (4-1)(4-1) = 9$

From the tables, we read $\chi^2_{[9,0.025]} = 19.023$. This indicates that $\chi^2_{cal} > \chi^2_{[9,0.025]}$. The data are significant. We therefore reject H_0 and conclude that there is an association between treatment regimens and the levels of recovery.

14.5 Use of MINITAB for Chi-Square Analysis

The MINITAB, the statistics software we have used in this text, can greatly simplify and remove the tedium involved in some of the calculations we make as well as the analyses we carry out. In this section, we discuss how MINITAB can be used to carry out the chi-square-based analyses which have been discussed in this chapter. The code is provided with each analysis, and can serve as a template for writing code that would analyze similar data.

In the next set of examples, we recall and analyze, using MINITAB, some data we had previously analyzed manually. Applications to data not previously analyzed will also be treated.

■ **EXAMPLE 14.7**   **Recall Example 14.1**
It is postulated that in human births, a boy is equally likely as a girl. 350 families with five children were surveyed for the distribution of girls. The following were the outcomes of the experiment.

Number of Girls	0	1	2	3	4	5
Observed	4	61	125	115	42	3

Test that the stated distribution is a good fit of the outcomes of this experiment at 1% level of significance. USE MINITAB.

Solution   To use MINITAB for this analysis, we enter the data into three columns in MINITAB (say $c1$–$c3$) as follows:

Number of Girls	Observed	Proportion
0	4	0.03125
1	61	0.15625
2	125	0.3125
3	115	0.3125
4	42	0.15625
5	3	0.03125

In the observed, we enter the observed values for the categories shown (in this case) for number of girls. In the proportion section, we enter the binomial probabilities for binomial (5, 0.5) for the values 0, 1, 2, 3, 4, and 5 under number of girls.

This means that in MINITAB, $c1$–$c3$ are dubbed "number of girls," "observed," and "proportion," respectively. The command to enter in MINITAB after setting up the data as described previously is:

```
MTB >  TChiSquare;
SUBC>    Observed 'observed';
SUBC>    Proportions c3;
SUBC>    GBar;
SUBC>    GChiSQ;
SUBC>     Pareto;
SUBC>    RTable.
```

The outcome of the above entries is:

Chi-Square Goodness-of-Fit Test for Observed Counts in Variable: observed

Category	Observed	Test Proportion	Expected	Contribution to Chi-Sq
1	4	0.03125	10.938	4.40036
2	61	0.15625	54.688	0.72864
3	125	0.31250	109.375	2.23214
4	115	0.31250	109.375	0.28929
5	42	0.15625	54.688	2.94350
6	3	0.03125	10.938	5.76036

N	DF	Chi-Sq	P-Value
350	5	16.3543	0.006

The analysis outputs include some graphics not presented. As we can see, the chi-square test statistic is the same as we calculated manually, and the p-value of 0.006 (which is less than level of significance of 0.01) supports our earlier decision to reject the null hypothesis.

■ **EXAMPLE 14.8**   **Recall Example 14.2**

A study was conducted in a conservation area regarding the health of bird nests in the forest. Bird nests in 81-square-mile plots were investigated to find if they have been ravaged by snakes. The number of nests ravaged in this way were collected and categorized as stated in the following frequency table:

Number of Nests	0	1	2	3	4	5	6
Counts	30	15	10	8	4	2	1

Test the hypothesis that the distribution of ravaged nests follows the Poisson distribution at 5% level of significance. Use MINITAB.

MINITAB Solution of EXAMPLE 14.2

This example requires the fitting of a Poisson distribution. The command in MINITAB is slightly different from the general case used in Example 14.1. We analyze the data using MINITAB. First, we enter the data in two columns in MINITAB $C1$ and $C2$, as follows:

X	Observed
0	25
1	31
2	10
3	8
4	4
5	2
6	0

The code in MINITAB for the analysis are given as follows:

```
MTB >  PGoodness C1;
SUBC>    Frequencies C2;
SUBC>    GBar;
SUBC>    GChiSQ;
SUBC>     Pareto;
SUBC>    RTable.
```

The output of the analysis is given as follows:

Goodness-of-Fit Test for Poisson Distribution:

```
Data column: X
Frequency column: observed

Poisson mean for X = 1.2625
```

X	Observed	Poisson Probability	Expected	Contribution to Chi-Sq
0	25	0.282946	22.6357	0.24696
1	31	0.357219	28.5775	0.20535
2	10	0.225495	18.0396	3.58293
3	8	0.094896	7.5916	0.02197
>=4	6	0.039445	3.1556	2.56388

```
N   N*  DF  Chi-Sq   P-Value
80  0   3   6.62109  0.085
1 cell(s) (20.00%) with expected value(s) less than 5.
```

Again, the analysis outputs include some graphics not presented. As is evident, the chi-square test statistic is the same as we calculated manually and the p-value of 0.085 which is higher than the level of significance of 0.05, supports our earlier decision not to reject the null hypothesis. We therefore conclude that the data is Poisson distributed with mean of 1.2625. Here, MINITAB comments on the cell with expected value less than 5, but proceeded with the analysis and did not merge categories. Evidently, the expected value e_5, 3.1556, the least of the expected value $>5(r/k) = 1$ and satisfies criterion (b) of page 2 for analysis to be carried out without any further merging of categories to reach expectation of 5 or more. This is the same decision we reached by manual analysis.

■ **EXAMPLE 14.8** Recall the claim that Mr. Zales sells cars according to the distribution in Example 5.3. Suppose 500 days for which he sold cars are taken at random and the following were observed for number of cars sold:

x	0	1	2	3	4	5
$p(x)$	0.083	0.215	0.35	0.176	0.112	0.064
Observed	50	117	200	66	40	27

Is the claim of Mr. Zales supported by the data obtained? Test at 1% level of significance. USE MINITAB.

MINITAB CODE

H_0: The number of cars sold per day is in the proportions stated by Mr. Zales as in the previous table.

Vs.

H_1: The number of cars sold per day is not in the proportions stated by Mr. Zales as in the previous table.

```
MTB > name c1 'category' c2 'observed' c3 'proportion'
MTB > set c1
DATA> 0 1 2 3 4 5
DATA> end
MTB > set c2
DATA> 50 117 200 66 40 27
DATA> end
MTB > set c3
DATA> 0.083 0.215 0.35 0.176 0.112 0.064
DATA> end
MTB > print c1-c3
```

Data Display

Row	category	observed	proportion
1	0	50	0.083
2	1	117	0.215
3	2	200	0.350
4	3	66	0.176
5	4	40	0.112
6	5	27	0.064

```
MTB > TChiSquare;
SUBC>    Observed c2;
SUBC>    Levels 'category';
SUBC>    Proportions c3;
SUBC>    GBar;
SUBC>    GChiSQ;
SUBC>      Pareto;
SUBC>    RTable.
```

Chi-Square Goodness-of-Fit Test for Observed Counts in Variable: Observed

Using category names in category

Category	Observed	Test Proportion	Expected	Contribution to Chi-Sq
0	50	0.083	41.5	1.74096
1	117	0.215	107.5	0.83953
2	200	0.350	175.0	3.57143
3	66	0.176	88.0	5.50000
4	40	0.112	56.0	4.57143

Continued.

Category	Observed	Test Proportion	Expected	Contribution to Chi-Sq
5	27	0.064	32.0	0.78125

N	DF	Chi-Sq	P-Value
500	5	17.0046	0.004

In view of the above output from the MINITAB analysis, the claim of Mr. Zales cannot be supported at 1% level of significance. This is because p-value < 0.01, the level of significance.

■ EXAMPLE 14.9 **Recall Example 14.5**

Four possible education policies were being proposed and to see whether preferences would be independent of the level of education attained by respondents, a single sample of size 310 was taken and the result of the survey was assigned to a two-way table as follows:

Highest Educ Level	Policy Favored			
	I	II	III	IV
College Grad	8	44	23	10
High School Grad	15	100	30	5
Grade-School Grad	15	40	10	10

Test the hypothesis that the policy favored is independent of the highest education level attained at 5% level of significance.

MINITAB analysis of Example 14.5

First, we enter the data into columns $C1$–$C5$ in MINITAB as in the columns of the following table:

Education Level	I	II	III	IV
College grad	8	44	23	10
High school grad	15	100	30	5
Grade-school grad	15	40	10	10

Next we enter the code:

```
MTB > XTabs;
SUBC> CSTA;
SUBC> Unstacked c2 - c5;
SUBC>   RLevels c1;
SUBC>   CName "Educational policies";
SUBC> ChiSquare;
SUBC> Counts;
SUBC> Expected.
```

Chi-Square Test for Association: Education level, Educational policies

```
Rows: Education level   Columns: Educational policies

                      I         II        III        IV        All
  COLLEGE GRAD        8         44         23        10         85
                   10.42      50.45      17.27      6.85

  HIGH SCHOOL GRAD   15        100         30         5        150
                   18.39      89.03      30.48     12.10

  GRADE-SCHOOL GRAD  15         40         10        10         75
                    9.19      44.52      15.24      6.05

  All                38        184         63        25        310
  Cell Contents:                              Count
                                        Expected count
```

```
Pearson Chi-Square = 19.384, DF = 6, P-Value = 0.004
Likelihood Ratio Chi-Square = 19.540, DF = 6, P-Value = 0.003
```

Again, the outputs and conclusion here agree with previous analysis. The p-value is less than 5%, the level of significance. This leads us to reject the null hypothesis of "there is no association between Education policies preferred and the level of Education attained" and accept that "there is an association between Education policies preferred and the level of Education attained" at 5% level of significance.

EXAMPLE 14.10 **Recall Example 14.6**

The following data show a sample of 224 patients taken from hospital records who were suffering from same disease; each of them was found to have been treated with one of four different treatment regimens; the frequencies observed for levels of recovery after 6 months were as recorded in the following. Do these data indicate that there is an association between the treatment regimens and the levels of recovery?

Treatment Regimens	Fully Recovered	Mildly Recovered	No Change	Worsened
I	17	19	6	5
II	23	35	6	5
III	10	14	15	11
IV	19	9	20	10

Use MINITAB and test at 2.5% level of significance.

Solution We first enter the data into five columns ($c1$–$c5$) in MINITAB as presented in the following:

Treatment	Fully recovered	Mildly recovered	No Change	Worsened
I	17	19	6	5
II	23	35	6	5
III	10	14	15	11
IV	19	9	20	10

We write the MINITAB code:

```
MTB > XTabs;
SUBC>    CSTA;
SUBC>    Unstacked c2 - c5;
SUBC>     RLevels c1;
SUBC>      CName "recovery levels";
SUBC>    ChiSquare;
SUBC>    Counts;
SUBC>    Expected.
```

Output:
Chi-Square Test for Association: Category, Recovery Levels

Rows: treatment   Columns: recovery levels

	Fully recovered	Mildly recovered	No Change	worsened	All
I	17	19	6	5	47
	14.48	16.16	9.86	6.50	
II	23	35	6	5	69
	21.25	23.72	14.48	9.55	
III	10	14	15	11	50
	15.40	17.19	10.49	6.92	
IV	19	9	20	10	58
	17.87	19.94	12.17	8.03	
All	69	77	47	31	224

Cell Contents: Count
                                   Expected count

```
Pearson Chi-Square = 33.866, DF = 9, P-Value = 0.000
Likelihood Ratio Chi-Square = 35.686, DF = 9, P-Value = 0.000
```

The output from MINITAB analysis shows a p-value < 0.0001, which is much less than the level of significance of 2.5%. We therefore reject the null hypothesis of "no association between treatment and recovery levels" and accept that such an association exists at 1% level of significance—the same was the result we reached by manual analysis carried out previously.

References

Cochran, William G. 1952. "The χ^2 Test of Goodness of Fit." *Annals of Mathematical Statistics* 23: 315-45.

Cochran, William G. 1954. "Some Methods of Strengthening the Common χ^2 Tests." *Biometrics* 10: 417-51.

Wackerly, D.D., W. Mendenhall III, and R.L. Scheaffer. 1996. *Mathematical Statistics With Applications.* Duxbury Press Corp., Belmont

Exercises—Chi-Square Analysis

14.1 1,200 bags, each containing 12 apples were inspected after 2 weeks after being stored under the same conditions. The number of rotten apples per bag and their frequencies are shown as follows:

Number of Rotten Apples	0	1	2	3	4	5	6	7
Frequency	308	426	300	118	43	2	2	1

Can we say that the rotten apples follow the binomial distribution with rotten proportion of 11.5%? Test at 5% level of significance.

14.2 The data show four treatment regimens applied to patients suffering from a disease and the levels of recovery observed:

Treatment Methods

Recovery-Level	A	B	C	D
Full	33	36	45	13
Mild	16	11	49	9
No Change	16	10	24	15
Worse	15	17	19	20

Is there an association between levels of recovery and the treatment regimen adopted? Test at 5% level of significance.

14.3 The following data were collected in a survey of drinking habits and marital status.

consumption of alcohol	marital status			
	married	divorced	separated	never married
heavy	65	113	90	44
regular	42	64	93	25
mild	55	51	56	27
teetotal	115	56	56	15

Test whether the drinking habits are independent from marital status. Test at 1% level of significance.

14.4   A sample of 270 voters was surveyed about their support of female presidency. They were assigned to the categories according to their political leanings as in the following table:

	Support of Free State Tuition		
Political Leaning	Yes	No	Undecided
Democratic	49	28	15
Republican	41	57	12
Independent	28	18	22

Is attitude to support for free state tuition independent of party leaning? Test at 5% level of significance.

14.5   A study was carried out to determine whether age of the driver is independent of the number of accidents he or she has had. The outcomes were assigned to a contingency table as shown in the following:

	Number of Accidents		
Age of Driver	1	2	≥3
16–21	66	50	10
22–35	49	21	15
≥35	35	18	18

Test for independence of age of driver and number of accidents they have had.

14.6   A survey was conducted to see whether there is an association between the type of orange preferred and the age of the consumer.

Age	Valencia	Navel	Blood	Mandarin
<30	20	25	17	30
30–50	15	41	13	10
>50	13	44	10	13

Test at 5% level of significance.

14.7   The following data show the number of times a sample of students were absent for a large introductory STAT class in a University:

Number of Times	0	1	2	3
Count	40	30	18	15

Is the data Poisson distributed? Test at 1% level of significance.

14.8   A survey of couples with serious marital problems was carried out in order to determine whether age of the couples is associated with the ultimate outcome of their relationships. The following data were obtained:

Outcome	Age		
	Above 50	25–50	<25
Stayed Married	45	40	22
Divorced	18	28	39
Separated	15	15	20

Are age and outcomes of marriage independent? Test at 5% level of significance.

14.9   Six flavors of a Greek yoghurt were subjected to market survey, and the preferences of the flavors are as follows:

Flavors	1	2	3	4	5	6
Preferences	40	46	38	40	24	22

Are the six flavors equally favored? Test at 1% level of significance.

14.10   Consider the data on five-child families of Example 14.2. Assume that probability of birth of a girl is 0.49. Test for goodness of fit at 5% level of significance.

CHAPTER 15

The Completely Randomized Design – One Way Analysis of Variance

15.0 One-Way Layout—An Introduction

In the previous sections of this text, we have tested hypotheses, in which we compared the means of two treatments or the means of two independent groups of observations. In making comparisons, we employed the t-test under different assumptions, particularly assuming normality for the distributions. Other assumptions had to do with the variances, whether they were equal or not. These assumptions helped us to determine the appropriate methods of analysis to employ.

Let us take samples from two independent normal populations. Let the sample of size n_1 observations be taken from population 1, and sample of size n_2 observations be taken from population 2. If both the variances of the two populations are unknown, but we have reason to believe that they are equal, we can compare their means by using the pooled variance t-test. The pooled variance serves as an estimate of the common variance for the two populations. The following test statistic can be used for all the tests to compare the means of the two populations:

$$t_c = \frac{\bar{Y}_1 - \bar{Y}_2}{s_p\sqrt{\left[\dfrac{1}{n_1} + \dfrac{1}{n_2}\right]}} \qquad (15.1)$$

$$s_p^2 = \frac{(n_1 - 1)s_1^2 + (n_2 - 1)s_2^2}{n_1 + n_2 - 2}; \text{ where } s_1^2 = \frac{\left[\displaystyle\sum_{i=1}^{n_1}(y_{1i} - \bar{y}_{1.})^2\right]}{n_1 - 1} \text{ and } s_2^2 = \frac{\left[\displaystyle\sum_{i=1}^{n_2}(y_{1i} - \bar{y}_{1.})^2\right]}{n_2 - 1}$$

Note that if we had taken samples of equal size, n from both then in that case, the pooled variance would be obtained as:

$$s_p^2 = \frac{(n - 1)s_1^2 + (n - 1)s_2^2}{2n - 2}; \text{ where } s_1^2 = \frac{\left[\displaystyle\sum_{i=1}^{n}(y_{1i} - \bar{y}_{1.})^2\right]}{n - 1} \text{ and } s_2^2 = \frac{\left[\displaystyle\sum_{i=1}^{n}(y_{1i} - \bar{y}_{1.})^2\right]}{n - 1}$$

We employed the previous method extensively in Chapters 12 and 13 in order to compare two means or make estimates when the populations met the assumptions. There is, however, an alternative way of looking at the analysis of samples collected in the circumstances described. The data can be analyzed by setting them in a one-way table and using the variation in the combined data set to determine whether the means are significantly different from each other.

As the name implies, the alternative approach requires that we lay out the entire data collected from both populations in a one-way table before analyzing them. We look at this alternative approach next, because it can also be extended to allow us to compare k means at once, where $k \geq 3$. For simplicity, we will apply this to two populations, and further restrict ourselves to discussing the method when the sample sizes taken from both populations are equal. Later we will consider the one-way layout when sample sizes are not equal.

The mathematical model suitable for the kind of design we have adopted in place of a pooled variance t-test expresses the typical observation (response) y_{ij} as a linear function of the grand mean μ, the treatment effects α_i, and the error term, ε_{ij} as shown in the following:

$$y_{ij} = \mu + \alpha_i + \varepsilon_{ij}; \quad i = 1, 2; \ j = 1, 2, \ldots, n. \tag{15.2}$$

The responses of the experiment could be presented in one-way layout as indicated in Table 15.1.

The method to be used for analysis is based on the total variation in the combined data set, as laid out in the following table. We will refer to the populations as groups 1 and 2 in the table.

■ **TABLE 15.1 Two Samples in One-Way Layout**

Groups	Sample Values					Totals
	1	2	3	...	n	
1	y_{11}	y_{12}	y_{13}	...	y_{1n}	$Y_{1.}$
2	y_{21}	y_{22}	y_{23}	...	y_{2n}	$Y_{2.}$
						$Y_{..}$

In the previous table, $y_{i.} = \sum_{j=1}^{n} y_{ij}$ the row total for row i, ($i = 1, 2$), referring to the two groups. Also, $y_{..} = \sum_{i=1}^{2} \sum_{j=1}^{n} y_{ij}$ is the overall total, the sum of all observations in the two samples from the two groups. The corresponding means are denoted are obtained as: $\overline{y}_{..} = \dfrac{\sum_{i=1}^{2} \sum_{j=1}^{n} y_{ij}}{2n}$ and $\overline{y}_{i.} = \dfrac{\sum_{j=1}^{n} y_{ij}}{n}$.

In order to carry out the comparison of means for both groups using this procedure called analysis of variance (ANOVA), we need to state the objective of our analysis as well as the assumptions we make about the sample values (called responses). In the model (15.2), μ is a parameter which is common to all observations y_{ij} from both groups and is called the grand mean, while α_i is the treatment effect of the ith group, a deviation for the ith group from the grand mean. The error component, ε_{ij} is assumed to be normally and independently distributed with mean zero and variance, σ^2, [i.e., $\varepsilon_{ij} \sim \text{NID}(0,\sigma^2)$] which means that σ^2 is the same across observations in both groups or for all groups being studied.

If each of the α_i is found in our test to be zero, then the grand mean is the mean of both groups. In other words, the means of all the observations made for each of the groups are the same. We would have found that there is no significant difference between both groups.

By the model of (15.2), a typical observation y_{ij} could be split into the grand mean which is common to all observations in the two groups, then α_i, the effect of group i, and a random error ε_{ij} which is $\text{NID}(0, \sigma^2)$. To test for difference between the means of the two groups, would mean the comparison of the effects of group 1 and group 2, namely, α_1 and α_2. It is therefore the difference or lack thereof between the two effects that would provide basis for any conclusion reached. This is because μ is the same, and ε_{ij} has zero mean, leaving only α_1 and α_2 as the only quantities in the model specifications that could be different for observations drawn from either groups. Under the fixed effects model, it is assumed that $\alpha_1 + \alpha_2 = 0$.

Our objective is to test the hypothesis that the means of the two groups are equal, against that they are not. However, the new hypothesis which is essentially the same, is stated in terms of α_i, that is the effects of either group one or group two, beyond the overall mean μ which is common to both groups:

$$H_0 : \alpha_1 = \alpha_2 = 0 \ \text{ vs. } \ H : \alpha_i \neq 0 \ \text{ for } i = 1 \text{ or } 2. \tag{15.3}$$

This objective is usually stated in the form of a hypothesis; in this case, the hypothesis we are to test is that there is no treatment effect, which can also be stated, in this particular case involving only two groups, as

$$H_0 : \mu_1 - \mu_2 = 0 \ \text{ vs. } \ H_1 : \mu_1 - \mu_2 \neq 0.$$

We demonstrate in the following section that one-way ANOVA actually uses pooled variance in its decisions about the hypothesis in (15.3). We show that variance is estimated the same way in both approaches and that both approaches reach the same conclusions. The total variation in the two samples from the two populations is usually encapsulated in the total sums of squares, SS_{total} defined as in (15.4):

$$SS_{\text{total}} = \sum_{i=1}^{2} \sum_{i=1}^{n} (y_{ij} - \bar{y}_{..})^2 \tag{15.4}$$

By adding and subtracting $\bar{y}_{i.}$ inside the parenthesis in (15.4), the definition of SS_{total} can be rewritten as:

$$SS_{\text{total}} = \sum_{i=1}^{2} \sum_{i=1}^{n} \left[(y_{ij} - \bar{y}_{i.}) + (\bar{y}_{i.} - \bar{y}_{..}) \right]^2$$

$$= n \sum_{i=1}^{2} (\bar{y}_{i.} - \bar{y}_{..})^2 + \sum_{i=1}^{2} \sum_{j=1}^{n} (y_{ij} - \bar{y}_{i.})^2 + \text{terms which sum to zero} \tag{15.5}$$

$$= SS_{\text{groups}} + SS_{\text{Error}}$$

In (15.5), we have split the total sum of squares into two component parts, namely, sum of squares groups, and sum of squares error.

The SS_{total}, SS_{groups}, and SS_{error} are chi-square distributed when divided by their respective degrees of freedom, to wit, $2n-1$, 1, and $2n-2$, degrees of freedom. This stems from the fact that the deviations, $(y_{ij} - \bar{y}_{..})$, $(y_{ij} - \bar{y}_{i.})$, and $(\bar{y}_{i.} - \bar{y}_{..})$, are estimates of normal variables, and each of them when squared would yield chi-square variables. Now, consider the error sum of squares, it is easily shown, as below, that it breaks down with a little algebra to the numerator in the estimation of the pooled variance, s_p^2:

$$SS_{Error} = \sum_{i=1}^{2}\sum_{j=1}^{n}(\bar{y}_{ij} - \bar{y}_{i.})^2 = \sum_{j=1}^{n}(\bar{y}_{1j} - \bar{y}_{1.})^2 + \sum_{j=1}^{n}(\bar{y}_{2j} - \bar{y}_{2.})^2$$

$$= (n-1)s_1^2 + (n-1)s_2^2$$

This is because $\sum_{j=1}^{n}(\bar{y}_{1j} - \bar{y}_1)^2$ and $\sum_{j=1}^{n}(\bar{y}_{2j} - \bar{y}_{2.})^2$ are the numerators in the definitions of the sample variances of Y_1 and Y_2, namely, s_1^2 and s_2^2.

The estimate of common variance used in the one-way ANOVA (the new method of analysis) divides the error sum of squares SS_{Error} by its degree of freedom. In this particular case when sample sizes are equal, the estimate of common variance is:

$$\hat{\sigma}^2 = \frac{\sum_{i=1}^{2}\sum_{j=1}^{n}(\bar{y}_{ij} - \bar{y}_{i.})^2}{2n-2} = \frac{(n-1)s_1^2 + (n-1)s_2^2}{2n-2} \tag{15.5}$$

The result (15.5) is the same as the estimate of the common variance s_p^2 which was used in pooled variance confidence intervals and t-tests discussed extensively in Chapters 12 and 13.

Statistically, it can be shown that $n\sum_{i=1}^{2}(\bar{y}_{i.} - \bar{y}_{..})^2$ divided by its degree of freedom, that is MS_{groups}, follows the chi-square distribution with 1 degree of freedom, while

$\sum_{i=1}^{2}\sum_{j=1}^{n}(y_{ij} - \bar{y}_{i.})^2$ divided by its degree of freedom $2n-2$, yields the MS_{Error} which

follows the chi-square distribution with $2n-2$ degrees of freedom. This is, however, outside the scope of this text.

To analyze the data from two independent normally distributed populations from which equal sample sizes were drawn using one-way ANOVA, we need to use the fact that the ratio of these two chi-square variables also follows the F-distribution with 1 and $2n-2$ degrees of freedom. Proof of this will not be pursued in this text.

The process of one-way ANOVA eventually leads to ANOVA table of the form:

Source	Sum of Squares	Degree of Freedom	Mean Square	F-Ratio
Groups	SS_{groups}	1	$SS_{groups}/1$	MS_{groups}/MS_{Error}
Error	SS_{Error}	$2n-2$	$SS_{Error}/(2n-2)$	
Total	SS_{total}	$2n-1$		

We typically will compare F-ratio with tabulated $F(1, 2n-2, \alpha)$ and reject the null hypothesis if the former is greater than the latter. Here, α is the level of significance at which we intend to test the hypothesis of (15.3).

Since we need the sums of squares for ANOVA, we suggest simpler formulas for calculating them as follows:

$$SS_{total} = \sum_{i=1}^{2} \sum_{j=1}^{n} y_{ij}^2 - \frac{y_{..}^2}{2n}$$

$$SS_{groups} = \sum_{i=1}^{2} y_{i.}^2 - \frac{y_{..}^2}{2n}$$

$$SS_{Error} = SS_{total} - SS_{groups}$$

These simpler forms of the formulas can easily be seen to have been arrived at by using algebra to manipulate the more complicated forms of the formulas. The latter forms make manual calculations less arduous.

To show that we would arrive at the same conclusion using both pooled variance t-test and one-way ANOVA, we reanalyze the data of Example 13.5 to which pooled variance t-test was previously applied, using one-way ANOVA:

■ **EXAMPLE 15.1   (Formerly EXAMPLE 13.5)**

You are a test engineer, and you wish to test the mileages attained by the same make of car run on two new formulations of biofuels.

Fuel A	Fuel B
29	38
30	33
31	32
31	37
28	35
27	39
22	40
24	38
25	37
28	43
40	36
38	32
33	31
19	42
30	33

Assuming that the mileages per gallon are drawn from two normally distributed populations with equal variance, test the hypothesis that the mean miles/gallon of Fuel B is greater than the mean miles/per gallon for Fuel A at 1% level of significance.

Solution   We rewrite the data in a one-way layout as follows:

Group (*i*)	Sample Number (*j* = 1, 2, 3, ... , 15)															Total $Y_{i.}$
	1	2	3	4	5	6	7	8	9	10	11	12	13	14	15	
1	29	30	31	31	28	27	22	24	25	28	40	38	33	19	30	435
2	38	33	32	37	35	39	40	38	37	43	36	32	31	42	33	536

$$y_{..} = y_{1.} + y_{2.} = 435 + 536 = 981$$

$$SS_{total} = 29^2 + 30^2 + ... + 42^2 + 33^2 - \frac{981^2}{2(15)} = 1028.3$$

$$SS_{group} = \frac{435^2 + 546^2}{15} - \frac{981^2}{2(15)} = 410.7$$

$$SS_{Error} = 1028.3 - 410.7 = 617.6$$

With the calculations done, we carry out ANOVA:

ANOVA

Source	Sum of Square	Degree of Freedom	Mean Square	*F*-Ratio
Groups	410.7	1	410.7	18.6198
Error	617.6	28	22.05714	
Total	1028.3	29		

From the tables of F (Appendix V, page 533), we find that $F(1,28,0.01) = 7.6356$. The *F*-ratio we obtained from the data is 18.6198, which is greater than then tabulated value of $F(1, 28, 0.01) = 7.6356$, so we reject the null hypothesis of (15.3) and accept the alternative that $\alpha_i \neq 0$ for $i = 1$ or 2; this is equivalent to rejecting the null hypothesis that $\mu_1 = \mu_2$ and accepting the alternative that $\mu_1 \neq \mu_2$ at 1% level of significance as we did in Example 13.5.

Note that MS_{Error} we used for ANOVA is the same value (22.01571) as s_p^2 that we used for pooled variance t-test for Example 13.5. Both analysis concluded that mean of group 1 is not equal to mean of group 2, although in Example 13.5 the analysis using t-test further showed that mean of group 2 is greater than the mean of group 1. In one-way ANOVA, one more step is usually taken to classify the means of the independent samples. We attend to that aspect later in this chapter after we have considered the use of one-way ANOVA to compare $k \geq 3$ means.

From the foregoing, both tests applied to the two sample comparison are equivalent. But note that one-way ANOVA is not designed primarily for comparing a pair of means but for k means, when $k \geq 3$. Consequently it does not immediately address which mean, of the k means, is greater than the others; further analysis would reveal such relationships.

Now that we have established the kinship between the pooled t-test we used for two-sample comparisons, and one-way ANOVA, let us look at the more general form of the one-way ANOVA.

15.1 The One-Way ANOVA or the Completely Randomized Design

The one-way ANOVA is not only applied to the comparison of k independent groups ($k \geq 3$), from which samples have been drawn, it is also used for analysis of designed experiments in which the object is to study the effects of a factor set at different levels on the responses (yields) of the experiment. Basically, this is a one factor factorial experiment. It is this form of one-way ANOVA that is referred to as the completely randomized design (CRD). A factor is an explanatory variable, a quantity which is expected to account for the responses or yields obtained in an experiment.

Thus, if a formal experiment is set up to study, say, the effect of temperature on time of completion of a chemical reaction; the phenomenon may be studied for different settings of temperature like 50°C, 60°C, 70°C, 80°C, and so on. For the process, a number of determinations of reaction times may be made at each of these temperatures, which represents the sampling process in situ. For such an experiment, in order to eliminate systematic variation, the order of the performances of the experiments for different settings of temperature is randomized.

Randomization is a deliberate, impersonal process which converts uncontrolled variation in an experiment (regardless of source) to random variation. In fact, randomization is one of the justifications that provide the statistical basis for the analysis of the responses of an experiment using ANOVA. Randomization also removes possible biases whose presence in the process generally makes the conclusions reached invariably not dependable. This type of experiment in which ANOVA is ultimately used to compare yields at the different temperature settings is an example of a CRD.

As stated earlier, we can use one-way ANOVA to compare independent groups which do not arise due to the settings of a single factor. A straight-forward example would be a case where we decide to study or test the mileages [per gallon] attained by samples of the same make of car which is run on k ($k \geq 3$) distinct formulations of biofuels for a specified quantity of the fuels. In this case, we are comparing miles per gallon for k distinct fuels (groups).

We extend the model for one-way ANOVA or CRD as follows: the observations (responses) y_{ij} is expressed as a linear function of the grand mean, the treatment effects, and the error term as shown in the model as follows:

$$y_{ij} = \mu + \alpha_i + \varepsilon_{ij}; \quad i = 1, 2, \ldots, k; \ j = 1, 2, \ldots, n. \tag{15.6}$$

For fixed effects model, then $\sum_{i=1}^{k} \alpha_i = 0$. The fixed effects assumption enables us to estimate α_i.

The responses of the experiment could be presented in an extended one-way layout involving k groups or k levels of factor rather than just two groups as indicated in the following table:

Groups	Sample Values					Totals
	1	2	3	...	n	
1	y_{11}	y_{12}	y_{13}	...	y_{1n}	$y_{1.}$
2	y_{21}	y_{22}	y_{23}	...	y_{2n}	$y_{2.}$
3	y_{31}	y_{32}	y_{33}	...	y_{3n}	$y_{3.}$
.	.	.	.	...	.	.
.	.	.	.	...	.	.
.	.	.	.	...	.	.
k	y_{k1}	y_{k2}	y_{k3}	...	y_{kn}	$y_{k.}$
						$y_{..}$

The splitting of total sums of squares for this general case of the CRD is as follows:

$$SS_{total} = \sum_{i=1}^{k} \sum_{i=1}^{n} (y_{ij} - \bar{y}_{..})^2 \tag{15.7}$$

$$SS_{total} = \sum_{i=1}^{k} \sum_{i=1}^{n} \left[(y_{ij} - \bar{y}_{i.}) + (\bar{y}_{i.} - \bar{y}_{..}) \right]^2$$

$$= n \sum_{i=1}^{k} (\bar{y}_{i.} - \bar{y}_{..})^2 + \sum_{i=1}^{k} \sum_{j=1}^{n} (y_{ij} - \bar{y}_{i.})^2 + \text{terms which sum to zero} \tag{15.8}$$

$$= SS_{groups} + SS_{Error}$$

In general, in one-way ANOVA, total variation is split into the ***between-group variation*** and the ***within-group variation.*** Thus, we usually compare the ***between-group variation*** with the ***within-group variation*** to assess whether there is a difference in the population means. The between group sums of squares (which is simply stated as sum of squares groups in this text) captures the variation occurring because of differences **between the groups**, while the **within group sums of squares** are residual, related to error—variation in the data which is free of the variations due to group differences.

Statistical theory shows that SS_{total}/σ^2 has a chi-square distribution with $k(n-1)$ degrees of freedom, and that SS_{error}/σ^2, has a chi-square distribution with $k(n-1)$ degrees of freedom. Similarly, *also,* SS_{groups}/σ^2 has a chi-square distribution with $k-1$ degrees of freedom.

Under the null hypothesis, $SS_{error}/k(n-1)$ is an unbiased estimate of σ^2, and $SS_{groups}/(k-1)$ is also an unbiased estimate of the same σ^2. By comparing these two measures of variance (OR spread of the data) with one another, we are able to detect if there are true differences among the underlying group population means. Their ratio is F-distributed, hence, if the null hypothesis is true, the F-ratio, the test statistic should be small, close to unity. When this ratio is large, it provides evidence that the differences between the groups are larger than one would expect and is used as a basis for supporting the alternative hypothesis.

We use the ratio of these two variance measures which follows the F-distribution with $k-1$ and $k(n-1)$ degrees of freedom to determine whether the means of the k groups are significantly different from each other. Thus, the hypotheses tested are:

$$H_0 : \mu_1 = \mu_2 = ... = \mu_k \text{ vs. } H_1 : \text{at least one } \mu_i \text{ is different}; i = 1,2,3...,k$$

The previous hypotheses can also be set in terms of the effects of the k groups or k levels of a factor:

$$H_0 : \alpha_1 = \alpha_2 = ... = \alpha_k = 0 \text{ vs. } H_1 : \text{at least one } \alpha_i \neq 0; i = 1,2,3...,k$$

If the variation between the group means is large, relative to the variation within the groups, then we would be in a position to detect significant differences among the sample means. If the variation between the sample means is small, relative to the variation within the samples, then there would be considerable overlap of observations in the different samples, and we would be unlikely to detect any differences among the population means.

The sums of squares required for the analysis, are presented in a form that makes them easy to use for manual calculation as follows:

$$SS_{\text{total}} = \sum_{i=1}^{k} \sum_{j=1}^{n} y_{ij}^2 - \frac{y_{..}^2}{kn}$$

$$SS_{\text{groups}} = \frac{1}{n} \sum_{i=1}^{k} y_{i.}^2 - \frac{y_{..}^2}{kn}$$

$$SS_{\text{Error}} = SS_{\text{total}} - SS_{\text{groups}}$$

Typical one-way ANOVA Table:

Source	Sum of Squares	Degree of Freedom	Mean Squares	F-Ratio
Between Groups	$SS_{\text{between groups}}$	$k-1$	$SS_{\text{between groups}}/(k-1)$	$(MS_{\text{between groups}})/MS_{\text{within groups}}$
Within Groups	$SS_{\text{within groups}}$	$k(n-1)$	$SS_{\text{within groups}}/k(n-1)$	
Total	SS_{total}	$kn-1$		

A **_large F_** provides evidence against the null hypothesis H_0 while a **_small F_** **_shows that the data_** supports the null hypothesis H_0. To calculate a *p-value* to test H_0, we compare the **_F_** we obtained from our data to the distribution it would have if H_0 were true, that is, an *F-distribution* with $(k-1)$ and $k(n-1)$ **_degrees of freedom_**. Note that F is always positive, so this is always a one-tailed test.

■ **EXAMPLE 15.2**   In order to study the effect of temperature on the yield of a chemical process, the yields of the process were studied at five settings of temperature as in the following table:

Temperature	1	2	3	4
40	30	36	37	40
50	41	43	48	36
60	43	47	47	61
70	43	45	61	64
80	41	46	52	62

Test the hypothesis that the mean yield is the same at the different temperature settings at 5% level of significance.

Solution

Temperature	1	2	3	4	Totals (y_i)
40	30	36	37	40	143
50	41	43	48	36	168
60	43	47	47	61	198
70	43	45	61	64	213
80	41	46	52	62	201

Overall total = 143 + 168 + 198 + 213 + 201 = 923 = $y_{..}$

The sums of squares required for ANOVA are obtained as follows:

$$SS_{total} = 30^2 + 36^2 + 37^2 + \ldots + 52^2 + 62^2 - \frac{923^2}{5(4)} = 1722.55$$

$$SS_{groups} = \frac{1}{4}(143^2 + 168^2 + 198^2 + 213^2 + 201^2) - \frac{923^2}{5(4)} = 815.3$$

$$SS_{Error} = 1722.55 - 815.3 = 907.25$$

Source	SS	df	MS	F	p-Value
Temperature	815.3	4	203.825	3.369937	0.037124
Error	907.25	15	60.48333		
Total	1722.55	19			

We can use the p-value to decide to reject or accept the null hypothesis. Testing at 5% level of significance, we reject the null hypothesis because p-value < 0.05. We can also read the table of F for the decision. From the tables $F(4,15,0.05) = 3.055$. The F-ratio $> F(4,15,0.05)$, so we reject the null hypothesis. There is significant temperature effect. The p-value in the earlier table is obtained very easily in a cell in EXCEL by typing **= FDIST(3.369937,4,15)**.

The earlier example reflects the type of one-way ANOVA called a CRD. The next example shows the use of one-way ANOVA to analyze the responses from independent samples. Four different methods of production of materials are being compared in the example:

EXAMPLE 15.3   Four methods for production of a fabric are being studied for the tensile strengths of the final products. Five measurements were made on material produced by each method, and the tensile strengths are recorded as follows:

Methods	1	2	3	4	5
1	24.95	26.44	28.57	23.17	26.34
2	30.41	29.23	28.87	32.95	35.87
3	26.49	28.5	33.13	32.36	28.95
4	30.66	29.72	30.14	30.44	28.73

Test at 5% level of significance for any difference in mean tensile strengths for the four different methods of production.

Solution   The hypotheses to be tested in this case are:

$$H_0 : \alpha_1 = \alpha_2 = \alpha_3 = \alpha_4 = 0 \quad \text{vs.} \quad H_1 : \text{at least one } \alpha_i \neq 0; \quad i = 1,2,3,4$$

Methods	Sample					Means
	1	2	3	4	5	
1	24.95	26.44	28.57	23.17	26.34	129.47
2	30.41	29.23	28.87	32.95	35.87	157.33
3	26.49	28.5	33.13	32.36	28.95	149.43
4	30.66	29.72	30.14	30.44	28.73	149.69

The sums of squares required for ANOVA are obtained as follows:

$$SS_{total} = 24.95^2 + 26.44^2 + 28.57^2 ... + 30.44^2 + 28.73^2 - \frac{585.92^2}{4(5)} = 168.9297$$

$$SS_{groups} = \frac{1}{5}(129.47^2 + 157.33^2 + 149.43^2 + 149.43^2 + 149.69^2) - \frac{585.92^2}{4(5)} = 85.21384$$

$$SS_{Error} = 168.9297 - 85.21384 = 83.71584$$

ANOVA

Source	SS	df	MS	F-Ratio	p-Value
Methods	85.21384	3	28.40461	5.428767	0.009075
Error	83.71584	16	5.23224		
Total	168.9297	19			

From the tables (Appendix V), $F(3,16, 0.05) = 3.23887$. Since *F-ratio* > $F(3,16,0.05)$ we reject the null hypothesis, H_0 and conclude that there is significant difference between the means of some of the four production methods at 5%. We can also say that there is a significant method effect. As usual we find the *p*-value by typing in a cell in EXCEL = *F*DIST(5.428767,3,16).

■ **EXAMPLE 15.4**   **Carry out a MINITAB analysis of data of Example 15.3**

MINITAB Solution   We first enter the data in four columns $C1$–$C4$, naming the columns as I–IV. Each column contains the samples taken for each level of the factor as follows:

I	II	III	IV
24.95	30.41	26.49	30.66
26.44	29.23	28.5	29.72
28.57	28.87	33.13	30.14
23.17	32.95	32.36	30.44
26.34	35.87	28.95	28.73

Then we enter this code:

```
MTB > OneWay;
SUBC>    Response C1 C2 C3 C4;
SUBC>    IType 0;
SUBC>    GMCI;
SUBC>    GIntPlot;
SUBC>    TMethod;
SUBC>    TFactor;
SUBC>    TANOVA;
SUBC>    TSummary;
SUBC>    TMeans;
SUBC>    Nodefault.
```

One-Way ANOVA: I, II, III, IV

```
Method
Null hypothesis              All means are equal
Alternative hypothesis       At least one mean is different
Significance level           α = 0.05
    Equal variances were assumed for the analysis.


Factor Information

Factor            Levels                    Values
Factor              4                     I, II, III, IV
```

```
Analysis of Variance

Source         DF        Adj SS      Adj MS      F-Value     P-Value
Factor          3         85.21      28.405       5.43        0.009
Error          16         83.72       5.232
Total          19        168.93

Model Summary

       S              R-sq            R-sq(adj)          R-sq(pred)
  2.28741           50.44%             41.15%              22.57%

Means

Factor         N          Mean      StDev              95% CI
I              5        25.894      1.998       (23.725, 28.063)
II             5        31.47       2.93         (29.30, 33.63)
III            5        29.89       2.78         (27.72, 32.05)
IV             5        29.938      0.762       (27.769, 32.107)

Pooled StDev = 2.28741
```

The ANOVA table compares to the one we obtained by manual analysis. As we can see, the *p*-value is 0.009, much less than 5%, the level of significance. We therefore reject H_0 that the means are equal, and accept the alternative that the means are not equal. The graph that follows, Figure 15.1 shows that mean of method I is different from the three means for II–IV.

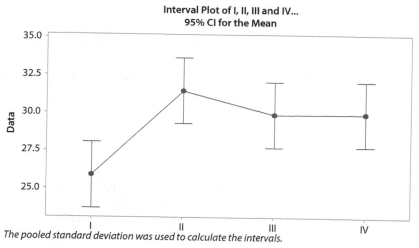

■ **FIGURE 15.1** Comparing the means for the four methods - 95% confidence interval plots

15.2 The Least Significant Difference Method

When ANOVA shows significance for group means in CRD, how do we determine which mean is larger or smaller? There are a number of ways for further analyzing the means to determine which ones are significantly different from each other. Some of the methods allow for us to group together the means which are not significantly different from each other. This type of analysis is often called a post-hoc analysis. Among those used extensively are the least significant difference (LSD) method, the Duncan's multiple range test, Tukey's honestly significant difference test, among a few others. Here, since this text is introductory, we introduce the simplest. For this text, we use the LSD method to achieve the purpose.

As we said, when we perform ANOVA in CRD, we use the design to compare k groups or k levels of a factor, we may wish to know whether any of the groups has mean that is greater than others. We may also wish to group the means to show the sets that belong together, and those that possibly stand out on their own.

Recall that in CRD, we described each observation by the model:

$$y_{ij} = \mu + \alpha_i + \varepsilon_{ij}$$

This model uses the assumption that $\varepsilon_{ij} \sim NID(0, \sigma^2)$. By this model, then the mean of the ith group is:

$$\mu_i = \mu + \alpha_i .$$

By same model, we can estimate the mean for the ith group by knowing that:

$$\hat{\mu} = \bar{y}_{..};$$
$$\hat{\alpha}_i = \bar{y}_{i.} - \bar{y}_{..}$$
$$\hat{\mu}_i = \hat{\mu} + \hat{\alpha}_i = \bar{y}_{i.}$$

From the foregoing we see that the sample mean for the ith $\bar{y}_{i.} \sim N\left(\mu + \alpha_i, \dfrac{\sigma^2}{n}\right)$.

If we consider two groups or two levels of a factor, the difference between their means, $\bar{y}_{i.} - \bar{y}_{j.}$ (where $i \neq j$, but $i,j = 1,2,...,k$), then:

$$\bar{y}_{i.} - \bar{y}_{j.} \sim N\left(0, \dfrac{2\sigma^2}{n}\right)$$

The above is true under the null hypothesis that states that $\mu_i = \mu_j$ is true.

More often than not, we do not know σ^2 so that we use t-distribution to make inference about difference between two means, and replace the σ^2 with the sample variance, s^2.

The LSD at α level of significance would then be defined in terms of the t-distribution which uses the mean square error as the estimate of s^2 and constructing the LSD following ANOVA as:

$$LSD = t\left[\frac{\alpha}{2}, k(n-1)\right] s \sqrt{\left(\frac{2}{n}\right)}$$

Thereafter, any two means within the design could be compared to see whether they are significantly different from each other. We show how this works practically by applying it to one of the comparisons we have carried out via CRD.

Consider Example 15.2, in which we rejected the null hypothesis that the means are equal, at $\alpha = 5\%$, here we use the same level of significance of 5% as α in order to calculate the LSD. We will use the LSD to explore the relationships between the means of the group. In that ANOVA the

$$MS_{Error} = 60.4833 \Rightarrow \sqrt{MS_{error}} = 7.7771 ;$$

which we use here as our best estimate of the sample standard deviation, s. From this we see that

$$LSD = t\left[\frac{\alpha}{2}, k(n-1)\right]s\sqrt{\left(\frac{2}{n}\right)} = t_{[0.025,15]}(7.7771)\sqrt{\frac{2}{5}}$$

$$= (2.131)(7.7771)\sqrt{\frac{2}{5}} = 10.482$$

The data from Example 15.2, is given in the following with the means of the groups:

Temperature	1	2	3	4	Totals ($y_{i.}$)	Means
40	30	36	37	40	143	35.75
50	41	43	48	36	168	42
60	43	47	47	61	198	49.5
70	43	45	61	64	213	53.25
80	41	46	52	62	201	50.25

Thus,

Means				
Temperature 1	Temperature 2	Temperature 3	Temperature 4	Temperature 5
35.75	42	49.5	53.25	50.25
$\bar{y}_{1.}$	$\bar{y}_{2.}$	$\bar{y}_{3.}$	$\bar{y}_{4.}$	$\bar{y}_{5.}$

The means at different levels of temperature can be arranged in order as follows:

$$\bar{y}_{1.} < \bar{y}_{2.} < \bar{y}_{3.} < \bar{y}_{5.} < \bar{y}_{4.}$$

The LSD = 10.482, therefore for any mean and one that is larger than it, they are significantly different if the difference between them is less than -10.482. The differences can be explored as follows:

$$\bar{y}_{1.} - \bar{y}_{2.} = 35.75 - 42 = -6.25 \ \text{(not significant)}$$
$$\bar{y}_{1.} - \bar{y}_{3.} = 35.75 - 49.5 = -13.75 \ \text{(significant)}$$
$$\bar{y}_{1.} - \bar{y}_{5.} = 35.75 - 50.25 = -14.50 \ \text{(significant)}$$
$$\bar{y}_{1.} - \bar{y}_{4.} = 35.75 - 53.25 = -17.50 \ \text{(significant)}$$

$$\bar{y}_{2.} - \bar{y}_{3.} = 42 - 49.5 = -7.5 \ \text{(not significant)}$$
$$\bar{y}_{2.} - \bar{y}_{5.} = 42 - 50.25 = -8.25 \ \text{(not significant)}$$
$$\bar{y}_{2.} - \bar{y}_{4.} = 42 - 53.25 = -11.25 \ \text{(significant)}$$

$$\bar{y}_{3.} - \bar{y}_{5.} = 49.5 - 50.25 = -0.75 \ \text{(not significant)}$$
$$\bar{y}_{3.} - \bar{y}_{4.} = 49.5 - 53.25 = -3.75 \ \text{(not significant)}$$

$$\bar{y}_{5.} - \bar{y}_{4.} = 50.25 - 53.25 = -3.0 \ \text{(not significant)}$$

By this analysis, we group the means for the levels of the factor as follows: $(\mu_1, \mu_2), (\mu_3, \mu_4, \mu_5)$.

15.3   The Unbalanced One-Way Layout

In treating one-way ANOVA, the discourse and examples have so far dealt with situations where the same number of measurements is made for each group or each level of the factor. With equal number of measurements as described, the design of the one-way layout is said to be balanced with a total of kn observations in the entire experiment. But in some designed experiments, the numbers of measurements for the different groups are different. Such a one-way layout is said to be unbalanced and the total number of observations in the entire study is $\sum_{i=1}^{k} n_i$.

The basic assumptions of the design for the unbalanced one-way layout are based on the general linear least-squares regression model:

$$y_{ij} = \mu + \alpha_i + \varepsilon_{ij}; \quad i = 1, 2, \ldots, k; \ j = 1, 2, \ldots, n_i \tag{15.9}$$

Thus, for group i ($i = 1, 2, \ldots, k$) in the design, the number of measurement made is designated as n_i. All other assumptions made for the balanced design apply.

These changes affect the formulas for the sums of squares which have to be calculated for ANOVA. The sums of squares total, groups, and error are obtained as follows for the unbalanced design:

$$SS_{\text{total}} = \sum_{i=1}^{k} \sum_{j=1}^{n_i} y_{ij}^2 - \frac{y_{..}^2}{\sum_{i=1}^{k} n_i}$$

$$SS_{\text{groups}} = \sum_{i=1}^{k} \frac{y_{i.}^2}{n_i} - \frac{y_{..}^2}{\sum_{i=1}^{k} n_i}$$

$$SS_{\text{Error}} = SS_{\text{total}} - SS_{\text{groups}}$$

In testing the hypothesis stated in (15.9), the following ANOVA table applies:

Source	Sum of Squares	Degree of Freedom	Mean Squares	F-Ratio
Between Groups	$SS_{\text{between groups}}$	$k-1$	$SS_{\text{between groups}}/(k-1)$	$(MS_{\text{between groups}})/MS_{\text{within groups}}$
Within Groups	$SS_{\text{within groups}}$	$\sum_{i=1}^{k} n_i - k$	$SS_{\text{within groups}}/\left(\sum_{i=1}^{k} n_i - k\right)$	
Total	SS_{total}	$\sum_{i=1}^{k} n_i - 1$		

To make decision based on a level of significance of α, the statistic, F-ratio is compared with $F\left(K-1, \sum_{i=1}^{k} n_i - k, \alpha\right)$ which is read from the table of F (Appendix V).

If F-ratio is greater than $F\left(k-1, \sum_{i=1}^{k} n_i - k, \alpha\right)$, the null hypothesis is rejected and conclusion is reached in favor of the alternative postulate.

Least Significance Difference for Unbalanced CRD

For post-hoc analysis when design is unbalanced, the LSD we used previously needs to be adjusted. The formula is changed as follows:

$$LSD = t\left[\frac{\alpha}{2}, \sum_{i=1}^{k} n_i - k\right] s \sqrt{\left(\frac{1}{n_i} + \frac{1}{n_j}\right)}$$

Again, the MS_{error} would be used as an estimate of s^2 and the rest of the LSD process proceeds as previously described for balanced design, except that for each pair of groups or levels of factor(i,j) the LSD will be different.

■ **EXAMPLE 15.5** Four different processes for making paper from sugar cane material were being studied to compare yield. The number of determinations of the yields was different for each process. The yields in percentage of input material are presented in the following.

Method										
I	75	46	74	75	76	48	68			
II	76	84	73	75	69	87	70	69	73	
III	80	90	96	88	76	64	71	90	78	79
IV	77	78	78	76	78	59	77	65		

Test whether the yields of the processes are significantly different at 5% level of significant.

Solutions

Process											Sample Size	Totals
I	75	46	74	75	76	48	68				7	462
II	76	84	73	75	69	87	70	69	73		9	676
III	80	90	96	88	76	64	71	90	78	79	10	812
IV	77	78	78	76	78	59	77	65			8	588
Total												2538

The sums of squares required for ANOVA are obtained as shown in the following using the modified formulas stated earlier:

$$SS_{total} = 75^2 + 76^2 + 74^2 ... + 77^2 + 65^2 - \frac{2538^2}{(7+9+10+8)} = 3587.76$$

$$SS_{process} = \frac{462^2}{7} + \frac{676^2}{9} + \frac{822^2}{10} + \frac{588^2}{8} - \frac{2538^2}{(7+9+10+8)} = 965.28$$

$$SS_{Error} = 3587.76 - 965.28 = 2622.49$$

ANOVA

Source	SS	df	MS	F	p-Value
Method	965.28	3	321.7586	3.860762	0.02278
Error	2622.49	30	87.4163		
Total	3587.76	33			

F-ratio is to be compared to $F(3, 30, 0.05)$. From the tables, $F(3,30, 0.05) = 2.92$. Since F-ratio $(3.860762) > F(3,30, 0.05)$, we reject H_0 and conclude that there is significant difference between the processes at 1% level of significance. Here p-value is obtained in a cell in EXCEL by typing **= FDIST(3.860762,3,30).**

EXAMPLE 15.6 Use LSD to determine which of the means in Example 15.5 are significantly different from each other.

Solution The data from Example 15.5, is given as follows:**Error! Not a valid link.**
$MS_{Error} = 87.4163 = s^2$; $s = \sqrt{87.4163} = 9.3497$; $t_{[0.025,30]} = 2.3596$

Now, $LSD = t\left[\dfrac{\alpha}{2}, \sum\limits_{i=1}^{k} n_i - k\right] s\sqrt{\left(\dfrac{1}{n_i} + \dfrac{1}{n_j}\right)}$

From this we see that for the combinations of i,j:

i,j	LSD
7,9	11.11794
7,10	10.87184
7,8	11.41772
9,10	10.13639
9,8	10.71978
10,8	10.46451

Thus,

Process 1	Process 2	Process 3	Process 4
16.25	19.2857	19	17
$\bar{y}_{1.}$	$\bar{y}_{2.}$	$\bar{y}_{3.}$	$\bar{y}_{4.}$

The means at different levels of temperature can be arranged in order as follows:

$\bar{y}_{1.} < \bar{y}_{4.} < \bar{y}_{2.} < \bar{y}_{3.}$

We use the LSD in the previous table to carry out the comparison as follows:

$\bar{y}_{1.} - \bar{y}_{4.} = 66 - 73.5 = -7.5 > -11.4177$ (not significant)
$\bar{y}_{1.} - \bar{y}_{2.} = 66 - 75.11 = -9.11 > -11.117$ (not significant)
$\bar{y}_{1.} - \bar{y}_{3.} = 66 - 81.20 = -15.20 < -10.8718$ (significant)

$\bar{y}_{4.} - \bar{y}_{2.} = 73.5 - 75.11 = -1.611 > -10.7198$ (not significant)
$\bar{y}_{4.} - \bar{y}_{3.} = 73.5 - 81.20 = -7.7 > -10.4645$ (not significant)

$\bar{y}_{2.} - \bar{y}_{3.} = 75.11 - 81.20 = -6.09 > -10.1364$ (not significant)

By this analysis, we group the means for the levels of the factors as follows:
$(\mu_1, \mu_4, \mu_2), (\mu_3)$.

■ **EXAMPLE 15.7**   MINITAB can also be used to analyze unbalanced one-way ANOVA. Let us use MINITAB to analyze the data of Example 15.5:

```
MTB > set c1
DATA> 75 46 74 75 76 48 68
DATA> end
MTB > set c2
DATA> 76 84 73 75 69 87 70 69 73
DATA> end
MTB > set c3
DATA> 80 90 96 88 76 64 71 90 78 79
DATA> end
MTB > set c4
DATA> 77 78 78 76 78 59 77 65
DATA> end
MTB > PRINT C1-c4
```

Data Display

Row	C1	C2	C3	C4
1	75	76	80	77
2	46	84	90	78
3	74	73	96	78
4	75	75	88	76
5	76	69	76	78
6	48	87	64	59
7	68	70	71	77
8		69	90	65
9		73	78	
10			79	

```
MTB > OneWay;
SUBC> Response C1 C2 C3 C4;
SUBC> IType 0;
SUBC> GMCI;
SUBC> GIntPlot;
SUBC> TMethod;
SUBC> TFactor;
SUBC> TANOVA;
SUBC> TSummary;
SUBC> TMeans;
SUBC> Nodefault.
```

One-Way ANOVA: I, II, III, IV
Method

Null hypothesis	All means are equal
Alternative hypothesis	At least one mean is different
Significance level	$\alpha = 0.05$

```
Equal variances were assumed for the analysis.

Factor Information
```

Factor	Levels	Values
Factor	4	C1, C2, C3, C4

```
Analysis of Variance
```

Source	DF	Adj SS	Adj MS	F-Value	P-Value
Facto	3	965.3	321.76	3.68	0.023
Error	30	2622.5	87.42		
Total	33	3587.8			

```
Model Summary
```

S	R-sq	R-sq(adj)	R-sq(pred)
9.34967	26.90%	19.60%	5.01%

```
Means
```

Factor	N	Mean	StDev	95% CI
C1	7	66.00	13.25	(58.78, 73.22)
C2	9	75.11	6.43	(68.75, 81.48)
C3	10	81.20	9.80	(75.16, 87.24)
C4	8	73.50	7.31	(66.75, 80.25)

```
Pooled StDev = 9.34967
```

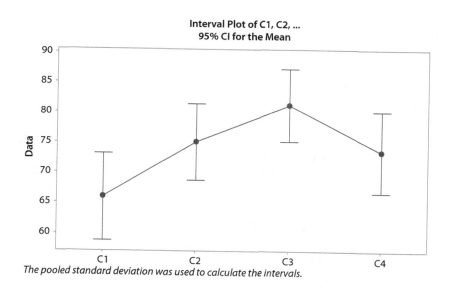

The pooled standard deviation was used to calculate the intervals.

■ **EXAMPLE 15.8** Four different methods for producing an item are being piloted. For each of the methods, four determinations were made for the number of products produced out of 1 kg of raw material.

Method I	Method I_1	Method I_2	Method I_3
34	42	23	37
38	35	35	45
26	42	32	41
33	35	15	37
32	46	33	41

Are the mean yields equal; test at 5% level of significance.

Solution   We read the data into MINITAB columns 1–4 using a READ statement.

```
MTB > name c1 'Method I' c2 'Method I_1' c3 'Method I_2' c4 'Method I_3'
MTB > read c1-c4
DATA> 34   42 23 37
DATA> 38   35 35 45
DATA> 26   42 32 41
DATA> 33   35 15 37
DATA> 32   46 33 41
DATA> end
5 rows read.
MTB > print c1-c4
```

Data Display

Row	Method I	Method I_1	Method I_2	Method I_3
1	34	42	23	37
2	38	35	35	45
3	26	42	32	41
4	33	35	15	37
5	32	46	33	41

```
MTB > OneWay;
SUBC> Response 'Method I' 'Method I_1' 'Method I_2' 'Method I_3';
SUBC> IType 0;
SUBC> GMCI;
SUBC> GIntPlot;
SUBC> TMethod;
SUBC> TFactor;
SUBC> TANOVA;
SUBC> TSummary;
SUBC> TMeans;
SUBC> Nodefault.
```

One-Way ANOVA: Method I, Method I_1, Method I_2, Method I_3

```
   Method

Null hypothesis               All means are equal
Alternative hypothesis        At least one mean is different
Significance level            α = 0.05


Equal variances were assumed for the analysis.

Factor Information

Factor  Levels  Values
Factor       4  Method I, Method I_1, Method I_2, Method I_3


Analysis of Variance


Source      DF    Adj SS    Adj MS    F-Value    P-Value
Factor       3     562.6    187.53       6.03      0.006
Error       16     497.2     31.08
Total       19    1059.8


Model Summary

S                   R-sq       R-sq(adj)     R-sq(pred)
5.57450           53.09%         44.29%         26.70%


Means

Factor           N        Mean       StDev        95% CI
Method I         5       32.60        4.34    (27.32, 37.88)
Method I_1       5       40.00        4.85    (34.72, 45.28)
Method I_2       5       27.60        8.41    (22.32, 32.88)
Method I_3       5       40.20        3.35    (34.92, 45.48)
                 Pooled StDev = 5.57450
```

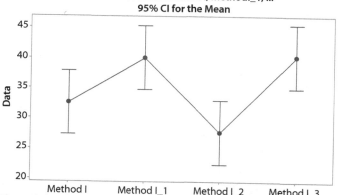

Interval Plot of Method I, MethodI_1, …
95% CI for the Mean

The pooled standard deviation was used to calculate the intervals.

The p-value of 0.006 for the ANOVA is the probability of obtaining the value of F of 6.06 or something more extreme. The more extreme the F-value, the less it lends support to the null hypothesis. This probability (p-value) is by far less than 0.05, the level of significance. We therefore reject the null hypothesis of the "equality of the means of the four methods" and accept the alternative that "at least one mean is different" at 5% level of significance.

Exercises

15.1 It is suspected that the means of the number of defectives produced by seven shifts of workers in a factory are not the same. Each of the shifts was observed randomly for four shift periods and the number of defectives produced are as follows:

Shifts	Samples of Number of Defectives for Four Periods			
	1	2	3	4
1	23	11	27	18
2	8	10	5	18
3	23	13	16	21
4	8	9	4	4
5	16	12	3	14
6	20	8	12	8
7	19	12	15	9

Test at 5% level of significance the claim that the mean number of defectives produced by shifts is not the same.

15.2 In a research project, the effect of temperature settings on the time it took to complete a process was being studied.

Temperature (°C)	1	2	3	4
20	11	10	12	8
30	11	12	13	10
40	14	15	15	11
50	21	16	20	20

Carry out ANOVA and test for differences in mean time between levels of temperature at 5% level of significance.

15.3 The following data was obtained from an experiment to study the yield of a process under five settings of the factor. Test the hypothesis to compare the means for the different levels of the factor at 1% level of significance.

Factor	1	2	3	4	5	6	7	8	9	10
1	10	12	10	11	12	9	11	13	10	9
2	14	14	14	12	16	10	14	15	10	15
3	17	13	15	12	16	10	14	15	10	15
4	19	15	19	14	23	12	17	17	15	19
5	20	19	23	16	20	18	21	22	18	20

15.4 Four types of forestry management techniques, which include different degrees of burning were being studied with respect to vegetation regrowth. The mean leaf length of regrown vegetation was used to compare the different management methods. Here are the samples of mean leaf lengths obtained from cropped leaves of new plants of the same age which grew after the forests were burned by the different methods. The lengths of the leaves are given in inches.

Management Method	Sample				
	1	2	3	4	5
I	2.244	2.559	2.480	2.441	2.323
II	2.559	2.283	2.716	2.244	2.244
III	1.929	2.047	2.244	2.008	2.086
IV	1.732	2.008	1.811	1.929	1.968

Test that the means are equal for the four management methods at 1% level of significance.

15.5 Pomegranates sourced from five countries were being studied for their yield of a certain extract with antioxidant properties. Five determinations of the yields for this extract are obtained from Pomegranate from each of the five countries. The yields in micrograms per pound of raw material were used.

Pomegranate Producing Countries	Sample				
	1	2	3	4	5
I	6.2	6.0	5.7	5.9	6.3
II	6.7	6.9	7.2	6.3	7.3
III	4.6	5.2	4.5	5.6	4.9
IV	5.8	5.0	5.6	5.8	5.0
V	6.0	5.6	5.3	6.6	6.0

Carry out ANOVA at 5% level of significance to test whether the means for the countries are equal.

15.6 Four green bean types are being studied to compare the mean time-to-first maturation (in days). Samples of sizes 6, 8, 5, and 6 were taken from bean types 1–4, respectively, and the times-to-first maturation for each bean type was observed under homogeneous conditions.

Type 1	Type 2	Type 3	Type 4
45	56	47	67
56	61	52	61
44	54	47	62
51	45	49	64
50	60	52	65
47	48		67
	53		
	55		

Analyze the data and test for differences in means at 1% level of significance.

15.7 Six varieties of melon were compared as sources for a compound used in making antiaging cream. The quantities of the extract obtained from equal weight of the melons in micrograms are presented as follows.

Varieties of Melon	Sample						
	1	2	3	4	5	6	7
Hales Best Jumbo	6.36	6.75	4.66	6.2	7.7	7.45	5.5
Hearts of Gold	3.1	4.25	5.25	3.35	5.3		
Ambrosia	4.65	5.12	4.35	5.26	4.49		
Athena	6.8	5.8	5.9	6.8	6.0	6.23	
Honey Bun Hybrid	5.0	5.26	5.33	5.56	5.0		
Sweet N Early Hybrid	5.71	5.8	5.75	6.0	5.77	5.92	

Compare the mean yields for the different varieties of melon, testing at 2.5% level of significance.

15.8 Four bottling processes are being tested, and are expected to fill each bottle with a beverage to the tune of 20 fluid ounces. The following samples of the amount of liquid with which each bottle was filled were recorded for the four processes. Compare the mean amount of liquid filled for the four processes. Analyze using one-way ANOVA. Use level of significance of 1%.

Filling Process	Sample						
	1	2	3	4	5	6	7
I	16	16	16	17			
II	20	19	19	19	20	20	18
III	20	18	18	20	19		
IV	18	18	16	17	16	17	

15.9 Five medications are being compared for use in palliative care for patients who have undergone a serious surgical procedure. 25 patients were randomly assigned in groups of 5–5 different palliatives. The following data were obtained from the patients who were treated, for time (in minutes) till pain subsided following the administration of the palliative. Test at 1% level of significance to compare mean times to relief.

Palliative	1	2	3	4	5
I	7.4	8.3	9.4	7	9.6
II	10.8	12.1	10	10.9	9.3
III	8.9	8.8	7.8	6.5	7.2
IV	6.8	6.8	6	7	6.5
V	9.1	9.5	7.8	8.5	9.4

15.10 Use LSD to determine the grouping of the means in Exercise 15.7.

15.11 Use LSD to determine the grouping of the means in Exercise 15.8.

CHAPTER 16

The Randomized Complete Block Design

16.0 The Randomized Complete Block Design

Often, in research, especially those involving designed experiments, one encounters a factor or a group of factors which may impinge on the responses of an experiment, but they are not primarily of interest to the researcher. Factors which act in this way are described as nuisance factors. Being ignorant of their existence or deliberately ignoring them, and carrying out the experiment without identifying and designing them into the study may affect the conclusions reached. If such a factor has a large effect, it could contribute to large error, and therefore, impede the ability of the designed experiment to detect any significant differences that may exist in the levels of the factors of primary interest in the research. In such experiments, the statistician would move to eliminate this possibility by actively designing such a factor or group of factors as *blocks* in the experiment. This way they make sure that if the effects of the nuisance factors are large, they would not inflate the size of the error sums of squares and impede the detection of significance of the effects of other factors in the design that are of primary interest.

Usually a **block** in an actual experiment would comprise a set of **experimental units** which are expected to be more homogeneous than the whole aggregate. An experimental unit on the other hand, is defined as the smallest subdivision of the experimental material so that any two of such units may receive different treatments in the experiment. The idea of blocking in an experiment is to include experimental units which are homogeneous in the same level of the blocking factor, while other sets of experimental units which are dissimilar with respect to homogeneity would be assigned to other levels of the blocking factor.

When **blocking** is applied to a basic comparative experiment, the process introduces **a randomization restriction** on the designed experiment. For instance, in the completely randomized design (CRD), that is the one-way layout, there is only one factor set at say k levels. There is no restriction imposed on randomization, which in CRD, is carried out across the entire experiment. In blocked experiments, all units are assigned to different levels of the block. Consequence is that **randomization is restricted, since it can only be carried out within the confines of each**

level of the block. The actual randomization is usually carried out at the design stages of the experiment in which the treatments are deliberately assigned haphazardly within the block.

There is a variety of designed experiments which are blocked—varying in the levels of the complexity of the blocking. The simplest experiment with blocking is the randomized complete block design (RCBD). There is only one blocking factor in this design. If the block were ignored, the data could easily be analyzed as coming from the CRD. The experimental unit in this design is all assigned into blocks, but they also receive different levels of treatment.

The RCBD is a design for comparing k treatments in b blocks. In the design of the experiment, the experimental units in each block must be homogeneous to provide comparative treatment for all the k treatments. Treatments are randomly distributed within each block, each treatment appearing only once in a block. Blocking as applied in this design is used **to control for the extraneous factor**. Examples of RCBD is found in industry if shifts of workers have varying skills in a factory, the shifts may become the blocks in studying production; in agriculture, if parcels of land are different (for instance loamy versus clayey, or sandy), parcels of land may become the blocks in studying yield. Batches of raw material may come from different sources and we may need to control for the differences. The batches become blocks in the design, and more.

16.1   Advantages and Disadvantages of RCBD

One advantage of RCBD is that it is a simple design that is easy to construct, and useful for comparing k treatments in the presence of an extraneous source of variability. Another is that it allows for any number of treatments, k which are being studied to be tested within each level of block.

One of the disadvantages of RCBD is that it requires that each level of the block be homogeneous: thus relatively small number of treatments can be studied with the design. Another is that RCBD controls only for one nuisance factor which is a source for extraneous variation: If more than one such factor exists, the ones not designed into the study would usually boost the error sum of squares and mask the ability of the design to detect significant treatment effect.

Since the block factor is a nuisance factor in which we are not primarily interested, should we test it in analysis of variance (ANOVA)? There is no harm done in testing for significance of block effect. If it is significant, then blocking is effective, else it is not. When block effect is not significant, it means that the mean square for block is close to mean square error in ANOVA. Under that scenario, blocking was ineffective and the attendant cost would be considered to have been wasted. RCBD would have been a relatively inefficient design, ergo, a CRD might as well have done the job better.

16.1.2 Model for RCBD and Analysis of Its Responses

Thus, responses from this experiment are classed in two ways: by the block they came from, and the level of treatment they received. The RCBD is an example of a two-way layout based on levels of the treatment and the block. In the linear least squares regression approach, the analysis of data from RCBD will eventually lead to ANOVA. The model suggested by the design would be of the form:

$$y_{ij} = \mu + \alpha_i + \beta_j + \varepsilon_{ij}; \ i = 1, 2, ..., k; \ j = 1, 2, ..., b \qquad (16.1)$$

y_{ij}, the responses of the experiment are made up of the grand mean, and the effects of treatment and block, plus a random error. The treatment (the factor of primary interest which is being studied) in this design has k levels, while the block (the nuisance factor, not of primary interest) has b levels. α_i and β_j are the respective effects of level i of treatment, and level j of block. They contribute these effects which either increase or detract from the μ which is the grand mean in the design. An effect can be positive or negative. The random error is normal, independently distributed with zero mean and constant variance, σ^2 which means that $\varepsilon_{ij} \sim NID(0, \sigma^2)$. For fixed effects model, the following assumptions are made, $\sum_{i=1}^{k} \alpha_i = 0$, and $\sum_{j=1}^{b} \beta_j = 0$. It is also possible to choose a random effects model for RCBD. If random effects are assumed for either the treatment or block, there is an understanding that the level of these factors are chosen randomly from many such levels, making the conclusions of the analysis transferrable to those levels of the factors not necessarily chosen for the experiment.

After the experiment, for easy understanding, the responses of the experiment are arranged in order in a table as follows:

Treatment	Block 1	2	3	...	b	Treatment Totals
1	y_{11}	y_{12}	y_{13}	...	y_{1b}	$Y_{1.}$
2	y_{21}	y_{22}	y_{23}	...	y_{2b}	$Y_{2.}$
3	y_{31}	y_{32}	y_{33}	...	y_{3b}	$Y_{3.}$
.	.	.	.	...	.	.
.	.	.	.	...	.	.
.	.	.	.	...	.	.
k	Y_{k1}	Y_{k2}	Y_{k3}	...	y_{kb}	$Y_{k.}$
Block Totals	$y_{.1}$	$y_{.2}$	$y_{.3}$	...	$y_{.b}$	$Y_{..}$

To be able to test hypotheses, we need to obtain the total sum of squares and the inherent partitions. The total sum of squares is given as:

$$SS_{\text{total}} = \sum_{i=1}^{k} \sum_{j=1}^{b} (y_{ij} - y_{..})^2$$

By adding and subtracting some quantities within the parentheses, this can be rewritten as:

$$SS_{total} = \sum_{i=1}^{k}\sum_{j=1}^{b}\left[y_{ij} - \bar{y}_{i.} + \bar{y}_{i.} - \bar{y}_{.j} + \bar{y}_{.j} - \bar{y}_{..} + \bar{y}_{..} - y_{..}\right]^2$$

$$= \sum_{i=1}^{k}\sum_{j=1}^{b}\left[\left(y_{ij} - \bar{y}_{i.} - \bar{y}_{.j} + \bar{y}_{..}\right) + \left(\bar{y}_{i.} - \bar{y}_{..}\right) + \left(\bar{y}_{.j} - y_{..}\right)\right]^2$$

By squaring, we obtain:

$$SS_{total} = \sum_{i=1}^{k}\sum_{j=1}^{b}(y_{ij} - \bar{y}_{i.} - \bar{y}_{.j} + \bar{y}_{..})^2 + \sum_{i=1}^{k}\sum_{j=1}^{b}(\bar{y}_{i.} - \bar{y}_{..})^2 + \sum_{i=1}^{k}\sum_{j=1}^{b}(\bar{y}_{.j} - y_{..})^2$$
$$+ \text{cross-product terms.}$$

The cross-product terms sum to zero. The result is the partition of the total sums of squares as follows:

$$SS_{total} = SS_{Error} + SS_{treatments} + SS_{block}$$

The formulas for these sums of squares which are easier to use for manual calculations are as follows:

$$SS_{total} = \sum_{i=1}^{k}\sum_{j=1}^{b}y_{ij}^2 - \frac{y_{..}^2}{kb}$$

$$SS_{block} = \frac{1}{k}\sum_{j=1}^{b}y_{.j}^2 - \frac{y_{..}^2}{kb}$$

$$SS_{treatment} = \frac{1}{b}\sum_{i=1}^{k}y_{i.}^2 - \frac{y_{..}^2}{kb}$$

$$SS_{Error} = SS_{total} - SS_{block} - SS_{treatment}$$

■ **EXAMPLE 16.1** Four different methods of extraction of an important compound were being studied. They were applied to raw materials obtained from four different sources. The amount of compound in micrograms extracted from a gram of raw material was recorded.

Method	Raw Material			
	A	B	C	D
1	5	7	10	8
2	8	10	12	8
3	10	14	12	14
4	10	14	14	9

Test that both treatment (methods) and raw material are significant at 5% level of significance.

Solutions

	Raw Material				
Method	A	B	C	D	Total
1	5	7	10	8	30
2	8	10	12	8	38
3	10	14	12	14	50
4	10	14	14	9	47
Total	33	45	48	39	165

$$SS_{total} = \sum_{i=1}^{k}\sum_{j=1}^{b} y_{ij}^2 - \frac{y_{..}^2}{kb} = 5^2 + 7^2 + 10^2 + \ldots + 14^2 + 14^2 + 9^2 - \frac{165^2}{4(4)} = 117.4375$$

$$SS_{block} = \frac{1}{k}\sum_{j=1}^{b} y_{.j}^2 - \frac{y_{..}^2}{kb} = \frac{33^2 + 45^2 + 48^2 + 39^2}{4} - \frac{165^2}{4(4)} = 33.1875$$

$$SS_{treatment} = \frac{1}{b}\sum_{i=1}^{k} y_{i.}^2 - \frac{y_{..}^2}{kb} = \frac{30^2 + 38^2 + 50^2 + 47^2}{4} - \frac{165^2}{4(4)} = 61.6875$$

$$SS_{Error} = SS_{total} - SS_{block} - SS_{treatment} = 117.4735 - 33.1875 - 61.6875 = 22.5625$$

Source	SS	df	MS	F-Ratio	p-Value
Method (treatment)	61.6875	3	20.5625	8.202216	0.006083
Raw material (block)	33.1875	3	11.0625	4.412742	0.036071
Error	22.5625	9	2.506944		
Total	117.4375				

From the tables, $F(3,9,0.01) = 6.9919$ and $F(3,9,.05) = 3.8625$, so the method(treatment) effect is significant at 1%; the raw material (block) effect is not significant at 1% level of significance, but it is significant at 5% level of significance. Note that the F-ratios are much less than the tabulated values for each significance level for which the factor or the block was declared significant. Every time we say that a test is significant, it means that the null hypothesis was rejected in favor of the alternative. The p-values for treatment and raw material were obtained by typing in different cells in EXCEL = **Fdist(8.202216,3,9)** and = **Fdist(4.412742,3,9)**, respectively.

■ **EXAMPLE 16.2** The yield of cassava for four different planting spacing were being studied using plots of one hectare in five different locations. The spacing being studied (a) 1×1 m, (b) 0.9×0.9 m, (c) 0.8×8 m, (d) 0.7×0.7 m were compared in the five locations. The responses of the experiment were the yield in metric tonnes per hectare. Same variety of cassava, cut and sown in similar manner was used in all the study. The result of the experiment is as follows:

	Locations				
Treatments	1	2	3	4	5
1	12.7	10.8	13.5	13.8	14.3
2	10.7	10	11	12.3	12.6
3	10.1	8.6	8.9	9.3	11.3
4	8.9	8	8.2	7.7	9.7

Analyze the data and determine whether there is significant difference in the mean yields for the different treatments and for different locations, testing at 1% level of significance.

Solutions

Treatments	Locations					Totals
	1	2	3	4	5	
1	12.7	10.8	13.5	13.8	14.3	65.1
2	10.7	10	11	12.3	12.6	56.6
3	10.1	8.6	8.9	9.3	11.3	48.2
4	8.9	8	8.2	7.7	9.7	42.5
Totals	42.4	37.4	41.6	43.1	47.9	212.4

The hypotheses tested for treatment is:

$$H_0 : \alpha_1 = \alpha_2 = \alpha_3 = \alpha_4 = 0 \text{ vs. } H_1 : \text{at least one } \alpha_i \neq 0$$

If the block factor is just a nuisance factor, it is only tested to show that it is significant, and that blocking was effective in removing a blocking effect that could have probably masked the ability to detect significant difference in treatment means.

For blocking, the hypothesis tested is:

$$H_0 : \beta_1 = \beta_2 = \beta_3 = \beta_4 = \beta_5 = 0 \text{ vs. } H_1 : \text{at least one } \beta_j \text{ is different}$$

The required sums of squares for ANOVA are obtained as follows:

$$SS_{total} = \sum_{i=1}^{k} \sum_{j=1}^{b} y_{ij}^2 - \frac{y_{..}^2}{kb} = 12.7^2 + 10.8^2 + 13.5^2 + \ldots + 7.7^2 + 9.7^2 - \frac{212.4^2}{4(5)} = 78.152$$

$$SS_{block} = \frac{1}{k} \sum_{j=1}^{b} y_{.j}^2 - \frac{y_{..}^2}{kb} = \frac{42.4^2 + 37.4^2 + 41.6^2 + 43.1^2 + 47.9^2}{4} - \frac{212.4^2}{4(5)} = 14.087$$

$$SS_{treatment} = \frac{1}{b} \sum_{i=1}^{k} y_{i.}^2 - \frac{y_{..}^2}{kb} = \frac{65.1^2 + 56.6^2 + 48.2^2 + 42.5^2}{5} - \frac{212.4^2}{4(5)} = 58.524$$

$$SS_{Error} = SS_{total} - SS_{block} - SS_{treatment} = 78.152 - 14.087 - 58.524 = 5.541$$

ANOVA

Source	SS	df	MS	F	p-Value
Treatment (spacing)	58.524	3	19.508	42.24797	0.00000118
Block (location)	14.087	4	3.52175	7.626963	0.002685
Error	5.541	12	0.46175		
Total	78.152				

From the tables $F(3,12,0.05) = 5.9525$ and $F(4,12,0.05) = 5.41195$, implying that both the treatment and block effects are significant at 5% level of significance. The p-values for treatment and location were obtained by typing in EXCEL cells = **Fdist(42.24797,3,12)** and = **Fdist(7.626963,4,9)**, respectively.

16.2 Missing Data in RCBD—An Approximate Analysis

Unforeseen circumstances may make some observations in RCBD to be missing. If only one of the observations in a RCBD is missing, we discuss here how it can be estimated. Even if more than one observation is missing, it is possible to modify the process we discuss here to have all of them estimated for ANOVA. The model of the RCBD is usually of the form:

$$y_{ij} = \mu + \gamma_i + \beta_j + \varepsilon_{ij}; \ i = 1, 2, ..., k; \ j = 1, 2, ..., b$$

There are b blocks and k treatments.

When a value is missing, it is desirable, if the principle of least squares is to be satisfied, to choose an estimate for the missing value in such a way as to minimize the sum of squares for error. To achieve the above, we minimize the sum of squares error in order to estimate the missing value, designated as x. If the observation in the ith treatment and jth block is missing, then by Least squares analysis, we find that we can estimate the missing value x, using the formula:

$$\hat{x} = \frac{ky_{i.}' + by_{.j}' - y_{..}'}{(k-1)(b-1)}$$

Where k = the number of treatments, and b = number of blocks.

$y_{i.}'$ = sum of all observations in ith row when the observation is missing

$y_{i.}'$ = sum of all observations in jth column when the observation is missing

$y_{..}'$ = sum of all observations in *the entire* study while the observation is missing

Now we utilize this result in a designed experiment with a missing value.

■ EXAMPLE 16.3   To compare five varieties of sorghum (A, B, C, D, and E), a randomized block design with four plots was used. The yield for one unit was misplaced and could not be found. Estimate the yield for the unit and complete the analysis.

Sorghum Varieties	Plots (Block)			
	1	2	3	4
A	5	10	7	8
B	7	x	8	8
C	9	16	11	12
D	10	14	11	9
E	6	12	7	8

Analysis

Sorghum Varieties	Plots (Block) 1	2	3	4	Total
A	5	10	7	8	30
B	7	x	8	8	23
C	9	16	11	12	48
D	10	14	11	9	44
E	6	12	7	8	33
Total	37	52	44	45	178

From the previous results,

$$\hat{x} = \frac{k y'_{i.} + b y'_{.j} - y'_{..}}{(k-1)(b-1)}$$

In the earlier example, the observation belonging to the treatment B and block 2 is missing. We estimate it as:

$$\hat{x} = \frac{5 y'_{2.} + 4 y'_{.2} - y'_{..}}{(5-1)(4-1)} = \frac{5(23) + 4(52) - 178}{(5-1)(4-1)} = 12.0833$$

Sorghum Varieties	Plots (Block) 1	2	3	4	Total
A	5	10	7	8	30
B	7	12.0833	8	8	35.0833
C	9	16	11	12	48
D	10	14	11	9	44
E	6	12	7	8	33
Total	37	66.0833	44	45	190.0833

In ANOVA we reduce degree of freedom of error and total by 1 because of the estimation of the missing value.

$$SS_{\text{total}} = 5^2 + 10^2 + 7^2 + \dots + 12^2 + 8^2 - \frac{190.0833^2}{5(4)} = 147.4233$$

$$SS_{\text{block}} = \frac{37^2 + 64.0833^2 + 44^2 + 45^2}{5} - \frac{190.0833^2}{5(4)} = 80.751$$

$$SS_{\text{sorghum}} = \frac{30^2 + 35.0833^2 + 48^2 + 44^2 + 33^2}{4} - \frac{190.0833^2}{5(4)} = 58.37639$$

$$SS_{\text{Error}} = 147.4233 - 58.37639 - 80.7512 = 8.295833$$

ANOVA

Source	SS	df	MS	F	p-Value
Sorghum	58.37639	4	14.5941	19.35129	0.000061183
Block	80.75104	3	26.91701	35.69107	0.000022942
Error	8.295833	11	0.754167		
Total	147.4233	18			

From the tables, we find that $F(0.01, 3,12) = 5.41$ and $F(0.01, 4,12) = 5.95$. Both the block and treatment (sorghum) are significant. The p-values for sorghum and block were obtained by typing in EXCEL cells = **Fdist(19.35129,4,11)** and = **Fdist(35.69107,3,11)**, respectively.

16.3 Using MINITAB to Carry Out ANOVA With Missing Data

Although we can estimate the missing observation in an RCBD as described in the previous section, we can always use software to carry out ANOVA. This method is recommended because it is based on the generalized linear model (GLM) which has been implemented in most major statistical software. The ANOVA using GLM is available on MINITAB. We will use MINITAB to analyze the data to show how it is done. The manual analysis in the previous section is at best a crude analysis in comparison to the one achieved by using GLM in MINITAB.

The original data with the missing value is used in MINITAB without estimating the missing value. MINITAB carries out all the analysis required. The data are read into MINITAB, assigning sorghum, block, and response to columns C1, C2, and C3, respectively, as shown in the following:

Sorghum	Block	Response
1	1	5
1	2	10
1	3	7
1	4	8
2	1	7
2	2	*
2	3	8
2	4	8
3	1	9
3	2	16
3	3	11
3	4	12

Continued.

Sorghum	Block	Response
4	1	10
4	2	14
4	3	11
4	4	9
5	1	6
5	1	12
5	3	7
5	4	8

Then the following code can be entered:

```
MTB >  GLM;
SUBC>  Response C3;
SUBC>  Nodefault;
SUBC>  Categorical C1 C2;
SUBC>  Terms C1 C2;
SUBC>  TMethod;
SUBC>  TAnova;
SUBC>  TSummary;
SUBC>  TCoefficients;
SUBC>  TEquation;
SUBC>  TFactor;
SUBC>  TDiagnostics 0.
```

The OUTPUT

```
Factor Information

Factor  Type   Levels Values
sorghum Fixed     5   1, 2, 3, 4, 5
block   Fixed     4   1, 2, 3, 4


Analysis of Variance
Source           DF  Adj SS  Adj MS  F-Value  P-Value
  sorghum         4   49.80  12.450    4.21    0.026
  block           3   41.92  13.972    4.73    0.024
Error            11   32.50   2.955
  Lack-of-Fit    10   14.50   1.450    0.08    0.994
  Pure Error      1   18.00  18.000
Total            18  140.42
```

The above analysis by GLM in MINITAB is different, and should not be surprising given what was said previously and the fact that the two analyses were premised differently.

16.3.1 Use of MINITAB for RCBD When Data Are Not Missing

■ **EXAMPLE 16.4** The data to be analyzed are the yields of an extract from five different varieties of a food (treatment) from a primary source, processed under five different conditions (block):

	Block				
Foods	I	II	III	IV	V
1	7	7	15	11	9
2	12	17	12	18	18
3	14	18	18	19	23
4	19	25	22	19	23
5	7	10	1	15	11

Use MINITAB to carry out ANOVA and test for difference between foods at 5% level of significance.

MINITAB Solution

```
MTB >   read c1-c3
DATA>   1   1   7
DATA>   1   2   7
DATA>   1   3   15
DATA>   1   4   11
DATA>   1   5   9
DATA>   2   1   12
DATA>   2   2   17
DATA>   2   3   12
DATA>   2   4   18
DATA>   2   5   18
DATA>   3   1   14
DATA>   3   2   18
DATA>   3   3   18
DATA>   3   4   19
DATA>   3   5   23
DATA>   4   1   19
DATA>   4   2   25
DATA>   4   3   22
DATA>   4   4   19
DATA>   4   5   23
DATA>   5   1   7
DATA>   5   2   10
DATA>   5   3   1
DATA>   5   4   15
DATA>   5   5   11
DATA>   end
25 rows read.
MTB >   print c1-c3
```

Data Display

Row	C1	C2	C3
1	1	1	7
2	1	2	7
3	1	3	15
4	1	4	11
5	1	5	9
6	2	1	12
7	2	2	17
8	2	3	12
9	2	4	18
10	2	5	18
11	3	1	14
12	3	2	18
13	3	3	18
14	3	4	19
15	3	5	23
16	4	1	19
17	4	2	25
18	4	3	22
19	4	4	19
20	4	5	23
21	5	1	7
22	5	2	10
23	5	3	1
24	5	4	15
25	5	5	11

```
MTB > name c1 'food' c2 'blocks' c3 'responses'
MTB > twoway c3 c1 c2
```

Two-Way ANOVA: Responses Versus Food, Blocks

Source	DF	SS	MS	F	P
food	4	602.8	150.7	13.83	0.000
blocks	4	86.8	21.7	1.99	0.145
Error	16	174.4	10.9		
Total	24	864.0			

S = 3.302   R-Sq = 79.81%   R-Sq(adj) = 69.72%

The p-value of less than 0.0001 for FOOD indicates that the factor FOOD is highly significant at 5% level of significance. The means for at least one of the levels of FOOD is different from the rest. The p-value of 0.145 for BLOCKS indicates that it is not significant at 5% level of significance. There is no significant block effect.

■ EXAMPLE 16.5

Ovens	Block				
	I	**II**	**III**	**IV**	**V**
1	21	16	20	15	18
2	12	10	17	13	12
3	11	12	16	11	5
4	4	10	11	7	3

Use MINITAB to carry out ANOVA and test at 1% level of significance.

MINITAB Solution

```
MTB > read c1-c3
DATA> 1 1   21
DATA> 1 2   16
DATA> 1 3   20
DATA> 1 4   15
DATA> 1 5   18
DATA> 2 1   12
DATA> 2 2   10
DATA> 2 3   17
DATA> 2 4   13
DATA> 2 5   12
DATA> 3 1   11
DATA> 3 2   12
DATA> 3 3   16
DATA> 3 4   11
DATA> 3 5   5
DATA> 4 1   4
DATA> 4 2   10
DATA> 4 3   11
DATA> 4 4   7
DATA> 4 5   3
DATA> end
20 rows read.
MTB > name c1 'ovens' c2 'block' c3 'y'
MTB > twoway c3 c1 c2
```

Two-Way ANOVA: y Versus Ovens, Block

```
Source   DF    SS      MS       F      P
ovens     3   312.4  104.133  16.53  0.000
block     4    89.2   22.300   3.54  0.040
Error    12    75.6    6.300
Total    19   477.2


S = 2.510    R-Sq = 84.16%   R-Sq(adj) = 74.92%
```

The p-value of less than 0.0001 for OVENS indicates that the factor OVENS is highly significant at 1% level of significance. The mean for at least one of the ovens is different from the rest. The p-value of 0.040 for BLOCK indicates that it is not significant at 1% level of significance. There is no significant block effect.

Exercises—Randomized Complete Block Design

16.1 The following data were collected from a study which compared four treatments in five blocks:

Treatment	Blocks				
	1	2	3	4	5
I	35	39	36	42	34
II	21	26	19	32	28
III	30	37	33	40	40
IV	24	22	23	35	33

Carry out ANOVA and test the hypothesis that the mean for the treatments are equal at 1% level of significance.

16.2 A research was to extract a type of resin from ginger uses three different laboratory facilities and ginger sourced from five countries. The study was to determine whether mean resin extracted from ginger from the five countries are different. The responses of the experiment are as follows:

Laboratories	Countries				
	1	2	3	4	5
I	12.5	14.9	12.6	11.8	12.4
II	12.7	17.6	11.9	13.2	15.3
III	13	15.7	10.3	14	16.4

Carry out ANOVA and test for significant country and laboratory effects at 5% level of significance.

16.3 Five food experts were chosen to taste four different formulations of a fast food to be marketed by a fast-food chain. Each of them blindly tasted each of the four formulations in random order and rated them in the scale 0–10. The following are the ratings:

Formulations	Experts				
	I	II	III	IV	V
I	9	7	8	7	8
II	7.5	6	7.5	6	7
III	6.5	4.3	5	5.5	6
IV	7	5	6	5	6

The formulations are the treatments that are of primary interest in the study. The experts are treated as the block factor in this study. Carry out ANOVA and test for significant treatment and block effects. Test at 1% level of significance. Use MINITAB.

16.4   An experiment was performed to study the effects of five fertilizer treatments on soybeans in six plots. The following data represent the yields of the soybeans in bushels per 0.3 acre. Write down the model for this design. Carry out ANOVA for the data at 2.5% level of significance if blocks or treatments or both are found significant.

Treatments	Blocks					
	I	II	III	IV	V	VI
I	34	36	61	24	43	42
II	30	30	20	20	30	30
III	29	44	51	26	36	27
IV	20	40	50	41	41	42
V	24	39	49	46	44	39

16.5   A study was conducted on the effect of four different diets on growth of broilers. One of the diets was generic. Seven chicks of the same age were used. The growth in the 4 weeks following were recorded as follows:

Treatments	Blocks						
	I	II	III	IV	V	VI	VII
I	134	136	161	142	134	151	157
II	120	120	130	130	130	136	136
III	129	144	151	136	120	144	145
IV	139	146	149	140	140	175	166

Test for diets and chicks effects in the study at 5% level of significance.

16.6   The following RCBD experiment resulted in the data below:

Block	Treatment			
	A	B	C	D
I	14	15	17	10
II	18	17	19	8
III	10	17	8	9
IV	8	18	10	8

Analyze the data to check whether there is significant treatment effect. Is blocking effective? Test at 5% level of significance.

16.7   The following RCBD experiment resulted in the data below:

Block	Treatment			
	A	B	C	D
I	14	14	8	12
II	9	12	6	10
III	8	11	8	7

Analyze the data to check whether there is significant treatment effect. Is blocking effective? Test at 5% level of significance.

CHAPTER 17

The Two-Way Layout

The Full Two Factor Factorial Experiment (A Two-Way Layout)

We had earlier considered a type of two-way layout, the randomized complete block design (RCBD). In that design, two factors were involved, but one of them was an extraneous factor which was not of primary interest in the study. That factor, usually called a BLOCK, was designed into the study in order to ensure that it did not mask the researcher's ability to detect the significance of the factor of primary interest (called treatment) by inflating the error sum of squares in analysis of variance (ANOVA). In RCBD, there was no expectation for the extraneous factor (block) to interact with the factor of interest.

Unlike in RCBD, in a full two-factor factorial experiment, the two factors set at multiple levels are of interest in the study. Thus, there are two factors A, set at a levels, and B set at b levels. Randomization is carried out within the levels of the factors and within the cells formed by the combination of levels of the factors. The purpose of randomization of the order of the performance of the experiments is to reduce experimental error. Unlike the RCBD, the full two-factor experiment must be replicated, so that when laid in a two-way table, within each cell of the design, there should be a minimum of two observations. When the design is balanced, there should be equal number of observations per cell.

The unbalanced two factor factorial experiment exists, but it is beyond the scope of the treatment of two-factor experiments in this text whose level is introductory. In RCBD, there is randomization restriction, obliging the designer of the experiment to randomize the treatments within each level of the block factor. Here, in two-factor factorial experiment, randomization is freely carried out for all levels of the factors A and B. A single replicate of this experiment requires ab units, so that for n replicates of the experiment, the total number of units used in the experiment is abn.

The two-way ANOVA is used for analyzing the responses of the designed experiment in which the levels of the two factors are compared. Testing for significance of the main effects of each of the factors compares the levels of the factors. In addition, testing for the significance of the interaction effect must be carried out. If a significant interaction exists, then the significance of the main effects does not mean much. In that context, the nature of the interaction between the two factors is studied subsequent to ANOVA. Further analysis of significant interaction effects was treated in Onyiah (2008).

In the model which applies to the two-factor factorial experiment, the observation (response) y_{ijk} is expressed as a linear function of the grand mean, the main effects of the two factors A and $B,$ their interaction effect, $AB,$ as well as the error term as shown in the following:

$$y_{ijk} = \mu + \alpha_i + \beta_j + (\alpha\beta)_{ij} + \varepsilon_{ijk}; \quad i = 1, 2, ..., a; \ j = 1, 2, ..., b; k = 1, 2, ..., n. \quad (17.1)$$

For the fixed effects model, then $\sum_{i=1}^{a} \alpha_i = 0; \sum_{i=1}^{b} \beta_j = 0;$ and $\sum_{i=1}^{a} (\alpha\beta)_{ij} = \sum_{j=1}^{b} (\alpha\beta)_{ij} = 0.$ The fixed effects model and its assumptions provide the basis for the estimation of each of the parameters $\alpha_i, \beta_j,$ and $(\alpha\beta)_{ij}.$ One of the assumptions in the model is that the error component, $\varepsilon_{ij} \sim NID(0, \sigma^2),$ that is to say that the error is normal, independently distributed, with zero mean and constant variance, $\sigma^2.$

After the experiments, the responses could be presented in a two-way layout involving a levels of factor A and b levels of factor B as indicated in Table 17.1.

■ **TABLE 17.1   A Two-Way Experimental Layout**

Factor A	Factor B					Totals for Levels of Factor A
	1	2	3	...	b	
1	$y111,$ $y112...y11n$	$y121,$ $y121...y12n$	$y131,$ $y131...y13n$	...	$y1b1,$ $y1b1...y1bn$	$y1..$
2	$y211,$ $y212...y21n$	$y221,$ $y221...y22n$	$y231,$ $y231...y23n$	...	$y2b1,$ $y2b1...y2bn$	$y2..$
3	$y311,$ $y312...y31n$	$y321,$ $y321...y32n$	$y331,$ $y331...y33n$	...	$y3b1,$ $y3b1...y3bn$	$y3...$
.	.	.	.	...	.	.
.	.	.	.	...	.	.
.	.	.	.	...	.	.
a	$ya11,$ $ya12...ya1n$	$ya21,$ $ya21...ya2n$	$ya31,$ $ya31...ya3n$	...	$yab1,$ $yab1...yabn$	$ya..$
Totals for Levels of Factor B	$y.1.$	$y.2.$	$y.3.$		$y.b.$	$y...$

The hypotheses tested are related to parameters specified in the model of 17.1. Two of them are about the main affects A and $B,$ while the other is about the interaction that may occur between the main effects A and $B,$ denoted by $AB.$

For the main effects $A,$ and $B,$ the set of hypotheses tested are stated in the following:

$$H_0 : \alpha_1 = \alpha_2 = ... = \alpha_a = 0 \text{ vs. } H_1 : \text{at least one } \alpha_i \neq 0; i = 1, 2, 3..., a$$
$$H_0 : \beta_1 = \beta_2 = ... = \beta_b = 0 \text{ vs. } H_1 : \text{at least one } \beta_j \neq 0; j = 1, 2, 3..., b$$

For interaction AB, we state similar hypothesis as:

$$H_0 : (\alpha\beta)_{ij} = 0 \text{ vs. } H_1 : \text{at least one } (\alpha\beta)_{ij} \neq 0; i = 1,2,3...,a; j = 1,2,3...,b$$

Next, we consider steps that would enable us to test the hypotheses stated. The above presentation of the data in a two-way table paves the way for clear calculations required for manual analysis of the responses of the experiment and which culminates in a two-way ANOVA.

In the other to perform the two-way ANOVA, we need to understand how to calculate some of the quantities stated in the earlier table as well as the formulas for the sums of squares to be stated. So we define the following quantities that will appear in the formulas:

The overall total for all combinations of levels of A and B is $y_{...} = \sum_{i=1}^{a}\sum_{j=1}^{b}\sum_{k=1}^{n} y_{ijk}$. This is the sum of all responses of the entire experiment.

The sum of all responses for ith level of factor A is $y_{i..} = \sum_{j=1}^{b}\sum_{k=1}^{n} y_{ijk}$. In a two-way layout this will be the sum of row i.

The sum of all responses for the jth level of factor B is $y_{.j.} = \sum_{i=1}^{a}\sum_{k=1}^{n} y_{ijk}$. In a two-way layout this will be the sum of column j.

The sum of all responses due to the combination of the ith and jth levels of A and B is $y_{ij.} = \sum_{k=1}^{n} y_{ijk}$. This is the cell total for treatment combination (i,j).

The total sum of squares is given as:

$$SS_{\text{total}} = \sum_{i=1}^{a}\sum_{j=1}^{b}\sum_{k=1}^{n} (y_{ijk} - \bar{y}_{...})^2 \tag{17.2}$$

As usual, this total sum of squares can be split, as we did for both CRD and RCBD, into its component parts. The split will show the component parts of the total sum of squares as shown in the following:

$$SS_{\text{total}} = SS_A + SS_B + SS_{AB} + SS_{\text{Error}} \tag{17.3}$$

For manual calculations, the sums of squares in (17.3) can be obtained by using the simplified forms of the formulas shown in (17.4):

$$SS_{\text{Total}} = \sum_{i=1}^{a}\sum_{j=1}^{b}\sum_{k=1}^{n} y_{ijk}^2 - \frac{y_{...}^2}{abn}$$

$$SS_A = \frac{1}{bn}\sum_{i=1}^{a} y_{i..}^2 - \frac{y_{...}^2}{abn}$$

$$SS_B = \frac{1}{an}\sum_{j=1}^{b} y_{.j.}^2 - \frac{y_{...}^2}{abn} \tag{17.4}$$

$$SS_{AB} = \frac{1}{n}\sum_{i=1}^{a}\sum_{j=1}^{b} y_{ij.}^2 - SS_A - SS_B - \frac{y_{...}^2}{abn}$$

$$SS_{\text{Error}} = \sum_{i=1}^{a}\sum_{j=1}^{b}\sum_{k=1}^{n} y_{ijk}^2 - SS_A - SS_B - SS_{AB}$$

The mean squares associated with some of the sums of squares in (17.4) are as follows:

$$MS_A = \frac{SS_A}{a-1}$$

$$MS = \frac{SS_B}{b-1}$$

$$MS_{AB} = \frac{SS_{AB}}{(a-1)(b-1)}$$

$$MS_{Error} = \frac{SS_{Error}}{ab(n-1)}$$

(17.5)

Statistical theory shows that, for the fixed effects model which was assumed as described in equation (17.1), each of these mean squares is chi-square distributed. The divisors of these sums of squares on the right-hand sides are their degrees of freedom. By token of this, the following ratios of two chi-square variables are distributed as F with the stated degrees of freedom:

$$\frac{MS_A}{MS_{Error}} \sim F[(a-1), ab(n-1)]$$

$$\frac{MS_B}{MS_{Error}} \sim F[(b-1), ab(n-1)]$$

$$\frac{MS_{AB}}{MS_{Error}} \sim F[(a-1)(b-1), ab(n-1)]$$

(17.6)

These ratios, which are F-distributed, shown in (17.6), form the bases for the two-way ANOVA. Their values are compared with the critical values of F for the specified level of significance for the test. The usual two-way ANOVA table is given as follows:

Source	Sum of Square	Degree of Freedom	Mean Square	F-Ratio
A	SS_A	$a-1$	$SS_A/(a-1)$	$[MS_A / MS_{Error}]$
B	SS_B	$b-1$	$SS_B/(b-1)$	$[MS_B / MS_{Error}]$
AB	SS_{AB}	$(a-1)(b-1)$	$SS_{AB}/(a-1)(b-1)$	$[MS_{AB} / MS_{Error}]$
Error	SS_{Error}	$ab(n-1)$	$SS_{Error}/ab(n-1)$	
Total	SS_{total}	$abn-1$		

The F-ratio for A in the earlier table is the test statistic for the hypothesis stated for main effect A; the ratio is compared to a tabulated value, $F[a-1, ab(n-1), \alpha]$. α is the level of significance specified for the hypothesis. The null hypothesis of "no effect of A" is rejected if data are significant, that is F-ratio is greater than the tabulated $F[a-1, ab(n-1), \alpha]$. Similarly, the F-ratio for B in the previous table is the test statistic for the hypothesis stated for main effect B and is used for testing that "there is no B effect"; this time, the F-ratio for B is compared with $F[b-1, ab(n-1), \alpha]$ when the level of significance is α. Rejection of the null hypothesis for B is carried out in the same manner as for A, that is when F-ratio $> F[b-1, ab(n-1), \alpha]$. By the same token, we reject the null hypothesis of "no interaction between A and B" when for AB in the above ANOVA, F-ratio $> F[(a-1)(b-1), ab(n-1), \alpha]$.

■ **EXAMPLE 17.1**   Four treatment regimens for three different types of cancer are being studied. Three treatments were applied, along with placebo which acted as the fourth. The measure is the change in percentage of hemoglobin levels measured at a specified period after treatment. The responses of the experiment are shown as follows:

Factor A (Three Types of Cancer)	Factor B (Placebo and Three Methods of Treatment)			
	I	II	III	IV
1	1.5, 1.2, −0.9, 1.8	−1.6, −1.7, −2.1, −1.8	−0.8, 0.4, −1.5, −1.7	−2, −1.4, 2, 2
2	2.7, 1.9, 2.0, 1.9	−2.7, 1.6, −2.1, 1.8	1, 2, 2.4, 3.3	2.4, 1.6, 2.2, 1.4
3	1.6, 2.9, 1.7, 1.7	−1.5, 0.5, −1.6, −1.5	1.7, −1.5, 2.8, 0.8	1.3, 2.5, −0.9, 1.2

Carry out a full two-way ANOVA for the data testing the hypotheses at 5% level of significance.

Solution   **The hypotheses to be tested in ANOVA are:**

$H_0 : \alpha_1 = \alpha_2 = \alpha_3 = 0$ vs. $H_1 :$ at least one $\alpha_i \neq 0$; $i = 1, 2, 3$.

$H_0 : \beta_1 = \beta_2 = \beta_3 = \beta_4 = 0$ vs. $H_1 :$ at least one $\beta_j \neq 0$; $j = 1, 2, 3, 4$.

$H_0 : (\alpha\beta)_{ij} = 0$ vs. $H_1 :$ at least one $(\alpha\beta)_{ij} \neq 0$; $i = 1, 2, 3$; $j = 1, 2, 3, 4$.

Factor A (Three Types of Cancer)	Factor B (Placebo and Three Methods of Treatment)				Totals
	I	II	III	IV	
1	1.5, 1.2, −0.9, 1.8 (3.6)	−1.6, −1.7, −2.1, −1.8 (−7.2)	−0.8, 0.4, −1.5, −1.7 (−3.6)	−2, −1.4, 2, 2 (0.6)	−6.6
2	2.7, 1.9, 2.0, 1.9 (8.5)	−2.7, 1.6, −2.1, 1.8 (−1.4)	1, 2, 2.4, 3.3 (8.7)	2.4, 1.6, 2.2, 1.4 (7.6)	23.4
3	1.6, 2.9, 1.7, 1.7 (7.9)	−1.5, 0.5, −1.6, −1.5 (−4.1)	1.7, −1.5, 2.8, 0.8 (3.8)	1.3, 2.5, −0.9, 1.2 (4.1)	11.7
Totals	20	−12.7	8.9	12.3	28.5

In the previous table, we have obtained the column totals, that is $y_{.j.}$ and the row totals, $y_{i..}$. The values in parenthesis in the cells are the cell totals, $y_{ij.}$.

Using the formulas in (17.4), we calculate the sums of squares needed for ANOVA as follows:

$$SS_{total} = 1.5^2 + 1.2^2 + (-0.9)^2 + 1.8^2 + \ldots + (-1.3)^2 + (-2.5)^2 + (-0.9)^2$$

$$+ 1.2^2 - \frac{28.5^2}{3(4)(4)} = 144.4681$$

$$SS_A = \frac{(-6.6)^2 + 23.4^2 + 11.7^2}{4(4)} - \frac{28.5^2}{3(4)(4)} = 49.0606$$

$$SS_B = \frac{20^2 + (-12.7)^2 + 8.9^2 + 12.3^2}{3(4)} - \frac{28.5^2}{3(4)(4)} = 28.5788$$

$$SS_{AB} = \frac{3.6^2 + (-7.2)^2 + (-3.6)^2 + 0.6^2 + \ldots + 3.8^2 + 4.1^2}{4} - 49.0606 - 28.5788$$

$$- \frac{28.5^2}{3(4)(4)} = 4.5013$$

$$SS_{Error} = 144.4681 - 49.0606 - 28.5788 - 4.5013 = 62.3275$$

ANOVA

Source	Sum of Squares	Degrees of Freedom	Mean Square	F-Ratio
A	49.0606	2	24.5303	14.1685
B	28.5788	3	9.5263	5.5023
AB	4.5012	6	0.7502	0.4333
Error	62.327	36	1.73132	
Total	144.4681	47		

From the tables, $F(2,36\ 0.05) = 3.259$; $F(3,36,0.05) = 2.866$ and $F(6,36,0.05) = 2.364$. The main effects A and B are highly significant at 5% level of significance. There is no significant interaction effect.

■ **EXAMPLE 17.2**

Cassava Variety	Fertilizer Type		
	I	II	III
T1	20.7, 21.2, 17.4	16.4, 17.3, 18.2	22, 23.1, 20.9
T2	13.6, 12.7, 14.5	10.3, 12.4, 10.5	14.3, 13.2, 13.5
T3	15.1, 15.7, 16.2	12.5, 14.2, 13.2	14.4, 14.8, 15.2
T4	14.0, 14.7, 13.2	11.5, 11.3, 12.2	13.4, 13.8, 14.2

The previous table contains the responses of a study of the yield of four different types of cassava under the application of three different fertilizers. The plots where this research was carried out were homogeneous. The responses represent the yields in metric tonnes per hectare. The experiment was replicated three times. Carry out ANOVA to determine whether the cassava variety, fertilizer type, and their interaction have significant effects on the responses. Test at 1% level of significance.

Solution

Cassava Variety	Fertilizer Type			Total
	I	**II**	**III**	**Total**
T1	20.7, 21.2, 17.4 (59.3)	16.4, 17.3, 18.2 (51.9)	22, 23.1, 20.9 (66)	177.2
T2	13.6, 12.7, 14.5 (40.8)	10.3, 12.4, 10.5 (33.2)	14.3, 13.2, 13.5 (41)	115
T3	15.1, 15.7, 16.2 (47)	12.5, 14.2, 13.2 (39.9)	14.4, 14.8, 15.2 (44.4)	131.3
T4	14.0, 14.7, 13.2 (41.9)	11.5, 11.3, 12.2 (35)	13.4, 13.8, 14.2 (41.4)	118.3
Total	189	160	192.8	541.80

The table shows the row totals (cassava variety) and the column totals (Fertilizer types). Cell totals are in parenthesis within each cell.

The hypothesis to be tested in ANOVA is:

$H_0 : \alpha_1 = \alpha_2 = \alpha_3 = \alpha_4 = 0$ vs. H_1 : at least one $\alpha_i \neq 0$; $i = 1,2,3,4$.

$H_0 : \beta_1 = \beta_2 = \beta_3 = 0$ vs. H_1 : at least one $\beta_j \neq 0$; $j = 1,2,3$.

$H_0 : (\alpha\beta)_{ij} = 0$ vs. H_1 : at least one $(\alpha\beta)_{ij} \neq 0$; $i = 1,2,3,4$; $j = 1,2,3$.

$$SS_{\text{total}} = 20.7^2 + 21.2^2 + \ldots + 13.4^2 + 13.8^2 + 14.2^2 - \frac{541.8^2}{4(3)(3)} = 361.35$$

$$SS_{\text{cassava}} = \frac{177.2^2 + 115^2 + 131.3^2 + 118.3^2 +}{(3)(3)} - \frac{541.8^2}{4(3)(3)} = 274.734$$

$$SS_{\text{Fertilizer}} = \frac{189^2 + 160^2 + 192.8^2}{(4)(3)} - \frac{541.8^2}{4(3)(2)} = 53.647$$

$$SS_{\text{Fetilizer*cassava}} = \frac{59.3^2 + 51.9^2 + 66^2 + \ldots + 41.9^2 + 35^2 + 41.4^2}{3}$$

$$- 274.734 - 53.647 - \frac{541.8^2}{4(3)(3)} = 11.169$$

$$SS_{\text{Error}} = 361.35 - 274.734 - 53.647 - 11.169 = 21.8$$

ANOVA

Source	Sum of Squares	Degrees of Freedom	Mean Square	F-Ratio
Cassava	274.734	3	91.578	100.82
Fertilizer	53.647	2	26.8235	29.53
Cassava*Fertilizer	11.169	6	1.8615	2.05
Error	21.800	24	0.9083	
Total	361.35			

From the tables, $F(3,24, 0.01) = 4.718$; $F(2,24,0.01) = 4.614$; and $F(6,24,0.01) = 3.667$. The main effects of cassava and fertilizer are highly significant at 1% level of significance. There is no significant interaction effect.

Reference

Onyiah, Leonard C. 2008. *Design and Analysis of Experiments Classical and Regression Approaches With SAS.* Boca Raton: Chapman and Hall/CRC.

Exercises—Two Factor Factorial Experiment (Two-Way Layout)

17.1 Five experts blindly rate each of three formulations of a cookie to be marketed by SmartFood™ Corp over 30; the ratings were repeated for each formulation. The ratings are as follows:

Experts	Formulations		
	1	2	3
1	20, 23	19, 21	27, 24
2	15, 18	16, 15	20, 25
3	20, 23	15, 20	24, 26
4	21, 18	19, 23	24, 25
5	20, 20	22, 25	22, 23

Carry out analysis as a full two-factor experiment, testing at 5% level of significance.

17.2 The effects of two factors A and B on the yield of a process were being studied. The data collected were arranged as in the two-way table in the following. Analyze the data, considering main effects and interaction in ANOVA, testing hypotheses at 1% level of significance.

Factor A	Factor B				
	I	II	III	IV	V
I	7.3, 10.2 (17.5)	11.8, 10.4 (22.2)	8.1, 10.7 (18.8)	10, 8.7 (18.7)	11.7, 11.1 (22.8)
II	9.8, 7.4 (17.2)	5.6, 11.1 (16.7)	9.5, 8.8 (18.3)	8.2, 7.7 (15.9)	8.6, 9.2 (17.8)
III	9.4, 7.9 (17.3)	9.6, 9.8 (19.4)	10.2, 10.2 (20.4)	10.8, 9.1 (19.9)	12.0, 10.5 (22.5)

17.3 The effects of four fertilizer treatments and three watering rates on the yield of vegetables grown randomly on contiguous plots of land were being studied. Plots were similarly prepared and randomization of allocation of assignments of combinations of the levels of the factors was thoroughly carried out.

Fertilizers	Watering Rates		
	I	II	III
1	30, 25, 29	32, 34, 29	25, 22, 23
2	31, 37, 33	31, 32, 30	28, 24, 25
3	33, 29, 24	35, 41, 31	30, 29, 29
4	31, 30, 32	33, 30, 31	26, 29, 25

Analyze the data as a full two-factor factorial experiment, testing for all effects at 5% level of significance.

17.4 A two-factor factorial experiment was performed and the responses were elicited and arranged in a two-way table as shown in the following.

Factor A	Factor B I	II	III	IV
1	5, 9, 10, 8	6, 7, 11, 8	8, 4, 5, 7	2, 4, 2, 2
2	7, 9, 10, 9	7, 6, 10, 8	1, 3, 4, 3	4, 6, 2, 4
3	6, 9, 7, 7	5, 5, 6, 5	7, 10, 8, 8	3, 5, 2, 2

Analyze the data as a full two factor factorial design and test for significance of all effects at 1% level of significance. USE ANOVA in MINITAB.

17.5 The following two-factor factorial experiment was performed and the following responses were elicited.

Material	Temperature 50	60	70	80
1	130, 155, 74, 180	34, 40, 80, 75	20, 70, 82, 58	120, 100, 101, 115
2	150, 188, 159, 126	136, 122, 106, 115	50, 60, 87, 86	114, 70, 54, 115
3	138, 110, 168, 160	74, 120, 150, 139	120, 96, 104, 82	108, 110, 109, 112

Analyze the data as a full two factor factorial design and test for significance of all effects at 1% level of significance. USE ANOVA in MINITAB.

17.6

Temperature	Plant I	II
I	7, −9, 6, 12, 11	19, 21, −4, 7, 18
II	4, 13, 10, 11, −14	29, 22, 11, 38, 21
III	15, 12, 11, 6, 9	13, 31, 50, 19, 28

The effects of two factors, temperature and plant type on the yield of a process were being studied. Three levels of temperature and two levels of plant type were studied. Carry out ANOVA, using MINITAB, testing for main effects and interaction at 5% level of significance.

17.7 Consider the following responses from an experiment with two factors A and B. Analyze the data and reach conclusions about the significance of the main effects and interaction at 1% level of significance.

Factor A	Factor B I	II	III
I	10, 10, 9	8.9, 8.5, 8.7	7.8, 7, 7.1
II	8.4, 8.8, 7.7	7, 5.6, 6.7	9, 8, 8

CHAPTER 18

Introductory Nonparametric Statistics

Introduction—Nonparametric Tests

Most of the statistical methods we studied up to Chapter 17 were applied to data which by nature are quantitative. Those methods we used are referred to as parametric methods. Under the parametric methods, data were assumed to follow some statistical distributions and inferences about the populations involved postulates that were about the parameters of those underlying distributions. Although the methods employed were generally speaking, robust, they were often sensitive to assumptions made on small samples and were limited in applications to quantitative data. In addition, the quantitative data were usually collected either on the interval or ratio scale.

Interval Scale

Most data have to do with measurements. Some measurements are made on an interval scale. The major features of data from an interval scale are that:

1. They can numerically be distinguished.
2. They can be ordered.
3. The differences among them have to be meaningful.

Ratio Scale

A ratio scale satisfies all the conditions of an interval scale, but in addition has a true origin.

A vast majority of analyses methods apply in the main, to quantitative data. However, a sizeable amount of data which are collected are qualitative. Examples include evaluations of products, quality of lectures, quality of taste, the proverbial shades of gray, and more. What is important for the scales used is that the evaluations reflect an ordering or ranking. Such data are defined in an **ordinal scale** because the distance between two evaluations is not important. Generally, these data would not qualify for analysis by parametric methods. Parametric methods often tie the use of an analysis method to an underlying distribution for the data. Data such as described previously may not qualify to be modeled by known distributions neither would they qualify for such analysis. In consequence, many distribution free methods have been developed for analyzing such data.

Most of these distribution-free methods are used for inference when data do not necessarily conform to the dictates required for parametric approaches. These approaches are covered under nonparametric statistics, which is a vast and growing field of statistical endeavor. We introduce some elements of the nonparametric approaches in this section.

Apart from the chi-square test, the rest of the tests we carried out elsewhere in the text rely on some basic assumptions about the distribution of the data: for instance, the underlying distribution of the data which should be analyzed by the t-test must be **normal**. The same principle applies to other tests. Further, we always had to make inference about some population value (usually called parameters) such as the population mean, variance, and others. We therefore estimated these parameters by some sample values (called statistics); thereafter based on the distribution of the statistics, we constructed a test statistic to enable us make inference about the population value (parameter) of interest.

Essentially, we can say that most of the inferential procedures we discussed in previous chapters were **parametric.** We do recognize, however, that many situations exist, when we may not know the underlying distributions of the data we have collected and therefore cannot establish a parametric procedure for inference about the data. In these situations, **nonparametric** procedures are adopted. Statistical theory in this field has developed enormously in recent years that equivalents of many parametric procedures now exist in the nonparametric form. Hence, it is possible to reach the same inference about a set of data using both approaches. For instance, one sample and two sample equivalents of the t-tests exist, will be discussed shortly, and can be applied even when *we know that the data conforms* **to the parametric approach**.

The nonparametric approaches have the added advantage that we can apply the procedures to data collected on the **ordinal** scale as opposed to the **ratio** scale on which parametric tests are mostly performed. In addition to these, the nonparametric tests are often based on less stringent assumptions than the parametric tests and are therefore, more widely applied. They are very useful for small samples and therefore are very useful for reducing the cost of an investigation. They are therefore applied in situations where human or destructive sampling restricts the size of a sample, for instance, in medicine.

That said, in some situations, due to the less restrictive assumptions for the nonparametric tests, their parametric equivalents may tend to be more **powerful** and would tend to reject the null hypothesis for a smaller size of type I error. However, the nonparametric tests would be correct for many situations since some of the assumptions made for the parametric equivalents may not apply to the dataset. Again, the nonparametric tests tend to require new tables for each test proposed, with less connections between the tables as was the case with their parametric equivalents. This disadvantage has to a certain extent been reduced by the fact that most of these tests can now be approximated by the normal distribution for reasonably large samples.

| 18.1 | **The Sign Tests** |

This test is used to study differences between pairs of observations. It is an equivalent of the parametric paired comparison test. Here, the signs of the differences between pairs of observations are used. The differences, D serve to test the hypothesis that the differences between the pairs have a median of zero. The equivalent of the above statement is that the signs + and − occur with **equal probability.**

We can therefore set up the hypothesis to be tested as:

$H_0 : p = 0.5$
$H_1 : p \neq 0.5$

In the hypothesis above, p is the probability of the + sign. The test statistic required for this test is:

$$\chi^2_{\text{cal}} = \frac{(n_1 - n_2)^2}{n_1 + n_2} \tag{18.1}$$

In (18.1), n_1 = number of plus (+) signs and n_2 = number of negative (−) signs, n = number of pairs of observations. The above test statistic is distributed as chi-square with $n-1$ degrees of freedom. So, we can compare the quantity calculated using the above test statistic with chi-square at n-1 degrees of freedom at half of the chosen level of significance for the test: that is if $\chi^2_{\text{cal}} > \chi^2_{\left[\frac{\alpha}{2}, n-1\right]}$, then reject the null hypothesis, and accept that the median of the differences is not zero.

The sign test can be modified for studying other situations:

1. For sample from a single populations, we can postulate that a given figure is the median, we subtract that postulated median from the observations. Count the positives and the negative, and calculate the test statistic above.
2. For the paired observations from populations A and B, we could ask whether A gives responses which are c units above the responses of B. In that case, we would then, for each pair of observations (a_j, b_i), subtract $b_j + c$ from a_i and calculate the test statistic and carry out the test.
3. For the paired observations from populations A and B, we can ask whether the response of A is $k\%$ better than B. In that case, for each pair of observations (a_j, b_i), we subtract $b_i + \dfrac{k}{100} b_i$ from a_i and apply the sign test to the differences as described previously.

■ **EXAMPLE 18.1** A doctor who is testing the effectiveness of a drug in reducing temperatures in patients who had undergone a certain operation, obtained the data on temperatures of patients before and after treatment as follows:

A	103.1	100	102.7	101.6	102.4	101	104	103	101
B	99.7	101	98.7	98.4	103	101	100	99	101.7

Is the treatment effective? Use α 5 5%.

Solution We calculate the differences and the signs as described previously:

Difference	3.4	−1	4	3.2	−0.6	0	4	4	−0.7
Sign	1	−	1	1	−	?	1	1	−

$n_1 = 5$, $n_2 = 3$, $n = 9$. The null hypothesis and alternatives are same as described in the previous section, namely,

$$H_0 : p = 0.5$$

vs.

$$H_1 : p \neq 0.5$$

$$\chi^2_{cal} = \frac{(5-3)^2}{5+3} = \frac{4}{8} = 0.5$$

From the tables, $\chi^2_{[7,0.025]} > \chi^2_{cal}$ which indicates that we should accept the null hypothesis; thus, showing that the drug was ineffective for lowering the temperatures of patients.

18.2 The Wilcoxon Signed-Rank Test

This is a nonparametric test which improves on the sign test for detecting actual differences between pairs of observations. The improvement is achieved because the magnitude of the differences between pairs is used along with the signs of the differences. As usual, no assumption is made about the underlying distribution of the data. We describe the procedure for this test:

Step 1: Rank all the differences between pairs of observations from the least to the largest without regard to the signs of the differences (or rank the absolute values of the differences). Assign to the smallest absolute difference the rank 1.

Step II: Assign to the ranks the signs of the original differences between each pair of observations. Note that if m (m . 2) differences are tied, then each of the m differences should be assigned the mean of the next m ranks.

Step III: Calculate the sums of the ranks which are negative and positive and call them $T-$ and $T+$ respectively and note that $T- + T+ = n(n+1)/2$; where n is the number of pairs of observations being used. Of $T-$ and $T+$, choose the smaller and call it T; this is the test statistic to be used for the test.

Step IV: Compare T with the critical region of the Table in Appendix VI. At the table, the values under 0.05, corresponds to the critical value for a two-sided test with two equal critical regions of size, 0.025. It is also the critical region for the test with one-sided alternative with level of significance α 5 0.05.

Step V: Decision rule: If T is less than the tabulated value, reject the null hypothesis of equality of the two populations which were paired; otherwise, accept the null hypothesis.

■ **EXAMPLE 18.2**   A new baby formula developed by Babymilk™ is believed to improve the growth of babies. A small study of this baby food was carried out using 14 babies of the same age. Their weekly weight gains before the formula were noted and their weights after 1 week of feeding on the baby formula are presented in the following:

| Before | 175 | 232 | 218 | 151 | 200 | 219 | 234 | 149 | 187 | 123 | 248 | 206 | 179 | 206 |
| After | 142 | 211 | 337 | 262 | 302 | 195 | 253 | 199 | 236 | 216 | 211 | 176 | 249 | 214 |

Has the formula improved the weekly weights gained by the babies?

Solution

(i)	Before	After	Differences (d_i)	Signed-Ranks
1	175	142	33	6
2	232	211	21	3
3	218	337	−119	−14
4	151	262	−111	−13
5	200	302	−102	−12
6	219	195	24	4
7	234	253	−18	−2
8	149	199	−50	−9
9	187	236	−49	−8
10	123	216	−93	−11
11	248	211	37	7
12	206	176	30	5
13	179	249	−70	−10
14	206	214	−8	−1

From the table, we see that $T- = 25$; $T+ = 80$, hence $T = 25$. Clearly, $n = 14 =>$ $T- + T+ = 14(14+1)/2 = 105$. We test the null hypothesis:

H_0: There is no difference between the gains in weight of the babies before and after the use of the baby formula.

Vs.

H_1: A difference exists between the gains in weight of the babies before and after the use of the new baby formula.

We now test the hypothesis at the level of significance of $\alpha = 5\%$. The test is two-tailed, implying that the critical regions are of sizes, 0.025. From the table (see Appendix VI), reading under $n = 14$, the tabulated $T = 21$. However, T from our data is 25. We therefore fail to reject the null hypothesis.

18.2.1 Normal Approximation for the Wilcoxon Signed-Rank Test

As we stated earlier, for larger values of n, it is possible to obtain a normal approximation for some of the nonparametric tests. When n is large (for instance, $n \geq 25$)

(generally for $n > 25$, the normal approximation can be used). The threshold of 25 is not universally accepted it can be lowered to 15 or even raised to about 50. If n is large, then for the Wilcoxon signed-rank test, we note that we can approximate the test by using the fact that T is normally distributed with mean, U_T:

$$U_T = \frac{n(n+1)}{4} \tag{18.2}$$

The variance is, $\sigma_T^2 = \dfrac{n(n+1)(2n+1)}{24}$ (18.3)

Thus, the test statistic is:

$$Z = \frac{(T - U_T)}{\sigma_T} \tag{18.4}$$

Z has a standard normal distribution with mean, 0 and variance, 1. The values of Z can be read off any table of the standard normal distribution in order to determine whether to reject or accept any hypothesis (See Table for Z in Appendix II).

■ **EXAMPLE 18.3**   A new method was used to teach pupils arithmetic in second term. It was decided to use a sample of the kids to evaluate the effectiveness of the new method in comparison to the old method used in previous terms. The pupils' scores in arithmetic in second term were obtained. Compare the scores and decide whether their first term scores are the same.

Pupils	Test1	Test2
1	20	21
2	67	29
3	37	25
4	42	32
5	57	38
6	39	20
7	43	34
8	35	23
9	41	24
10	39	25
11	47	32
12	53	33
13	38	41
14	60	35
15	37	37
16	59	33
17	67	41
18	43	35
19	64	37

Continued.

Pupils	Test1	Test2
20	41	35
21	21	43
22	29	42
23	35	49
24	59	62
25	29	42

Solution   Here, we use the normal approximation to test the hypothesis

H_0: There is no difference between the performances of the pupils in the two terms.

Vs.

H_1: There is a difference in the performances of the pupils in the two tests.

To carry out the test, we rank the absolute differences in the performances of the pupils (i.e., we rank the differences ignoring their signs but using the sizes) and then we obtain the sum of the ranks for positive and negative differences ($T+$ and $T-$).

Pupils	Test1	Test2	Differences	Signed Ranks
1	20	21	−1	1
2	67	29	38	24
3	37	25	12	9.5
4	42	32	10	7.5
5	57	38	19	16.5
6	39	20	19	16.5
7	43	34	9	6
8	35	23	12	9.5
9	41	24	17	15
10	39	25	14	12.5
11	47	32	15	14
12	53	33	20	18
13	38	41	−3	2.5
14	60	35	25	20
15	37	37	0	No rank
16	59	33	26	21.5
17	67	41	26	21.5
18	43	35	8	5
19	64	37	27	23
20	41	35	6	4

Continued.

Pupils	Test1	Test2	Differences	Signed Ranks
21	21	43	−22	19
22	53	43	10	7.5
23	35	49	−14	12.5
24	59	62	−3	2.5
25	29	42	−13	11

Using the sum of positive ranks, $T+ = 251.5$; and $T- = 48.5$.

We know from previous discussion of the normal approximation for the Wilcoxon signed-rank test, which yielded the equations (18.2–18.4), that $T \sim N(U_T, \sigma_T^2)$, application to the above data, which shows that:

$$U_T = \frac{24(24+1)}{4} = 150; \sigma_T = \sqrt{\frac{24(24+1)(49+1)}{24}} = 35.35534;$$

$$Z = \frac{251.5 - 150}{35.35534} = 2.87$$

The data are significant at $\alpha = 1\%$ level of significance. This is due to the fact that from the table of the normal distribution we see that when $\alpha = 1\%$, we would reject the null hypothesis because Z is outside the interval $[-2.575, 2.575]$. If we had used $T- = 48.5$, we would have found that $Z = -2.87$, enabling us to arrive at the same decision.

18.3 The Wilcoxon Rank-Sum Test

In the previous sections, we considered those nonparametric tests which apply to paired observations. The Wilcoxon rank-sum test which we discuss in this section is developed for comparing two independent samples which may or may not be of equal size. It is often referred to as the nonparametric analogue of the t-test. The test is in general for unpaired observations in two groups, one of which contains n_1 observations, and the other, n_2, where $n_1 \geq n_2$. Like other nonparametric tests, there is no assumption about the distribution of the two samples involved in this test, it is therefore distribution-free. Unlike the t-test, this test does not require that we assume that the two samples are from populations which are at least approximately normally distributed. The null hypothesis under this test states that the two samples come from the same parent population, which is a stronger requirement than the t-test which simply compares the means of the two populations, except that the t-test also requires that the variances of both populations be equal. The test is equivalent to another nonparametric test called the Mann-Whitney U-test. We discuss the procedure for the test:

Step I: Given two samples with n_1 and n_2 observations, rank observations from both samples together, assigning the lowest rank to the lowest observation. If there are m tied observations, assign the mean of the next m ranks to each of them.

Step II: Add ranks for the smaller sample and call it T.

Step III: Obtain $T' = n_1(n_1 + n_2 + 1) - T$

Step IV: Choose the smaller of T and T' and compare with the tabulated value under n_1 and n_2 in the table (Appendix VI).

Step V: Decision Rule: if the calculated smaller sum (which may be T or T') is smaller than the tabulated value, reject the null hypothesis. Otherwise, accept the null hypothesis. For a two-tailed test, note that the tabulated value at the level of significance, $\alpha = 0.05$ is equivalent of two critical regions of size, 0.025.

■ **EXAMPLE 18.4**   Two different production processes for the manufacture of a soft drink were being studied. The hourly yield of the processes (in cu liters of drink) was recorded for 10 hours (the data are given in the following). Are the two processes the same? Test at 5% level of significance.

Process 1	Process 2
1133	1200
1422	1150
1288	1300
1317	1370
1080	1410
1344	1580
1506	1650
1752	1650
1924	1750
1783	2000

Solution   Here we test the hypothesis: H_0: The yields of the two production processes are equal against the alternative. H_1: The yields of the two production processes are different.

Rank(Process)1)	Process 1	Rank (Process 2)	Process 2
2	1,133	4	1,200
11	1,422	3	1,150
5	1,288	6	1,300
7	1,317	9	1,370
1	1,080	10	1,410
8	1,344	13	1,580
12	1,506	14.5	1,650
17	1,752	14.5	1,650
19	1,924	16	1,750
18	1,783	20	2,000
Total = 100			Total = 110

From the ranks, we see that $T = 100$. We calculate T', that is, $T' = n_1(n_1 + n_2 + 1) - T = 210 - 100 = 110$.

From the tables (Appendix VII), we see that under $n_1 = n_2 = 10$, the value of the critical T is 78. Hence $T' >$ tabulated T, so we accept the null hypothesis of equality of the two production processes.

18.3.1 Normal Approximation of the Wilcoxon Rank-Sum Test

Although we can make inferences by comparing values of T or T' with appropriate tabulated values, we can reach the same decisions by making use of the fact T has a sampling distribution which can be approximated by the normal curve for most values of n. T is known to be approximately normally distributed with mean,

$$T_m = \frac{n_1(n_1 + n_2 + 1)}{2} \tag{18.5}$$

Its variance is:

$$\sigma_T^2 = \frac{n_1 n_2(n_1 + n_2 + 1)}{12} \tag{18.6}$$

This implies that:

$$Z = \frac{T - \dfrac{n_1(n_1 + n_2 + 1)}{2}}{\sqrt{\dfrac{n_1 n_2(n_1 + n_2 + 1)}{12}}} \sim N(0,1) \tag{18.7}$$

The values of Z can be read from the table of the standard normal random variable. Thus, Z is the test statistic and the decision rule depends on the alternative hypotheses to be tested. We use examples to demonstrate the applications of this test for different alternative hypotheses.

■ **EXAMPLE 18.5** **A Typical Test With Two-Sided Alternative**
The times to burn out of two filaments used in an electronic equipment are being studied and are recorded in hours. Test the hypothesis that lifetimes of the filaments are equal.

Type I Filament	Type II Filament
650	640
810	560
820	710
570	830
660	590
820	650
670	690
590	740
750	820

Continued.

Type I Filament	Type II Filament
700	720
820	
680	
650	

Solution We test H_0: The lifetimes of the filaments are the same against. H_1: The lifetimes of the filaments are not the same. First we rank the two sets of observations together and use average ranks in cases of tied observations.

Observation	Type of Filament	Rank
560	2	1
570	1	2
590	2	3.5
590	1	3.5
640	2	5
650	1	7
650	1	7
650	2	7
660	1	9
670	1	10
680	1	11
690	2	12
700	1	13
710	2	14
720	2	15
740	2	16
750	1	17
810	1	18
820	1	20.5
820	1	20.5
820	1	20.5
820	2	20.5
830	2	23

From the tables for $n_1 = 13$, $n_2 = 10$, the critical T does not exist, so we can use the approximate normal distribution for this test. We know that $T \sim N\left[\dfrac{n_1(n_1+n_2+1)}{2}, \dfrac{n_1 n_2(n_1+n_2+1)}{12}\right]$. So that by choosing $n_1 = 13$, thereby making $n_2 = 10$, then total ranks for type 1 filaments, $T = 159$. This means that $E(T) = \dfrac{13(13+10+1)}{2} = 156$, $\sigma_T^2 = \dfrac{10 \times 13 \times (13+10+1)}{12} = 260$, and the test statistic for the hypothesis is:

$$Z = \frac{150-156}{\sqrt{260}} = 0.18605$$

At the level of significance, $\alpha = 1\%$, the critical value for Z is 2.575, so that we reject H_0 if Z lies outside the interval $[-2.575, 2.575]$. On that basis, we could not reject H_0. Hence, the lifetimes of the two types of filaments are deemed to be the same.

It is instructive to note that we could have come to exactly the same decision by assuming that $n_1 = 10$, making the total ranks for filaments type II, $T = 117$, with the attendant mean of $E(T) = \dfrac{n_1(n_1+n_2+1)}{2} = \dfrac{10(13+10+1)}{2} = 120$. The variance remains the same, $\sigma_T^2 = \dfrac{10 \times 13 \times (13+10+1)}{12} = 260$, so that $Z = \dfrac{117-126}{\sqrt{260}} = -0.18605$. We see that the calculated Z leads us to exactly the same conclusion as we reached earlier. The above decisions were reached because our alternative hypothesis is two-sided.

Suppose that we wanted to test H_0 against a one-sided alternative, H_1: Lifetime of filament 1 is greater than the lifetime of filament 2. Then, we could have used $n_1 = 13$, and the attendant $Z = 0.18605$ for our decision. In that case, if Z value had been greater than 2.575, we would have had to accept that H_1, meaning that the alternative hypothesis is true.

18.4 The Kruskal–Wallis Test (A Nonparametric Equivalent of One-Way ANOVA)

In experimentation which subsequently leads to one-way analysis of variance (ANOVA), there are situations where the normality assumptions of the model may not be satisfied. An alternative method which could be used for comparing k groups of observations or responses of k treatments without the requirement that the samples be normally distributed is the test attributed to Kruskal and Wallis (1952). The test is **nonparametric**. It is an extension of the Mann–Whitney test. This means that we make no assumptions about the distributions of the data generated in the experiment. Furthermore, we estimate no parameters in the process of our analysis. **The null hypothesis in this case is that all the k groups are identical which is tested against an alternative that some groups contain observations that are larger than others.** This test was developed as an alternative to the completely randomized design (one-way layout) and is therefore sensitive enough to detect differences in the means of the groups or treatments. We can therefore think of what we do in this this test as if we are comparing the means of k groups or treatments. We can therefore regard this test as the **nonparametric analogue** of the one-way ANOVA.

To carry out the analysis of the responses of an experiment using the Kruskal–Wallis test, we first rank all the observations in all the k groups in ascending order of magnitude, that is, we assign the smallest rank to the smallest observation, and the highest rank to the highest observation. For each observation y_{ij}, we assign the rank, r_{ij}. If ties occur, we assign the average of the ranks to the tied observations and assign the next higher rank to the next observation that is larger than the tied observations. Let $N = \sum_{i=1}^{k} n_i$. and $r_{i.}$ be the sum of the ranks for the ith group or ith treatment, which contains, n_i observations, then the test statistic is:

$$W = \frac{1}{s^2}\left[\sum_{i=1}^{k}\frac{r_i^2}{n_i} - \frac{N(N+1)}{4}\right] \tag{18.8}$$

The variance s^2 used in (18.8) is:

$$s^2 = \frac{1}{N-1}\left[\sum_{i=1}^{k}\sum_{j=1}^{n_i} r_{ij}^2 - \frac{N(N+1)^2}{4}\right] \tag{18.9}$$

We can easily see that s^2 is the variance of the ranks, because $(N + 1)/2$ is the mean of the numbers $1, 2, \ldots, N$. W, the test statistic is approximately distributed as chi-square with k-1 degrees of freedom. We therefore compare W with $\chi^2_{[\alpha,n-1]}$ and reject the null hypothesis if $W > \chi^2_{[\alpha,n-1]}$. If there are no ties, it can be shown that (17.9) simplifies to yield:

$$s^2 = \frac{N(N+1)}{12} \tag{18.10}$$

which in turn simplifies the formula for W in (17.8) so that

$$W = \frac{12}{N(N+1)}\sum_{i=1}^{k}\frac{r_i^2}{n_i} - 3(N+1) \tag{18.11}$$

■ **EXAMPLE 18.6** A nutritionist used four different methods to determine the nutrient content of a food. The percentage contents according to the four methods are:

Method 1	Method II	Method III	Method IV
48.2	51.1	50.1	47.6
50.8	50.3	54.2	48.2
49.5	52.3	52.2	47.2
49.7	51.9	53.8	48.3
		53.3	

Analyze the data using Kruskal–Wallis test at $\alpha = 5\%$ level of significance.

Analysis

In the following table, we assign ranks to all the observations in the experiment.

Method 1 (Ranks)	Method II (Ranks)	Method III (Ranks)	Method IV (Ranks)
48.2 (3.5)	51.1 (11)	50.1 (8)	47.6 (2)
50.8 (10)	50.3 (9)	54.2 (17)	48.2 (3.5)
49.5 (6)	52.3 (14)	52.2 (13)	47.2 (1)
49.7 (7)	51.9 (12)	53.8 (16)	48.3 (5)
		53.3 (15)	

The ranks are in the brackets and two observations are tied. $N = 4 + 4 + 5 + 4 = 17$. The total ranks for the different methods are:

Methods	I	II	III	IV
Total Rank	26.5	46	69	11.5

We test the hypothesis: H_0: The four methods generate equal measurements for nutrient content against the alternative: H_1: The measurements generated by some methods are different.

We need to obtain $s^2 = \dfrac{1}{N-1}\left[\sum_{i=1}^{k}\sum_{j=1}^{n_i} r_{ij}^2 - \dfrac{N(N+1)^2}{4}\right]$ and hence

$$W = \frac{1}{s^2}\left[\sum_{i=1}^{k} \frac{r_i^2}{n_i} - \frac{N(N+1)}{4}\right]$$

$$s^2 = \frac{1}{17-1}\left[3^2 + 10^2 + ... + 1^2 + 5^2 - \frac{17(18)^2}{4}\right] = 25.4687$$

$$W = \frac{1}{25.4687}\left[\frac{26.5^2}{4} + \frac{46^2}{4} + \frac{69^2}{4} + \frac{11.5^2}{4} - \frac{17(18)^2}{4}\right] = 21.629$$

But $W < \chi^2_{[0.05,16]} = 26.30$, so we accept the null hypothesis, H_0, and reject the alternative, H_1.

Reference

Kruskal, William H., and W. Allen Wallis. 1952. "Use of Ranks in One-Criterion Analysis of Variance." *Journal of the American Statistical Association* 47 (260): 583–621.

Exercises (Nonparametic Statistics)

18.1   Reanalyze Example 13.5, page 388

You are a test engineer, and you wish to test the mileages attained by the same make of car run on two new formulations of biofuels. The following are mileages recorded for the two types of fuel, each having been studied with 15 cars. In the study, cars were chosen at random from a large pool and assigned to the fuels in at random.

Fuel A	Fuel B
29	38
30	33
31	32
31	37
28	35
27	39
22	40
24	38
25	37
28	43
40	36
38	32
33	31
19	42
30	33

If normality could not be assumed for the data, test the hypothesis that the data on Fuel A and B are from the same or equal populations at 1% level of significance.

18.2   A teacher teaches the same statistical theory to two samples of students, and administers the same test at the end of the course. The scores of the students in the test are as follows:

Method I	Method II
77	72
88	82
62	98
66	99
98	84

Continued.

Method I	Method II
75	98
97	90
65	91
99	81
59	76
88	90

It is not certain that the samples are from normally distributed populations. Do the scores from the test come from same or equal populations? Test at 5% level of significance.

18.3 Recall data of Example 13.11:

The times used by a sample of 18 chemistry students from School A and a sample of 15 chemistry students from School B to complete a titration experiment are as follows:

A	2.3	6.7	3.8	5	4.9	6.1	4.4	5.2	3.9	4.8	4.6	5.7	5.3	4.7	4.2	5.7	4.8	4.7
B	6.7	7.3	4.4	8.3	6.2	4.3	5.5	3.2	8.4	7.3	5.5	4.8	4.9	6.7	7.5			

Test at 5% level of significance that data for both schools are from same or equal population, against that they are not.

18.4 High low-density lipoprotein (LDL) type of cholesterol is bad for our health. A larger sample of LDL levels (mg/dL) for some patients were obtained before and after they had been put on a cholesterol lowering treatment for 3 months. The observed values of LDL for each individual in the study are presented as follows:

LDL Before	LDL After
137	145
169	174
170	173
167	170
177	180
145	147
177	179
156	155
136	133
126	122
174	170
145	140
156	150
137	129

Continued.

LDL Before	LDL After
165	156
143	133
166	154
122	108
121	105
160	141
153	132
163	141
166	143
165	140
177	152
171	146

Test at 1% level of significance whether the two sets of LDLs are from the same or equal population.

18.5 A measure of performance on a machine was made for a sample of individuals for 2 consecutive weeks to assess the effect of repeated use. The data are shown in the following:

x1	x2
6.54	8.83
7.43	9.11
7.45	10.09
7.58	10.36
7.92	11.78
12.93	10.88
8.39	13.91
8.88	14.77
9.08	15.79
16.6	11.61
10.39	17.17
11.39	18
12.57	10.6
13.2	10.92
14.46	9.66
6.45	8.08
7.34	9.23
7.36	10.21

Continued.

x1	x2
7.49	6.78
7.83	12.09
7.88	12.36
13.47	8.3
14.3	10.4
8.99	15.26
9.42	16.67
17.9	14.3
11.3	10.44

Do the data suggest any difference in performance due to repeated use? Test at 5% level of significance.

18.6   Example 15.2 reanalyzed

In order to study the effect of temperature on the yield of a chemical process, the process, the yields of the process were studied at five settings of temperature as in the following table:

Temperature	1	2	3	4
40	30	36	37	40
50	41	43	48	36
60	43	47	47	61
70	43	45	61	64
80	41	46	52	62

If normality assumptions cannot be used here, test the hypothesis that the mean yield is the same at the different temperature settings at 5% level of significance.

18.7   Example 15.3 reanalyzed

Four methods for production of a fabric are being studied for the tensile strengths of the final products. Five measurements were made on material produced by each method, and the tensile strengths are recorded as follows:

Methods	1	2	3	4	5
1	24.95	26.44	28.57	23.17	26.34
2	30.41	29.23	28.87	32.95	35.87
3	26.49	28.5	33.13	32.36	28.95
4	30.66	29.72	30.14	30.44	28.73

Test at 5% level of significance for any difference in mean tensile strengths for the four different methods of production if it is not assumed that data are normal.

18.8   The cost of tuning up the heating system for single family houses in four cities A, B, C, and D which are close to each other were surveyed. The following amounts charged in dollars are:

A	B	C	D
100	144	210	178
123	186	234	190
89	177	189	177
156	103	204	223
210	110	278	197
134	134	216	103
167		145	
220			

Are the differences in the charges in the four cities significant at 5% level of significance?

18.9   Thirty-one tasters rate two flavors of the same cookie on some criteria with maximum score total of 100. The ratings are as follows:

Individual	Cookie 1	Cookie2
1	91	69
2	94	69
3	65	81
4	70	62
5	82	74
6	81	80
7	50	70
8	62	79
9	75	81
10	77	82
11	76	78
12	82	73
13	78	74
14	91	86
15	80	74
16	74	64
17	77	66
18	81	69
19	92	83
20	70	52

Continued.

Individual	Cookie 1	Cookie2
21	83	66
22	77	56
23	70	91
24	77	90
25	66	56
26	92	79
27	83	69
28	90	75
29	91	75
30	93	76
31	85	76

Can the cookies be considered to be similarly rated? Test at 5% level of significance.

18.10   Recall data of example 13.13.

Individual	Chocolate A	Chocolate B
1	9	7
2	7	6
3	9	3
4	8	5
5	7	7
6	9	6
7	8	8
8	9	5
9	7	9
10	7	8
11	8	6
12	9	7

Analyze the data using sign test. Test at 5% level of significance.

Appendix I—Table of Random Digits

Rows	1–5	6–10	11–15	16–20	21–25	26–30	31–35	36–40	41–45	46–50	50–55	56–60
1	30120	13850	81903	56587	39129	94727	78226	95207	76354	38718	01669	33878
2	69696	81799	27328	33287	35476	56650	57330	02079	82972	82287	75634	55692
3	17784	00005	25584	51364	02330	44697	39343	75351	34890	87080	64632	49892
4	35821	49630	87686	53852	56806	57379	26797	94186	19280	64342	78912	96759
5	75763	40570	04655	30679	00398	95493	91902	11365	83316	07188	27891	31001
6	87138	59718	98465	91803	82212	98364	92735	35410	78370	12608	21966	61052
7	11654	48648	36173	91029	30989	50408	12641	21246	16638	36266	85720	71967
8	41754	43010	16749	73270	70472	54849	40547	08786	29933	99319	62931	49428
9	20754	31665	87389	81339	79106	92621	89831	06127	45080	82142	94014	68695
10	11198	19216	82884	95655	63248	20721	32433	75861	77861	55700	83148	89194
11	19976	38351	52917	64368	80502	23253	88104	10826	06580	92391	49517	42460
12	30418	63656	60308	20088	86266	04800	71798	70611	16329	84639	31713	15717
13	03813	55452	12072	65137	68304	68835	96418	38610	10270	28783	91173	93877
14	36635	26966	00037	47517	11038	92559	13220	62684	25891	07377	84848	06536
15	10433	21728	44543	76598	55639	67408	07560	53371	80692	92056	56721	64087
16	53496	62915	73129	53761	90505	45837	85376	49571	96246	58693	41725	90269
17	65705	73230	23348	93900	98303	58430	86614	72905	54098	44340	32081	50058
18	77435	12048	82677	30529	60360	29855	43089	37825	28210	59090	14250	06752
19	20070	92043	24037	83583	25355	55480	11193	54184	18361	51270	98499	61398
20	55342	49934	11082	54818	79576	36977	37721	41914	39538	30179	49244	36056
21	07567	03612	81928	05111	43492	75714	97754	64447	25100	72756	02198	95592
22	68140	15564	89299	88433	55768	65473	32746	61493	19649	80919	38101	52314
23	90081	94223	88193	70464	63503	40140	97620	01805	92242	40242	64402	30683
24	63887	28878	95742	07029	52663	93720	89502	55551	41135	99199	38865	45337
25	06718	36267	04006	90154	86114	47543	82086	45230	58218	39799	06650	80884
26	02050	84631	38553	67622	11489	26119	62581	96146	93108	44378	32050	14470
27	19352	65919	06096	82444	20818	64342	87826	37332	77672	90399	32535	61002
28	91882	47249	86196	69169	22571	71419	58379	91656	44157	28443	16095	09186
29	21890	49206	28120	56209	21158	24639	17042	37034	86757	47673	02127	66913
30	71266	90921	67705	63770	41962	83737	63284	14900	54646	50966	47298	59165

Rows	1–5	6–10	11–15	16–20	21–25	26–30	31–35	36–40	41–45	46–50	50–55	56–60
						Columns						
31	06910	70975	81644	87240	93841	47206	50645	82994	72155	43492	03554	02371
32	67192	42127	36387	54886	50590	75099	43463	87116	04279	70709	94437	33201
33	13710	33875	80634	36230	55774	66277	52248	05794	93379	68338	07635	97764
34	92866	39641	43754	21920	10555	87509	68595	86524	54432	17106	85324	76684
35	73805	58967	53850	02887	24026	99077	88445	80076	03387	18983	01881	85633
36	82184	00144	36534	27061	20621	36213	89024	20944	83539	18079	04923	62871
37	74241	20930	55156	32173	74289	18739	89341	18749	95080	92976	77994	47274
38	98135	25411	49569	54907	69233	02176	23906	84356	56102	25131	65558	05969
39	79131	93858	23598	82938	85525	44416	73168	49378	02654	66410	84477	49229
40	14834	33594	82125	23681	02892	63118	96853	35632	33014	44085	66664	29900
41	94539	90867	63392	56092	72658	83677	31880	01922	52559	13759	49878	70038
42	94082	41975	02433	55758	41188	86952	35027	40867	82724	55949	03205	55231
43	63571	75636	64594	24715	74438	29793	60304	40566	79428	31355	49690	50023
44	23735	26942	09467	86699	46783	68004	49807	28067	58183	27562	04510	38564
45	23058	41489	09596	48957	78581	37003	02835	70991	00637	56205	55988	86716
46	04611	22913	53999	22825	29524	35725	32667	75573	63286	62191	19604	67222
47	27280	44306	61397	06112	84833	35907	84332	56427	18291	16725	05287	13393
48	43281	83120	04399	67542	04021	71163	61568	36141	08417	61576	47224	31488
49	81145	49102	26158	22946	56398	20575	58013	80801	52669	28443	02232	47913
50	02472	93734	25928	54042	23864	16400	12609	66281	97155	58773	79130	14579
51	21841	00021	49649	55396	49026	05092	86167	79487	97287	70064	58844	31775
52	61578	24025	05690	19185	65233	31338	68038	10503	09360	27241	83379	13417
53	80794	08643	61406	70009	31906	54962	02415	05169	24386	57836	03376	21700
54	80104	02474	11481	54034	73147	01883	05586	88572	81821	52034	82960	01114
55	80191	95373	08098	83779	39557	38779	79934	50648	51626	45459	55074	48268
56	45850	94026	49830	11281	83396	76894	63339	60038	44412	39201	10852	50561
57	62608	26249	74921	75064	81342	32050	81234	96110	96774	05872	71230	98739
58	78611	82103	22524	87069	57691	71013	11912	79660	14745	69357	40463	58575
59	42695	67012	53281	85270	03758	20525	56774	77059	51607	67027	73280	91777
60	91505	75468	25148	60872	09303	29766	46042	19720	82766	24178	96874	48006

Appendix II—Table of the Standard Normal Distribution

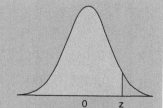

Z	0	0.01	0.02	0.03	0.04	0.05	0.06	0.07	0.08	0.09	Z
−3.7	0.0001	0.0001	0.0001	0.0001	0.0001	0.0001	0.0001	0.0001	0.0001	0.0001	−3.7
−3.6	0.0002	0.0002	0.0002	0.0001	0.0001	0.0001	0.0001	0.0001	0.0001	0.0001	−3.6
−3.5	0.0002	0.0002	0.0002	0.0002	0.0002	0.0002	0.0002	0.0002	0.0002	0.0002	−3.5
−3.4	0.0003	0.0003	0.0003	0.0003	0.0003	0.0003	0.0003	0.0003	0.0003	0.0002	−3.4
−3.3	0.0005	0.0005	0.0005	0.0004	0.0004	0.0004	0.0004	0.0004	0.0004	0.0004	−3.3
−3.2	0.0007	0.0007	0.0006	0.0006	0.0006	0.0006	0.0006	0.0005	0.0005	0.0005	−3.2
−3.1	0.0010	0.0009	0.0009	0.0009	0.0008	0.0008	0.0008	0.0008	0.0007	0.0007	−3.1
−3.0	0.0014	0.0013	0.0013	0.0012	0.0012	0.0011	0.0011	0.0011	0.0010	0.0010	−3.0
−2.9	0.0019	0.0018	0.0018	0.0017	0.0016	0.0016	0.0015	0.0015	0.0014	0.0014	−2.9
−2.8	0.0026	0.0025	0.0024	0.0023	0.0023	0.0022	0.0021	0.0021	0.0020	0.0019	−2.8
−2.7	0.0035	0.0034	0.0033	0.0032	0.0031	0.0030	0.0029	0.0028	0.0027	0.0026	−2.7
−2.6	0.0047	0.0045	0.0044	0.0043	0.0042	0.0040	0.0039	0.0038	0.0037	0.0036	−2.6
−2.5	0.0062	0.0060	0.0059	0.0057	0.0055	0.0054	0.0052	0.0051	0.0049	0.0048	−2.5
−2.4	0.0082	0.0080	0.0078	0.0076	0.0073	0.0071	0.0070	0.0068	0.0066	0.0064	−2.4
−2.3	0.0107	0.0104	0.0102	0.0099	0.0096	0.0094	0.0091	0.0089	0.0087	0.0084	−2.3
−2.2	0.0139	0.0136	0.0132	0.0129	0.0126	0.0122	0.0119	0.0116	0.0113	0.0110	−2.2
−2.1	0.0179	0.0174	0.0170	0.0166	0.0162	0.0158	0.0154	0.0150	0.0146	0.0143	−2.1
−2.0	0.0228	0.0222	0.0217	0.0212	0.0207	0.0202	0.0197	0.0192	0.0188	0.0183	−2.0
−1.9	0.0287	0.0281	0.0274	0.0268	0.0262	0.0256	0.0250	0.0244	0.0239	0.0233	−1.9
−1.8	0.0359	0.0352	0.0344	0.0336	0.0329	0.0322	0.0314	0.0307	0.0301	0.0294	−1.8
−1.7	0.0446	0.0436	0.0427	0.0418	0.0409	0.0401	0.0392	0.0384	0.0375	0.0367	−1.7
−1.6	0.0548	0.0537	0.0526	0.0516	0.0505	0.0495	0.0485	0.0475	0.0465	0.0455	−1.6
−1.5	0.0668	0.0655	0.0643	0.0630	0.0618	0.0606	0.0594	0.0582	0.0571	0.0559	−1.5
−1.4	0.0808	0.0793	0.0778	0.0764	0.0749	0.0735	0.0722	0.0708	0.0694	0.0681	−1.4
−1.3	0.0968	0.0951	0.0934	0.0918	0.0901	0.0885	0.0869	0.0853	0.0838	0.0823	−1.3
−1.2	0.1151	0.1131	0.1112	0.1094	0.1075	0.1057	0.1038	0.1020	0.1003	0.0985	−1.2

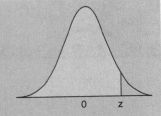

Z	0	0.01	0.02	0.03	0.04	0.05	0.06	0.07	0.08	0.09	Z
−1.1	0.1357	0.1335	0.1314	0.1292	0.1271	0.1251	0.1230	0.1210	0.1190	0.1170	−1.1
−1.0	0.1587	0.1563	0.1539	0.1515	0.1492	0.1469	0.1446	0.1423	0.1401	0.1379	−1.0
−0.9	0.1841	0.1814	0.1788	0.1762	0.1736	0.1711	0.1685	0.1660	0.1635	0.1611	−0.9
−0.8	0.2119	0.2090	0.2061	0.2033	0.2005	0.1977	0.1949	0.1922	0.1894	0.1867	−0.8
−0.7	0.2420	0.2389	0.2358	0.2327	0.2297	0.2266	0.2236	0.2207	0.2177	0.2148	−0.7
−0.6	0.2743	0.2709	0.2676	0.2644	0.2611	0.2579	0.2546	0.2514	0.2483	0.2451	−0.6
−0.5	0.3085	0.3050	0.3015	0.2981	0.2946	0.2912	0.2877	0.2843	0.2810	0.2776	−0.5
−0.4	0.3446	0.3409	0.3372	0.3336	0.3300	0.3264	0.3228	0.3192	0.3156	0.3121	−0.4
−0.3	0.3821	0.3783	0.3745	0.3707	0.3669	0.3632	0.3594	0.3557	0.3520	0.3483	−0.3
−0.2	0.4207	0.4168	0.4129	0.4091	0.4052	0.4013	0.3974	0.3936	0.3897	0.3859	−0.2
−0.1	0.4602	0.4562	0.4522	0.4483	0.4443	0.4404	0.4364	0.4325	0.4286	0.4247	−0.1
0.0	0.5000	0.4960	0.4920	0.4880	0.4840	0.4801	0.4761	0.4721	0.4681	0.4641	0.0
0.1	0.5398	0.5438	0.5478	0.5517	0.5557	0.5596	0.5636	0.5675	0.5714	0.5754	0.1
0.2	0.5793	0.5832	0.5871	0.5910	0.5948	0.5987	0.6026	0.6064	0.6103	0.6141	0.2
0.3	0.6179	0.6217	0.6255	0.6293	0.6331	0.6368	0.6406	0.6443	0.6480	0.6517	0.3
0.4	0.6554	0.6591	0.6628	0.6664	0.6700	0.6736	0.6772	0.6808	0.6844	0.6879	0.4
0.5	0.6915	0.6950	0.6985	0.7019	0.7054	0.7088	0.7123	0.7157	0.7190	0.7224	0.5
0.6	0.7258	0.7291	0.7324	0.7357	0.7389	0.7422	0.7454	0.7486	0.7518	0.7549	0.6
0.7	0.7580	0.7612	0.7642	0.7673	0.7704	0.7734	0.7764	0.7794	0.7823	0.7852	0.7
0.8	0.7881	0.7910	0.7939	0.7967	0.7996	0.8023	0.8051	0.8079	0.8106	0.8133	0.8
0.9	0.8159	0.8186	0.8212	0.8238	0.8264	0.8289	0.8315	0.8340	0.8365	0.8389	0.9
1.0	0.8413	0.8438	0.8461	0.8485	0.8508	0.8531	0.8554	0.8577	0.8599	0.8621	1.0
1.1	0.8643	0.8665	0.8686	0.8708	0.8729	0.8749	0.8770	0.8790	0.8810	0.8830	1.1
1.2	0.8849	0.8869	0.8888	0.8907	0.8925	0.8944	0.8962	0.8980	0.8997	0.9015	1.2
1.3	0.9032	0.9049	0.9066	0.9082	0.9099	0.9115	0.9131	0.9147	0.9162	0.9177	1.3
1.4	0.9192	0.9207	0.9222	0.9236	0.9251	0.9265	0.9279	0.9292	0.9306	0.9319	1.4
1.5	0.9332	0.9345	0.9357	0.9370	0.9382	0.9394	0.9406	0.9418	0.9430	0.9441	1.5
1.6	0.9452	0.9463	0.9474	0.9485	0.9495	0.9505	0.9515	0.9525	0.9535	0.9545	1.6
1.7	0.9554	0.9564	0.9573	0.9582	0.9591	0.9599	0.9608	0.9616	0.9625	0.9633	1.7

Z	0	0.01	0.02	0.03	0.04	0.05	0.06	0.07	0.08	0.09	Z
1.8	0.9641	0.9649	0.9656	0.9664	0.9671	0.9678	0.9686	0.9693	0.9700	0.9706	1.8
1.9	0.9713	0.9719	0.9726	0.9732	0.9738	0.9744	0.9750	0.9756	0.9762	0.9767	1.9
2.0	0.9773	0.9778	0.9783	0.9788	0.9793	0.9798	0.9803	0.9808	0.9812	0.9817	2.0
2.1	0.9821	0.9826	0.9830	0.9834	0.9838	0.9842	0.9846	0.9850	0.9854	0.9857	2.1
2.2	0.9861	0.9865	0.9868	0.9871	0.9875	0.9878	0.9881	0.9884	0.9887	0.9890	2.2
2.3	0.9893	0.9896	0.9898	0.9901	0.9904	0.9906	0.9909	0.9911	0.9913	0.9916	2.3
2.4	0.9918	0.9920	0.9922	0.9925	0.9927	0.9929	0.9931	0.9932	0.9934	0.9936	2.4
2.5	0.9938	0.9940	0.9941	0.9943	0.9945	0.9946	0.9948	0.9949	0.9951	0.9952	2.5
2.6	0.9953	0.9955	0.9956	0.9957	0.9959	0.9960	0.9961	0.9962	0.9963	0.9964	2.6
2.7	0.9965	0.9966	0.9967	0.9968	0.9969	0.9970	0.9971	0.9972	0.9973	0.9974	2.7
2.8	0.9974	0.9975	0.9976	0.9977	0.9977	0.9978	0.9979	0.9980	0.9980	0.9981	2.8
2.9	0.9981	0.9982	0.9983	0.9983	0.9984	0.9984	0.9985	0.9985	0.9986	0.9986	2.9
3.0	0.9987	0.9987	0.9987	0.9988	0.9988	0.9989	0.9989	0.9989	0.9990	0.9990	3.0
3.1	0.9990	0.9991	0.9991	0.9991	0.9992	0.9992	0.9992	0.9992	0.9993	0.9993	3.1
3.2	0.9993	0.9993	0.9994	0.9994	0.9994	0.9994	0.9994	0.9995	0.9995	0.9995	3.2
3.3	0.9995	0.9995	0.9996	0.9996	0.9996	0.9996	0.9996	0.9996	0.9996	0.9997	3.3
3.4	0.9997	0.9997	0.9997	0.9997	0.9997	0.9997	0.9997	0.9997	0.9998	0.9998	3.4
3.5	0.9998	0.9998	0.9998	0.9998	0.9998	0.9998	0.9998	0.9998	0.9998	0.9998	3.5
3.6	0.9998	0.9999	0.9999	0.9999	0.9999	0.9999	0.9999	0.9999	0.9999	0.9999	3.6
3.7	0.9999	0.9999	0.9999	0.9999	0.9999	0.9999	0.9999	0.9999	0.9999	0.9999	3.7

Appendix III—Table of the t-Distribution

df	t.10	t.05	t.025	t.01	t.005	df
1	3.078	6.314	12.706	31.821	63.657	1
2	1.886	2.920	4.303	6.965	9.925	2
3	1.638	2.353	3.182	4.541	5.841	3
4	1.533	2.132	2.776	3.747	4.604	4
5	1.476	2.015	2.571	3.365	4.032	5
6	1.440	1.943	2.447	3.143	3.707	6
7	1.415	1.895	2.365	2.998	3.499	7
8	1.397	1.860	2.306	2.896	3.355	8
9	1.383	1.833	2.262	2.821	3.250	9
10	1.372	1.812	2.228	2.764	3.169	10
11	1.363	1.796	2.201	2.718	3.106	11
12	1.356	1.782	2.179	2.681	3.055	12
13	1.350	1.771	2.160	2.650	3.012	13
14	1.345	1.761	2.145	2.624	2.977	14
15	1.341	1.753	2.131	2.602	2.947	15
16	1.337	1.746	2.120	2.583	2.921	16
17	1.333	1.740	2.110	2.567	2.898	17
18	1.330	1.734	2.101	2.552	2.878	18
19	1.328	1.729	2.093	2.539	2.861	19
20	1.325	1.725	2.086	2.528	2.845	20
21	1.323	1.721	2.080	2.518	2.831	21
22	1.321	1.717	2.074	2.508	2.819	22
23	1.319	1.714	2.069	2.500	2.807	23
24	1.318	1.711	2.064	2.492	2.797	24
25	1.316	1.708	2.060	2.485	2.787	25
26	1.315	1.706	2.056	2.479	2.779	26

df	t.10	t.05	t.025	t.01	t.005	df
27	1.314	1.703	2.052	2.473	2.771	27
28	1.313	1.701	2.048	2.467	2.763	28
29	1.311	1.699	2.045	2.462	2.756	29
30	1.310	1.697	2.042	2.457	2.750	30
31	1.309	1.696	2.040	2.453	2.744	31
32	1.309	1.694	2.037	2.449	2.738	32
33	1.308	1.692	2.035	2.445	2.733	33
34	1.307	1.691	2.032	2.441	2.728	34
35	1.306	1.690	2.030	2.438	2.724	35
36	1.306	1.688	2.028	2.434	2.719	36
37	1.305	1.687	2.026	2.431	2.715	37
38	1.304	1.686	2.024	2.429	2.712	38
39	1.304	1.685	2.023	2.426	2.708	39
40	1.303	1.684	2.021	2.423	2.704	40
41	1.303	1.683	2.020	2.421	2.701	41
42	1.302	1.682	2.018	2.418	2.698	42
43	1.302	1.681	2.017	2.416	2.695	43
44	1.301	1.680	2.015	2.414	2.692	44
45	1.301	1.679	2.014	2.412	2.690	45
46	1.300	1.679	2.013	2.410	2.687	46
47	1.300	1.678	2.012	2.408	2.685	47
48	1.299	1.677	2.011	2.407	2.682	48
49	1.299	1.677	2.010	2.405	2.680	49
50	1.299	1.676	2.009	2.403	2.678	50
51	1.298	1.675	2.008	2.402	2.676	51
52	1.298	1.675	2.007	2.400	2.674	52
53	1.298	1.674	2.006	2.399	2.672	53
54	1.297	1.674	2.005	2.397	2.670	54

df	t.10	t.05	t.025	t.01	t.005	df
55	1.297	1.673	2.004	2.396	2.668	55
56	1.297	1.673	2.003	2.395	2.667	56
57	1.297	1.672	2.002	2.394	2.665	57
58	1.296	1.672	2.002	2.392	2.663	58
59	1.296	1.671	2.001	2.391	2.662	59
60	1.296	1.671	2.000	2.390	2.660	60
61	1.296	1.670	2.000	2.389	2.659	61
62	1.295	1.670	1.999	2.388	2.657	62
63	1.295	1.669	1.998	2.387	2.656	63
64	1.295	1.669	1.998	2.386	2.655	64
65	1.295	1.669	1.997	2.385	2.654	65
66	1.295	1.668	1.997	2.384	2.652	66
67	1.294	1.668	1.996	2.383	2.651	67
68	1.294	1.668	1.995	2.382	2.650	68
69	1.294	1.667	1.995	2.382	2.649	69
70	1.294	1.667	1.994	2.381	2.648	70
71	1.294	1.667	1.994	2.380	2.647	71
72	1.293	1.666	1.993	2.379	2.646	72
73	1.293	1.666	1.993	2.379	2.645	73
74	1.293	1.666	1.993	2.378	2.644	74
75	1.293	1.665	1.992	2.377	2.643	75
76	1.293	1.665	1.992	2.376	2.642	76
77	1.293	1.665	1.991	2.376	2.641	77
80	1.292	1.664	1.990	2.374	2.639	80
85	1.292	1.663	1.988	2.371	2.635	85
95	1.291	1.661	1.985	2.366	2.629	95
100	1.290	1.660	1.984	2.364	2.626	100
150	1.287	1.655	1.976	2.351	2.609	150

Appendix IV—Table of the Upper α-Percentile of the Chi-Square Distribution

df	Values of α								
	0.99	0.975	0.95	0.9	0.1	0.05	0.025	0.01	0.005
1	0.000	0.001	0.004	0.016	2.706	3.841	5.024	9.210	7.879
2	0.020	0.051	0.103	0.211	4.605	5.991	7.378	13.277	10.597
3	0.115	0.216	0.352	0.584	6.251	7.815	9.348	16.812	12.838
4	0.297	0.484	0.711	1.064	7.779	9.488	11.143	18.475	14.860
5	0.554	0.831	1.145	1.610	9.236	11.070	12.833	21.666	16.750
6	0.872	1.237	1.635	2.204	10.645	12.592	14.449	23.209	18.548
7	1.239	1.690	2.167	2.833	12.017	14.067	16.013	26.217	20.278
8	1.646	2.180	2.733	3.490	13.362	15.507	17.535	27.688	21.955
9	2.088	2.700	3.325	4.168	14.684	16.919	19.023	29.141	23.589
10	2.558	3.247	3.940	4.865	15.987	18.307	20.483	30.578	25.188
11	3.053	3.816	4.575	5.578	17.275	19.675	21.920	33.409	26.757
12	3.571	4.404	5.226	6.304	18.549	21.026	23.337	34.805	28.300
13	4.107	5.009	5.892	7.042	19.812	22.362	24.736	36.191	29.819
14	4.660	5.629	6.571	7.790	21.064	23.685	26.119	38.932	31.319
15	5.229	6.262	7.261	8.547	22.307	24.996	27.488	40.289	32.801
16	5.812	6.908	7.962	9.312	23.542	26.296	28.845	41.638	34.267
17	6.408	7.564	8.672	10.085	24.769	27.587	30.191	42.980	35.718
18	7.015	8.231	9.390	10.865	25.989	28.869	31.526	44.314	37.156
19	7.633	8.907	10.117	11.651	27.204	30.144	32.852	46.963	38.582
20	8.260	9.591	10.851	12.443	28.412	31.410	34.170	48.278	39.997
21	8.897	10.283	11.591	13.240	29.615	32.671	35.479	49.588	41.401
22	9.542	10.982	12.338	14.041	30.813	33.924	36.781	50.892	42.796
23	10.196	11.689	13.091	14.848	32.007	35.172	38.076	53.486	44.181
24	10.856	12.401	13.848	15.659	33.196	36.415	39.364	54.776	45.559
25	11.524	13.120	14.611	16.473	34.382	37.652	40.646	56.061	46.928
26	12.198	13.844	15.379	17.292	35.563	38.885	41.923	57.342	48.290
27	12.879	14.573	16.151	18.114	36.741	40.113	43.195	58.619	49.645
28	13.565	15.308	16.928	18.939	37.916	41.337	44.461	59.893	50.993

df				Values of α				
	0.99	**0.975**	**0.95**	**0.9**	**0.1**	**0.05**	**0.025**	**0.01**	**0.005**
29	14.256	16.047	17.708	19.768	39.087	42.557	45.722	62.428	52.336
30	14.953	16.791	18.493	20.599	40.256	43.773	46.979	63.691	53.672
40	22.164	24.433	26.509	29.051	51.805	55.758	59.342	77.386	66.766
50	29.707	32.357	34.764	37.689	63.167	67.505	71.420	92.010	79.490
60	37.485	40.482	43.188	46.459	74.397	79.082	83.298	105.202	91.952
70	45.442	48.758	51.739	55.329	85.527	90.531	95.023	118.236	104.215
80	53.540	57.153	60.391	64.278	96.578	101.879	106.629	131.141	116.321
90	61.754	65.647	69.126	73.291	107.565	113.145	118.136	143.940	128.299
100	70.065	74.222	77.929	82.358	118.498	124.342	129.561	156.648	140.169

Appendix V—Table of the Upper α-Percentile of the F-Distribution

Denominator df	F-distribution ($\alpha = 0.05$)												
	Numerator Degree of Freedom												
	1	2	3	4	5	6	7	8	9	10	11	12	13
1	161.4	199.5	215.7	224.6	230.2	234	236.8	238.9	240.5	241.9	243	243.9	244.7
2	18.51	19	19.16	19.25	19.3	19.33	19.49	19.37	19.38	19.4	19.4	19.41	19.49
3	10.13	9.552	9.277	9.117	9.013	8.941	8.667	8.845	8.812	8.786	8.763	8.745	8.667
4	7.709	6.944	6.591	6.388	6.256	6.163	6.041	6.041	5.999	5.964	5.936	5.912	6.041
5	6.608	5.786	5.409	5.192	5.05	4.95	4.95	4.818	4.772	4.735	4.704	4.678	4.95
6	5.987	5.143	4.757	4.534	4.387	4.284	4.534	4.147	4.099	4.06	4.027	4	4.534
7	5.591	4.737	4.347	4.12	3.972	3.866	4.12	3.726	3.677	3.637	3.603	3.575	4.12
8	5.318	4.459	4.066	3.838	3.687	3.581	3.838	3.438	3.388	3.347	3.313	3.284	3.838
9	5.117	4.256	3.863	3.633	3.482	3.374	3.863	3.23	3.179	3.137	3.102	3.073	3.863
10	4.965	4.103	3.708	3.478	3.326	3.217	3.708	3.072	3.02	2.978	2.943	2.913	3.708
11	4.844	3.982	3.587	3.357	3.204	3.095	3.587	2.948	2.896	2.854	2.818	2.788	3.587
12	4.747	3.885	3.49	3.259	3.106	2.996	3.49	2.849	2.796	2.753	2.717	2.687	3.49
13	4.667	3.806	3.411	3.179	3.025	2.915	3.411	2.767	2.714	2.671	2.635	2.604	3.411
14	4.6	3.739	3.344	3.112	2.958	2.848	3.344	2.699	2.646	2.602	2.565	2.534	3.344
15	4.543	3.682	3.287	3.056	2.901	2.79	3.287	2.641	2.588	2.544	2.507	2.475	3.287
16	4.494	3.634	3.239	3.007	2.852	2.741	3.239	2.591	2.538	2.494	2.456	2.425	3.239
17	4.451	3.592	3.197	2.965	2.81	2.699	3.197	2.548	2.494	2.45	2.413	2.381	3.197
18	4.414	3.555	3.16	2.928	2.773	2.661	3.16	2.51	2.456	2.412	2.374	2.342	3.16
19	4.381	3.522	3.127	2.895	2.74	2.628	3.127	2.477	2.423	2.378	2.34	2.308	3.127
20	4.351	3.493	3.098	2.866	2.711	2.599	3.098	2.447	2.393	2.348	2.31	2.278	3.098
21	4.325	3.467	3.072	2.84	2.685	2.573	3.072	2.42	2.366	2.321	2.283	2.25	3.072
22	4.301	3.443	3.049	2.817	2.661	2.549	3.049	2.397	2.342	2.297	2.259	2.226	3.049
23	4.279	3.422	3.028	2.796	2.64	2.528	3.028	2.375	2.32	2.275	2.236	2.204	3.028
24	4.26	3.403	3.009	2.776	2.621	2.508	3.009	2.355	2.3	2.255	2.216	2.183	3.009
25	4.242	3.385	2.991	2.759	2.603	2.49	2.991	2.337	2.282	2.236	2.198	2.165	2.991
26	4.225	3.369	2.975	2.743	2.587	2.474	3.369	2.321	2.265	2.22	2.181	2.148	3.369
27	4.21	3.354	2.96	2.728	2.572	2.459	2.96	2.305	2.25	2.204	2.166	2.132	2.96
28	4.196	3.34	2.947	2.714	2.558	2.445	3.34	2.291	2.236	2.19	2.151	2.118	3.34
29	4.183	3.328	2.934	2.701	2.545	2.432	2.934	2.278	2.223	2.177	2.138	2.104	2.934
30	4.171	3.316	2.922	2.69	2.534	2.421	3.316	2.266	2.211	2.165	2.126	2.092	3.316

Denominator df	F-distribution ($\alpha = 0.025$)												
	Numerator Degree of Freedom												
	1	2	3	4	5	6	7	8	9	10	11	12	13
1	647.8	799.5	864.2	899.6	921.8	937.1	948.2	956.7	963.3	968.6	973	976.7	979.8
2	38.51	39	39.17	39.25	39.3	39.33	39.36	39.37	39.39	39.4	39.41	39.41	39.42
3	17.44	16.04	15.44	15.1	14.88	14.73	14.62	14.54	14.47	14.42	14.37	14.34	14.3
4	12.22	10.65	9.979	9.605	9.364	9.197	9.074	8.98	8.905	8.844	8.794	8.751	8.715
5	10.01	8.434	7.764	7.388	7.146	6.978	6.853	6.757	6.681	6.619	6.568	6.525	6.488
6	8.813	7.26	6.599	6.227	5.988	5.82	5.695	5.6	5.523	5.461	5.41	5.366	5.329
7	8.073	6.542	5.89	5.523	5.285	5.119	4.995	4.899	4.823	4.761	4.709	4.666	4.628
8	7.571	6.059	5.416	5.053	4.817	4.652	4.529	4.433	4.357	4.295	4.243	4.2	4.162
9	7.209	5.715	5.078	4.718	4.484	4.32	4.197	4.102	4.026	3.964	3.912	3.868	3.831
10	6.937	5.456	4.826	4.468	4.236	4.072	3.95	3.855	3.779	3.717	3.665	3.621	3.583
11	6.724	5.256	4.63	4.275	4.044	3.881	3.759	3.664	3.588	3.526	3.474	3.43	3.392
12	6.554	5.096	4.474	4.121	3.891	3.728	3.607	3.512	3.436	3.374	3.321	3.277	3.239
13	6.414	4.965	4.347	3.996	3.767	3.604	3.483	3.388	3.312	3.25	3.197	3.153	3.115
14	6.298	4.857	4.242	3.892	3.663	3.501	3.38	3.285	3.209	3.147	3.095	3.05	3.012
15	6.2	4.765	4.153	3.804	3.576	3.415	3.293	3.199	3.123	3.06	3.008	2.963	2.925
16	6.115	4.687	4.077	3.729	3.502	3.341	3.219	3.125	3.049	2.986	2.934	2.889	2.851
17	6.042	4.619	4.011	3.665	3.438	3.277	3.156	3.061	2.985	2.922	2.87	2.825	2.786
18	5.978	4.56	3.954	3.608	3.382	3.221	3.1	3.005	2.929	2.866	2.814	2.769	2.73
19	5.922	4.508	3.903	3.559	3.333	3.172	3.051	2.956	2.88	2.817	2.765	2.72	2.681
20	5.871	4.461	3.859	3.515	3.289	3.128	3.007	2.913	2.837	2.774	2.721	2.676	2.637
21	5.827	4.42	3.819	3.475	3.25	3.09	2.969	2.874	2.798	2.735	2.682	2.637	2.598
22	5.786	4.383	3.783	3.44	3.215	3.055	2.934	2.839	2.763	2.7	2.647	2.602	2.563
23	5.75	4.349	3.75	3.408	3.183	3.023	2.902	2.808	2.731	2.668	2.615	2.57	2.531
24	5.717	4.319	3.721	3.379	3.155	2.995	2.874	2.779	2.703	2.64	2.586	2.541	2.502
25	5.686	4.291	3.694	3.353	3.129	2.969	2.848	2.753	2.677	2.613	2.56	2.515	2.476
26	5.659	4.265	3.67	3.329	3.105	2.945	2.824	2.729	2.653	2.59	2.536	2.491	2.451
27	5.633	4.242	3.647	3.307	3.083	2.923	2.802	2.707	2.631	2.568	2.514	2.469	2.429
28	5.61	4.221	3.626	3.286	3.063	2.903	2.782	2.687	2.611	2.547	2.494	2.448	2.409
29	5.588	4.201	3.607	3.267	3.044	2.884	2.763	2.669	2.592	2.529	2.475	2.43	2.39
30	5.568	4.182	3.589	3.25	3.026	2.867	2.746	2.651	2.575	2.511	2.458	2.412	2.372

Denominator df	\multicolumn{13}{c}{F-distribution ($\alpha=0.01$) Numerator Degree of Freedom}

Denominator df	1	2	3	4	5	6	7	8	9	10	11	12	13
1	4052	5000	5403	5625	5764	5859	5928	5981	6022	6056	6083	6106	6126
2	98.5	99	99.17	99.25	99.3	99.33	99.36	99.37	99.39	99.4	99.41	99.42	99.42
3	34.12	30.82	29.46	28.71	28.24	27.91	27.67	27.49	27.35	27.23	27.13	27.05	26.98
4	21.2	18	16.69	15.98	15.52	15.21	14.98	14.8	14.66	14.55	14.45	14.37	14.31
5	16.26	13.27	12.06	11.39	10.97	10.67	10.46	10.29	10.16	10.05	9.963	9.888	9.825
6	13.75	10.92	9.78	9.148	8.746	8.466	8.26	8.102	7.976	7.874	7.79	7.718	7.657
7	12.25	9.547	8.451	7.847	7.46	7.191	6.993	6.84	6.719	6.62	6.538	6.469	6.41
8	11.26	8.649	7.591	7.006	6.632	6.371	6.178	6.029	5.911	5.814	5.734	5.667	5.609
9	10.56	8.022	6.992	6.422	6.057	5.802	5.613	5.467	5.351	5.257	5.178	5.111	5.055
10	10.04	7.559	6.552	5.994	5.636	5.386	5.2	5.057	4.942	4.849	4.772	4.706	4.65
11	9.646	7.206	6.217	5.668	5.316	5.069	4.886	4.744	4.632	4.539	4.462	4.397	4.342
12	9.33	6.927	5.953	5.412	5.064	4.821	4.64	4.499	4.388	4.296	4.22	4.155	4.1
13	9.074	6.701	5.739	5.205	4.862	4.62	4.441	4.302	4.191	4.1	4.025	3.96	3.905
14	8.862	6.515	5.564	5.035	4.695	4.456	4.278	4.14	4.03	3.939	3.864	3.8	3.745
15	8.683	6.359	5.417	4.893	4.556	4.318	4.142	4.004	3.895	3.805	3.73	3.666	3.612
16	8.531	6.226	5.292	4.773	4.437	4.202	4.026	3.89	3.78	3.691	3.616	3.553	3.498
17	8.4	6.112	5.185	4.669	4.336	4.102	3.927	3.791	3.682	3.593	3.519	3.455	3.401
18	8.285	6.013	5.092	4.579	4.248	4.015	3.841	3.705	3.597	3.508	3.434	3.371	3.316
19	8.185	5.926	5.01	4.5	4.171	3.939	3.765	3.631	3.523	3.434	3.36	3.297	3.242
20	8.096	5.849	4.938	4.431	4.103	3.871	3.699	3.564	3.457	3.368	3.294	3.231	3.177
21	8.017	5.78	4.874	4.369	4.042	3.812	3.64	3.506	3.398	3.31	3.236	3.173	3.119
22	7.945	5.719	4.817	4.313	3.988	3.758	3.587	3.453	3.346	3.258	3.184	3.121	3.067
23	7.881	5.664	4.765	4.264	3.939	3.71	3.539	3.406	3.299	3.211	3.137	3.074	3.02
24	7.823	5.614	4.718	4.218	3.895	3.667	3.496	3.363	3.256	3.168	3.094	3.032	2.977
25	7.77	5.568	4.675	4.177	3.855	3.627	3.457	3.324	3.217	3.129	3.056	2.993	2.939
26	7.721	5.526	4.637	4.14	3.818	3.591	3.421	3.288	3.182	3.094	3.021	2.958	2.904
27	7.677	5.488	4.601	4.106	3.785	3.558	3.388	3.256	3.149	3.062	2.988	2.926	2.871
28	7.636	5.453	4.568	4.074	3.754	3.528	3.358	3.226	3.12	3.032	2.959	2.896	2.842
29	7.598	5.42	4.538	4.045	3.725	3.499	3.33	3.198	3.092	3.005	2.931	2.868	2.814
30	7.562	5.39	4.51	4.018	3.699	3.473	3.304	3.173	3.067	2.979	2.906	2.843	2.789

Appendix VI—Table for Wilcoxon Signed-Rank Test

	Level of Significance (α)				
n	0.1	0.05	0.02	0.01	0.001
5	0				
6	2	0			
7	3	2	0		
8	5	3	1	0	
9	8	5	3	1	
10	10	8	5	3	
11	13	10	7	5	0
12	17	13	9	7	1
13	21	17	12	9	2
14	25	21	15	12	4
15	30	25	19	15	6
16	35	29	23	19	8
17	41	34	27	23	11
18	47	40	32	27	14
19	53	46	37	32	18
20	60	52	43	37	21
21	67	58	49	42	25
22	75	65	56	48	30
23	83	73	62	54	35
24	91	81	69	61	40
25	100	89	77	68	45
26	110	98	84	75	51
27	119	107	92	83	57
28	130	116	101	91	64
29	140	126	110	100	71
30	151	137	120	109	78
31	163	147	130	118	86
32	175	159	140	128	94
33	187	170	151	138	102
34	200	182	162	148	111
35	213	195	173	159	120

	Level of Significance (α)				
n	0.1	0.05	0.02	0.01	0.001
36	227	208	185	171	130
37	241	221	198	182	140
38	256	235	211	194	150
39	271	249	224	207	161
40	286	264	238	220	172
41	302	279	252	233	183
42	319	294	266	247	195
43	336	310	280	261	207
44	353	327	296	276	220
45	371	343	312	291	233
46	389	361	328	307	246
47	407	378	345	322	260
48	426	396	362	339	274
49	446	415	379	355	289
50	466	434	397	373	304

level of significance		n1 = smaller sample	n2 = larger sample size																	
			1	2	3	4	5	6	7	8	9	10	11	12	13	14	15	16	17	18
two-sided	0.2	1																		
one-sided	0.1	2																		
		3		3	7															
		4		3	7	13														
		5		4	8	14	20													
		6		4	9	15	22	30												
		7		4	10	16	23	32	41											
		8		5	11	17	25	34	44	55										
		9	1	5	11	19	27	36	46	58	70									
		10	1	6	12	20	28	38	49	60	73	87								
		11	1	6	13	21	30	40	51	63	76	91	106							
		12	1	7	14	22	32	42	54	66	80	94	110	127						
		13		7	15	23	33	44	56	69	83	98	114	131	149					
		14	1	7	16	25	35	46	59	72	86	102	118	136	154	174				
		15	1	8	16	26	37	48	61	75	90	106	123	141	159	179	200			
		16	1	8	17	27	38	50	64	78	93	109	127	145	165	185	206	229		
		17	1	9	18	28	40	52	66	81	97	113	131	150	170	190	212	235	259	
		18	1	9	19	30	42	55	69	84	100	117	135	155	175	196	218	242	266	291
		19	2	10	20	31	43	57	71	87	103	121	139	159	180	202	224	248		

level of significance		n1 = smaller sample size	n2 = larger sample size																	
			1	2	3	4	5	6	7	8	9	10	11	12	13	14	15	16	17	18
two-sided	0.1	1																		
one-sided	0.05	2																		
		3			6															
		4		6	7	13														
		5		3	7	12	19													
		6		3	8	13	20	28												

level of significance	n1 = smaller sample size	1	2	3	4	5	6	7	8	9	10	11	12	13	14	15	16	17	18
								n2 = larger sample size											
	7		3	8	14	21	29	39											
	8		4	9	15	23	31	41	51										
	9		4	9	16	24	33	43	54	66									
	10		4	10	17	26	35	45	56	69	82								
	11		4	11	18	27	37	47	59	72	86	100							
	12		5	11	19	28	38	49	62	75	89	104	120						
	13		5	12	20	30	40	52	64	78	92	108	125	149					
	14		5	13	21	31	42	54	67	81	96	112	129	147	166				
	15		6	13	22	33	44	56	69	84	99	116	133	152	171	192			
	16		6	14	24	34	46	58	72	87	103	120	138	156	176	197	219		
	17		6	15	28	40	52	66	81	97	113	131	150	170	190	212	235	259	
	18		7	15	26	37	49	63	77	93	110	127	146	166	187	208	231	255	280
	19	1	7	16	27	38	51	65	80	96	113	131	150	171	192	214	237		

level of significance	n1 = larger sample size	1	2	3	4	5	6	7	8	9	10	11	12	13	14	15	16	17	18
								n2 = smaller sample size											
two-sided 0.05	1																		
one-sided 0.025	2																		
	3																		
	4				10														
	5			6	11	17													
	6			7	12	18	26												
	7			7	13	20	27	36											
	8		3	8	14	21	29	38	49										
	9		3	8	14	22	31	40	51	62									
	10		3	9	15	23	32	42	53	65	78								
	11		3	9	16	24	34	44	55	68	81	96							
	12		4	10	17	26	35	46	58	71	84	99	115						
	13		4	10	18	27	37	48	60	73	88	103	119	136					
	14		4	11	19	28	38	50	62	76	91	106	123	141	160				
	15		4	11	20	29	40	52	65	79	94	110	127	145	164	184			
	16		4	12	21	30	42	54	67	82	97	113	131	150	161	190	211		
	17		5	12	21	32	43	56	70	84	100	117	135	154	174	195	217	240	

level of significance	n1 = larger sample size																	
		n2 = smaller sample size																
		1	2	3	4	5	6	7	8	9	10	11	12	13	14	15	16	17	18
	18		5	13	22	33	45	58	72	87	113	121	139	158	179	200	222	246	270
	19		5	13	23	34	46	60	74	90	107	124	143	163	182	205	228			

level of significance		n1 = larger sample size	**n2 = smaller sample size**																
			1	2	3	4	5	6	7	8	9	10	11	12	13	14	15	16	17	18
two-sided	0.01	1																		
one-sided	0.005	2																		
		3																		
		4																		
		5					15													
		6				10	16	13												
		7				10	16	24	32											
		8				11	17	25	34	43										
		9			6	11	18	26	35	45	56									
		10			6	12	19	27	37	47	58	71								
		11			6	12	20	28	38	49	61	73	87							
		12			7	13	21	30	40	51	63	76	90	105						
		13			7	14	22	31	44	53	64	79	93	109	125					
		14			7	14	22	32	43	54	67	81	96	112	129	147				
		15			8	15	23	33	44	56	69	84	99	115	133	151	171			
		16			8	15	24	34	46	58	72	86	102	119	130	155	175	196		
		17			8	16	25	36	47	60	74	89	105	122	140	159	180	201	223	
		18			8	16	26	37	49	62	76	92	108	125	144	163	184	206	228	252
		19		3	9	17	27	38	50	64	78	94	111	129	147	168	189	210		

Solutions to Exercises

Exercise 1.1

Label	Grade 9	Expense	Gender
02	Sven	26	M
09	Lona	27	F
20	Anne	19	F
21	Brianna	22	F
24	Henry	11	M
26	Lu	29	F
36	Cassie	13	F
38	Brian	13	M
40	Erin	17	F
41	Briahn	18	M

$$\overline{X} = \frac{195}{10} = 19.5; \hat{p} = \frac{6}{10} = 0.6$$

Exercise 1.3

Label	Grade 7	Expense	Gender
05	Matthew	8	M
07	Derek	17	M
08	Judith	24	F
15	Lauren	10	F
16	Graham	20	M
20	Mark	23	M
29	Robin	17	F
33	Scott	18	M
40	Brice	23	M
49	Jeremy	20	M

$$\overline{X} = \frac{180}{10} = 18.0; \hat{p} = \frac{3}{10} = 0.3$$

Exercise 1.5

From stratum I, using row 11, column 10.

Label	Grade 9	Expense	Gender
03	Scot	30	M
08	Crystal	20	F
20	Anne	19	F
04	Miles	25	M
07	Ramon	18	M
17	Lance	23	M
11	Paul	21	M
29	Margaret	18	F
13	Matt	30	M

Stratum II (Grade 8) using row 16, column 9.

Label	Grade 8	Expense	Gender
15	Amanda	10	F
12	Carrie	6	F
37	Alberto	24	M
24	Collin	20	M
41	Hadiza	17	F
02	Hoe	24	F
32	Cory	9	M
30	Amy	10	F

Stratum III (Grade 7) using row 14, column 5.

Label	Grade 7	Expense	Gender
00	Sarah	6	F
03	Daniel	13	M
17	Amy	9	F
11	Travis	23	M
25	Fatah	22	F
13	Nicholas	6	M
22	Sandra	5	F

$$\overline{X}_w = \frac{50}{150}(22.6667) + \frac{50}{150}(15) + \frac{50}{150}(12) = 16.5556$$

$$\hat{p}_m = \frac{50}{150}\left(\frac{6}{9}\right) + \frac{50}{150}\left(\frac{3}{8}\right) + \frac{50}{150}\left(\frac{3}{7}\right) = 0.4901$$

Exercise 1.7

Row 17, column 7 (Grades 7–9)

Label	Grade 7–9	Expense	Gender
023	Ashley	27	F
009	Lona	27	F
147	Kimberly	20	F
098	Humphrey	15	M
058	Courtney	21	F
036	Cassie	13	F
029	Margaret	18	F
006	Phil	25	M
037	Allison	13	F
111	Travis	23	M

$$\overline{X} = \frac{202}{10} = 20.2; \hat{p}_m = \frac{3}{10} = 0.3$$

Exercise 1.9

Simple random sample of size 11, row 7, column 15.

Label	Grade 9	Expense	Gender
029	Margaret	18	F
040	Erin	17	F
061	Rachel	25	F
069	Brent	21	M
072	David	10	M
082	Cory	9	M
087	Alberto	24	M
101	Ben	17	M
132	Bobbi	12	F
142	Paul	17	M
146	Amanda	23	F

$$\overline{X} = \frac{193}{11} = 17.55; \hat{p}_f = \frac{5}{11} = 0.450$$

Exercise 1.11

1 in 15 systematic sampling; labeling up to k; 00, 01, 02, . . . , 14. Therefore $a = 00 \Rightarrow 1$

Label	Grade 7–9	Expense	Gender
1	Jan	21	F
16	Dennis	11	M
31	Penny	20	F
46	Kirstin	14	F
61	Kelly	12	F
76	Devan	7	F
91	Adil	18	M
106	Matthew	8	M
121	Mark	23	M
136	Joshua	11	M

$$\overline{X} = \frac{145}{10} = 14.5; \hat{p}_m = \frac{5}{10} = 0.50$$

Exercise 1.13

Biomedics Aged 40–44 (Stratum I)	Investment	Gender	Biomedics Aged 35–39 (Stratum II)	Investment	Gender
08	8	M	01	9.5	M
13	4.7	M	02	22.7	M
15	14.9	F	04	16	F
17	9.1	F	13	8	M
19	9.7	F	18	14.6	F
20	14.8	F	20	9.8	F

$$\text{weighted mean} = \overline{X} = \frac{21}{42}\left(\frac{61.2}{6}\right) + \frac{21}{42}\left(\frac{80.6}{6}\right) = 11.81667$$

$$\text{weighted proportion of females} = \hat{p}_{\text{weighted}} = \frac{21}{42}\left(\frac{4}{6}\right) + \frac{21}{42}\left(\frac{3}{6}\right) = 0.5833$$

Exercise 1.15

We assign labels to the classes as follows:

Class	S193s1-01	S193s1-02	S193s1-03	S193s1-04	S193s1-05	S193s2-01	S193s2-02
Class-Size	58	56	60	60	57	55	60
Labels	00	01	02	03	04	05	06
Class	S193s2-03	S193s2-04	S193s2-05	S193sum1-01	S193sum1-02		
Class-Size	57	60	60	56	60		
Labels	07	08	09	10	11		

Using row 14 and column 16 of the table of random digits, the selected clusters are:

Label	Cluster	Size
04	S193s1-05	57
06	S193s2-02	60
07	S193s2-03	57
10	S193sum1-01	56
Total		230

Exercise 1.17

Using row 16, column 17, the selected clusters are:

Label	Church Name	Size
24	St. Joseph	239
02	St Peter	110
30	City Assembly	504
23	Stanley Memorial	622
00	Blaine Assembly	537
Total		**2012**

Exercise 1.19

Label: 000, 001, 002, . . . , 119
Using row 14, column 19, the simple random sample is:

Label	Weights (X)	Label	Weights (X)
103	38	36	55
73	76	29	46
104	51	6	34
75	52	37	37
87	44	111	33
50	44	82	41
57	39	56	34
23	66	51	52
9	49	114	45
98	37	100	53
58	43	**Total**	969

$$\text{mean weight} = \bar{x} = \frac{\sum_{i=1}^{21} x_i}{21} = \frac{969}{21} = 46.14286$$

Exercise 1.21

Using row 12, column 14, we obtain a 1 in 6 systematic sample:
Labels: 0, 1, 2, 3, 4, 5
First label chosen is: 0

Chosen sample is:

Chosen 1 in Systematic Sample			
Serial Number	**First Name**	**Weight**	**Height**
0	Nicole	126	64
6	Abby	170	66
12	Brianna	176	69
18	Krystal	145	61
24	Patricia	188	63
30	Pamela	172	71
36	Phil	216	57
42	Sharon	186	73
48	Joy	172	66
54	Holly	145	67
60	Blaine	180	69
66	Chad	214	63
72	Matt	167	66
78	Juan	191	62

$$\text{mean weight} = \bar{x} = \frac{\sum_{i=1}^{13} x_i}{14} = \frac{2448}{14} = 174.86$$

$$\text{mean height} = \bar{y} = \frac{\sum_{i=1}^{13} y_i}{14} = \frac{917}{14} = 65.5$$

Exercise 2.1

Classes	Class Boundaries	Class Midpoints	Class Frequencies	Frequencies, Cumulative	Relative Frequencies	Relative Cumulative Frequencies
30–39	29.5–39.5	34.5	7	7	7/132	7/132
40–49	39.5–49.5	44.5	29	36	29/132	36/132
50–59	49.50–59.50	54.5	45	81	45/132	81/132
60–69	59.50–69.50	64.5	33	114	33/132	114/132
70–79	69.50–79.50	74.5	13	127	13/132	127/132
80–89	79.50–89.50	84.5	3	130	3/132	130/132
90–99	89.90–99.90	94.5	2	132	2/132	132/132

Exercise 2.3

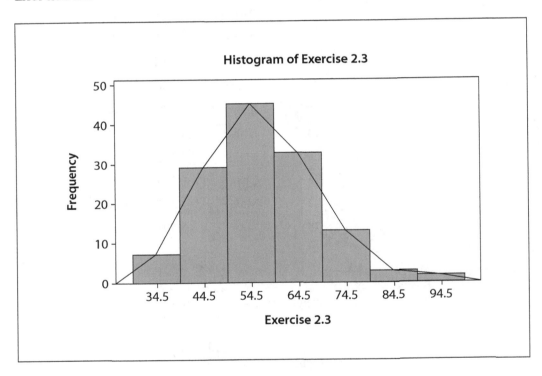

Exercise 2.5

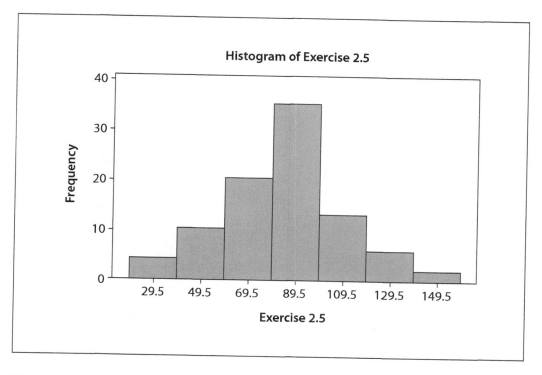

Exercise 2.7

Classes	Class Boundaries	Class Midpoints	Class Frequencies	Frequencies, Cumulative	Relative Frequencies	Relative Cumulative Frequencies
125–134	124.5–134.5	129.5	2	2	2/81	2/81
135–144	134.5–144.5	139.5	1	3	1/81	3/81
145–154	144.5–154.5	149.5	5	8	5/81	8/81
155–164	154.5–164.5	159.5	6	14	6/81	14/81
165–174	164.5–174.5	169.5	25	39	25/81	39/81
175–184	174.5–184.5	179.5	15	54	15/81	54/81
185–194	184.5–194.5	189.5	18	72	18/81	72/81
195–204	194.5–204.5	199.5	3	75	3/81	75/81
205–214	204.5–214.5	209.5	4	79	4/81	79/81
215–224	214.5–224.5	219.50	2	81	2/81	81/81

Exercise 2.9

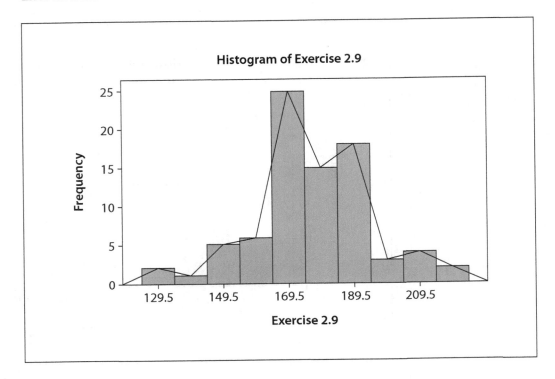

Exercise 2.11

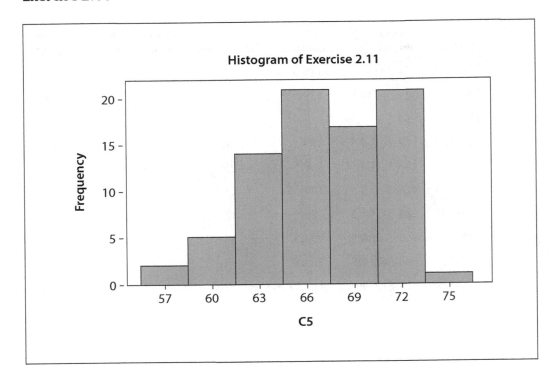

Exercise 2.13

Classes	Class Boundaries	Class Midpoints	Class Frequencies	Frequencies, Cumulative	Relative Frequencies	Relative Cumulative Frequencies
5–9	4.5–9.5	7	17	17	17/150	17/150
10–14	9.5–14.5	12	35	52	35/150	52/150
15–19	14.5–19.5	17	43	95	43/150	95/150
20–24	19.5–24.5	22	40	135	40/150	135/150
55–29	24.5–29.5	27	13	148	13/150	148/150
30–34	29.5–34.5	32	2	150	2/150	150/150

Exercise 2.15

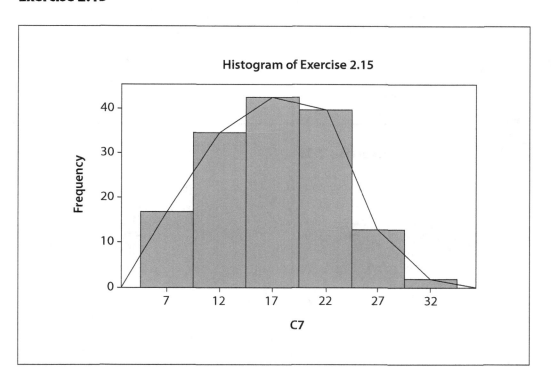

Exercise 2.17

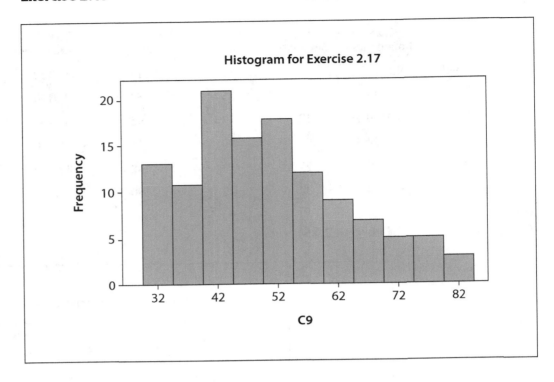

Exercise 2.19

```
Stem-and-leaf of Exercise 2.19  N = 132
Leaf Unit = 1.0

   4    3 0113
   7    3 557
  16    4 003333444
  36    4 5555555566677778899999
  47    5 00112223344
 (34)   5 5555555666677777888888889999999999
  51    6 0111111122222223344444444
  26    6 66677999
  18    7 01111234
  10    7 56789
   5    8 12
   3    8 8
   2    9 01
```

Exercise 2.21

```
Stem-and-leaf of Exercise 2.21 N = 81
Leaf Unit = 1.0

   1    12 6
   2    13 0
   8    14 355779
  13    15 56789
  22    16 057788999
 (24)   17 000001112222233445668888
  35    18 00002224668889
  21    19 0000122223446
   8    20 04
   6    21 224466
```

Exercise 2.23

```
Stem-and-leaf of Exercise 2.23 N = 150
Leaf Unit = 1.0

   1     0 5
  12     0 66666677777
  17     0 88999
  29     1 000000111111
  45     1 2222222333333333
  57     1 444444455555
 (21)    1 666666677777777777777
  72     1 88888888888999999
  55     2 00000000001111111111
  35     2 2222233333333
  22     2 4444444555555
   9     2 677
   6     2 8999
   2     3 00
```

Exercise 3.1

Age

    30 29 21 33 24 28 57 28 48 60 20 19 53 34 26 18 26
    25 48 24 48 28 52 32 46 20 23 26 27 46 37 27 25 55
    27 53 50 18 36 52 29 31 17 24 37 48 23 45 21 22

Array:

    17 18 18 19 20 20 21 21 22 23 23 24 24 24 25 25 26
    26 26 27 27 27 28 28 28 29 29 30 31 32 33 34 36 37
    37 45 46 46 48 48 48 48 50 52 52 53 53 55 57 60

Five number summary:
Minimum = 17 $Q_1 = 24$ Median = $(28 + 29)/2 = 28.50$ $Q_3 = 46$
Maximum = 60

Exercise 3.3

Hemoglobin

    13.50 15.90 13.90 15.85 13.55 10.95 12.50 15.60 13.05 12.20
    14.25 11.50 14.00 14.05 11.40 14.65 15.40 12.80 14.30 14.70
    14.90 15.20 14.90 11.50 14.45 14.05 12.20 15.35 14.15 14.50
    11.10 14.85 14.60 15.60 15.45 14.60 12.75 15.20 15.65 15.30
    12.75 13.10 11.40 13.30 13.70 15.10 13.55 15.95 11.90 14.20

Array

    10.95 11.10 11.40 11.40 11.50 11.50 11.90 12.20 12.20 12.50
    12.75 12.75 12.80 13.05 13.10 13.30 13.50 13.55 13.55 13.70
    13.90 14.00 14.05 14.05 14.15 14.20 14.25 14.30 14.45 14.50
    14.60 14.60 14.65 14.70 14.85 14.90 14.90 15.10 15.20 15.20
    15.30 15.35 15.40 15.45 15.60 15.60 15.65 15.85 15.90 15.95

Five number summary:
Minimum = 10.95 $Q_1 = 12.80$ Median = $(14.15 + 14.20)/2 = 14.175$
$Q_3 = 15.10$ Maximum = 15.95

Exercise 3.5

Variable↓	n	Total	Mean
Age	50	1676	33.52
Red BCC	50	231.56	4.6312
Hemoglobin	50	695.3	13.906
Platelets	50	14579	291.58

Exercise 3.7

Red BCC	Red BCC²	Red BCC	Red BCC²
4.39	19.2721	4.4	19.36
5.18	26.8324	4.88	23.8144
4.88	23.8144	4.22	17.8084
5.94	35.2836	4.36	19.0096
4.48	20.0704	4.59	21.0681
3.58	12.8164	3.63	13.1769
		Total→ **54.53**	**252.327**

$$\overline{X} = \frac{54.53}{12} = 4.544167; \; s^2 = \frac{\sum_{i=1}^{n} X_i - n(\overline{X})^2}{n-1} = \frac{252.3267 - 12(4.544167)^2}{12-1}$$

$$= 0.412117$$

$$s = \sqrt{0.412117} = 0.641964$$

Exercise 3.9

	Platelets	Platelets²
	224	50176
	264.5	69960.25
	360	129600
	384.5	147840.3
	364.5	132860.3
	468	219024
	171	29241
	328.5	107912.3
	323.5	104652.3
	306.5	93942.25
	264.5	69960.25
	233	54289
	254.5	64770.25
	267	71289
Total→	**4214**	**1345517**

$$\overline{X} = \frac{4214}{14} = 301; \; s^2 = \frac{\sum_{i=1}^{n} X_i^2 - n(\overline{X})^2}{n-1} = \frac{1345517 - 14(301)^2}{14-1} = 5931$$

$$s = \sqrt{5931} = 77.013$$

Exercise 3.11

Solution:

32 33 33 35 38 39 41 42 42 43 45 47 47 48 49 52
53 54 56 57 58 61 64 65 65 73 74 78 80 82 85 89

$$\sum_{i=1}^{n} X_i = 1760; \quad \sum_{i=1}^{n} X_i^2 = 105202; \quad \overline{X} = \frac{1760}{32} = 55;$$

$$s^2 = \frac{\sum_{i=1}^{n} X_i^2 - n(\overline{X})^2}{n-1} = \frac{105202 - 32(55)^2}{32-1} = 271.032 \Rightarrow s = \sqrt{271.032} = 16.4631$$

Exercise 3.13

Solution:
Using row 7, column 10 to select the sample, we write labels 00, 01, . . . , 10. The first value found was 02, which translated to serial numbers is equal to 3.
The resulting sample is shown in the table below:

Serial Number	Name	Weight	Height	Weight	Weight^2	Height	Height^2
3	Peter	171	70	171	29241	70	4900
14	Tyler	212	71	212	44944	71	5041
25	Pat	188	63	188	35344	63	3969
36	Gordon	180	73	180	32400	73	5329
47	Jamie	169	71	169	28561	71	5041
58	Shirley	172	67	172	29584	67	4489
69	Andy	184	62	184	33856	62	3844
80	Eduardo	192	63	192	36864	63	3969
				1468	**270794**	**540**	**36582**

For weight:

$$\overline{X} = \frac{1468}{8} = 183.5; \quad s^2 = \frac{\sum_{i=1}^{n} X^2 - n(\overline{X})^2}{n-1} = \frac{270794 - 8(183.5)^2}{8-1} = 202.2857$$

$$s = \sqrt{202.2857} = 14.2227$$

For height:

$$\overline{X} = \frac{540}{8} = 67.5; \quad s^2 = \frac{\sum_{i=1}^{n} X^2 - n(\overline{X})^2}{n-1} = \frac{36582 - 8(67.5)^2}{8-1} = 18.85714$$

$$s = \sqrt{18.85714} = 4.342481$$

Exercise 3.15

$$\overline{X} = \frac{513}{11} = 46.6364;$$

$$s^2 = \frac{\sum_{i=1}^{n} X_i^2 - n(\overline{X})^2}{n-1} = \frac{25339 - 11(46.6364)^2}{11-1} = 141.455$$

$$s = \sqrt{141.455} = 11.8935$$

Exercise 3.17

$$\overline{x} = \frac{\sum_{i=1}^{21} x_i}{21} = \frac{969}{21} = 46.14286$$

$$s^2 = \frac{\sum_{i=1}^{n} X_i^2 - n(\overline{X})^2}{n-1} = \frac{47023 - 21(46.14286)^2}{21-1} = 115.528$$

$$s = \sqrt{115.528} = 10.7484$$

Exercise 3.19

$$\text{mean weight} = \overline{x} = \frac{\sum_{i=1}^{20} x_i}{20} = \frac{3637}{20} = 181.85$$

$$\text{mean height} = \overline{y} = \frac{\sum_{i=1}^{20} y_i}{20} = \frac{1324}{20} = 66.20$$

$$s_x^2 = \frac{671693 - 20(181.85)^2}{20-1} = 542.345 \Rightarrow SD = \sqrt{542.345} = 23.883$$

$$s_y^2 = \frac{87934 - 20(66.20)^2}{20-1} = 15.0105 \Rightarrow \sqrt{15.0105} = 3.87434$$

Exercise 3.21

$$\text{mean weight} = \overline{x} = \frac{\sum_{i=1}^{14} x_i}{14} = \frac{2448}{14} = 174.8571$$

$$\text{mean height} = \overline{y} = \frac{\sum_{i=1}^{14} y_i}{14} = \frac{917}{14} = 65.5$$

$$s_x^2 = \frac{436132 - 14(174.8571)^2}{14 - 1} = 621.6703 \Rightarrow SD = \sqrt{621.6703} = 24.9333$$

$$s_y^2 = \frac{60297 - 14(66.20)^2}{14 - 1} = 17.96154 \Rightarrow \sqrt{17.96154} = 4.238106$$

Exercises on grouped data

3.22 Obtain the median for the data set.

3.23 Obtain the first quartile for the data set.

3.24 Obtain the third quartile for the data set.

3.25 Obtain the mean, variance, and standard deviation for the data set.

Class	Frequency
0–4	2
4–9	5
10–14	7
15–19	9
20–24	12
25–29	14
29–34	10
35–39	8
40–44	5
45–49	3

Exercise 3.23

There are $n = 75$ observations, so that $n/4 = 18.75$. This value is found in the class 15–19 because the cumulative frequency attains and exceed 18.75 in that class. That is the first quartile class; $\frac{1}{4} \sum_{i=1}^{8} f_i = 75/2 = 18.75$

$$(\Sigma f_i)_{bQ_1c} = 2 + 5 + 7 = 14$$

For Q_1

$$L_{Q_1} = 14.5; (\Sigma f_i)_{bQ_1c} = 14; f_{Q_1} = 9; c = 5;$$

$$Q_1 = 14.5 + \frac{\left(\frac{75}{4} - 14\right) \times 5}{9} = 17.13889$$

Exercise 3.25

$$\sum_{i=1}^{9} f_i X_i = 5414; \sum_{i=1}^{9} f_i = 108 \text{ and mean} = \overline{X} = \frac{5414}{108} = 50.1296$$

$$\sum_{i=1}^{k} f_i x_i^2 = 282312;$$

$$s^2 = \frac{1}{\left(\sum_{i=1}^{k} f_i - 1\right)} \left[\sum_{i=1}^{k} f_i x_i^2 - \left(\sum_{i=1}^{k} f_i\right) \bar{x}^2 \right]$$

$$= \frac{1}{108 - 1} \left[282312 - 108(50.1296)^2 \right] = 101.9643$$

$$s = \sqrt{101.9643} = 10.09774$$

classes	Frequency
26–30	4
31–35	6
36–40	10
41–45	12
46–50	20
51–55	24
56–60	15
61–65	10
66–70	7

3.26 Obtain the median for the data set.
3.27 Obtain the first quartile for the data set.
3.28 Obtain the third quartile for the data set.
3.29 Obtain the mean, variance, and standard deviation for the data set.

Solutions of 3.22–3.25 will depend on the table below:
3.22

Classes	Boundaries	Mid-Points	Freq.	Cum. Freq.	$f_i x_i$	X^2	$f_i x_i^2$
26–30	25.5–30.5	28	4	4	112	784	3136
31–35	30.5–35.5	33	6	10	198	1089	6534
36–40	35.5–40.5	38	10	20	380	1444	14440
41–45	40.5–45.5	43	12	32	516	1849	22188
46–50	45.5–50.5	48	20	52	960	2304	46080
51–55	50.5–55.5	53	24	76	1272	2809	67416
56–60	55.5–60.5	58	15	91	870	3364	50460
61–65	60.5–65.5	63	10	101	630	3969	39690
66–70	65.5–70.5	68	7	108	476	4624	32368
			108		**5414**	**22236**	**282312**

Exercise 3.27

There are $n = 108$ observations, so that $n/4 = 27$. This value is found in the class 41–45 because the cumulative frequency attains and exceed 27 in that class. That is the first quartile class; $\dfrac{1}{4}\sum_{i=1}^{9} f_i = 108/4 = 27$; $(\Sigma f_i)_{bmc} = 4 + 6 + 10 = 20$;

For Q_1

$$L_{Q_1} = 40.5; (\Sigma f_i)_{bQ_1c} = 20; f_{Q_1} = 12; c = 5;$$

$$Q_1 = 40.5 + \frac{\left(\dfrac{108}{4} - 20\right) \times 5}{12} = 40.5 + 2.91667 = 43.41667$$

Exercise 3.29

$$\sum_{i=1}^{9} f_i X_i = 5415; \sum_{i=1}^{9} f_i = 108 \text{ and mean} = \overline{X} = \frac{5415}{108} = 50.129629;$$

$$\sum_{i=1}^{k} f_i x_i^2 = 282312;$$

$$s^2 = \frac{1}{\left(\sum_{i=1}^{k} f_i - 1\right)}\left[\sum_{i=1}^{k} f_i x_i^2 - \left(\sum_{i=1}^{k} f_i\right)\overline{x}^2\right]$$

$$= \frac{1}{108 - 1}[282312 - 108(50.129629)^2] = 101.96435$$

$$s = \sqrt{101.96435} = 10.097739$$

Classes	Boundaries	Class Mid Points	f_i	$f_i X_i$	x_i^2	$f_i x_i^2$	$x_i - \overline{x}$	$f_i|x_i - \overline{x}|$
20–29	19.5–29.5	24.5	3	73.5	600.25	1800.75	-36.9697	110.9091
30–39	29.5–39.5	34.5	5	172.5	1190.25	5951.25	-26.9697	134.8485
40–49	39.5–49.5	44.5	8	356	1980.25	15842	-16.9697	135.7576
50–59	49.5–59.5	54.5	15	817.5	2970.25	44553.8	-6.9697	104.5455
60–69	59.5–69.5	64.5	13	838.5	4160.25	54083.3	3.030303	39.39394
70–79	69.5–79.5	74.5	10	745	5550.25	55502.5	13.0303	130.303
80–89	79.5–89.5	84.5	8	676	7140.25	57122	23.0303	184.2424
90–99	89.5–99.5	94.5	4	378	8930.25	35721	33.0303	132.1212
			66	4057	32522	270576.5		972.1212

Exercise 3.31

There are $n = 66$ observations, so that $n/4 = 16.5$. This value is found in the class 50–59 because the cumulative frequency attains and exceed 16.5 in that class. That is the first quartile class; $\dfrac{1}{4}\sum_{i=1}^{8} f_i = 66/4 = 16.50$; $(\Sigma f_i)_{\text{bmc}} = 3 + 5 + 8 = 16$;
For Q_1

$$L_{Q_1} = 49.5; \ (\Sigma f_i)_{bQ_1c} = 16; \ f_{Q_1} = 15; \ c = 10;$$

$$Q_1 = 49.5 + \frac{\left(\dfrac{66}{4} - 16\right) \times 10}{15} = 49.5 + 0.3333 = 49.8333$$

Exercise 3.33

$$\sum_{i=1}^{8} f_i X_i = 4057; \quad \sum_{i=1}^{8} f_i = 66 \text{ and mean} = \overline{X} = \frac{4057}{66} = 61.4697;$$

$$\sum_{i=1}^{k} f_i x_i^2 = 270576.5$$

$$s^2 = \frac{1}{\left(\sum_{i=1}^{k} f_i - 1\right)} \left[\sum_{i=1}^{k} f_i x_i^2 - \left(\sum_{i=1}^{k} f_i\right)\overline{x}^2 \right]$$

$$= \frac{1}{66 - 1}[270576.5 - 66(61.4697)^2] = 326.0606$$

$$s = \sqrt{326.0606} = 18.05715$$

Exercise 4.1

(a) $P(5 \text{ chosen with at least 2 women}) = P(2 \text{ women, 3 men})$
$+ P(3 \text{ women, 2 men}) + P(4 \text{ women, 1 man}) + P(5 \text{ women, no man})$

$$= \frac{\dbinom{6}{2}\dbinom{8}{3} + \dbinom{6}{3}\dbinom{8}{2} + \dbinom{6}{4}\dbinom{8}{1} + \dbinom{6}{5}\dbinom{8}{0}}{\dbinom{14}{5}}$$

$$= \frac{15 \times 56 + 20 \times 28 + 15 \times 8 + 6 \times 1}{2002} = 0.762238$$

(b) Anna is chosen with certainty, so only 4 more will be chosen, ensuring there are at least two women from the rest $\Rightarrow P(5$ chosen with at least 2 women which must include Anna$) =$

$P($Anna,1 woman,3 men$) + P($Anna, 2 women,2 men$) + P($Anna, 3 women, 1 man$) + P($Anna, 4 women, no man$)$

$$= \frac{\binom{5}{1}\binom{8}{3} + \binom{5}{2}\binom{8}{2} + \binom{5}{3}\binom{8}{1} + \binom{5}{4}\binom{8}{0}}{\binom{13}{5}}$$

$$= \frac{5 \times 56 + 10 \times 28 + 20 \times 8 + 5 \times 1}{1287} = 0.501166$$

Exercise 4.3

Solution:

The pairs are:
  (John, Jeremy), (John, Marcus), (John, Phil), (John, Justin),(John, Tyler), (John, Ray), (John, Duncan), (Jeremy, Marcus), (Jeremy, Phil), (Jeremy, Justin), (Jeremy, Tyler),(Jeremy, Ray), (Jeremy, Duncan) (Marcus, Phil),(Marcus, Justin), (Marcus, Tyler), (Marcus, Ray), (Marcus, Duncan), (Phil, Justin), (Phil, Tyler) (Phil, Ray), (Phil, Duncan), (Justin, Tyler), (Justin, Ray), (Justin, Duncan), (Tyler, Ray), (Tyler, Duncan), (Ray, Duncan)

(a) (John, Phil), (Marcus, Phil), (Jeremy, Phil), (Phil, Justin), (Phil, Tyler), (Phil, Ray), (Phil, Duncan)
(b) $P($Phil is chosen$) = 7/28 = \frac{1}{4}$
(c) $P($Phil or Justin$) = P($Phil$) + P($Justin$) - P($Phil, Justin$) = 1/4 + 1/4 - 1/28 = 13/28$.

Exercise 4.5

(a) $P(2$ chosen: 1 man, 1 woman$) = \dfrac{\binom{6}{1}\binom{8}{1}}{\binom{14}{2}} = \dfrac{48}{91} = 0.527473$

(b) $P(2$ chosen: 0 man, 2 women$) = \dfrac{\binom{6}{2}\binom{8}{0}}{\binom{14}{2}} = \dfrac{15}{91} = 0.164835$

(c) $P(2$ chosen: 2 men, 0 women$) = \dfrac{\binom{6}{0}\binom{8}{2}}{\binom{14}{2}} = \dfrac{28}{91} = 0.307692$

Exercise 4.7

(a) {(1, 1), (1, 2), (1, 3), (1, 4), (1, 5), (1, 6), (2, 1), (2, 2), (2, 3),
    (2, 4), (2, 5), (2, 6), (3, 1), (3, 2), (3, 3), (3, 4), (3, 5), (3, 6),
    (4, 1), (4, 2), (4, 3), (4, 4), (4, 5), (4, 6), (5, 1), (5, 2), (5, 3),
    (5, 4), (5, 5), (5, 6), (6, 1), (6, 2), (6, 3), (6, 4), (6, 5), (6, 6)}

(b) E_1 = {(2, 1), (3, 1), (3, 2), (4, 1), (4, 2), (4, 3), (5, 1), (5, 2),
    (5, 3), (5, 4), (6, 1), (6, 2), (6, 3), (6, 4), (6, 5)} $\Rightarrow P(E_1)$ = 15/36

(c) E_2 = {(3, 6), (4, 5), (4, 6), (5, 4), (5, 5),
    (5, 6), (6, 3), (6, 4), (6, 5), (6, 6)} $\Rightarrow P(E_2)$ = 10/36

(d) {(1, 1), (1, 2), (1, 3), (1, 4), (1, 5), (1, 6), (2, 1), (2, 2), (2, 3),
    (2, 4), (2, 5), (2, 6), (3, 2), (3, 3), (3, 4), (3, 5), (3, 6), (4, 3),
    (4, 4), (4, 5), (4, 6), (5, 4), (5, 5), (5, 6), (6, 5), (6, 6)} $\Rightarrow P(E_3)$ = 26/36

(e) (i) $E_1 \cap E_2$ = {(5, 4), (6, 3), (6, 4), (6, 5)} $\Rightarrow P(E_1 \cap E_2)$ = 4/36

    (ii) $E_1 \cap E_3$ = {(2, 1), (3, 2), (4, 3), (5, 4), (6, 5), (6, 6)} $\Rightarrow P(E_1 \cap E_3)$
        = 6/36

    (iii) $E_2 \cap E_3$ = {(3, 6), (4, 5), (5, 4), (5, 4), (5, 5), (6, 5), (6, 6)}
        $E_1 \cap E_2 = E_1 \Rightarrow P(E_1 \cap E_2)$ = 7/36

Exercise 4.9

(a) S = {BBBB, BBBG, BBGB, BBGG, BGBB, BGBG, BGGB, BGGG,
    GBBB, GBBG, GBGB, GBGG, GGBB, GGBG, GGGB, GGGG}

(b) E_1 = {BBBB, BBBG, BBGB, BBGG, BGBB, BGBG,
    BGGB, GBBB, GBBG, GBGB, GGBB} $\Rightarrow P(E_1)$ = 11/16

(c) E_2 = {S − GGGG} $\Rightarrow P(E_2)$ = 15/16

(d) E_3 = {BBGG, BGBG, BGGB, GBBG, BGGG, BGBG,
    GBGG, GGBB, GGBG, GGGG} $\Rightarrow P(E_3)$ = 10/16

(e) $E_1 \cap E_2 = E_1 \Rightarrow P(E_1 \cap E_2)$ = 11/16

(f) $E_1 \cup E_2 = P(E_1) + P(E_2) - P(E_1 \cap E_2) = \dfrac{11}{16} + \dfrac{15}{16} - \dfrac{11}{16} = \dfrac{15}{16}$

(g) $E_1 \cap E_3$ = {BBGG, BGBG, BGGB,

    GBBG, GBGB, GGBB} $\Rightarrow P(E_1 \cap E_3) = \dfrac{6}{16}$

Exercise 4.10

(a) $E_1 \cup E_3 = P(E_1) + P(E_3) - P(E_1 \cap E_3) = \dfrac{11}{16} + \dfrac{10}{16} - \dfrac{6}{16} = \dfrac{15}{16} = 1$

(b) $E_2 \cap E_3$ = {E_3 − GGGG} $\Rightarrow P(E_2)$ = 9/16

(c) $E_2 \cup E_3 = P(E_2) + P(E_3) - P(E_2 \cap E_3) = \dfrac{15}{16} + \dfrac{10}{16} - \dfrac{9}{16} = \dfrac{16}{16} = 1$

(d) $E_1 \cap E_2 = E_1 \Rightarrow E_1 \cap E_3 = E_1 \cap E_2 \cap E_3 \Rightarrow P(E_1 \cap E_2 \cap E_3)$ = 6/16

(e) $P(E_1 \cup E_2 \cup E_3) = P(E_1) + P(E_2) + P(E_3) - P(E_1 \cap E_2) - P(E_1 \cap E_3)$
    $- P(E_2 \cap E_3) + p(P(E_1 \cap E_2 \cap E_3))$

    $= \dfrac{11}{16} + \dfrac{15}{16} + \dfrac{10}{16} - \dfrac{11}{16} - \dfrac{6}{16} - \dfrac{9}{16} + \dfrac{6}{16} = \dfrac{16}{16} = 1.$

Exercise 4.11

Let G stand for graduated.

Total probability is $P(G) = P(G \cap P_1) + P(G \cap P_2) + P(G \cap P_3)$

But, $P(G | P_1) = \dfrac{P(P_1 \cap G)}{P(P_1)} \Rightarrow P(P_1 \cap G) = P(P_1)P(G | P_1)$

Similarly, $P(G | P_2) = \dfrac{P(P_2 \cap G)}{P(P_2)} \Rightarrow P(P_2 \cap G) = P(P_2)P(G | D_2)$

$P(G | P_3) = \dfrac{P(P_3 \cap G)}{P(P_3)} \Rightarrow P(P_3 \cap G) = P(P_3)P(G | D_3)$

Total probability $= P(G) = P(P_1)P(G | P_1) + P(P_2)P(G | P_2) + P(P_3)P(G | P_3)$

$P(P_1) = 0.33; P(P_2) = 0.36; P(P_3) = 0.31$

$P(G | P_1) = 0.80; P(G | P_2) = 0.85; P(G | P_2) = 0.86$

Total probability $= P(G) = 0.80 \times 0.33 + 0.36 \times 0.85 + 0.31 \times 0.86 = 0.8366$

(i) $P(P_1 | G) = \dfrac{P(P_1 \cap G)}{P(G)} = \dfrac{P(P_1)P(G | P_1)}{P(G)}$

$= \dfrac{P(P_1)P(G | P_1)}{P(P_1)P(G | P_1) + P(P_2)P(G | P_2) + P(P_3)P(G | P_3)}$

$= \dfrac{0.80 \times 0.33}{0.80 \times 0.33 + 0.36 \times 0.85 + 0.31 \times 0.86} = 0.315563$

(ii) $P(P_2 | G) = \dfrac{0.85 \times 0.36}{0.80 \times 0.33 + 0.36 \times 0.85 + 0.31 \times 0.86} = 0.365766$

(iii) $P(P_2 | G) = \dfrac{0.86 \times 0.31}{0.80 \times 0.33 + 0.36 \times 0.85 + 0.31 \times 0.86} = 0.318671$

Exercise 4.13

(a) $P(2 \text{ white balls, 2 blue balls, and 2 yellow ball are chosen})$

$= \dfrac{\binom{8}{2}\binom{7}{2}\binom{9}{2}}{\binom{24}{6}} = \dfrac{28 \times 21 \times 36}{134596} = 0.157271$

(b) $P(\text{All balls are the same color}) = P(\text{all balls are white})$
$+ P(\text{all balls are blue}) + P(\text{all balls are yellow})$

$= \dfrac{\binom{8}{6} + \binom{7}{6} + \binom{9}{6}}{\binom{24}{6}} = \dfrac{28 + 21 + 84}{134596} = 0.000884$

Exercise 4.15

(a) P(two balls of different colors are selected) $= \dfrac{\dbinom{4}{2}\dbinom{9}{2}}{\dbinom{13}{4}} = \dfrac{6(36)}{715} = 0.302098$

(b) P(at least two yellow balls are selected) $= \dfrac{\dbinom{4}{2}\dbinom{9}{2} + \dbinom{4}{1}\dbinom{9}{3} + \dbinom{4}{0}\dbinom{9}{4}}{\dbinom{13}{4}}$

$= \dfrac{6(36) + 4(84) + 1(126)}{715} = 0.948252$

Exercise 4.17

Let D stand for defective:

Total probability that an item is defective is $P(D) = P(D \cap A) + P(D \cap B) + P(D \cap C)$

But, $P(D \mid A) = \dfrac{P(A \cap D)}{P(A)} \Rightarrow P(A \cap D) = P(A)P(D \mid A)$

Similarly, $P(D \mid B) = \dfrac{P(B \cap D)}{P(B)} \Rightarrow P(B \cap D) = P(B)P(D \mid B)$

$P(D \mid C) = \dfrac{P(C \cap D)}{P(C)} \Rightarrow P(C \cap D) = P(C)P(G \mid C)$

Total probability $= P(D) = P(A)P(D \mid A) + P(B)P(D \mid B) + P(C)P(G \mid C)$

$P(A) = 0.25$; $P(B) = 0.25$; $P(C) = 0.50$

$P(D \mid A) = 0.08$; $P(D \mid B) = 0.07$; $P(D \mid C) = 0.04$;

Total probability $= P(G) = 0.25 \times 0.08 + 0.25 \times 0.07 + 0.5 \times 0.04 = 0.0575$

(i) $P(A \mid D) = \dfrac{P(A \cap D)}{P(D)} \Rightarrow P(A \cap D)$

$= \dfrac{P(A)P(D \mid A)}{P(A)P(D \mid A) + P(B)P(D \mid B) + P(C)P(G \mid C)}$

$= \dfrac{0.25 \times 0.08}{0.25 \times 0.08 + 0.25 \times 0.07 + 0.50 \times 0.04} = 0.347826$

(ii) $P(B \mid D) = \dfrac{0.25 \times 0.07}{0.25 \times 0.08 + 0.25 \times 0.07 + 0.50 \times 0.04} = 0.304348$

(iii) $P(C \mid D) = \dfrac{0.50 \times 0.04}{0.25 \times 0.08 + 0.25 \times 0.07 + 0.50 \times 0.04} = 0.347826$

Exercise 4.19

Let C stand for car of the brand is crashed; F = females; M = males:

Total probability that a car of the make is crashed is $P(C) = P(F \cap C) + P(M \cap C)$

But, $P(C \mid F) = \dfrac{P(F \cap C)}{P(F)} \Rightarrow P(F \cap F) = P(F)P(C \mid F)$

Similarly, $P(C \mid M) = \dfrac{P(M \cap C)}{P(M)} \Rightarrow P(M \cap C) = P(M)P(C \mid M)$

Total probability that orange is available and would be bought = $P(C) = P(F)P(C \mid F) + P(M)P(C \mid M)$

$P(F) = 0.601; P(M) = 0.399;$

$P(C \mid F) = 0.77; P(C \mid M) = 0.84;$

Total probability of crashing that type of car = $P(C) = 0.601 \times 0.77 + 0.399 \times 0.84 = 0.79793$

(i) $P(F \mid C) = \dfrac{P(F)P(F \mid C)}{P(F)P(F \mid C) + P(M)P(C \mid M)}$

$= \dfrac{0.601 \times 0.77}{0.601 \times 0.77 + 0.399 \times 0.84} = 0.579963$

(ii) $P(M \mid C) = \dfrac{0.399 \times 0.84}{0.601 \times 0.77 + 0.399 \times 0.84} = 0.420037$

Exercise 4.21

(a) $P(A_1 \cup A_2) = P(A_1) + P(A_2) - P(A_1 \cap A_2) = \dfrac{7}{15} + \dfrac{6}{15} - \dfrac{2}{15} = \dfrac{11}{15};$

(b) $P(A_1 \cup A_3) = P(A_1) + P(A_3) - P(A_1 \cap A_3) = \dfrac{7}{15} + \dfrac{4}{15} - \dfrac{1}{15} = \dfrac{10}{15};$

(c) $P(A_1 \cup A_4) = P(A_1) + P(A_4) - P(A_1 \cap A_4) = \dfrac{7}{15} + \dfrac{5}{15} - \dfrac{2}{15} = \dfrac{10}{15};$

(d) $P(A_2 \cup A_3) = P(A_2) + P(A_3) - P(A_2 \cap A_3) = \dfrac{6}{15} + \dfrac{4}{15} - \dfrac{1}{15} = \dfrac{9}{15};$

(e) $(A_2 \cup A_4) \Rightarrow P(A_2) + P(A_4) - P(A_2 \cap A_4) = \dfrac{6}{15} + \dfrac{5}{15} - \dfrac{2}{15} = \dfrac{9}{15};$

(f) $(A_3 \cup A_4) \Rightarrow P(A_3) + P(A_4) - P(A_3 \cap A_4) = \dfrac{4}{15} + \dfrac{5}{15} - \dfrac{2}{15} = \dfrac{7}{15};$

Exercise 4.23

Solution:
$S = \{2, 3, 4, 5, 6, 7, 8, 9, 10, 11, 12, 14\}$

$A = \{2, 4, 6, 8, 10, 12, 14\}; B = \{2, 4, 7, 8, 9, 11, 12\}; C = \{5, 6, 8, 10, 11, 12\};$

$P(A) = \dfrac{7}{12}; P(B) = \dfrac{7}{12}; P(C) = \dfrac{6}{12};$

$A \cap B = \{2, 4, 8, 12\} \Rightarrow P(A \cap B) = \dfrac{4}{12};$

$B \cap C = \{8, 11, 12\} \Rightarrow P(B \cap C) = \dfrac{3}{12};$

$A \cap C = \{6, 8, 10, 12\} \Rightarrow P(A \cap C) = \dfrac{4}{12};$

$A \cap B \cap C = \{8, 12\} \Rightarrow P(A \cap B \cap C) = \dfrac{2}{12};$

Exercise 5.1

X	0	1	2	3	4	5	Sum
P(x)	0.15	0.2	0.25	0.18	0.14	0.08	1
$x \cdot p(x)$	0	0.2	0.5	0.54	0.56	0.4	2.2
$x^2 \cdot p(x)$	0	0.2	1	1.62	2.24	2	7.06

$EX = \mu = 2.2 = \text{mean}$

$\mu = EX = 2.2; EX^2 = \sum_x x^2 \cdot p(x) = 7.06$

$\sigma^2 = EX^2 - \mu^2 = 7.06 - 2.2^2 = 2.22 \Rightarrow \sigma = \sqrt{2.22} = 1.48997$

Exercise 5.3

X	0	1	2	3	4	5	6	7	Sum
P(x)	0.08	0.17	0.22	0.24	0.12	0.08	0.05	0.04	0.92
$x \cdot p(x)$	0	0.17	0.44	0.72	0.48	0.4	0.3	0.28	2.79
$x^2 \cdot p(x)$	0	0.17	0.88	2.16	1.92	2	1.8	1.96	10.89

$\mu = EX = 2.79; EX^2 = \sum_x x^2 \cdot p(x) = 10.89$

$\sigma^2 = EX^2 - \mu^2 = 10.59 - 2.79^2 = 3.1059 \Rightarrow \sigma = \sqrt{3.1059} = 1.76236$

Exercise 5.5

	0	1	2	3	4	5	Sum
	0.03125	0.15625	0.3125	0.3125	0.15625	0.03125	1
$x \cdot p(x)$	0	0.15625	0.625	0.9375	0.625	0.15625	2.5
$x^2 \cdot p(x)$	0	0.15625	1.25	2.8125	2.5	0.78125	7.5

$$\mu = EX = 2.5; \ EX^2 = \sum_x x^2 \cdot p(x) = 7.5$$

$$\sigma^2 = EX^2 - \mu^2 = 7.5 - 2.5^2 = 1.25 \Rightarrow \sigma = \sqrt{1.25} = 1.118$$

Exercise 5.7

(a) $\binom{13}{5} = \dfrac{13!}{5!(13-5)!} = \dfrac{13!}{5!8!} = \dfrac{13 \times 12 \times 11 \times 10 \times 9 \times 8!}{5!8!}$

$\qquad = \dfrac{13 \times 12 \times 11 \times 10 \times 9}{5 \times 4 \times 3 \times 2 \times 1} = 1287$

(b) $\binom{9}{2} = \dfrac{9!}{2!(9-2)!} = \dfrac{9!}{2!7!} = \dfrac{9 \times 8 \times 7!}{2!7!} = \dfrac{9 \times 8}{2 \times 1} = 36$

(c) $\binom{11}{4} = \dfrac{11!}{4!(11-4)!} = \dfrac{11!}{4!7!} = \dfrac{11 \times 10 \times 9 \times 8 \times 7!}{4!7!} = \dfrac{11 \times 10 \times 9 \times 8}{4 \times 3 \times 2 \times 1}$

$\qquad = 330$

Exercise 5.9

(a) $(0.36)^4(0.64)^{11} = 0.000123934$
(b) $(0.55)^3(0.45)^7 = 0.0006211692$
(c) $(0.341)^6(0.659)^8 = 0.000559255$

Exercise 5.11

(a) $\binom{13}{5}(0.6)^5(0.4)^8 = 0.065587$

(b) $\binom{9}{2}(0.35)^2(0.65)^7 = 0.216188$

(c) $\binom{11}{4}(0.41)^4(0.59)^7 = 0.232067$

Exercise 5.13

The relevant Poisson distribution is:

$$p(x) = \frac{4^x e^{-4}}{x!}; x = 0, 1, 2, \ldots$$

(a) $P(4) = \dfrac{4^3 e^{-4}}{4!} = 0.195367$

(b) $P(X \geq 3) = 1 - P(X \leq 2) = 1 - P(0) - P(1) - P(2)$
$$= 1 - 0.018316 - 0.073263 - 0.146525 = 0.7619$$

(c) $P(2 \leq X \leq 5) = 0.146525 + 0.195367 + 0.195367 + 0.156297 = 0.69355$

Exercise 5.15

The relevant binomial distribution is:

$$p(x) = \binom{12}{x}(0.45)^x(0.55)^{9-x}; x = 0, 1, 2, \ldots, 12$$

(a) $P(X > 4) = 1 - P(X \leq 4) = 1 - \left[\binom{12}{0}(0.45)^0(0.55)^{12-0} \right.$

$$+ \binom{12}{1}(0.45)^5(0.55)^{10-5} + \binom{12}{2}(0.45)^2(0.55)^{12-2}$$

$$\left. + \binom{12}{3}(0.45)^3(0.55)^{12-3} + \binom{12}{4}(0.45)^4(0.55)^{12-4} \right]$$

$$= 1 - (0.000766 + 0.007523 + 0.033853 + 0.092326 + 0.169964)$$
$$= 1 - 0.30443 = 0.69557$$

(b) $P(0) = \binom{12}{0}(0.45)^0(0.55)^{12-0} = 0.000766$

(c) For black flies, the relevant binomial is:
$$p(x) = \binom{12}{x}(0.55)^x(0.45)^{12-x}; x = 0, 1, 2, \ldots, 12.$$

$$\binom{12}{3}(0.55)^3(0.45)^{12-3} + \binom{12}{4}(0.55)^4(0.45)^{12-4} + \binom{12}{5}(0.55)^5(0.45)^{12-5}$$

$$+ \binom{12}{6}(0.55)^6(0.45)^{12-6}$$

$$= 0.0923 + 0.076165 + 0.148945 + 0.212385 = 0.52982$$

Exercise 5.17

The relevant binomial distribution is:

$$p(x) = \binom{15}{x}(0.08)^x(0.92)^{15-x}; \ x = 0, 1, 2, \ldots, 15$$

(a) $P(X \le 3) = \binom{15}{0}(0.08)^0(0.92)^{15-0} + \binom{15}{1}(0.08)^1(0.92)^{15-1}$

$+ \binom{15}{2}(0.08)^2(0.92)^{15-2} + \binom{15}{3}(0.08)^3(0.92)^{15-3}$

$= 0.286297 + 0.373431 + 0.227306 + 0.085652 = 0.97269$

(b) $P(5) = \binom{15}{5}(0.08)^5(0.92)^{15-5} = 0.004274$

(c) mean $= np = 15 \times 0.08 = 1.2$ variance $= 1.104$ $SD = \sqrt{1.104} = 1.0507$

Exercise 5.19

$\lambda = 2 =$ the mean number of defectives produced per hour

(a) $P(0) = \dfrac{2^0 e^{-2}}{0!} = 0.1353$

(b) $P(X \le 2) = P(0) + P(1) + P(2) = \dfrac{2^0 e^{-2}}{0!} + \dfrac{2^1 e^{-2}}{1!} + \dfrac{2^2 e^{-2}}{2!}$

$= 0.13533 + 0.27067 + 0.27067 = 0.67668$

(c) $P(X = 4) = \dfrac{2^4 e^{-2}}{4!} = 0.09022$

Exercise 5.21

Solutions:
$\lambda = 3.9$, therefore the relevant Poisson distribution is:

$$p(x) = \frac{3.9^x e^{-3.9}}{x!}; \ x = 0, 1, 2\ldots$$

$P(X = 3) = p(3) = \dfrac{3.9^3 e^{-3.9}}{3!} = 0.20012$

$P(0 \le x \le 4) = p(0) + p(1) + p(2) + p(3) + p(4)$

$= \dfrac{3.9^0 e^{-3.9}}{0!} + \dfrac{3.9^1 e^{-3.9}}{1!} + \dfrac{3.9^2 e^{-3.9}}{2!} + \dfrac{3.9^3 e^{-3.9}}{3!} + \dfrac{3.9^4 e^{-3.9}}{4!}$

$= e^{-3.9}\left(1 + 3.9 + \dfrac{3.9^2}{2} + \dfrac{3.9^3}{6} + \dfrac{3.9^4}{24}\right)$

$= 0.02024 + 0.07894 + 0.15394 + 0.20112 + 0.019512$

$= 0.64836$

$$P(4 \le x \le 6) = p(4) + p(5) + p(6)$$

$$= \frac{3.9^4 e^{-3.9}}{4!} + \frac{3.9^5 e^{-3.9}}{5!} + \frac{3.9^6 e^{-3.5}}{6!}$$

$$= \frac{3.9^4 e^{-3.9}}{24}\left(1 + \frac{3.9}{5} + \frac{3.9^2}{30}\right)$$

$$= 0.195119 + 0.152193 + 0.098925 = 0.44623$$

$$P(x \ge 3) = 1 - P(x \le 2) = 1 - p(0) - p(1) - p(2)$$

$$= 1 - \frac{3.9^0 e^{-3.9}}{0!} - \frac{3.9^1 e^{-3.9}}{1!} + \frac{3.9^2 e^{-3.9}}{2!}$$

$$= 1 - 0.02024 - 0.078943 - 0.15394$$

$$= 1 - 0.25313 = 0.74687$$

Exercise 5.23

x	P(x)
0	0.002479
1	0.014873
2	0.044618
3	0.089235
4	0.133853
5	0.160623
6	0.160623

$\lambda = 6 =$ mean number of cars arriving at toll gate per minute.

$$p(x) = \frac{6^x e^{-6}}{x!}; x = 0, 1, 2, \dots,$$

(a) $P(0) = \dfrac{6^0 e^{-6}}{0!} = 0.002479$

(b) $P(X < 3) = P(0) + P(1) + P(2) = \dfrac{6^0 e^{-6}}{0!} + \dfrac{6^1 e^{-6}}{1!} + \dfrac{6^2 e^{-6}}{2!}$

$$= 0.002479 + 0.014873 + 0.044618 = 0.06197$$

(c) $P(X \ge 4) = 1 - P(X \le 3) = 1 - [P(0) + P(1) + P(2) + P(3)]$

$$= 1 - (0.002479 + 0.014873 + 0.044618 + 0.08925)$$

$$= 1 - 0.151205 = 0.848795$$

Exercise 5.25

$\lambda = 2.6 =$ mean number of errors per page of document

$$p(x) = \frac{2.6^x e^{-2.6}}{x!}; x = 0, 1, 2, \ldots,$$

(a) $P(X \le 3) = \dfrac{2.6^0 e^{-2.6}}{0!} + \dfrac{2.6^1 e^{-2.6}}{1!} + \dfrac{2.6^2 e^{-2.6}}{2!} + \dfrac{2.6^3 e^{-2.6}}{3!} = 0.7360$

(b) $P(2 \le X \le 5) = P(2) + P(3) + P(4) + P(5)$

$$= \frac{2.6^2 e^{-2.6}}{2!} + \frac{2.6^3 e^{-2.6}}{3!} + \frac{2.6^4 e^{-2.6}}{4!} + \frac{2.6^5 e^{-2.6}}{5!}$$

$$= 0.2510 + 0.2176 + 0.1414 + 0.0735 = 0.6836$$

(c) $P(0) = \dfrac{2.6^0 e^{-2.6}}{0!} = 0.07427$

Exercise 6.1

In general, $f(x) = \begin{cases} \dfrac{1}{b-a}; a \le x \le b \\ 0 \quad \text{elsewhere} \end{cases}$

In this case $a = 259$, and $b = 294$. The relevant-distribution is:

$$f(x) = \begin{cases} \dfrac{1}{294 - 259}; 259 \le x \le 294 \\ 0 \quad \text{elsewhere} \end{cases}$$

(a) $P(X \ge 277) = \dfrac{277 - 259}{35} = \dfrac{18}{35}$

(b) $P(266 \le X \le 280) = \dfrac{280 - 266}{35} = \dfrac{14}{35}$

(c) $P(X \le 270) = \dfrac{270 - 259}{35} = \dfrac{11}{35}$

Exercise 6.3

In general, $f(x) = \begin{cases} \dfrac{1}{b-a}; a \le x \le b \\ 0 \quad \text{elsewhere} \end{cases}$

In this case $a = 19.80$, and $b = 20.2$. The relevant-distribution is:

$$f(x) = \begin{cases} \dfrac{1}{20.2 - 19.8} = \dfrac{1}{0.40}; 19.8 \le x \le 20.2 \\ 0 \quad \text{elsewhere} \end{cases}$$

(a) minimum $= 19.8$ median $=$ mean $= \dfrac{19.8 + 20.2}{2} = 20$ maximum $= 20.2$

If Q_1 is the first quartile, then

$\dfrac{Q_1 - 19.8}{0.40} = 0.25 \Rightarrow Q_1 - 19.8 = 0.40(0.25) \Rightarrow Q_1 = 19.8 + 0.1 = 19.9$

If Q_3 is the third quartile, then

$\dfrac{Q_3 - 20}{0.40} = 0.25 \Rightarrow Q_3 - 20 = 0.40(0.25) \Rightarrow Q_3 = 20 + 0.1 = 20.1$

Five number summary: min $= 19.8$, $Q_1 = 19.9$ median $= 20$ $Q_3 = 20.1$ maximum $= 20.2$

(b) $P(X > 20) = \dfrac{20.2 - 20}{0.40} = 0.50$

(c) if a_1 is the lower 10th percentile $\Rightarrow \dfrac{a_1 - 19.8}{0.40} = 0.1 \Rightarrow a_1 - 19.8 =$ $0.1(0.40) \Rightarrow 19.8 + 0.04 = 19.84$

if a_2 is the upper 10th percentile $\Rightarrow \dfrac{20.2 - a_2}{0.40} = 0.1 \Rightarrow 20.2 - a_2 =$ $0.1(0.40) \Rightarrow a_2 = 20.2 - 0.04 = 20.16$

Exercise 6.5

(a) In general, $f(x) = \begin{cases} \dfrac{1}{b - a}; & a \le x \le b \\ 0 & \text{elsewhere} \end{cases}$

In this case $a = 2$, and $b = 36$. The relevant-distribution is:

$f(x) = \begin{cases} \dfrac{1}{36 - 2} = \dfrac{1}{34}; & 2 \le x \le 36 \\ 0 & \text{elsewhere} \end{cases}$

(b) $P(X > 28) = \dfrac{36 - 28}{34} = \dfrac{8}{34} = 0.235294$

(c) if a_2 is the upper 30 percentile $\Rightarrow \dfrac{36 - a_2}{34} = 0.3 \Rightarrow 36 - a_2 =$ $34(0.30) \Rightarrow a_2 = 36 - 34(0.30) = 25.8$

(d) if a_1 is the lower 40 percentile $\Rightarrow \dfrac{a_1 - 2}{34} = 0.40 \Rightarrow a_1 - 2 =$ $34(0.40) \Rightarrow a_1 = 2 + 34(0.40) = 15.6$

Exercise 6.7

We use $Z = \dfrac{\overline{X} - \mu}{\sigma}$:

(a) $(X = 87) \Rightarrow Z = \dfrac{87 - 80}{\sqrt{20}} = 1.565$

(b) $(X = 88.95) \Rightarrow Z = \dfrac{88.95 - 80}{\sqrt{20}} = 2.00$

(c) $(X = 72) \Rightarrow Z = \dfrac{72 - 80}{\sqrt{20}} = 1.79$

(d) $(X = 68) \Rightarrow Z = \dfrac{68 - 80}{\sqrt{20}} = -2.68$

(e) $(X = 93) \Rightarrow Z = \dfrac{93 - 80}{\sqrt{20}} = 2.91$

Exercise 6.9

By symmetry:

(i) $P(-Z_1 \leq Z \leq Z_1) = 0.78$
$\Rightarrow 1 - 2P(Z < -Z_1) = 0.78 \Rightarrow 2P(Z < -Z_1) = 1 - 0.78 = 0.22$
$\Rightarrow P(Z < -Z_1) = 0.11 \Rightarrow Z_1 \approx 1.23$

(ii) $P(-1.25 \leq Z \leq Z_1) = 0.8403$
$\Rightarrow P(Z < Z_1) = P(Z < -1.25) + 0.8403 = 0.1056 + 0.8403 = 0.9459$
$\Rightarrow Z_1 \approx 1.61$

(iii) $P(-Z_1 \leq Z \leq 1.68) = 0.5367$
$\Rightarrow P(Z < 1.68) = 0.5367 + P(Z < -Z_1) \Rightarrow 0.9535 - 0.5367)$
$= P(Z < -Z_1) = 0.4168 \Rightarrow Z_1 \approx 0.21$

Exercise 6.11

(a) $P(50 < X < 77) = P\left(\dfrac{50 - 66}{\sqrt{53}} < Z < \dfrac{77 - 66}{\sqrt{53}}\right)$
$= P(-2.20 < Z < 1.51) = 0.9345 - 0.0139 = 0.9206$

(b) $P(X < 47) = P\left(Z < \dfrac{47 - 66}{\sqrt{53}}\right) = P(Z < -2.61) = 0.0045$

(c) $P(X > 82) = P\left(Z > \dfrac{82 - 66}{\sqrt{53}}\right) = P(Z > 2.20) = 1 - 0.9861 = 0.0139$

Exercise 6.13

(a) $P(X < 70) = P\left(Z < \dfrac{70 - 64}{\sqrt{16}}\right) = P(Z < 1.50) = 0.9332$

(b) $P(60 < X < 72) = P\left(\dfrac{60 - 64}{\sqrt{16}} < Z < \dfrac{72 - 64}{\sqrt{16}}\right)$
$= P(-1.0 < Z < 2.0) = 0.9773 - 0.1587 = 0.8186$

(c) $P(X > 57) = P\left(Z > \dfrac{57 - 64}{\sqrt{16}}\right) = P(Z > -1.75) = 1 - P(Z < -1.75)$

$\qquad\qquad = 1 - 0.0401 = 0.9599$

Exercise 6.15

(a) $P(Y > 97) = P\left(Z > \dfrac{97 - 100}{\sqrt{112}}\right) = P(Z > -0.28) = 1 - 0.3897 = 0.6103$

(b) $P(94 < Y < 107) = P\left(\dfrac{94 - 100}{\sqrt{112}} < Z < \dfrac{107 - 100}{\sqrt{112}}\right)$

$\qquad\qquad = P(-0.57 < Z < 0.66) = 0.7454 - 0.2843 = 0.4611$

(c) $P(104 < Y < 118) = P\left(\dfrac{104 - 100}{\sqrt{112}} < Z < \dfrac{118 - 100}{\sqrt{112}}\right)$

$\qquad\qquad = P(0.38 < Z < 1.7) = 0.9554 - 0.6480 = 0.3074$

(d) $P(Y < 89) = P\left(Z < \dfrac{89 - 100}{\sqrt{112}}\right) = P(Z < -1.04) = 0.1492$

Exercise 6.17

(a) $P(86 < Y < 94) = P\left(\dfrac{86 - 80}{6} < Z < \dfrac{94 - 80}{6}\right) = P(1 < Z < 2.33)$

$\qquad\qquad = 0.9901 - 0.8413 = 0.1488$

(b) $P(72 < Y < 90) = P\left(\dfrac{72 - 80}{6} < Z < \dfrac{90 - 80}{6}\right) = P(-1.33 < Z < 1.67)$

$\qquad\qquad = 0.9525 - 0.0918 = 0.8607$

(c) $P(65 < Y < 72) = P\left(\dfrac{65 - 80}{6} < Z < \dfrac{72 - 80}{6}\right)$

$\qquad\qquad = P(-2.5 < Z < -1.33) = 0.0918 - 0.0062 = 0.0856$

(d) $P(68 < Y < 92) = P\left(\dfrac{68 - 80}{6} < Z < \dfrac{92 - 80}{6}\right) = P(-2 < Z < 2)$

$\qquad\qquad = 0.9772 - 0.0228 = 0.9544$

Exercise 6.19

(a) $P(81 < Y < 99) = P\left(\dfrac{81 - 90}{6.5} < Z < \dfrac{99 - 90}{6.5}\right)$

$\qquad\qquad = P(-1.385 < Z < 1.385) = 0.91695 - 0.08305 = 0.8339$

(b) $P(77 < Y < 98) = P\left(\dfrac{77 - 90}{6.5} < Z < \dfrac{98 - 90}{6.5}\right) = P(-2.0 < Z < 1.23)$

$\qquad\qquad = 0.8907 - 0.0228 = 0.8679$

(c) $P(80 < Y < 102) = P\left(\dfrac{80 - 90}{6.5} < Z < \dfrac{102 - 90}{6.5}\right)$

$\qquad\qquad = P(-1.54 < Z < 1.85) = 0.9678 - 0.0618 = 0.9060$

(d) $P(70 < Y < 89) = P\left(\dfrac{70 - 90}{6.5} < Z < \dfrac{89 - 90}{6.5}\right)$

$\qquad\qquad = P(-3.07 < Z < -0.15) = 0.4404 - 0.0011 = 0.4393$

Exercise 6.21

$$\mu = 595; \sigma^2 \Rightarrow \sigma = 105$$

(i) $Y_{0.96} = \mu + \sigma Z_{0.96} \Rightarrow Y_{0.96} = 595 + 105(Z_{0.96}) = 595 + 105(1.75)$
$$= 778.75$$
upper 66th percentile $\Rightarrow$ lower 34th percentile

(ii) $Y_{0.34} = \mu + \sigma Z_{0.34} \Rightarrow Y_{0.34} = 595 + 105(Z_{0.34}) = 595 + 105(-0.41)$
$$= 551.95$$

Exercise 6.23

$$\mu = 578; \sigma^2 = 10201 \Rightarrow \sigma = 103$$
upper 20th percentile $\Rightarrow$ lower 80th percentile

(i) $Y_{0.72} = \mu + \sigma Z_{0.80} \Rightarrow Y_{0.80} = 578 + 103(Z_{0.80}) = 578 + 103(0.84)$
$$= 664.52$$
upper 70th percentile:

(ii) $Y_{0.70} = \mu + \sigma Z_{70} \Rightarrow Y_{0.70} = 578 + 103(Z_{0.70}) = 578 + 103(0.52)$
$$= 631.56$$
upper 5th percentile $\Rightarrow$ lower 95th percentile

(iii) $Y_{0.95} = \mu + \sigma Z_{0.95} \Rightarrow Y_{0.95} = 578 + 103(Z_{0.95}) = 578 + 103(1.65)$
$$= 747.44$$

Exercise 6.25

$$\mu = 19; \sigma^2 = 12.24 \Rightarrow \sigma = 3.4986$$
lower 96th percentile, upper 4th percentile

(i) $Y_{0.96} = \mu + \sigma Z_{0.96} \Rightarrow Y_{0.96} = 19 + 3.4986(Z_{0.96}) = 19 + 3.4986(1.75)$
$$= 25.1226$$
median:

(ii) $Y_{0.5} = \mu + \sigma Z_{0.5} \Rightarrow Y_{0.5} = 100 + 3.4986(Z_{0.5}) = 19 + 15(0.0) = 19$
lower 85th percentile $\Rightarrow$ upper 15th percentile

(iii) $Y_{0.85} = \mu + \sigma Z_{0.85} \Rightarrow Y_{0.85} = 19 + 3.4986(Z_{0.85})$
$$= 19 + 3.4986(+1.04) = 22.639$$

Exercise 6.27

(a) $Y_{0.90} = 64 + \sqrt{7.9} \times Z_{0.90} = 64 + \sqrt{7.9} \times 1.28 = 67.5977$

(b) $Y_{0.94} = 64 + \sqrt{7.9} \times Z_{0.94} = 64 + \sqrt{7.9} \times (1.56) = 68.3847$

(c) $Y_{0.8} = 64 + \sqrt{7.9} \times Z_{0.8} = 64 + \sqrt{7.9} \times (0.84) = 66.3610$

Exercise 6.29

(a) $Y_{0.70} = 16 + \sqrt{1.33} \times Z_{0.70} = 16 + \sqrt{1.33} \times 0.52 = 16.5997$

(b) $Y_{0.95} = 16 + \sqrt{1.33} \times Z_{0.95} = 16 + \sqrt{27.8} \times (1.645) = 17.8971$

(c) $Y_{0.88} = 16 + \sqrt{1.33} \times Z_{0.88} = 16 + \sqrt{1.33} \times 1.175 = 17.3551$

Exercise 6.31

Solution:

$n = 730; p = 0.66 \Rightarrow np = 730(0.66) = 481.8$

$npq = 730(0.66)(0.34) = 163.812; \sqrt{npq} = 12.79891$

(a) Let X be the number of women who support the proposed reservation, then P(number of women who support the reservation is more than 500).

$$P(X > 500) = P\left(Z > \frac{500.5 - 481.8}{12.79891}\right) = P(Z > 1.46) = 1 - 0.9279$$
$$= 0.0721$$

(b) Let X be the number of women who support the reservation, then P(number of women who support the reservation is between 460 and 480).

$$P(460 < X \leq 490) = P\left(\frac{460.5 - 481.8}{12.79891} \leq Z \leq \frac{490.5 - 481.8}{12.79891}\right)$$
$$= P(-1.66 \leq Z \leq 0.68) = 0.7517 - 0.0485$$
$$= 0.7032$$

(c) Let X be the number of women, then P(less than 464 number of women support the reservation).

$$P(X < 464) = P\left(Z < \frac{463.5 - 481.8}{12.79891}\right) = P(Z < -1.43) = 0.0764$$

Exercise 6.33

$n = 500; p = 0.43 \Rightarrow np = 500(0.43) = 215$

$npq = 500(0.43)(0.57) = 122.55; \sqrt{npq} = 11.07023$

(a) Let X be the number of times the head turns up, then P(the number of times the head turns up is greater 182):

$$P(X > 182) = P\left(Z > \frac{182.5 - 215}{11.07023}\right) = P(Z < -2.94)$$
$$= 1 - p(Z < -2.94) = 1 - 0.0016 = 0.9984$$

(b) Let X be the number of times the head turns up, then P(number of times the head turns up is less than 233) is:

$$P(X < 233) = P\left(Z \le \frac{232.5 - 215}{11.07023}\right) = P(Z \le 1.58) = 0.9430$$

(c) Let X be the number of times the head turns up, then P(the number of times the head turns up, that is between 185 and 235) is:

$$P(185 \le X \le 235) = P\left(\frac{184.5 - 215}{11.07023} \le Z \le \frac{235.5 - 215}{11.07023}\right)$$
$$= P(-2.755 \le Z \le 1.85) = 0.9678 - 0.00295$$
$$= 0.96485$$

Exercise 6.35

$n = 1500;\ np = 1500(0.51) = 765;\ npq = 1500(0.51)(0.49) = 374.85;$
$\sqrt{npq} = 19.36104;$

(a) P(that between 720 and 805 have binged on alcohol) $= P(720 < X < 805)$
$$= P\left(\frac{719.5 - 765}{19.36104} \le Z \le \frac{805.5 - 765}{19.36104}\right) = P(-2.35 < Z < 2.09)$$
$$= 0.9817 - 0.0094 = 0.9723$$
(b) P(less than 790 have binged on alcohol) $= P(X < 790)$
$$= P\left(Z < \frac{789.5 - 765}{19.36104}\right) = P(Z < 1.265) = 0.5(0.8962 + 0.898)$$
$$= 0.8971$$
(c) P(greater than 767 have binged on alcohol) $= P(X > 767)$
$$= \left(Z > \frac{767.5 - 765}{19.36104}\right) = P(Z > 0.13) = 1 - P(Z \le 0.13)$$
$$= 1 - 0.5517 = 0.4483$$

Exercise 6.37

$n = 900;\ np = 1000(0.88) = 792;$
$npq = 1000(0.88)(0.12) = 95.04 \sqrt{npq} = 9.74885$

(a) P(greater than 770 would support female president) $= P(X > 770)$
$$= P\left(Z > \frac{770.5 - 792}{9.74885}\right) = P(Z > -2.21) = 1 - P(Z < -2.21)$$
$$= 1 - 0.0136 = 0.9864$$
(b) P(between 790 and 822 support female presidency)
$$= P\left(Z < \frac{789.5 - 792}{9.74885} < Z < \frac{822.5 - 792}{9.74885}\right) = P(-0.26 < Z < 3.13)$$
$$= 0.9991 - 0.3974 = 0.6017$$
(c) P(less than 820 support female presidency) $= P(X < 820)$
$$= \left(Z < \frac{819.5 - 792}{9.74885}\right) = P(Z < 2.82) = 0.9976$$

Exercise 6.39

$n = 1000; p = 0.314; np = 1000(0.314) = 314;$
$npq = 1000(0.314)(0.686) = 215.404; \sqrt{npq} = 14.67665$

(a) P(less than 287 had gestation period which exceeded 9 months)

$$= P(X < 287) = P\left(Z < \frac{286.5 - 314}{14.67665}\right) = P(Z < -1.87) = 0.0307;$$

(b) P(greater than 340 had gestation period between)

$$= P(X > 340) = \left(Z > \frac{340.5 - 314}{14.67665}\right) = 1 - P(Z < 1.81)$$

$$= 1 - 0.9649 = 0.0351$$

(c) P(between 286 and 336 had a gestation period of more than 9 months)

$$= P(286 \le X \le 336) = P\left(\frac{285.5 - 314}{14.67665} < Z < \frac{336.5 - 314}{14.677498}\right)$$

$$= P(-1.94 < Z < 1.53) = 0.9370 - 0.0262 = 0.9108$$

Exercise 6.41

$\lambda = 40$; We approximate the Poisson by the normal and adjust values with continuity correction:

$$X \sim N(\lambda, \lambda) \Rightarrow Z = \frac{X - \lambda}{\sqrt{\lambda}}$$

(a) $P(X \le 55) = P\left(Z \le \dfrac{55.5 - 40}{\sqrt{40}}\right) = P(Z \le 2.45) = 0.9929$

(b) $P(33 < X < 51) = P\left(\dfrac{33.5 - 40}{\sqrt{40}} < Z < \dfrac{50.5 - 40}{\sqrt{40}}\right)$

$$= P(-1.03 < Z < 1.66) = 0.9515 - 0.1515 = 0.80$$

(c) $P(X > 35) = P\left(Z > \dfrac{35.5 - 40}{\sqrt{40}}\right) = P(Z > -0.71) = 1 - P(Z < -0.71)$

$$= 1 - 0.2389 = 0.7611$$

Exercise 6.43

$\lambda = 52$; We approximate the Poisson by the normal and adjust values with continuity correction:

$$X \sim N(\lambda, \lambda) \Rightarrow Z = \frac{X - \lambda}{\sqrt{\lambda}}$$

(a) $P(X < 40) = P\left(Z < \dfrac{39.5 - 52}{\sqrt{52}}\right) = P(Z < -1.73) = 0.0418$

(b) $P(55 < X < 68) = P\left(\dfrac{55.5 - 52}{\sqrt{52}} < Z < \dfrac{67.5 - 52}{\sqrt{52}}\right)$

$$= P(0.485 < Z < 2.15) = 0.9842 - \dfrac{1}{2}(0.6844 + 0.6879)$$

$$= 0.29805$$

(c) $P(X > 37) = P\left(Z > \dfrac{37.5 - 52}{\sqrt{52}}\right) = P(Z > -2.01) = 1 - P(Z < -2.01)$

$$= 1 - 0.0222 = 0.9778$$

Exercise 6.45

$\lambda = 849$; We approximate the Poisson by the normal and adjust values with continuity correction:

$$X \sim N(\lambda, \lambda) \Rightarrow Z = \dfrac{X - \lambda}{\sqrt{\lambda}}$$

(a) $P(X < 878) = P\left(Z < \dfrac{877.5 - 849}{\sqrt{849}}\right) = P(Z < 0.98) = 0.8365$

(b) $P(800 \le X \le 900) = P\left(\dfrac{799.5 - 849}{\sqrt{849}} < Z < \dfrac{900.5 - 849}{\sqrt{849}}\right)$

$$= P(-1.70 < Z < 1.77) = 0.9616 - 0.0446 = 0.9170$$

(c) $P(X > 790) = P\left(Z > \dfrac{790.5 - 849}{\sqrt{849}}\right) = P(Z > -2.01)$

$$= 1 - P(Z < -2.01) = 1 - 0.0222 = 0.9778$$

Exercise 7.1

$Y \sim U(7, 24); n = 35$

In general, $\overline{Y} \sim N\left(\dfrac{a + b}{2}, \dfrac{(b - a)^2}{12n}\right)$

So $\overline{Y} \sim N\left[\dfrac{24 + 7}{2}, \dfrac{(24 - 7)^2}{12(35)}\right] \Rightarrow \overline{Y} \sim N[15.5, 0.688095]$

Exercise 7.3

$\lambda = 44, n = 50;$

In general, $\overline{Y} \sim N\left(\lambda, \dfrac{\lambda}{n}\right)$

So $\overline{Y} \sim N\left(44, \dfrac{44}{50}\right) \Rightarrow \overline{Y} \sim N(44, 0.88)$

Exercise 7.5

$\lambda = 35, n = 36;$

In general, $\overline{Y} \sim N\left(\lambda, \dfrac{\lambda}{n}\right)$

So $\overline{Y} \sim N\left(35, \dfrac{35}{36}\right) \Rightarrow \overline{Y} \sim N(35, 0.97222)$

Exercise 7.7

$p = 0.3, n = 40$ sample size $m = 89$

In general, $\overline{Y} \sim N\left(np, \dfrac{npq}{m}\right)$

So $\overline{Y} \sim N\left(40(0.3), \dfrac{40(0.3)(0.7)}{89}\right) \Rightarrow \overline{Y} \sim N(12, 0.094382)$

Exercise 7.9

$X \sim N(100, 560); n = 50;$

$Z = \dfrac{\sqrt{n}(\overline{X} - \mu)}{\sigma}; \mu = 100; \sigma = \sqrt{560} = 23.66$

(a) $P(\overline{X} > 92) = P\left[Z > \dfrac{\sqrt{50}(92 - 100)}{23.66}\right] = P(Z > -2.39) = 1 - 0.0084$
$= 0.9916$

(b) $P(95 \le \overline{X} \le 108) = P\left[\dfrac{\sqrt{50}(95 - 100)}{23.66} \le Z \le \dfrac{\sqrt{50}(108 - 100)}{23.66}\right]$
$= P(-1.49 \le Z \le 2.39) = 0.9916 - 0.0681 = 0.9235$

(c) $P(\overline{X} < 110) = P\left[Z < \dfrac{\sqrt{50}(110 - 100)}{23.66}\right] = P(Z < 2.99) = 0.9986$

Exercise 7.11

$X \sim N(60, 235); n = 50;$

$Z = \dfrac{\sqrt{n}(\overline{X} - \mu)}{\sigma}; \mu = 60; \sigma = \sqrt{235} = 15.33$

(a) $P(\overline{X} < 55) = P\left[Z < \dfrac{\sqrt{50}(55 - 60)}{15.33}\right] = P(Z < -2.31) = 0.0104$

(b) $P(58 \le \overline{X} \le 66) = P\left[\dfrac{\sqrt{50}(58 - 60)}{15.33} \le Z \le \dfrac{\sqrt{50}(66 - 60)}{15.33}\right]$
$= P(-0.92 < Z < 2.77) = 0.9972 - 0.1788 = 0.8184$

(c) $P(\overline{X} > 65) = P\left[Z > \dfrac{\sqrt{50}(65 - 60)}{15.33}\right] = P(Z > 2.31) = 1 - 0.9896$
$= 0.0104$

Exercise 7.13

$X \sim N(609, 10201); n = 100;$

$Z = \dfrac{\sqrt{n}(\overline{X} - \mu)}{\sigma}; \mu = 609; \sigma = \sqrt{10201} = 101;$

(a) $P(\overline{X} < 580) = P\left[Z < \dfrac{\sqrt{100}(580 - 595)}{101}\right] = P(Z < -2.87) = 0.0021$

(b) $P(612 \leq \overline{X} \leq 620) = P\left[\dfrac{\sqrt{100}(612 - 620)}{101} \leq Z \leq \dfrac{\sqrt{100}(590 - 595)}{101}\right]$

$= P(0.30 \leq Z \leq 1.09) = 0.8621 - 0.6179 = 0.2442$

(c) $P(\overline{X} > 605) = P\left[Z > \dfrac{\sqrt{100}(605 - 609)}{101}\right] = P(Z > -0.40)$

$= 1 - P(Z < -0.40) = 1 - 0.3446 = 0.6554$

Exercise 7.15

$X \sim N(300, 7680); n = 100;$

$Z = \dfrac{\sqrt{n}(\overline{X} - \mu)}{\sigma}; \mu = 300; \sigma = \sqrt{7680} = 87.6353;$

(a) $P(\overline{X} < 280) = P\left[Z < \dfrac{\sqrt{100}(280 - 300)}{87.6353}\right] = P(Z < -2.28) = 0.0113$

(b) $P(290 \leq \overline{X} \leq 315) = P\left[\dfrac{\sqrt{100}(290 - 300)}{87.6353} \leq Z \leq \dfrac{\sqrt{100}(315 - 300)}{87.6353}\right]$

$= P(-1.14 \leq Z \leq 1.71) = 0.9564 - 0.1271 = 0.8293$

(c) $P(\overline{X} > 308) = P\left[Z > \dfrac{\sqrt{100}(308 - 300)}{87.6353}\right] = P(Z > 0.91)$

$= 1 - 0.8186 = 0.1814$

Exercise 7.17

$X \sim N(4290, 1732800); n = 40;$

$Z = \dfrac{\sqrt{n}(\overline{X} - \mu)}{\sigma}; \mu = 4290; \sigma = \sqrt{1732800} = 1316.3586$

(a) $P(\overline{X} < 3650) = P\left[Z < \dfrac{\sqrt{40}(3650 - 4290)}{1316.3586}\right] = P(Z < -3.07) = 0.0011$

(b) $P(4000 \leq \overline{X} \leq 4600)$

$= P\left[\dfrac{\sqrt{40}(4000 - 4290)}{1316.3586} \leq Z \leq \dfrac{\sqrt{40}(4600 - 4290)}{1316.3586}\right]$

$= P(-1.39 \leq Z \leq 1.49) = 0.9319 - 0.0832 = 0.8487$

(c) $P(\overline{X} > 3900) = P\left[Z > \dfrac{\sqrt{40}(3900 - 4290)}{1316.3586}\right] = P(Z > -1.87)$

$= 1 - P(Z < -1.87) = 1 - 0.0307 = 0.9693$

Exercise 7.19

$X \sim N(10569, 1760700); n = 40;$

$Z = \dfrac{\sqrt{n}(\overline{X} - \mu)}{\sigma}; \mu = 10569; \sigma = \sqrt{1760700} = 1326.9137$

(a) $P(10069 \leq \overline{X} \leq 11069)$

$= P\left[\dfrac{\sqrt{40}(10069 - 10569)}{1326.9137} \leq Z \leq \dfrac{\sqrt{40}(11069 - 10569)}{1326.9137} \right]$

$= P(-2.38 \leq Z \leq 2.38) = 0.9913 - 0.0087 = 0.9826$

(b) $P(\overline{X} > 11019) = P\left[Z > \dfrac{\sqrt{40}(11019 - 10569)}{1326.9137} \right] = P(Z > 2.145)$

$= 1 - 0.98765 = 0.01235$

(c) $P(\overline{X} < 10324) = P\left[Z < \dfrac{\sqrt{40}(10324 - 10569)}{1326.9137} \right] = P(Z < -1.17)$

$= 0.1210$

Exercise 7.21

$n = 260; p = 0.31 \ \sigma_{\hat{p}} = \sqrt{\dfrac{0.31(1 - 0.31)}{260}} = 0.028683$

(a) $P(\hat{p} < 0.24) = P\left(Z < \dfrac{0.24 - 0.31}{0.028683} \right) = P(Z < -2.44) = 0.0073$

(b) $P(0.27 < \hat{p} < 0.35) = P\left(\dfrac{0.27 - 0.31}{0.028683} < Z < \dfrac{0.35 - 0.31}{0.028683} \right)$

$= P(-1.39 < Z < 1.39) = 0.9177 - 0.0823 = 0.8354$

(c) $P(p > 0.36) = P\left(Z > \dfrac{0.36 - 0.31}{0.028683} \right) = P(Z > 1.74) = 1 - 0.9591$

$= 0.0409$

Exercise 7.23

$n = 1000; p = 0.53 \ \sigma_{\hat{p}} = \sqrt{\dfrac{0.53(1 - 0.53)}{1000}} = 0.015783$

(a) $P(\hat{p} > 0.5) = P\left(Z > \dfrac{0.5 - 0.53}{0.015783} \right) = P(Z > -1.9) = 1 - P(Z < -1.9)$

$= 1 - 0.0287 = 0.9713$

(b) $P(0.49 < \hat{p} < 0.57) = P\left(\dfrac{0.49 - 0.53}{0.015783} < Z < \dfrac{0.57 - 0.53}{0.015783} \right)$

$= P(-2.53 < Z < 2.53) = 0.9943 - 0.0057 = 0.9886$

(c) $P(p < 0.565) = P\left(Z < \dfrac{0.565 - 0.53}{0.015783} \right) = P(Z < 2.22) = 0.9868$

Exercise 7.25

$$n = 500; p = 0.17 \; \sigma_{\hat{p}} = \sqrt{\frac{0.17(1 - 0.17)}{500}} = 0.016799$$

(a) $P(\hat{p} > 0.145) = P\left(Z > \dfrac{0.145 - 0.17}{0.016799}\right) = P(Z > -1.49)$
$$= 1 - P(Z < -1.49) = 1 - 0.0681 = 0.9319$$

(b) $P(p < 0.20) = P\left(Z < \dfrac{0.20 - 0.17}{0.016799}\right) = P(Z < 1.79) = 0.9633$

(c) $P(0.15 < \hat{p} < 0.205) = P\left(\dfrac{0.15 - 0.17}{0.016799} < Z < \dfrac{0.205 - 0.17}{0.016799}\right)$
$$= P(-1.19 < Z < 2.08) = 0.9812 - 0.1170$$
$$= 0.8642$$

Exercise 7.27

$$n = 407; p = 0.14; \sigma_{\hat{p}} = \sqrt{\frac{0.14(1 - 0.14)}{407}} = 0.0172$$

(a) $P(\hat{p} < 0.1) = P\left(Z < \dfrac{0.1 - 0.14}{0.0172}\right) = P(Z < -2.33) = 0.0099$

(b) $P(0.09 < \hat{p} < 0.12) = P\left(\dfrac{0.09 - 0.14)}{0.0172} < Z < \dfrac{0.12 - 0.14}{0.0172}\right)$
$$= P(-2.91 < Z < -1.16) = 0.1230 - 0.0018$$
$$= 0.1212$$

(c) $P(p > 0.17) = P\left(Z > \dfrac{0.17 - 0.14}{0.0172}\right) = P(Z > 1.74) = 1 - P(Z < 1.74)$
$$= 1 - 0.9591 = 0.0409$$

Exercise 7.29

$$n = 250; p = 0.25; \sigma_{\hat{p}} = \sqrt{\frac{0.25(1 - 0.25)}{250}} = 0.027386$$

(a) $P(\hat{p} > 0.28 = P\left(Z > \dfrac{0.28 - 0.25}{0.027386}\right) = P(Z > 1.1) = 1 - 0.8643$
$$= 0.1357$$

(b) $P(p < 0.20) = P\left(Z < \dfrac{0.20 - 0.25}{0.027386}\right) = P(Z < -1.83) = 0.0336$

(c) $P(0.21 < \hat{p} < 0.32) = P\left(\dfrac{0.21 - 0.25)}{0.027386} < Z < \dfrac{0.32 - 0.25}{0.027386}\right)$
$$= P(-1.46 < Z < 2.56) = 0.9948 - 0.0721 = 0.9227$$

Exercise 8.1

Confidence intervals for single population mean using t-distribution

$$\bar{x} = \frac{246}{40} = 6.15; \quad s^2 = \frac{\sum_{i=1}^{n} x_i^2 - n\bar{x}^2}{n-1} = \frac{2074 - 40(6.15)^2}{40-1} = 14.38718;$$

$$s = \sqrt{14.38718} = 3.793044$$

$$100(1 - \alpha)\% \text{ CI} = \left[\bar{x} - t_{(n-1,\,\alpha/2)} \times \frac{s}{\sqrt{n}}, \bar{x} + t_{(n-1,\,\alpha/2)} \times \frac{s}{\sqrt{n}} \right]$$

(a) To find the 95% = $100(1 - 0.05)\%$ CI $\Rightarrow \alpha = 0.05$; $\alpha/2 = 0.025$.
From the tables we see that $t_{(40-1,\,0.025)} = 2.023$, so that:

$$6.15 \pm 2.023 \times \frac{3.793044}{\sqrt{40}} = [4.9367, 7.3633].$$

$$\text{Margin of error} = 2.023 \times \frac{3.793044}{\sqrt{40}} = 1.2133$$

(b) To find the 99% = $100(1 - 0.01)\%$ CI $\Rightarrow \alpha = 0.01$; $\alpha/2 = 0.005$.
From the tables we see that $t_{(40-1,\,0.005)} = 2.708$, so that:

$$6.15 \pm 2.708 \times \frac{3.793044}{\sqrt{40}} = [4.5259, 7.7741].$$

$$\text{Margin of error} = 2.708 \times \frac{3.793044}{\sqrt{40}} = 1.6241$$

Exercise 8.3

$$\bar{x} = \frac{1100}{36} = 30.5556; \quad s^2 = \frac{\sum_{i=1}^{n} X_i^2 - n\bar{X}^2}{n-1} = \frac{40334 - 36(30.5556)^2}{36-1} = 192.0825$$

$$s = \sqrt{192.0825} = 13.83938$$

$$100(1 - \alpha)\% \text{ CI} = \left[\bar{x} - t_{(n-1,\,\alpha/2)} \times \frac{s}{\sqrt{n}}, \bar{x} + t_{(n-1,\,\alpha/2)} \times \frac{s}{\sqrt{n}} \right]$$

(a) To find the 90% = $100(1 - 0.1)\%$ CI $\Rightarrow \alpha = 0.1$; $\alpha/2 = 0.05$.
From the tables we see that $t_{(36-1,\,0.05)} = 1.69$, so that:

$$30.5556 \pm 1.69 \times \frac{13.83938}{\sqrt{36}} = [26.6575, 34.4537].$$

(b) To find the 98% = $100(1 - 0.02)\%$ CI $\Rightarrow \alpha = 0.02$; $\alpha/2 = 0.01$.
From the tables we see that $_{(36-1,\,0.01)} = 2.438$, so that:

$$30.5556 \pm 2.438 \times \frac{13.83938}{\sqrt{36}} = [24.9322, 36.179].$$

Exercise 8.5

$$\bar{x} = \frac{610.88}{35} = 17.4537;$$

$$s^2 = \frac{\sum\limits_{i=1}^{n} X_i^2 - n\bar{X}^2}{n-1} = \frac{10900.02 - 35(17.4537)^2}{35 - 1} = 6.99685$$

$$s = \sqrt{6.99685} = 2.64516$$

$$100(1 - \alpha)\% \ \text{CI} = \left[\bar{x} - t_{(n-1,\,\alpha/2)} \times \frac{s}{\sqrt{n}}, \bar{x} + t_{(n-1,\,\alpha/2)} \times \frac{s}{\sqrt{n}}\right]$$

(a) To find the 95% = 100(1 − 0.05)% CI ⇒ α = 0.10; α/2 = 0.05.
From the tables we see that $t_{(35-1,\,0.025)} = 2.032$, so that:
$$17.4537 \pm 2.032 \times \frac{2.64516}{\sqrt{35}} = [16.5452, 18.3622].$$

(b) To find the 99% = 100(1 − 0.01)% CI ⇒ α = 0.01; α/2 = 0.005.
From the tables we see that $t_{(35-1,\,0.005)} = 2.728$, so that:
$$17.4537 \pm 2.728 \times \frac{2.64516}{\sqrt{35}} = [16.2340, 18.67343].$$

Exercise 8.7

$$\bar{x} = \frac{2928}{37} = 79.1351; \ s^2 = \frac{\sum\limits_{i=1}^{n} X_i^2 - n\bar{X}^2}{n-1} = \frac{233732 - 37(79.1351)^2}{37 - 1} = 56.2312$$

$$s = \sqrt{56.2312} = 7.4987$$

$$100(1 - \alpha)\% \ \text{CI} = \left[\bar{x} - t_{(n-1,\,\alpha/2)} \times \frac{s}{\sqrt{n}}, \bar{x} + t_{(n-1,\,\alpha/2)} \times \frac{s}{\sqrt{n}}\right]$$

(a) To find the 90% = 100(1 − 0.1)% CI ⇒ α = 0.1; α/2 = 0.05.
From the tables we see that $t_{(37-1,\,0.05)} = 2.028$, so that:
$$79.1351 \pm 2.028 \times \frac{7.4987}{\sqrt{37}} = [76.6350, 81.6352].$$

(b) To find the 99% = 100(1 − 0.01)% CI ⇒ α = 0.01; α/2 = 0.005.
From the tables we see that $t_{(37-1,\,0.005)} = 2.719$, so that:
$$79.1351 \pm 2.719 \times \frac{7.4987}{\sqrt{37}} = [75.7832, 82.4870].$$

Exercise 8.9

$$\bar{x} = \frac{168.7}{33} = 5.1121; \, s^2 = \frac{\sum\limits_{i=1}^{n} X_i^2 - n\bar{X}^2}{n-1} = \frac{868.49 - 33(5.1121)^2}{33-1} = 0.18985$$

$$s = \sqrt{0.18985} = 0.43572$$

$$100(1-\alpha)\% \, \text{CI} = \left[\bar{x} - t_{(n-1,\,\alpha/2)} \times \frac{s}{\sqrt{n}}, \bar{x} + t_{(n-1,\,\alpha/2)} \times \frac{s}{\sqrt{n}}\right]$$

(a) To find the 95% = $100(1 - 0.05)\%$ CI $\Rightarrow \alpha = 0.05$; $\alpha/2 = 0.025$.
     From the tables we see that $t_{(33-1,\,0.025)} = 2.037$, so that:

$$5.1121 \pm 2.037 \times \frac{0.43572}{\sqrt{33}} = [4.9576, 5.2666].$$

(b) To find the 98% = $100(1 - 0.02)\%$ CI $\Rightarrow \alpha = 0.02$; $\alpha/2 = 0.01$.
     From the tables we see that $t_{(33-1,\,0.01)} = 2.738$, so that:

$$5.1121 \pm 2.738 \times \frac{0.43572}{\sqrt{33}} = [4.9044, 5.3198].$$

Exercise 8.11

$$\bar{x} = \frac{419}{16} = 26.1875; \, s^2 = \frac{\sum\limits_{i=1}^{n} X_i^2 - n\bar{X}^2}{n-1} = \frac{12059 - 16(26.1875)^2}{16-1} = 72.4292$$

$$s = \sqrt{72.4292} = 8.5105$$

$$100(1-\alpha)\% \, \text{CI} = \left[\bar{x} - t_{(n-1,\,\alpha/2)} \times \frac{s}{\sqrt{n}}, \bar{x} + t_{(n-1,\,\alpha/2)} \times \frac{s}{\sqrt{n}}\right]$$

(a) To find the 95% = $100(1 - 0.05)\%$ CI $\Rightarrow \alpha = 0.05$; $\alpha/2 = 0.025$.
     From the tables we see that $t_{(15,\,0.025)} = 2.131$, so that:

$$26.1875 \pm 2.131 \times \frac{8.5105}{\sqrt{16}} = [21.6535, 30.7215].$$

$$\text{Margin of error} = 2.131 \times \frac{8.5105}{\sqrt{16}} = 4.5339$$

(b) To find the 98% = $100(1 - 0.02)\%$ CI $\Rightarrow \alpha = 0.02$; $\alpha/2 = 0.01$.
     From the tables we see that $t_{(15,\,0.01)} = 2.602$, so that:

$$26.1875 \pm 2.602 \times \frac{8.5105}{\sqrt{16}} = [20.6514, 31.7236].$$

$$\text{Margin of error} = 2.602 \times \frac{8.5105}{\sqrt{16}} = 5.5361$$

Exercise 8.13

$$\bar{x} = \frac{301}{17} = 17.70588; \quad s^2 = \frac{\sum\limits_{i=1}^{n} X_i^2 - n\bar{X}^2}{n-1} = \frac{429.7 - 21(17.70588)^2}{17-1} = 17.4707$$

$$s = \sqrt{17.4707} = 4.23918$$

$$100(1-\alpha)\% \text{ CI} = \left[\bar{x} - t_{(n-1,\,\alpha/2)} \times \frac{s}{\sqrt{n}}, \bar{x} + t_{(n-1,\,\alpha/2)} \times \frac{s}{\sqrt{n}}\right]$$

(a) To find the 95% = $100(1 - 0.05)\%$ CI $\Rightarrow \alpha = 0.05$; $\alpha/2 = 0.025$.
From the tables we see that $t_{(16,\,0.025)} = 2.12$, so that

$$17.70588 \pm 2.12 \times \frac{4.23917}{\sqrt{17}} = [15.5262, 19.8856].$$

Margin of error = $2.12 \times \dfrac{4.23917}{\sqrt{17}} = 2.1797$

(b) To find the 90% = $100(1 - 0.10)\%$ CI $\Rightarrow \alpha = 0.10$; $\alpha/2 = 0.05$.
From the tables we see that $t_{(16,\,0.05)} = 1.746$, so that

$$17.70588 \pm 1.746 \times \frac{4.23917}{\sqrt{17}} = [15.9107, 19.5010].$$

Margin of error = $1.746 \times \dfrac{4.23917}{\sqrt{17}} = 1.7952$

Exercise 8.15

$$\bar{x} = \frac{4302}{12} = 358.5; \quad s^2 = \frac{\sum\limits_{i=1}^{n} X_i^2 - n\bar{X}^2}{n-1} = \frac{1549962 - 12(358.5)^2}{12-1} = 699.5455$$

$$s = \sqrt{699.5455} = 26.44892$$

$$100(1-\alpha)\% \text{ CI} = \left[\bar{x} - t_{(n-1,\,\alpha/2)} \times \frac{s}{\sqrt{n}}, \bar{x} + t_{(n-1,\,\alpha/2)} \times \frac{s}{\sqrt{n}}\right]$$

(a) To find 80% CI, we see that 80% CI = $100(1 - 0.20)\%$ CI $\Rightarrow \alpha = 0.20$; $\alpha/2 = 0.10$.
From the tables we obtain $t_{(15,\,0.10)} = 1.363$, so that:

$$358.5 \pm 1.363 \times \frac{26.44892}{\sqrt{12}} = [348.0933, 368.9067].$$

(b) To find 95% CI, we see that 95% CI = $100(1 - 0.05)\%$ CI $\Rightarrow \alpha = 0.05$; $\alpha/2 = 0.025$.
From the tables we see that $t_{(15,\,0.025)} = 2.201$, so that:

$$358.5 \pm 2.201 \times \frac{26.44892}{\sqrt{12}} = [341.695, 375.305].$$

Exercise 8.17

$$\bar{x} = \frac{9473}{11} = 861.1818; \; s^2 = \frac{\sum\limits_{i=1}^{n} X_i^2 - n\bar{X}^2}{n-1} = \frac{8168335 - 11(861.1818)^2}{11-1} = 1035.998$$

$$s = \sqrt{1035.998} = 32.18692$$

$$100(1-\alpha)\% \; CI = \left[\bar{x} - t_{(n-1,\,\alpha/2)} \times \frac{s}{\sqrt{n}}, \bar{x} + t_{(n-1,\,\alpha/2)} \times \frac{s}{\sqrt{n}} \right]$$

(a) To find 95% CI, we know that 95% CI $= 100(1 - 0.05)\%$ CI $\Rightarrow \alpha = 0.05$; $\alpha/2 = 0.025$.
From the tables we obtain $t_{(10,\,0.025)} = 2.228$, so that:
$$886.1818 \pm 2.228 \times \frac{32.18692}{\sqrt{11}} = [839.5587, 882.8039].$$

(b) To find 98% CI, we see that 98% CI $= 100(1 - 0.02)\%$ CI $\Rightarrow \alpha = 0.02$; $\alpha/2 = 0.01$.
From the tables we see that $t_{(10,\,0.01)} = 2.764$, so that:
$$886.1818 \pm 2.764 \times \frac{32.18692}{\sqrt{11}} = [834.3601, 888.0057].$$

Exercise 8.19

$$\bar{x} = \frac{350}{13} = 26.9231; \; s^2 = \frac{\sum\limits_{i=1}^{n} X_i^2 - n\bar{X}^2}{n-1} = \frac{10834 - 19(26.9231)^2}{13-1} = 117.5769$$

$$s = \sqrt{117.5769} = 10.8433$$

$$100(1-\alpha)\% \; CI = \left[\bar{x} - t_{(n-1,\,\alpha/2)} \times \frac{s}{\sqrt{n}}, \bar{x} + t_{(n-1,\,\alpha/2)} \times \frac{s}{\sqrt{n}} \right]$$

(a) To find 80% CI, we know that 80% CI $= 100(1 - 0.20)\%$ CI $\Rightarrow \alpha = 0.20$; $\alpha/2 = 0.10$.
From the tables we obtain $t_{(12,\,0.10)} = 1.356$, so that:
$$26.9231 \pm 1.356 \times \frac{10.8433}{\sqrt{13}} = [22.844, 31.002].$$

(b) To find 95% CI, we see that 95% CI $= 100(1 - 0.05)\%$ CI $\Rightarrow \alpha = 0.05$; $\alpha/2 = 0.025$.
From the tables we see that $t_{(12,\,0.025)} = 2.179$, so that:
$$26.9231 \pm 2.179 \times \frac{10.8433}{\sqrt{13}} = [20.3706, 33.4756].$$

Exercise 8.21

$$\bar{x} = \frac{148}{17} = 8.7059; \quad s^2 = \frac{\sum\limits_{i=1}^{n} X_i^2 - n\bar{X}^2}{n-1} = \frac{1336 - 17(8.7059)^2}{17 - 1} = 2.9706$$

$$s = \sqrt{2.9706} = 1.72354$$

$$100(1-\alpha)\% \text{ CI} = \left[\bar{x} - t_{(n-1,\,\alpha/2)} \times \frac{s}{\sqrt{n}}, \bar{x} + t_{(n-1,\,\alpha/2)} \times \frac{s}{\sqrt{n}}\right]$$

(a) To find 90% CI, we know that 90% CI $= 100(1 - 0.10)\%$ CI $\Rightarrow \alpha = 0.10$; $\alpha/2 = 0.05$.
From the tables we obtain $t_{(16,\,0.05)} = 1.746$, so that

$$8.7059 \pm 1.746 \times \frac{1.72354}{\sqrt{17}} = [7.97604, 9.4358].$$

(b) To find 98% CI, we see that 98% CI $= 100(1 - 0.02)\%$ CI $\Rightarrow \alpha = 0.02$; $\alpha/2 = 0.01$.
From the tables we see that $t_{(16,\,0.01)} = 2.583$, so that

$$8.7059 \pm 2.583 \times \frac{1.72354}{\sqrt{17}} = [7.6262, 9.7856].$$

Confidence Intervals about a Single Population Proportion

Exercise 8.23

$$\hat{p} = \frac{720}{1030} = 0.699029;$$

$$se = \sqrt{\frac{\hat{p}(1-\hat{p})}{n}} = \sqrt{\frac{0.699029(1 - 0.6999029)}{1030}} = 0.014292$$

(i) confidence level $= 90\%$
$90\% = 100(1 - 0.10)\% \Rightarrow \alpha = 0.10 \Rightarrow \alpha/2 = 0.05$
90% CI: $\hat{p} \pm Z_{0.05} \times se; \; Z_{0.05} = 1.645$
$0.699029 \pm 1.645 \times 0.014292 = (0.675519, 0.722539)$
margin of error $= 1.645 \times 0.014292 = 0.02351$

(ii) confidence level $= 97\%$
$97\% = 100(1 - 0.03)\% \Rightarrow \alpha = 0.03 \Rightarrow \alpha/2 = 0.015$
97% CI: $\hat{p} \pm Z_{0.015} \times se; \; Z_{0.015} = 2.17$
$0.699029 \pm 2.17 \times 0.014292 = (0.668016, 0.730043)$
margin of error $= 2.17 \times 0.014292 = 0.031014$

Exercise 8.25

$$\hat{p} = \frac{128}{895} = 0.143017$$

$$se = \sqrt{\frac{\hat{p}(1-\hat{p})}{n}} = \sqrt{\frac{0.143017(1 - 0.143017)}{895}} = 0.011702$$

(i) confidence level = 80%

80% = 100(1 − 0.20)% ⟹ $\alpha = 0.20 \Rightarrow \alpha/2 = 0.10$

80% CI: $\hat{p} \pm Z_{0.10} \times se$; $Z_{0.10} = 1.28$

$0.143017 \pm 1.28 \times 0.011702 = (0.128038, 0.157996)$

margin of error = $1.28 \times 0.011702 = 0.014979$

(ii) confidence level = 85%

85% = 100(1 − 0.15)% ⟹ $\alpha = 0.15 \Rightarrow \alpha/2 = 0.075$

85% CI: $\hat{p} \pm Z_{0.075} \times se$; $Z_{0.075} = 1.44$

$0.143017 \pm 1.44 \times 0.011702 = (0.126166, 0.159868)$

margin of error = $1.44 \times 0.011702 = 0.016851$

Exercise 8.27

$$\hat{p} = \frac{1032}{1500} = 0.688; \; se = \sqrt{\frac{\hat{p}(1 - \hat{p})}{n}} = \sqrt{\frac{0.688(1 - 0.688)}{1500}} = 0.011963$$

(i) confidence level = 91%

91% = 100(1 − 0.09)% ⟹ $\alpha = 0.09 \Rightarrow \alpha/2 = 0.045$;

91% CI: $\hat{p} \pm Z_{0.045} \times se$; $Z_{0.045} = 1.7$

$0.688 \pm 1.7 \times 0.011963 = (0.66766, 0.708337)$

margin of error = $1.7 \times 0.011963 = 0.0203371$

(ii) confidence level = 95%

95% = 100(1 − 0.05)% ⟹ $\alpha = 0.05 \Rightarrow \alpha/2 = 0.025$

95% CI: $\hat{p} \pm Z_{0.025} \times se$; $Z_{0.025} = 1.96$

$0.688 \pm 1.96 \times 0.011963 = (0.664553, 0.711448)$

margin of error = $1.96 \times 0.011963 = 0.023447$

Exercise 8.29

$$\hat{p} = \frac{804}{1006} = 0.799205$$

$$se = \sqrt{\frac{\hat{p}(1 - \hat{p})}{n}} = \sqrt{\frac{0.69(1 - 0.69)}{1200}} = 0.01263$$

(i) confidence level = 92%

92% = 100(1 − 0.08)% ⟹ $\alpha = 0.08 \Rightarrow \alpha/2 = 0.04$

92% CI: $\hat{p} \pm Z_{0.04} \times se$; $Z_{0.04} = 1.75$

$0.799205 \pm 1.75 \times 0.01263 = (0.777102, 0.821307)$

margin of error = $1.75 \times 0.01263 = 0.022103$

(ii) confidence level = 94%

94% = 100(1 − 0.06)% ⟹ $\alpha = 0.06 \Rightarrow \alpha/2 = 0.03$

94% CI: $\hat{p} \pm Z_{0.03} \times se$; $Z_{0.03} = 1.88$

$0.799205 \pm 1.88 \times 0.01263 = (0.77546, 0.822949)$

margin of error = $1.88 \times 0.01263 = 0.023745$

Exercise 8.31

$$\hat{p} = \frac{94}{107} = 0.878505$$

$$se = \sqrt{\frac{\hat{p}(1 - \hat{p})}{n}} = \sqrt{\frac{0.878505(1 - 0.878505)}{107}} = 0.031583$$

(i) confidence level = 92%
92% = 100(1 − 0.08)% ⟹ $\alpha = 0.08 \Rightarrow \alpha/2 = 0.04$
92% CI: $\hat{p} \pm Z_{0.04} \times se$; $Z_{0.04} = 1.75$
$0.878505 \pm 1.75 \times 0.031583 = (0.8232348, 0.9337753)$
margin of error = $1.75 \times 0.031583 = 0.0552703$

(ii) confidence level = 99%
99% = 100(1 − 0.01)% ⟹ $\alpha = 0.01 \Rightarrow \alpha/2 = 0.005$
99% CI: $\hat{p} \pm Z_{0.005} \times se$; $Z_{0.025} = 2.575$
$0.878505 \pm 2.575 \times 0.031583 = (0.797178, 0.959832)$
margin of error = $2.575 \times 0.031583 = 0.081326$

Testing Hypothesis about the Single Population Mean Using the t-distribution

Exercise 9.1

$$\overline{X} = \frac{\sum_{i=1}^{16} X_i}{16} = \frac{1900}{16} = 118.75;$$

$$s^2 = \frac{\sum_{i=1}^{n} X_i^2 - n\overline{X}^2}{n - 1} = \frac{244688 - 16(118.75)^2}{16 - 1} = 1270.867$$

$s = 35.64922$

$H_0: \mu = 100$ vs. $H_1: \mu \neq 100$;

$$t_{cal} = \frac{118.75 - 100}{35.64922/\sqrt{16}} = 2.1038$$

Level of significance, $\alpha = 0.01$, then from the tables, we read $t_{(15, 0.005)} = 2.947$

We accept the null hypothesis and conclude that the mean time people spent on this website is one hundred minutes at a 1% level of significance.

p-Value Approach:
We can use EXCEL to obtain the p-value by entering in any cell:

= tdist(2.1038, 15, 2)

The output from Excel is 0.052677, ⟹ p-value = $P[t_{(15)} < 2.1038] = P[t_{(33)} > 2.1038] = 0.052677$.

The p-value (0.052677) is greater than the level of significance, $\alpha = 0.01$, so we accept the null hypothesis and conclude that the mean time people spent on this website is one hundred minutes at a 1% level of significance.

Exercise 9.3

$$\overline{X} = \frac{\sum\limits_{i=1}^{17} X_i}{17} = \frac{518}{17} = 30.471; \; s^2 = \frac{\sum\limits_{i=1}^{n} X_i^2 - n\overline{X}^2}{n-1} = \frac{15894 - 17(30.471)^2}{17-1} = 6.8897$$

$$s = 2.6248$$

$$H_0: \mu = 32 \text{ vs. } H_1: \mu < 32$$

$$t_{cal} = \frac{30.471 - 32}{2.6248/\sqrt{17}} = -2.40$$

Level of significance, $\alpha = 0.05$, then from the tables, we read $t_{(16,\, 0.05)} = 1.746$.

Here, we reject the null hypothesis at the 5% level of significance; we conclude that mean miles/gallon is less than thirty-two.

p-Value Approach:
Level of significance is $\alpha = 0.05$. We obtain the *p*-value by entering the following in any cell in EXCEL:

$$= \text{tdist}(2.40, 16, 1)$$

The EXCEL output is $0.01446 \Rightarrow p\text{-value} = P[t_{(16)} < -2.40] = 0.01446$. The *p*-value is less than the level of significance, so we reject the null hypothesis at the 5% level of significance; we conclude that mean miles/gallon is less than thirty-two.

Exercise 9.5

$$\overline{x} = \frac{94}{21} = 4.47619; \; s^2 = \frac{\sum\limits_{i=1}^{n} X_i^2 - n\overline{X}^2}{n-1} = \frac{429.7 - 21(4.47619)^2}{21-1} = 0.446905$$

$$s = \sqrt{0.446905} = 0.668509$$

$$H_0: \mu = 4.2 \text{ vs. } H_1: \mu \neq 4.2;$$

$$t_c = \frac{\overline{X} - \mu_0}{s/\sqrt{n}} = \frac{4.47619 - 4.2}{0.668509/\sqrt{21}} = 1.90$$

From the tables, $t_{(20,\, 0.025)} = 2.086$ because $\alpha = 0.05$.

Since $t_c < t_{(20,\, 0.025)}$ we accept H_0 and conclude that the mean concentration of the compound is probably 4.2 at 5% level of significance.

p-Value Approach:
Level of significance is $\alpha = 0.05$. We obtain the *p*-value by entering the following in any cell in EXCEL:

$$= \text{tdist}(1.90, 20, 2)$$

The EXCEL output is $0.071948 \Rightarrow p\text{-value} = P[t_{(20)} > 1.90] + P[t_{(20)} < -90] = 0.071948$. The *p*-value is greater than the level of significance, we accept H_0 and conclude that the mean concentration of the compound is 4.2 at 5% level of significance.

Exercise 9.7

$$\overline{x} = \frac{22840}{13} = 2218.462;$$

$$s^2 = \frac{\sum\limits_{i=1}^{n} X_i^2 - n\overline{X}^2}{n - 1} = \frac{64083200 - 13(2218.462)^2}{13 - 1} = 8564.103$$

$$s = \sqrt{8564.103} = 92.54244$$

$H_0 : \mu = 2150$ vs. $H_1 : \mu > 2150$;

$$t_c = \frac{\overline{X} - \mu_0}{s/\sqrt{n}} = \frac{2218.462 - 2150}{92.54244/\sqrt{13}} = 2.6673$$

From the tables, $t_{(12, 0.025)} = 2.179$ because $\alpha = 0.025$.

Since $t_c > t_{(12, 0.025)}$ we reject H_0 and conclude that the mean property tax for houses whose prices are in the vicinity of $200,000 at Stearns County is greater than $2150 at a 2.5% level of significance.

p-Value Approach:
Level of significance is $\alpha = 0.025$. We obtain the *p*-value by entering the following in any cell in EXCEL:

$$= \text{tdist}(2.6673, 12, 1)$$

The EXCEL output is $0.010256 \Rightarrow p\text{-value} = P[t_{(12)} > 2.6673] = 0.010256$. The *p*-value is less than the level of significance; we reject H_0 and conclude that the mean property tax for houses whose prices are in the vicinity of $200,000 at Stearns County is greater than $2150 at 2.5% level of significance.

Exercise 9.9

Solution:

$$\overline{x} = \frac{261.5}{16} = 16.34375;$$

$$s^2 = \frac{\sum\limits_{i=1}^{n} X_i^2 - n\overline{X}^2}{n - 1} = \frac{4684.35 - 16(16.34375)^2}{16 - 1} = 27.36396$$

$$s = \sqrt{27.36396} = 5.231057$$

$H_0 : \mu = 13$ vs. $H_1 : \mu > 13$;

$$t_c = \frac{\overline{X} - \mu_0}{s/\sqrt{n}} = \frac{16.34375 - 13}{5.231057/\sqrt{16}} = 2.557$$

From the tables, $t_{(15, 0.025)} = 2.131$ because $\alpha = 0.025$.

Since $t_c > t_{(15, 0.025)}$ we reject H_0 and conclude that the mean waiting times to checkout at this popular supermarket is greater than thirteen minutes at a 5% level of significance.

p-Value Approach:
Level of significance is $\alpha = 0.025$. We obtain the *p*-value by entering the following in any cell in EXCEL:

$$= \text{tdist}(2.557, 15, 1)$$

The EXCEL output is $0.015729 \Rightarrow$ *p*-value $= P[t_{(15)} > 2.557] = 0.015729$. The *p*-value is less than the level of significance; we reject H_0 and conclude that the mean waiting times to checkout at this popular supermarket is greater than thirteen minutes at a 2.5% level of significance.

Exercise 9.11

Solution:
First, we set up the null hypothesis and the alternative. In this case, the alternative should be two-sided as the dispute states that it is not equal. The disputed idea is put in the alternative hypothesis:

$$H_0{:}\mu = 3.65 \text{ vs. } H_1{:}\mu \neq 3.65$$

The test statistic is obtained as:

$$\sum_{i=1}^{34} X_i = 127.59 \Rightarrow \overline{X} = \frac{1}{34}\sum_{i=1}^{34} X_i = \frac{127.59}{34} = 3.753$$

$$s^2 = \frac{\sum_{i=1}^{n} X_i^2 - n\overline{X}^2}{34 - 1} = \frac{480.2793 - 34(3.753)^2}{34 - 1} = 0.04179$$

$$s = \sqrt{0.04179} = 0.204425$$

$$t_c = \frac{\sqrt{34}(3.753 - 3.65)}{0.204425/\sqrt{34}} = 2.93$$

Classical test: Using $\alpha = 1\%$, we need two rejection regions, each of size $\alpha/2$. We read from the table of $t_c \sim t(34 - 1, \alpha/2)$ that $t(33, \alpha/2)$ and $-t(33, 0.005) = -2.733$; $t(33, 0.005) = 2.733$. These are the two critical values needed to set up the rejection and acceptance regions. Since $t_c > t(33, 0.005)$, we reject the null hypothesis and conclude that the mean GPA of accepted students is not equal to 3.65 at a 1% level of significance.

p-Value Approach:
We use EXCEL to obtain the *p*-value by entering in any cell:

$$= \text{tdist}(2.93, 33, 2)$$

The output from Excel is 0.006108, implying that *p*-value $= P[t_{(33)} < -2.93] + P[t_{(33)} > 2.93] = 0.006108$.

The *p*-value (0.006108) is less than the level of significance, $\alpha = 0.01$, so we reject the null hypothesis and conclude that the mean GPA of students accepted in the program is not equal to 3.65 at a 1% level of significance.

Exercise 9.13

Solution:

$$\overline{x} = \frac{204.16}{39} = 5.2349;$$

$$s^2 = \frac{\sum\limits_{i=1}^{n} X_i^2 - n\overline{X}^2}{n-1} = \frac{1091.452 - 39(5.2349)^2}{39 - 1} = 0.597394$$

$$s = \sqrt{0.597394} = 0.772913$$

$$H_0\!:\mu = 4.9 \text{ vs. } H_1\!:\mu \neq 4.9$$

$$t_c = \frac{5.2349 - 4.9}{0.772913/\sqrt{39}} = 2.71$$

Classical test: Using $\alpha = 5\%$, we need one rejection region to the right and to the left, each of size $\alpha/2$. We read from the table of $t_c \sim t_{(39-1,\,\alpha/2)}$ that $t_c \sim t_{(38,\,\alpha/2)}$; $t_{(38,\,0.025)} = -2.024$ and $t_{(38,\,0.025)} = 2.024$. These are the two critical values needed to set up the rejection and acceptance regions. Since $t_c > t_{(38,\,0.025)}$, we reject the null hypothesis and conclude that the mean percentage nutrient content of the fruit is not equal to 4.9 at the 5% level of significance.

p-Value Approach:
We use EXCEL to obtain the *p*-value by entering in any cell:

$$= \text{tdist}(2.71, 38, 2)$$

The EXCEL output is $0.010039 \Rightarrow p\text{-value} = P[t_{(34)} > 2.71] + P[t_{(34)} < -2.71] = 0.010039$.

The *p*-value (0.010039) is less than the level of significance, $\alpha = 0.05$, so we reject the null hypothesis and conclude that the mean percentage nutrient content of the fruit is not equal to 4.9 at the 5% level of significance.

Exercise 9.15

Solution:

$$\overline{x} = \frac{7724}{37} = 208.7568;$$

$$s^2 = \frac{\sum\limits_{i=1}^{n} X_i^2 - n\overline{X}^2}{n-1} = \frac{1639848 - 37(208.7568)^2}{37 - 1} = 761.4114$$

$$s = \sqrt{761.4114} = 27.59368$$

$$H_0\!:\mu = 200 \text{ vs. } H_1\!:\mu > 200$$

$$t_c = \frac{208.7568 - 200}{27.59368/\sqrt{37}} = 1.93$$

Classical test: Using $\alpha = 5\%$, we need one rejection region to the right of size α. We read from the table of $t_c \sim t_{(37-1,\,\alpha)}$ that $t_c \sim t_{(36,\,\alpha)}$; $t_{(36,\,0.05)} = 1.688$. This is the critical value needed to set up the rejection and acceptance regions. Since $t_c > t_{(36,\,0.05)}$, we reject the null hypothesis and conclude that the mean cost of the extraction of a wisdom tooth is \$200 against that it is greater, at a 5% level of significance.

p-Value Approach:
The test is one-sided to the right. **We use EXCEL to obtain the *p*-value by entering in any cell:**

$$= \text{tdist}(1.93, 36, 1)$$

EXCEL output is 0.030548. The *p*-value (0.030548) is less than the level of significance $\alpha = 0.05$, so we reject the null hypothesis and conclude that the mean cost of the extraction of a wisdom tooth is greater than \$200 at a 5% level of significance.

Exercise 9.17

Solution:

$$\bar{x} = \frac{17740}{34} = 521.765;$$

$$s^2 = \frac{\sum_{i=1}^{n} X_i^2 - n\bar{X}^2}{n-1} = \frac{9409200 - 34(521.765)^2}{34-1} = 4638.90$$

$$s = \sqrt{4638.90} = 68.1118$$

$$H_0{:}\mu = 495 \text{ vs. } H_1{:}\mu \neq 495$$

$$t_c = \frac{521.765 - 495}{68.1118/\sqrt{34}} = 2.29$$

Classical test: Using $\alpha = 2\%$, we need two rejection regions to the right and left, each of size $\alpha/2$. We read from the table of $t_c \sim t_{(34-1,\,\alpha/2)}$ that $t_c \sim t_{(33,\,\alpha/2)}$; $-t_{(33,\,0.01)} = -2.445$ and $-t_{(33,\,0.01)} = 2.445$. These are the critical values needed to set up the rejection and acceptance regions. Since $t_c < t_{(33,\,0.01)}$, we accept the null hypothesis and conclude that the mean water consumption is 500 milliliters at a 2% level of significance.

p-Value Approach:
The test is two-sided to the left and right. **We use EXCEL to obtain the *p*-value by entering in any cell:**

$$= \text{tdist}(2.29, 33, 2)$$

The Excel output is 0.02835. The *p*-value (0.02835) is greater than the level of significance, $\alpha = 0.02$, so we accept the null hypothesis, and conclude that the mean water consumption is 500 milliliters at a 2% level of significance.

Exercise 9.19

Solution:

$$H_0: \mu = 12.5 \text{ vs. } H_1: \mu < 12.5$$

The test statistic is obtained as

$$\sum_{i=1}^{42} X_i = 511 \Rightarrow \overline{X} = \frac{1}{42} \sum_{i=1}^{42} X_i = \frac{511}{42} = 12.1667$$

$$s^2 = \frac{\sum_{i=1}^{n} X_i^2 - n\overline{X}^2}{42 - 1} = \frac{6295.22 - 42(12.1667)^2}{42 - 1} = 1.90291$$

$$s = \sqrt{1.90291} = 1.37946$$

$$t_c = \frac{\sqrt{42}(12.1667 - 12.5)}{1.37946} = -1.57$$

Classical test: Using $\alpha = 5\%$, we need one rejection region of size α. We read from the table of $t_c \sim t_{(42-1, \, \alpha)}$ that $t_c \sim t_{(41, \, \alpha)}$; $-t_{(41, \, 0.05)} = -1.683$. This is the critical value needed to set up the rejection and acceptance regions. Since $t_c > -t_{(41, \, 0.05)}$, we accept the null hypothesis and conclude that the mean life span of LRs is 12.5 at a 5% level of significance.

p-Value Approach:
The test is one-sided to the left. **We use EXCEL to obtain the *p*-value by entering in any cell:**

$$= \text{tdist}(1.57, 41, 1)$$

The Excel output is 0.062051. The *p*-value (0.062051) is greater than the level of significance, $\alpha = 0.05$, so we accept the null hypothesis and conclude that the mean life span of LRs is 12.5 at a 5% level of significance.

Solutions to Exercises on Testing Hypothesis for a Single Population Proportion

Exercise 9.21

$$H_0: p_L = 0.39 \text{ vs. } H_1: p_L > 0.39$$

$$\hat{p}_L = \frac{66}{156} = 0.423077; \; \sigma_L = \sqrt{\frac{0.39(1 - 0.39)}{156}} = 0.039051$$

$$Z_{cal} = \frac{0.423077 - 0.39}{0.039051} = 0.847 \approx 0.85$$

$$\text{l.o.s.} = 0.025 \Rightarrow Z_{0.025} = 1.96$$

Since $Z_{cal} < Z_\alpha$ we accept H_0 and conclude that the proportion of patients of Dr. Lenny who did not fully recover is 39%.

$$p\text{-value} = P(Z > 0.85) = 1 - 0.8023 = 0.1977$$

p-value > 0.025, so we accept H_0 and conclude that the proportion of patients of Dr. Lenny who did not fully recover is 39%.

Exercise 9.23

$H_0 : p_L = p_0$ vs. $H_1 : p_L > p_0$

$H_0 : p_L = 0.27$ vs. $H_1 : p_L > 0.27$

$$\hat{p}_L = \frac{178}{520} = 0.342308; \quad \sigma_L = \sqrt{\frac{0.27(1 - 0.27)}{520}} = 0.019469$$

$$Z_c = \frac{0.342308 - 0.27}{0.019469} = 3.71$$

$$\alpha = 0.02 \Rightarrow Z_{0.02} = 2.05$$

Since $Z_c > Z_\alpha$ we reject H_0 and conclude that the proportion of eggs that hatched is more than 27%.

$$p\text{-value} = P(Z > 3.71) = 1 - 0.99999 = 0.00001$$

p-value < 0.02 so we reject H_0 and conclude that the proportion of eggs that hatched is greater than 27%.

Exercise 9.25

$H_0 : p = 0.05$ vs. $H_1 : p > 0.05$

$$\hat{p} = \frac{90}{1500} = 0.06; \quad \sigma = \sqrt{\frac{0.05(1 - 0.05)}{1500}} = 0.005627$$

$$Z_c = \frac{0.06 - 0.05}{0.005627} = 1.777 \approx 1.78$$

$$\alpha = 0.01 \Rightarrow Z_{0.01} = 2.33$$

Since $Z_c < Z_\alpha$ and falls within the acceptance region, we accept H_0 and conclude process is operating as specified, that is the proportion of defectives is 0.05.

Exercise 9.27

$H_0{:}p = 0.54$ vs. $H_1{:}p > 0.54$

$\hat{p} = \dfrac{276}{747} = 0.585987; \sigma = \sqrt{\dfrac{0.54(1 - 0.54)}{470}} = 0.022965$

$Z_c = \dfrac{0.585987 - 0.54}{0.022965} = 2.003$

$\alpha = 0.03 \Rightarrow Z_\alpha = Z_{0.03} = 1.88$

Since $Z_c > Z_\alpha$ and falls within the rejection region, we reject H_0 and conclude that the proportion of patients treated with D_1 who are allergic is greater than 0.54.

Using p-Value

$P\text{-value} = P(Z > 2.003) = 0.0228.$

Since p-value $< \alpha$, so we reject H_0 and conclude proportion that is allergic to D_1 formulation is greater than 0.54 at 0.03 level of significance.

Exercise 9.29

$H_0{:}p = 0.40$ vs. $H_1{:}p < 0.40$

$\hat{p} = \dfrac{470}{1235} = 0.380567; \sigma = \sqrt{\dfrac{0.40(1 - 0.40)}{1235}} = 0.01394$

$Z_{cal} = \dfrac{0.380567 - 0.40}{0.01394} = -1.39$

$\alpha = 0.03 \Rightarrow Z_\alpha = Z_{0.03} = 1.88$

Since $Z_{cal} > -Z_{\alpha/2}$ and falls within the acceptance region, we accept H_0 and conclude that the proportion of part time students have not decreased from 40% at 3% level of significance.

The Solutions of Regression Exercises

Exercise 10.1

x	y	xy	x^2	y^2	$\hat{y}$	$y - \hat{y}$	$(y - \hat{y})^2$
12	8	96	144	64	6.852941	1.147059	1.315744
14	9	126	196	81	8.397059	0.602941	0.363538
11	6	66	121	36	6.080882	−0.08088	0.006542
11	4	44	121	16	6.080882	−2.08088	4.330071
10	3	30	100	9	5.308824	−2.30882	5.330666
9	8	72	81	64	4.536765	3.463235	11.994
8	3	24	64	9	3.764706	−0.76471	0.584775
6	2	12	36	4	2.220588	−0.22059	0.048659
7	4	28	49	16	2.992647	1.007353	1.01476
8	3	24	64	9	3.764706	−0.76471	0.584775
96	**50**	**522**	**976**	**308**	**50**	**3.55E-15**	**25.57353**

$$n = 10; \overline{X} = \frac{96}{10} = 9.6; \overline{Y} = \frac{50}{10} = 5.0; \sum_{i=1}^{10} x_i y_i = 522; \sum_{i=1}^{10} x_i^2 = 976;$$

$$\sum_{i=1}^{10} y_i^2 = 308; S_{xx} = \sum_{i=1}^{10} x_i^2 - n\overline{x}^2 = 976 - 10(9.6)^2 = 54.40;$$

$$S_{yy} = \sum_{i=1}^{10} y_i^2 - n\overline{y}^2 = 308 - 10(5)^2 = 58;$$

$$S_{xy} = \sum_{i=1}^{10} x_i y_i - n\overline{x}\overline{y} = 522 - 10(9.6)(5) = 42;$$

$$\hat{\beta}_1 = \frac{S_{xy}}{S_{xx}} = \frac{42}{54.40} = 0.772059;$$

$$\hat{\beta}_0 = \overline{y} - \hat{\beta}_1\overline{x} = 5.0 - 0.772059(9.6) = -2.41176$$

The fitted line is: $\hat{y} = -2.41176 + 0.772059x$

$$r = \frac{S_{xy}}{\sqrt{S_{xx} \cdot S_{yy}}} = \frac{42}{\sqrt{54.40 \times 58}} = 0.747715$$

$$SSE = \sum_{i=1}^{n} (y_i - \hat{y}_i)^2 = 25.57628.$$

Thus $\hat{\sigma}^2 = \dfrac{SSE}{n-2} = \dfrac{25.57628}{10-2} = 3.197035 \Rightarrow \hat{\sigma} = \sqrt{3.197035} = 1.788025$

Next, we test that the slope is zero, against that it is not, that is, $H_0{:}\beta_1 = 0$ vs. $H_1{:}\beta_1 \neq 0$

$$t_{cal} = \frac{\hat{\beta}_1}{\hat{\sigma}/\sqrt{S_{xx}}} = \frac{0.772089}{1.788025/\sqrt{54.40}} = 3.18488$$

From the tables, $t_{(8, 0.025)} = 2.306$ we see that fitted line is significant. We therefore reject H_0 and conclude that $\beta_1 \neq 0$.

Exercise 10.3

x	y	xy	x^2	y^2	$\hat{y}$	$y - \hat{y}$	$(y - \hat{y})^2$
3.7	650	2405	13.69	422500	650.2208	−0.22079	0.048746
2	490	980	4	240100	478.0189	11.9811	143.5467
3.3	580	1914	10.89	336400	609.7027	−29.7027	882.2501
2.8	530	1484	7.84	280900	559.0551	−29.0551	844.1978
3.5	640	2240	12.25	409600	629.9617	10.03826	100.7666
3.15	615	1937.25	9.9225	378225	594.5084	20.49159	419.9052
3.7	650	2405	13.69	422500	650.2208	− 0.22079	0.048746
2.2	498	1095.6	4.84	248004	498.2779	−0.27795	0.077255
3.3	590	1947	10.89	348100	609.7027	−19.7027	388.1962
2.8	550	1540	7.84	302500	559.0551	−9.05508	81.99452
3.3	630	2079	10.89	396900	609.7027	20.2973	411.9806
3.2	625	2000	10.24	390625	599.5732	25.42683	646.5235
36.95	**7048**	**22026.85**	**116.9825**	**4176354**	**7048**	**−6.8E-13**	**3919.536**

$$n = 12; \quad \overline{X} = \frac{36.95}{12} = 3.079167; \quad \overline{Y} = \frac{7048}{12} = 587.3333; \quad \sum_{i=1}^{12} x_i y_i = 22026.85;$$

$$\sum_{i=1}^{12} x_i^2 = 116.9825; \quad \sum_{i=1}^{12} y_i^2 = 4176354;$$

$$S_{xx} = \sum_{i=1}^{12} x_i^2 - n\overline{x}^2 = 116.9825 - 12(3.079167)^2 = 3.207292;$$

$$S_{yy} = \sum_{i=1}^{12} y_i^2 - n\overline{y}^2 = 4176354 - 12(587.333)^2 = 36828.67;$$

$$S_{xy} = \sum_{i=1}^{12} x_i y_i - n\overline{xy} = 22026.85 - 12(3.079167)(587.333) = 324.8833;$$

$$\hat{\beta}_1 = \frac{S_{xy}}{S_{xx}} = \frac{324.8833}{3.207292} = 101.2952;$$

$$\hat{\beta}_0 = \overline{y} - \hat{\beta}_1\overline{x} = 587.3333 - 101.2952(3.079167) = 275.4285$$

The fitted line is: $\hat{y} = 275.4285 + 101.2952x$

$$r = \frac{S_{xy}}{\sqrt{S_{xx} \cdot S_{yy}}} = \frac{324.8833}{\sqrt{3.207292 \times 36828.67}} = 0.94529$$

Now, $SSE = \sum_{i=1}^{n} (y_i - \hat{y}_i)^2 = 3919.536.$ $\hat{\sigma}^2 = \frac{SSE}{n-2} = \frac{3919.536}{12-2} = 391.9536 \Rightarrow$
$\hat{\sigma} = \sqrt{391.9536} = 19.79782$

Next, we test that the slope is zero, against that it is not, that is, $H_0:\beta_1 = 0$ vs. $H_1:\beta_1 \neq 0$

$$t_{cal} = \frac{\hat{\beta}_1}{\hat{\sigma}/\sqrt{S_{xx}}} = \frac{101.2952}{19.79782/\sqrt{3.207292}} = 9.163065$$

From the tables, $t_{(10, 0.005)} = 3.169$ we see that fitted line is significant. We therefore reject H_0 and conclude that $\beta_1 \neq 0$ at 1% level of significance.

Exercise 10.5

Tree	1	2	3	4	5	6
X	1.8	1.9	1.8	2.4	5.1	3.1
Y	22	32.5	23.6	40.8	63.2	27.9
XY	39.6	61.75	42.48	97.92	322.32	86.49
x^2	3.24	3.61	3.24	5.76	26.01	9.61
Y^2	484	1056.25	556.96	1664.64	3994.24	778.41
$\hat{y}$	28.276	28.8269	28.276	31.5817	46.4572	35.4383
$y - \hat{y}$	−6.27599	3.67307	−4.67599	9.21834	16.7428	−7.53827
$(y - \hat{y})^2$	39.388	13.4914	21.8649	84.9778	280.322	56.8256
$\hat{y} - \bar{y}$	−16.914	−16.3631	−16.914	−13.6083	1.26717	−9.75173
$(\hat{y} - \bar{y})^2$	286.084	267.75	286.084	185.187	1.60573	95.0962

Tree	7	8	9	10	Sum
X	5.5	5.1	8.3	13.7	**48.7**
Y	41.4	44.3	63.5	92.7	**451.9**
XY	227.7	225.93	527.05	1269.99	**2901.23**
x^2	30.25	26.01	68.89	187.69	**364.31**
Y^2	1713.96	1962.49	4032.25	8593.29	**24836.5**
$\hat{y}$	48.661	46.4572	64.0874	93.8384	**451.9**
$y - \hat{y}$	−7.26095	−2.15717	−0.58741	−1.13844	**2.8E-14**
$(y - \hat{y})^2$	52.7214	4.6534	0.34506	1.29606	**555.886**
$\hat{y} - \bar{y}$	3.47095	1.26717	18.8974	48.6484	**0**
$(\hat{y} - \bar{y})^2$	12.0475	1.60573	357.112	2366.67	**3859.24**

$$n = 10; \overline{X} = \frac{48.7}{10} = 4.87; \overline{Y} = \frac{451.9}{10} = 45.19;$$

$$\sum_{i=1}^{10} x_i y_i = 2901.23; \sum_{i=1}^{10} x_i^2 = 364.31; \sum_{i=1}^{10} y_i^2 = 24836.5$$

$$S_{xx} = \sum_{i=1}^{10} x_i^2 - n\overline{x}^2 = 364.31 - 10(4.87)^2 = 127.141$$

$$S_{yy} = \sum_{i=1}^{10} y_i^2 - n\overline{y}^2 = 24836.5 - 10(45.19)^2 = 4415.129$$

$$S_{xy} = \sum_{i=1}^{10} x_i y_i - n\overline{xy} = 2901.23 - 10(45.19)(4.87) = 700.477$$

$$\hat{\beta}_1 = \frac{S_{xy}}{S_{xx}} = \frac{700.477}{127.141} = 5.50945$$

$$\hat{\beta}_0 = \overline{y} - \hat{\beta}_1\overline{x} = 45.19 - 5.50945(4.87) = 18.35898$$

The fitted line is: $\hat{y} = 18.35898 + 5.50945x$

$$SSE = 555.8859 \; \hat{\sigma}^2 = \frac{SSE}{n-2} = \frac{555.8859}{10-2} = 69.4874 \Rightarrow \hat{\sigma} = \sqrt{69.4874}$$

$$= 8.335811$$

Next, we test that the slope is zero, against that it is not, that is, $H_0:\beta_1 = 0$ vs. $H_1:\beta_1 \neq 0$

$$t_{cal} = \frac{\hat{\beta}_1}{\hat{\sigma}/\sqrt{S_{xx}}} = \frac{5.50945}{8.335811/\sqrt{127.141}} = 7.452521$$

From the tables, $t_{(8, 0.025)} = 2.306$ we see that fitted line is significant. We therefore reject H_0 and conclude that $\beta_1 \neq 0$.

Exercise 10.7

Height (ins.)	Weight (lbs.)	xy	x^2	y^2	$\hat{y}$	$y - \hat{y}$	$(y - \hat{y})^2$	$\hat{y} - \bar{y}$	$(\hat{y} - \bar{y})^2$	xy
63	154	9702	3969	23716	150.9793	3.020657	9.12437	148.3214	−3.72066	13.84329
65	145	9425	4225	21025	153.9559	−8.95587	80.20758	139.3214	−0.74413	0.553732
70	166	11620	4900	27556	161.3972	4.602817	21.18592	160.3214	6.697183	44.85226
66	150	9900	4356	22500	155.4441	−5.44413	29.63857	144.3214	0.744131	0.553732
58	145	8410	3364	21025	143.538	1.461972	2.137362	139.3214	−11.162	124.5896
69	164	11316	4761	26896	159.9089	4.09108	16.73693	158.3214	5.20892	27.13285
67	156	10452	4489	24336	156.9324	−0.93239	0.869359	150.3214	2.232394	4.983585
64	158	10112	4096	24964	152.4676	5.532394	30.60739	152.3214	−2.23239	4.983585
65	154	10010	4225	23716	153.9559	0.044131	0.001948	148.3214	−0.74413	0.553732
68	155	10540	4624	24025	158.4207	−3.42066	11.7009	149.3214	3.720657	13.84329
655	**1547**	**101487**	**43009**	**239759**	**1547**	**1.42E-13**	**202.2103**	**1490.214**	**−2.8E-14**	**235.8897**

$$n = 10; \overline{X} = \frac{655}{10} = 65.50; \overline{Y} = \frac{1547}{10} = 154.7;$$

$$\sum_{i=1}^{10} x_i y_i = 101487; \sum_{i=1}^{10} x_i^2 = 43009; \sum_{i=1}^{10} y_i^2 = 239759;$$

$$S_{xx} = \sum_{i=1}^{10} x_i^2 - n\overline{x}^2 = 239759 - 10(65.50)^2 = 106.50$$

$$S_{yy} = \sum_{i=1}^{10} y_i^2 - n\overline{y}^2 = 43009 - 10(154.7)^2 = 438.1$$

$$S_{xy} = \sum_{i=1}^{15} x_i y_i - n\overline{xy} = 101487 - 10(65.50)(154.7) = 158.50$$

$$\hat{\beta}_1 = \frac{S_{xy}}{S_{xx}} = \frac{158.50}{106.5} = 1.488263$$

$$\hat{\beta}_0 = \overline{y} - \hat{\beta}_1 \overline{x} = 154.7 - 1.488263(65.5) = 57.21878$$

The fitted line is: $\hat{y} = 57.21878 + 1.488263x$

$$SSE = 202.2103;$$

$$\hat{\sigma}^2 = \frac{SSE}{n-2} = \frac{202.2103}{10-2} = 25.27629 \Rightarrow \hat{\sigma} = \sqrt{25.27629} = 5.027553$$

Next, we test that the slope is zero, against that it is not, that is, $H_0 : \beta_1 = 0$ vs. $H_1 : \beta_1 \neq 0$

$$t_{cal} = \frac{\hat{\beta}_1}{\hat{\sigma}/\sqrt{S_{xx}}} = \frac{1.488263}{5.027553/\sqrt{106.5}} = 3.054906$$

From the tables, $t_{(13, 0.01)} = 2.65$ we see that fitted line is significant. We therefore reject H_0 and conclude that $\beta_1 \neq 0$.

Exercise 10.9

x	y	xy	x^2	y^2	$\hat{y}$	$y - \hat{y}$	$(y - \hat{y})^2$
5.66	11.96	67.6936	32.0356	143.0416	10.483	1.477004	2.181541
15.6	12	187.2	243.36	144	12.11164	−0.11164	0.012464
10.51	11.44	120.2344	110.4601	130.8736	11.27766	0.162343	0.026355
6.24	11.44	71.3856	38.9376	130.8736	10.57803	0.861972	0.742996
1.76	9.61	16.9136	3.0976	92.3521	9.84399	−0.23399	0.054751
3.94	11.31	44.5614	15.5236	127.9161	10.20118	1.108822	1.229486
4.28	9.99	42.7572	18.3184	99.8001	10.25689	−0.26689	0.071228
2.98	9.97	29.7106	8.8804	99.4009	10.04388	−0.07388	0.005459
15.3	11.29	172.737	234.09	127.4641	12.06249	−0.77249	0.596738
12.28	11.67	143.3076	150.7984	136.1889	11.56767	0.102332	0.010472
2.34	8.84	20.6856	5.4756	78.1456	9.939022	−1.09902	1.207848
2.74	8.85	24.249	7.5076	78.3225	10.00456	−1.15456	1.33301
83.63	**128.37**	**941.4356**	**868.4849**	**1388.379**	**128.37**	**1.24E-14**	**7.472349**

$$n = 12; \overline{X} = \frac{83.63}{12} = 6.96917; \overline{Y} = \frac{128.37}{12} = 10.6975;$$

$$\sum_{i=1}^{12} x_i y_i = 941.4356; \sum_{i=1}^{12} x_i^2 = 868.4849; \sum_{i=1}^{12} y_i^2 = 1388.379$$

$$S_{xx} = \sum_{i=1}^{13} x_i^2 - n\overline{x}^2 = 868.4849 - 12(6.96917)^2 = 285.6535$$

$$S_{yy} = \sum_{i=1}^{13} y_i^2 - n\overline{y}^2 = 1388.379 - 12(10.6975)^2 = 15.14103$$

$$S_{xy} = \sum_{i=1}^{n} x_i y_i - n\overline{xy} = 941.4356 - 12(6.96917)(10.6975) = 46.80368$$

$$\hat{\beta}_1 = \frac{S_{xy}}{S_{xx}} = \frac{46.80368}{285.6535} = 0.163848$$

$$\hat{\beta}_0 = \overline{y} - \hat{\beta}_1\overline{x} = 128.37/12 - 0.163848(83.63/12) = 9.555618$$

The fitted line is: $\hat{y} = 9.55568 + 0.163848x$

$$SSE = 7.47235;$$

$$\hat{\sigma}^2 = \frac{SSE}{n-2} = \frac{7.47235}{12-2} = 0.747235 \Rightarrow \hat{\sigma} = \sqrt{0.747235} = 0.864428$$

Next, we test that the slope is zero, against that it is not, that is, $H_0: \beta_1 = 0$ vs. $H_1: \beta_1 \neq 0$

$$t_{cal} = \frac{\hat{\beta}_1}{\hat{\sigma}/\sqrt{S_{xx}}} = \frac{0.163848}{0.864428/\sqrt{285.6535}} = 3.203551$$

From the tables, $t_{(10,\,0.01)} = 2.764$ we see that fitted line is significant. We therefore reject H_0 and conclude that $\beta_1 \neq 0$.

Exercise 10.11

x	y	xy	x^2	y^2	$\hat{y}$	$y - \hat{y}$	$(y - \hat{y})^2$	$(\hat{y} - \overline{y})$	$(\hat{y} - \overline{y})^2$
1.08	81	87.48	1.1664	6561	84.12367	−3.12367	9.7573	−416.676	173619.2
5.66	290	1641.4	32.0356	84100	310.4851	−20.4851	419.6384	−190.351	36219.76
15.7	466	7316.2	246.49	217156	806.7009	−340.701	116077.1	305.9009	93575.36
6.95	435	3023.25	48.3025	189225	374.2419	60.7581	3691.547	−126.558	16016.95
10.32	475	4902	106.5024	225625	540.8004	−65.8004	4329.693	40.0004	1600.032
4.88	225	1098	23.8144	50625	271.9344	−46.9344	2202.842	−228.866	52379.46
15.35	959	14720.65	235.6225	919681	789.4026	169.5974	28763.29	288.6026	83291.46
15.91	848	13491.68	253.1281	719104	817.0799	30.92006	956.0501	316.2799	100033
5.56	340	1890.4	30.9136	115600	305.5427	34.45731	1187.306	−195.257	38125.41
9.29	485	4505.65	86.3041	235225	489.8938	−4.8938	23.94923	−10.9062	118.9452
12.09	643	7773.87	146.1681	413449	628.2807	14.71932	216.6583	127.4807	16251.33
15.85	875	13868.75	251.2225	765625	814.1145	60.88549	3707.043	313.3145	98165.98
19.54	1071	20927.34	381.8116	1147041	996.4887	74.51134	5551.94	495.6887	245707.3
1.83	99	181.17	3.3489	9801	121.1916	−22.1916	492.4664	−379.608	144102.5
2.65	220	583	7.0225	48400	161.7192	58.28083	3396.655	−339.081	114975.8
142.7	**7512**	**96010.84**	**1853.853**	**5147218**	**7512**	**6.68E-13**	**171026**	**7E-05**	**1214182**

$$n = 15;\ \overline{X} = \frac{142.66}{15} = 9.51067;\ \overline{Y} = \frac{7512}{15} = 500.8;$$

$$\sum_{i=1}^{15} x_i y_i = 96010.84;\ \sum_{i=1}^{15} x_i^2 = 1853.853;\ \sum_{i=1}^{15} y_i^2 = 11240.182$$

$$S_{xx} = \sum_{i=1}^{15} x_i^2 - n\overline{x}^2 = 1853.858 - 15(9.51067)^2 = 497.0615$$

$$S_{yy} = \sum_{i=1}^{15} y_i^2 - n\overline{y}^2 = 5147218 - 15(500.8)^2 = 1385208$$

$$S_{xy} = \sum_{i=1}^{15} x_i y_i - n\overline{xy} = 96010.84 - 15(9.51067)(500.8) = 24566.71$$

$$\hat{\beta}_1 = \frac{S_{xy}}{S_{xx}} = \frac{24566.71}{497.0615} = 49.42389$$

$$\hat{\beta}_0 = \overline{y} - \hat{\beta}_1\overline{x} = 500.8 - 49.42389(9.510667) = 30.74587$$

The fitted line is: $\hat{y} = 30.74587 + 49.42389x$

$$SSE = 171026;$$

$$\hat{\sigma}^2 = \frac{SSE}{n-2} = \frac{171026}{15-2} = 13155.84 \Rightarrow \hat{\sigma} = \sqrt{13155.85} = 114.6989$$

Next, we test that the slope is zero, against that it is not, that is, $H_0:\beta_1 = 0$ vs. $H_1:\beta_1 \neq 0$

$$t_{cal} = \frac{\hat{\beta}_1}{\hat{\sigma}/\sqrt{S_{xx}}} = \frac{049.42389}{114.6989/\sqrt{497.0615}} = 9.60689$$

From the tables, $t_{(14, 0.005)} = 2.977$ we see that fitted line is significant. We therefore reject H_0 and conclude that $\beta_1 \neq 0$.

Exercise 10.13

x	y	xy	x^2	y^2	$\hat{y}$	$y - \hat{y}$	$(y - \hat{y})^2$	$\hat{y} - \bar{y}$	$(\hat{y} - \bar{y})^2$
2	70	140	4	4900	70.49864	−0.49864	0.248644	53.57007	2869.753
2	75	150	4	5625	70.49864	4.501357	20.26222	53.57007	2869.753
2.5	72	180	6.25	5184	74.92036	−2.92036	8.528514	57.99179	3363.048
2.5	78	195	6.25	6084	74.92036	3.079638	9.48417	57.99179	3363.048
2.75	80	220	7.5625	6400	77.13122	2.868778	8.229889	60.20265	3624.359
3	81	243	9	6561	79.34208	1.657919	2.748694	62.41351	3895.446
3	70	210	9	4900	79.34208	−9.34208	87.27449	62.41351	3895.446
3.5	85	297.5	12.25	7225	83.7638	1.236199	1.528188	66.83523	4466.948
3	80	240	9	6400	79.34208	0.657919	0.432857	62.41351	3895.446
3.25	70	227.5	10.5625	4900	81.55294	−11.5529	133.4704	64.62437	4176.309
3.5	87	304.5	12.25	7569	83.7638	3.236199	10.47298	66.83523	4466.948
3.5	85	297.5	12.25	7225	83.7638	1.236199	1.528188	66.83523	4466.948
3.75	90	337.5	14.0625	8100	85.97466	4.025339	16.20336	69.04609	4767.362
4	90	360	16	8100	88.18552	1.81448	3.292336	71.25695	5077.553
42.25	1113	3402.5	132.4375	89173	1113	−8.5E-14	303.705	876	55198.37

$$n = 14; \quad \overline{X} = \frac{42.25}{14} = 3.018; \quad \overline{Y} = \frac{1113}{14} = 79.5;$$

$$\sum_{i=1}^{14} x_i y_i = 3402; \quad \sum_{i=1}^{14} x_i^2 = 132.4375; \quad \sum_{i=1}^{14} y_i^2 = 89173;$$

$$S_{xx} = \sum_{i=1}^{14} x_i^2 - n\bar{x}^2 = 132.4375 - 14(3.018)^2 = 4.933;$$

$$S_{yy} = \sum_{i=1}^{14} y_i^2 - n\bar{y}^2 = 89173 - 14(79.5)^2 = 689.5$$

$$S_{xy} = \sum_{i=1}^{14} x_i y_i - n\overline{xy} = 3402.5 - 14(3.018)(79.5) = 43.625$$

$$\hat{\beta}_1 = \frac{S_{xy}}{S_{xx}} = \frac{43.625}{4.933} = 8.843439$$

$$\hat{\beta}_0 = \bar{y} - \hat{\beta}_1\bar{x} = 79.5 - 8.843439(3.018) = 52.81176$$

The fitted line is: $\hat{y} = 52.81176 + 8.843439x$

$$SSE = 303.705;$$

$$\hat{\sigma}^2 = \frac{SSE}{n-2} = \frac{303.705}{14-2} = 25.30875 \Rightarrow \hat{\sigma} = \sqrt{25.30875} = 5.03078$$

Next, we test that the slope is zero, against that it is not, that is, $H_0:\beta_1 = 0$ vs. $H_1:\beta_1 \neq 0$

$$t_{cal} = \frac{\hat{\beta}_1}{\hat{\sigma}/\sqrt{S_{xx}}} = \frac{8.843439}{5.03078/\sqrt{4.933}} = 3.904298$$

From the tables, $t_{(10,\,0.01)} = 2.764$ we see that fitted line is significant. We therefore reject H_0 and conclude that $\beta_1 \neq 0$.

Exercise 10.15

x	y	xy	x^2	y^2	$\hat{y}$	$y - \hat{y}$	$(y - \hat{y})^2$	$y - \bar{y}$	$(y - \bar{y})^2$
3.2	7.4	23.68	10.24	54.76	6.561562	0.838438	0.702978	1.634289	2.670901
3	6.5	19.5	9	42.25	6.6242	−0.1242	0.015426	1.696928	2.879563
7.2	5.5	39.6	51.84	30.25	5.308797	0.191203	0.036559	0.381524	0.145561
6	6.2	37.2	36	38.44	5.684626	0.515374	0.26561	0.757354	0.573584
7	6.4	44.8	49	40.96	5.371435	1.028565	1.057946	0.444162	0.19728
5.4	4.3	23.22	29.16	18.49	5.872541	−1.57254	2.472885	0.945268	0.893532
8	4	32	64	16	5.058244	−1.05824	1.11988	0.130971	0.017153
12.2	3.7	45.14	148.84	13.69	3.74284	−0.04284	0.001835	−1.18443	1.402881
12.5	3.3	41.25	156.25	10.89	3.648883	−0.34888	0.121719	−1.27839	1.634281
14.7	3.1	45.57	216.09	9.61	2.959862	0.140138	0.019639	−1.96741	3.870706
13.4	3.8	50.92	179.56	14.44	3.36701	0.43299	0.18748	−1.56026	2.434418
92.6	54.2	402.88	949.98	289.78	54.2	−1.8E-15	6.001957	7.55E-15	16.71986

$$n = 11; \overline{X} = \frac{92.6}{11} = 8.418182; \overline{Y} = \frac{54.2}{11} = 4.927273;$$

$$\sum_{i=1}^{14} x_i y_i = 402.88; \quad \sum_{i=1}^{14} x_i^2 = 949.98; \quad \sum_{i=1}^{14} y_i^2 = 289.78;$$

$$S_{xx} = \sum_{i=1}^{11} x_i^2 - n\overline{x}^2 = 949.98 - 11(8.418182)^2 = 170.4564;$$

$$S_{yy} = \sum_{i=1}^{14} y_i^2 - n\overline{y}^2 = 289.78 - 11(4.927273)^2 = 22.72182$$

$$S_{xy} = \sum_{i=1}^{14} x_i y_i - n\overline{xy} = 402.88 - 11(9.714286)(4.927273) = -53.3855$$

$$\hat{\beta}_1 = \frac{S_{xy}}{S_{xx}} = \frac{-53.3855}{170.4564} = -0.31319$$

$$\hat{\beta}_0 = \overline{y} - \hat{\beta}_1 \overline{x} = 4.927273 - (-0.31319)(8.418182) = 7.563774$$

The fitted line is: $\hat{y} = 7.563774 - 0.313194x$

$$SSE = 6.001957;$$

$$\hat{\sigma}^2 = \frac{SSE}{n-2} = \frac{6.001957}{11-2} = 0.666884 \Rightarrow \hat{\sigma} = \sqrt{0.666884} = 0.81663$$

Next, we test that the slope is zero, against that it is not, that is, $H_0:\beta_1 = 0$ vs. $H_1:\beta_1 \neq 0$

$$t_{cal} = \frac{\hat{\beta}_1}{\hat{\sigma}/\sqrt{S_{xx}}} = \frac{-0.313194}{0.81663/\sqrt{170.4564}} = -5.00716$$

From the tables, $t_{(10,\ 0.025)} = 2.201$ we see that fitted line is significant. We therefore reject H_0 and conclude that $\beta_1 \neq 0$.

Exercise 11.1

x	y	xy	x^2	y^2	$\hat{y}$	$y - \hat{y}$	$(y - \hat{y})^2$
12	8	96	144	64	6.852941	1.147059	1.315744
14	9	126	196	81	8.397059	0.602941	0.363538
11	6	66	121	36	6.080882	−0.08088	0.006542
11	4	44	121	16	6.080882	−2.08088	4.330071
10	3	30	100	9	5.308824	−2.30882	5.330666
9	8	72	81	64	4.536765	3.463235	11.994
8	3	24	64	9	3.764706	−0.76471	0.584775
6	2	12	36	4	2.220588	−0.22059	0.048659
7	4	28	49	16	2.992647	1.007353	1.01476
8	3	24	64	9	3.764706	−0.76471	0.584775
96	50	522	976	308	50	3.55E-15	25.57353

$$n = 10; \overline{X} = \frac{96}{10} = 9.6; \overline{Y} = \frac{50}{10} = 5.0; \sum_{i=1}^{10} x_i y_i = 522; \sum_{i=1}^{10} x_i^2 = 976;$$

$$\sum_{i=1}^{10} y_i^2 = 308;$$

$$S_{xx} = \sum_{i=1}^{10} x_i^2 - n\bar{x}^2 = 976 - 10(9.6)^2 = 54.40;$$

$$S_{yy} = \sum_{i=1}^{10} y_i^2 - n\bar{y}^2 = 308 - 10(5)^2 = 58;$$

$$S_{xy} = \sum_{i=1}^{10} x_i y_i - n\overline{xy} = 522 - 10(9.6)(5) = 42;$$

$$\hat{\beta}_1 = \frac{S_{xy}}{S_{xx}} = \frac{42}{54.40} = 0.772059;$$

$$\hat{\beta}_0 = \bar{y} - \hat{\beta}_1\bar{x} = 5.0 - 0.772059(9.6) = -2.41176$$

The fitted line is: $\hat{y} = -2.41176 + 0.772059x$

$$r = \frac{S_{xy}}{\sqrt{S_{xx} \cdot S_{yy}}} = \frac{42}{\sqrt{54.40 \times 58}} = 0.747715$$

Next, we test that the population correlation coefficient, ρ, is zero, against that it is not, that is: $H_0 : \rho = 0$ vs. $H_1 : \rho \neq 0$;
The test statistic is:

$$t_{cal} = \frac{r}{\sqrt{\dfrac{1 - r^2}{n - 2}}} = \frac{0.747715}{\sqrt{\dfrac{1 - 0.747715^2}{10 - 2}}} = 3.184927.$$

From the tables, $t_{(8, 0.025)} = 2.306$. We reject H_0 and conclude that $\rho \neq 0$ at a 5% level of significance.

Now $SSE = \sum_{i=1}^{n} (y_i - \hat{y}_i)^2 = 25.57628$. Thus,

$$R^2 = \frac{S_{yy} - SSE}{S_{yy}} = \frac{58 - 25.57353}{58} = 0.559077$$

Exercise 11.3

x	y	xy	x^2	y^2	$\hat{y}$	$y - \hat{y}$	$(y - \hat{y})^2$
3.7	650	2405	13.69	422500	650.2208	−0.22079	0.048746
2	490	980	4	240100	478.0189	11.9811	143.5467
3.3	580	1914	10.89	336400	609.7027	−29.7027	882.2501
2.8	530	1484	7.84	280900	559.0551	−29.0551	844.1978
3.5	640	2240	12.25	409600	629.9617	10.03826	100.7666
3.15	615	1937.25	9.9225	378225	594.5084	20.49159	419.9052
3.7	650	2405	13.69	422500	650.2208	−0.22079	0.048746
2.2	498	1095.6	4.84	248004	498.2779	−0.27795	0.077255
3.3	590	1947	10.89	348100	609.7027	−19.7027	388.1962
2.8	550	1540	7.84	302500	559.0551	−9.05508	81.99452
3.3	630	2079	10.89	396900	609.7027	20.2973	411.9806
3.2	625	2000	10.24	390625	599.5732	25.42683	646.5235
36.95	**7048**	**22026.85**	**116.9825**	**4176354**	**7048**	**−6.8E-13**	**3919.536**

$$n = 12; \overline{X} = \frac{36.95}{12} = 3.079167; \overline{Y} = \frac{7048}{12} = 587.3333; \sum_{i=1}^{12} x_i y_i = 22026.85;$$

$$\sum_{i=1}^{12} x_i^2 = 116.9825; \sum_{i=1}^{12} y_i^2 = 4176354;$$

$$S_{xx} = \sum_{i=1}^{12} x_i^2 - n\overline{x}^2 = 116.9825 - 12(3.079167)^2 = 3.207292;$$

$$S_{yy} = \sum_{i=1}^{12} y_i^2 - n\overline{y}^2 = 4176354 - 12(587.333)^2 = 36828.67;$$

$$S_{xy} = \sum_{i=1}^{12} x_i y_i - n\overline{xy} = 22026.85 - 12(3.079167)(587.333) = 324.8833;$$

$$\hat{\beta}_1 = \frac{S_{xy}}{S_{xx}} = \frac{324.8833}{3.207292} = 101.2952;$$

$$\hat{\beta}_0 = \overline{y} - \hat{\beta}_1\overline{x} = 587.3333 - 101.2952(3.079167) = 275.4285$$

The fitted line is: $\hat{y} = 275.4285 + 101.2952x$

$$r = \frac{S_{xy}}{\sqrt{S_{xx} \cdot S_{yy}}} = \frac{324.8833}{\sqrt{3.207292 \times 36828.67}} = 0.94529$$

Next, we test that the population correlation coefficient, ρ, is zero, against that it is not, that is: $H_0: \rho = 0$ vs. $H_1: \rho > 0$;

The test statistic is:

$$t_{cal} = \frac{r}{\sqrt{\frac{1-r^2}{n-2}}} = \frac{0.94529}{\sqrt{\frac{1-0.94529^2}{12-2}}} = 9.163067.$$

From the tables, $t_{(10,\,0.01)} = 2.764$. We reject H_0 and conclude that $\rho \neq 0$ at a 1% level of significance.

Now, $SSE = \sum_{i=1}^{n}(y_i - \hat{y}_i)^2 = 3919.536$. Thus,

$$R^2 = \frac{S_{yy} - SSE}{S_{yy}} = \frac{36828.67 - 3919.536}{38628.67} = 0.893574$$

Exercise 11.5

Tree	1	2	3	4	5	6
x	1.8	1.9	1.8	2.4	5.1	3.1
y	22	32.5	23.6	40.8	63.2	27.9
xy	39.6	61.75	42.48	97.92	322.32	86.49
x^2	3.24	3.61	3.24	5.76	26.01	9.61
y^2	484	1056.25	556.96	1664.64	3994.24	778.41
$y_i - \hat{y}_i$	−6.276	3.67307	−4.676	9.21834	16.7428	−7.5383
$(y_i - \hat{y}_i)^2$	39.388	13.4914	21.8649	84.9778	280.322	56.8256

Tree	7	8	9	10	Total
x	5.5	5.1	8.3	13.7	**48.7**
y	41.4	44.3	63.5	92.7	**451.9**
xy	227.7	225.93	527.05	1269.99	**2901.23**
x^2	30.25	26.01	68.89	187.69	**364.31**
y^2	1713.96	1962.49	4032.25	8593.29	**24836.49**
$y_i - \hat{y}_i$	−7.261	−2.1572	−0.5874	−1.1384	**3.55E-14**
$(y_i - \hat{y}_i)^2$	52.7214	4.6534	0.34506	1.29606	**555.8859**

$$n = 10; \quad \overline{X} = \frac{48.7}{10} = 4.87; \quad \overline{Y} = \frac{451.9}{10} = 45.19; \quad \sum_{i=1}^{10} x_i y_i = 2901.23;$$

$$\sum_{i=1}^{10} x_i^2 = 364.31; \quad \sum_{i=1}^{10} y_i^2 = 24836.5;$$

$$S_{xx} = \sum_{i=1}^{10} x_i^2 - n\overline{x}^2 = 364.31 - 10(4.87)^2 = 127.141;$$

$$S_{yy} = \sum_{i=1}^{10} y_i^2 - n\bar{y}^2 = 24836.5 - 10(45.19)^2 = 4415.29;$$

$$S_{xy} = \sum_{i=1}^{10} x_i y_i - n\overline{xy} = 2901.23 - 10(45.19)(4.87) = 700.477$$

$$\hat{\beta}_1 = \frac{S_{xy}}{S_{xx}} = \frac{700.477}{127.141} = 5.50945;$$

$$\hat{\beta}_0 = \bar{y} - \hat{\beta}_1 \bar{x} = 45.19 - 5.50945(4.87) = 18.35898$$

The fitted line is: $\hat{y} = 18.35898 + 5.50945x$

$$r = \frac{S_{xy}}{\sqrt{S_{xx} \cdot S_{yy}}} = \frac{700.477}{\sqrt{127.141 \times 4415.29}} = 0.934931$$

Next, we test that the population correlation coefficient, ρ, is zero, against that it is not, that is: $H_0:\rho = 0$ vs. $H_1:\rho \neq 0$;
The test statistic is:

$$t_{cal} = \frac{r}{\sqrt{\dfrac{1 - r^2}{n - 2}}} = \frac{0.934931}{\sqrt{\dfrac{1 - 0.934931^2}{10 - 2}}} = 7.45144.$$

From the tables, $t_{(8, 0.025)} = 2.306$. We reject H_0 and conclude that $\rho \neq 0$ at a 5% level of significance.

Now, $SSE = \sum_{i=1}^{n} (y_i - \hat{y}_i)^2 = 555.8859$. Thus,

$$R^2 = \frac{S_{yy} - SSE}{S_{yy}} = \frac{4415.129 - 555.8859}{4415.129} = 0.874095$$

Exercise 11.7

Height (ins.)	Weight (lbs.)	xy	x^2	y^2	$\hat{y}$	$y - \hat{y}$	$(y - \hat{y})^2$
63	154	9702	3969	23716	150.9793	3.020657	9.12437
65	145	9425	4225	21025	153.9559	−8.95587	80.20758
70	166	11620	4900	27556	161.3972	4.602817	21.18592
66	150	9900	4356	22500	155.4441	−5.44413	29.63857
58	145	8410	3364	21025	143.538	1.461972	2.137362
69	164	11316	4761	26896	159.9089	4.09108	16.73693
67	156	10452	4489	24336	156.9324	−0.93239	0.869359
64	158	10112	4096	24964	152.4676	5.532394	30.60739
65	154	10010	4225	23716	153.9559	0.044131	0.001948
68	155	10540	4624	24025	158.4207	−3.42066	11.7009
655	**1547**	**101487**	**43009**	**239759**	**1547**	**1.42E-13**	**202.2103**

$$n = 10; \ \overline{X} = \frac{655}{10} = 65.50; \ \overline{Y} = \frac{1547}{10} = 154.7; \ \sum_{i=1}^{10} x_i y_i = 101487;$$

$$\sum_{i=1}^{10} x_i^2 = 43009; \ \sum_{i=1}^{10} y_i^2 = 239759;$$

$$S_{xx} = \sum_{i=1}^{10} x_i^2 - n\overline{x}^2 = 43009 - 10(65.50)^2 = 106.50;$$

$$S_{yy} = \sum_{i=1}^{10} y_i^2 - n\overline{y}^2 = 239759 - 10(154.7)^2 = 438.1$$

$$S_{xy} = \sum_{i=1}^{15} x_i y_i - n\overline{xy} = 101487 - 10(65.50)(154.7) = 158.50$$

$$\hat{\beta}_1 = \frac{S_{xy}}{S_{xx}} = \frac{158.50}{106.5} = 1.488263$$

$$\hat{\beta}_0 = \overline{y} - \hat{\beta}_1 \overline{x} = 154.7 - 1.488263(65.5) = 57.21878$$

Fitted line is: $\hat{y} = 57.21878 + 1.488263x$

$$r = \frac{S_{xy}}{\sqrt{S_{xx}S_{yy}}} = \frac{158.50}{\sqrt{106.50 \times 438.1}} = 0.733783$$

We test the ρ is zero, against that it greater than zero, that is: $H_0 : \rho = 0$ vs. $H_1 : \rho > 0$;

The test statistic is:

$$t_{cal} = \frac{r}{\sqrt{\dfrac{1 - r^2}{n - 2}}} = \frac{0.733783}{\sqrt{\dfrac{1 - 0.733783^2}{10 - 2}}} = 3.054906.$$

From the tables, $t_{(8, 0.01)} = 2.896$. We reject H_0 and conclude that ρ is probably greater than zero at a 1% level of significance.

With $S_{yy} = 438.1$; $SSE = \sum_{i=1}^{n} (y_i - \hat{y}_i)^2 = 202.2103$. Thus,

$$R^2 = \frac{S_{yy} - SSE}{S_{yy}} = \frac{438.1 - 202.2102}{438.1} = 0.538438.$$

Exercise 11.9

x	y	xy	x^2	y^2	$\hat{y}$	$y - \hat{y}$	$(y - \hat{y})^2$
5.66	11.96	67.6936	32.0356	143.0416	10.483	1.477004	2.181541
15.6	12	187.2	243.36	144	12.11164	−0.11164	0.012464
10.51	11.44	120.2344	110.4601	130.8736	11.27766	0.162343	0.026355
6.24	11.44	71.3856	38.9376	130.8736	10.57803	0.861972	0.742996
1.76	9.61	16.9136	3.0976	92.3521	9.84399	−0.23399	0.054751
3.94	11.31	44.5614	15.5236	127.9161	10.20118	1.108822	1.229486
4.28	9.99	42.7572	18.3184	99.8001	10.25689	−0.26689	0.071228
2.98	9.97	29.7106	8.8804	99.4009	10.04388	−0.07388	0.005459
15.3	11.29	172.737	234.09	127.4641	12.06249	−0.77249	0.596738
12.28	11.67	143.3076	150.7984	136.1889	11.56767	0.102332	0.010472
2.34	8.84	20.6856	5.4756	78.1456	9.939022	−1.09902	1.207848
2.74	8.85	24.249	7.5076	78.3225	10.00456	−1.15456	1.33301
83.63	128.37	941.4356	868.4849	1388.379	128.37	1.24E-14	7.472349

$$n = 12; \overline{X} = \frac{83.63}{12} = 6.96917; \overline{Y} = \frac{128.37}{12} = 10.6975; \sum_{i=1}^{12} x_i y_i = 941.4356;$$

$$\sum_{i=1}^{12} x_i^2 = 868.4849; \sum_{i=1}^{12} y_i^2 = 1388.379$$

$$S_{xx} = \sum_{i=1}^{13} x_i^2 - n\overline{x}^2 = 868.4849 - 12(6.96917)^2 = 285.6535;$$

$$S_{yy} = \sum_{i=1}^{13} y_i^2 - n\overline{y}^2 = 1388.379 - 12(10.6975)^2 = 15.14103$$

$$S_{xy} = \sum_{i=1}^{n} x_i y_i - n\overline{xy} = 941.4356 - 12(6.96917)(10.6975) = 46.80368;$$

$$\hat{\beta}_1 = \frac{S_{xy}}{S_{xx}} = \frac{46.80368}{285.6535} = 0.163848;$$

$$\hat{\beta}_0 = \overline{y} - \hat{\beta}_1 \overline{x} = 128.37/12 - 0.163848(83.63/12) = 9.555618;$$

The fitted line is: $\hat{y} = 9.55568 + 0.163848x$

$$r = \frac{S_{xy}}{\sqrt{S_{xx}S_{yy}}} = \frac{46.80368}{\sqrt{285.6535 \times 15.14103}} = 0.711616$$

We test the hypothesis that $H_0:\rho = 0$ vs. $H_1:\rho \neq 0$
The test statistics is:

$$t_c = \frac{r}{\sqrt{\frac{1 - r^2}{n - 2}}} = \frac{0.711616}{\sqrt{\frac{1 - 0.711616^2}{12 - 2}}} = 3.20355$$

From the tables, $t_{(10, 0.025)} = 2.228$; we reject H_0 and conclude that $\rho \neq 0$ at a 5% level of significance.

With $S_{yy} = 15.14103$ and $SSE = \sum_{i=1}^{n} (y_i - \hat{y}_i)^2 = 7.47235$. Thus,

$$R^2 = \frac{S_{yy} - SSE}{S_{yy}} = \frac{15.14103 - 7.47235}{15.14103} = 0.506483$$

Exercise 11.11

x	y	xy	x^2	y^2	$\hat{y}$	$y - \hat{y}$	$(y - \hat{y})^2$
1.08	81	87.48	1.1664	6561	84.12367	−3.12367	9.7573
5.66	290	1641.4	32.0356	84100	310.4851	−20.4851	419.6384
15.7	466	7316.2	246.49	217156	806.7009	−340.701	116077.1
6.95	435	3023.25	48.3025	189225	374.2419	60.7581	3691.547
10.32	475	4902	106.5024	225625	540.8004	−65.8004	4329.693
4.88	225	1098	23.8144	50625	271.9344	−46.9344	2202.842
15.35	959	14720.65	235.6225	919681	789.4026	169.5974	28763.29
15.91	848	13491.68	253.1281	719104	817.0799	30.92006	956.0501
5.56	340	1890.4	30.9136	115600	305.5427	34.45731	1187.306
9.29	485	4505.65	86.3041	235225	489.8938	−4.8938	23.94923
12.09	643	7773.87	146.1681	413449	628.2807	14.71932	216.6583
15.85	875	13868.75	251.2225	765625	814.1145	60.88549	3707.043
19.54	1071	20927.34	381.8116	1147041	996.4887	74.51134	5551.94
1.83	99	181.17	3.3489	9801	121.1916	−22.1916	492.4664
2.65	220	583	7.0225	48400	161.7192	58.28083	3396.655
142.66	7512	96010.84	1853.853	5147218	7512	6.68E-13	171026

$$n = 15; \overline{X} = \frac{142.66}{15} = 9.51067; \overline{Y} = \frac{7512}{15} = 500.8; \sum_{i=1}^{15} x_i y_i = 96010.84;$$

$$\sum_{i=1}^{15} x_i^2 = 1853.853; \sum_{i=1}^{15} y_i^2 = 11240.182$$

$$S_{xx} = \sum_{i=1}^{15} x_i^2 - n\overline{x}^2 = 1853.858 - 15(9.51067)^2 = 497.0615;$$

$$S_{yy} = \sum_{i=1}^{15} y_i^2 - n\overline{y}^2 = 5147218 - 15(500.8)^2 = 1385208$$

$$S_{xy} = \sum_{i=1}^{15} x_i y_i - n\overline{xy} = 96010.84 - 15(9.51067)(500.8) = 24566.71;$$

$$\hat{\beta}_1 = \frac{S_{xy}}{S_{xx}} = \frac{24566.71}{497.0615} = 49.42389$$

$$\hat{\beta}_0 = \overline{y} - \hat{\beta}_1 \overline{x} = 500.8 - 49.42389(9.510667) = 30.74587$$

The fitted line is: $\hat{y} = 30.74587 + 49.42389x$

$$r = \frac{S_{xy}}{\sqrt{S_{xx}S_{yy}}} = \frac{24566.71}{\sqrt{497.0615 \times 1385208}} = 0.936234$$

$$\hat{\beta}_1 = \frac{S_{xy}}{S_{xx}} = \frac{24566.71}{497.0615} = 49.42389$$

$$\hat{\beta}_0 = \overline{y} - \hat{\beta}_1 \overline{x} = 500.8 - 49.42389(9.510667) = 30.74587$$

We test the hypothesis that $H_0: \rho = 0$ vs. $H_1: \rho > 0$
  The test statistics is:

$$t_c = \frac{r}{\sqrt{\frac{1 - r^2}{n - 2}}} = \frac{0.936234}{\sqrt{\frac{1 - 0.936234^2}{15 - 2}}} = 9.60689$$

From the tables, $t_{(13, 0.05)} = 1.771$; we reject H_0 and conclude that $\rho \neq 0$ at a 5% level of significance.

With $S_{yy} = 1385208$ and $SSE = \sum_{i=1}^{n} (y_i - \hat{y}_i)^2 = 171026$. Thus,

$$R^2 = \frac{S_{yy} - SSE}{S_{yy}} = \frac{1385208 - 171026}{1385208} = 0.876534.$$

Exercise 11.13

x	y	xy	x^2	y^2	$\hat{y}$	$y - \hat{y}$	$(y - \hat{y})^2$
2	70	140	4	4900	70.49864	−0.49864	0.248644
2	75	150	4	5625	70.49864	4.501357	20.26222
2.5	72	180	6.25	5184	74.92036	−2.92036	8.528514
2.5	78	195	6.25	6084	74.92036	3.079638	9.48417
2.75	80	220	7.5625	6400	77.13122	2.868778	8.229889
3	81	243	9	6561	79.34208	1.657919	2.748694
3	70	210	9	4900	79.34208	−9.34208	87.27449
3.5	85	297.5	12.25	7225	83.7638	1.236199	1.528188
3	80	240	9	6400	79.34208	0.657919	0.432857
3.25	70	227.5	10.5625	4900	81.55294	−11.5529	133.4704
3.5	87	304.5	12.25	7569	83.7638	3.236199	10.47298
3.5	85	297.5	12.25	7225	83.7638	1.236199	1.528188
3.75	90	337.5	14.0625	8100	85.97466	4.025339	16.20336
4	90	360	16	8100	88.18552	1.81448	3.292336
42.25	**1113**	**3402.5**	**132.4375**	**89173**	**1113**	**−8.5E-14**	**303.705**

$$n = 14; \overline{X} = \frac{42.25}{14} = 3.018; \overline{Y} = \frac{1113}{14} = 79.5; \sum_{i=1}^{14} x_i y_i = 3402;$$

$$\sum_{i=1}^{14} x_i^2 = 132.4375; \sum_{i=1}^{14} y_i^2 = 89173;$$

$$S_{xx} = \sum_{i=1}^{14} x_i^2 - n\overline{x}^2 = 132.4375 - 14(3.018)^2 = 4.933;$$

$$S_{yy} = \sum_{i=1}^{14} y_i^2 - n\overline{y}^2 = 89173 - 14(79.5)^2 = 689.5$$

$$S_{xy} = \sum_{i=1}^{14} x_i y_i - n\overline{xy} = 3402.5 - 14(3.018)(79.5) = 43.625$$

$$\hat{\beta}_1 = \frac{S_{xy}}{S_{xx}} = \frac{43.625}{4.933} = 8.843439$$

$$\hat{\beta}_0 = \overline{y} - \hat{\beta}_1\overline{x} = 79.5 - 8.843439(3.018) = 52.81176$$

The fitted line is: $\hat{y} = 52.81176 + 8.843439x$

$$r = \frac{S_{xy}}{\sqrt{S_{xx}S_{yy}}} = \frac{43.625}{\sqrt{4.933 \times 689.5}} = 0.748901$$

We test the hypothesis that $H_0{:}\rho = 0$ vs. $H_1{:}\rho \neq 0$

The test statistics is:

$$t_c = \frac{r}{\sqrt{\frac{1 - r^2}{n - 2}}} = \frac{0.748091}{\sqrt{\frac{1 - 0.748091^2}{14 - 2}}} = 3.90433$$

From the tables, $t_{(12, 0.005)} = 3.055$; we reject H_0 and conclude that $\rho \neq 0$ at 1% level of significance.

With $S_{yy} = 689.5$ and $SSE = \sum_{i=1}^{n}(y_i - \hat{y}_i)^2 = 303.7095$. Thus

$$R^2 = \frac{S_{yy} - SSE}{S_{yy}} = \frac{689.5 - 303.7095}{689.5} = 0.559552$$

Exercise 11.15

x	y	xy	x^2	y^2	$\hat{y}$	$y - \hat{y}$	$(y - \hat{y})^2$
3.2	7.4	23.68	10.24	54.76	6.561562	0.838438	0.702978
3	6.5	19.5	9	42.25	6.6242	−0.1242	0.015426
7.2	5.5	39.6	51.84	30.25	5.308797	0.191203	0.036559
6	6.2	37.2	36	38.44	5.684626	0.515374	0.26561
7	6.4	44.8	49	40.96	5.371435	1.028565	1.057946
5.4	4.3	23.22	29.16	18.49	5.872541	−1.57254	2.472885
8	4	32	64	16	5.058244	−1.05824	1.11988
12.2	3.7	45.14	148.84	13.69	3.74284	−0.04284	0.001835
12.5	3.3	41.25	156.25	10.89	3.648883	−0.34888	0.121719
14.7	3.1	45.57	216.09	9.61	2.959862	0.140138	0.019639
13.4	3.8	50.92	179.56	14.44	3.36701	0.43299	0.18748
92.6	**54.2**	**402.88**	**949.98**	**289.78**	**54.2**	**1.33E-15**	**6.001957**

$$n = 11; \overline{X} = \frac{92.6}{11} = 8.418182; \overline{Y} = \frac{54.2}{11} = 4.927273; \sum_{i=1}^{11} x_i y_i = 402.88;$$

$$\sum_{i=1}^{11} x_i^2 = 949.98; \sum_{i=1}^{11} y_i^2 = 289.78;$$

$$S_{xx} = \sum_{i=1}^{11} x_i^2 - n\overline{x}^2 = 949.98 - 11(8.418182)^2 = 170.4564;$$

$$S_{yy} = \sum_{i=1}^{11} y_i^2 - n\overline{y}^2 = 289.78 - 11(4.927273)^2 = 22.72182$$

$$S_{xy} = \sum_{i=1}^{11} x_i y_i - n\overline{xy} = 402.88 - 11(8.418182)(4.927273) = -53.3855$$

$$\hat{\beta}_1 = \frac{S_{xy}}{S_{xx}} = \frac{-53.3855}{170.4564} = -0.31319$$

$$\hat{\beta}_0 = \bar{y} - \hat{\beta}_1\bar{x} = 4.927273 - (-0.31319)(8.418182) = 7.563774$$

The fitted line is: $\hat{y} = 7.563774 - 0.31319x$

$$r = \frac{S_{xy}}{\sqrt{S_{xx}S_{yy}}} = \frac{-53.3855}{\sqrt{22.72182 \times 170.4564}} = -0.857817$$

We test the hypothesis that $H_0: \rho = 0$ vs. $H_1: \rho \neq 0$
The test statistics is:

$$t_c = \frac{r}{\sqrt{\dfrac{1 - r^2}{n - 2}}} = \frac{-0.857817}{\sqrt{\dfrac{1 - (-0.857817)^2}{11 - 2}}} = -5.007156$$

From the tables, $t_{(10,\,0.005)} = 3.25$; we reject H_0 and conclude that $\rho \neq 0$ at a 1% level of significance.

With $S_{yy} = 22.72182$ and $SSE = \displaystyle\sum_{i=1}^{n}(y_i - \hat{y}_i)^2 = 6.001957$. Thus

$$R^2 = \frac{S_{yy} - SSE}{S_{yy}} = \frac{22.72182 - 6.001957}{22.72182} = 0.735851.$$

Exercise 12.1

$$n_1 = 400; n_2 = 300; p_1 = 0.7; p_2 = 0.6; \sigma_{\hat{p}_1 - \hat{p}_2} = \sqrt{\frac{0.7(1-0.7)}{400} + \frac{0.6(1-0.6)}{300}} = 0.036401$$

(a) $P(\hat{p}_1 - \hat{p}_2 > 0.059) = p\left(Z > \dfrac{0.059 - (0.7 - 0.6)}{0.036401}\right) = P(Z > -1.13)$

$$= 1 - 0.1292 = 0.8708.$$

(b) $P(0.04 < \hat{p}_1 - \hat{p}_2 < 0.15) = p\left(\dfrac{0.04 - (0.7 - 0.6)}{0.036401} < Z < \dfrac{0.15 - (0.7 - 0.6)}{0.036401}\right)$

$$= P(-1.65 < Z < 1.37) = 0.9147 - 0.0495 = 0.8652$$

(c) $P(\hat{p}_1 - \hat{p}_2 < 0.08) = P\left(Z < \dfrac{0.08 - (0.7 - 0.6)}{0.036401}\right) = P(Z < -0.55) = 0.2912.$

Exercise 12.3

(a) $Z = \dfrac{\bar{X}_1 - \bar{X}_2 - (\mu_1 - \mu_2)}{\sigma_{\bar{x}_1 - \bar{x}_2}}$

$$\mu_1 = 114; \sigma_1^2 = 560; n_1 = 50; \mu_2 = 102; \sigma_2^2 = 481; n_2 = 42;$$

$$\sigma_{\bar{x}_1 - \bar{x}_2} = \sqrt{\frac{\sigma_1^2}{n_1} + \frac{\sigma_2^2}{n_2}} = \sqrt{\frac{560}{50} + \frac{481}{42}} = 4.759452$$

$$P[(\bar{X}_1 - \bar{X}_2) > 20] = P\left(Z > \frac{20 - (114 - 102)}{4.759452}\right)$$

$$= p(Z > 1.68) = 1 - 0.9535 = 0.0464$$

(b) $\mu_1 = 114; \sigma_1^2 = 560; n_1 = 50; \mu_2 = 102; \sigma_2^2 = 481; n_2 = 42;$

$$\sigma_{\bar{x}_1 - \bar{x}_2} = \sqrt{\frac{\sigma_1^2}{n_1} + \frac{\sigma_2^2}{n_2}} = \sqrt{\frac{560}{50} + \frac{481}{42}} = 4.759452$$

$$P[3 < (\bar{X}_1 - \bar{X}_2) < 15] = P\left(\frac{3 - (114 - 102)}{4.759452} < Z < \frac{15 - (114 - 102)}{4.759452}\right)$$

$$= p(-1.68 < Z < 0.84) = 0.7996 - 0.0465 = 0.7531$$

(c) $\mu_1 = 114; \sigma_1^2 = 560; n_1 = 50; \mu_2 = 102; \sigma_2^2 = 481; n_2 = 42;$

$$\sigma_{\bar{x}_1 - \bar{x}_2} = \sqrt{\frac{\sigma_1^2}{n_1} + \frac{\sigma_2^2}{n_2}} = \sqrt{\frac{560}{50} + \frac{481}{42}} = 4.759452$$

$$P[(\bar{X}_1 - \bar{X}_2) > 20] = P\left(Z > \frac{20 - (114 - 102)}{4.759452}\right)$$

$$= p(Z > 1.89) = 1 - 0.9706 = 0.0294$$

Exercise 12.5

$$n_1 = 200; n_2 = 200; p_1 = 0.6; p_2 = 0.4; \sigma_{\hat{p}_1 - \hat{p}_2} = \sqrt{\frac{0.6(1 - 0.6)}{200} + \frac{0.4(1 - 0.6)}{200}} = 0.04899;$$

(a) $P(\hat{p}_1 - \hat{p}_2 > 0.27) = p\left(Z > \frac{0.27 - (0.6 - 0.4)}{0.04899}\right) = P(Z > 1.43) = 1 - 0.9236 = 0.0764.$

(b) $P(0.14 < \hat{p}_1 - \hat{p}_2 < 0.25) = p\left(\frac{0.14 - (0.6 - 0.4)}{0.04899} < Z < \frac{0.25 - (0.6 - 0.4)}{0.04899}\right)$

$$= P(-1.22 < Z < 1.02) = 0.9485 - 0.1113 = 0.8373$$

(c) $P(\hat{p}_1 - \hat{p}_2 < 0.28) = P\left(Z < \frac{0.28 - (0.6 - 0.4)}{0.04899}\right) = P(Z < 1.63) = 0.9485.$

Exercise 12.7

$$n_1 = 750, n_2 = 560; x_1 = 257, x_2 = 149;$$

$$\hat{p}_1 = \frac{257}{750} = 0.342667; \hat{p}_2 = \frac{149}{560} = 0.266071$$

$$[(\hat{p}_1 - \hat{p}_2) - Z_{\frac{\alpha}{2}}\hat{\sigma}_{\hat{p}_1 - \hat{p}_2}, (\hat{p}_1 - \hat{p}_2) + Z_{\frac{\alpha}{2}}\hat{\sigma}_{\hat{p}_1 - \hat{p}_2}]$$

where $\hat{\sigma}_{\hat{p}_1 - \hat{p}_2} = \sqrt{\dfrac{\hat{p}_1(1 - \hat{p}_1)}{n_1} + \dfrac{\hat{p}_2(1 - \hat{p}_2)}{n_2}}$

$$[(0.342667 - 0.266071) - Z_{\frac{0.01}{2}}\hat{\sigma}_{\hat{p}_1 - \hat{p}_2}, (0.342667 - 0.266071) - Z_{\frac{0.01}{2}}\hat{\sigma}_{\hat{p}_1 - \hat{p}_2}]$$

where $\hat{\sigma}_{\hat{p}_1 - \hat{p}_2} = \sqrt{\dfrac{0.342667(1 - 0.342667)}{750} + \dfrac{0.266071(1 - 0.266071)}{560}} = 0.025476$

The 99% confidence interval means that $\alpha = 0.01$ and $\alpha/2 = 0.005 \Rightarrow Z_{0.005} = 2.575$. Thus, $100(1 - 0.01)\%$ CI is: $(0.076595 - 2.575 \times 0.025476, 0.076595 + 2.575 \times 0.025476) = (0.010994, 0.142197)$.

Exercise 12.9

$n_1 = 609, n_2 = 806; x_1 = 211, x_2 = 217;$

$$\hat{p}_1 = \frac{211}{609} = 0.34647; \hat{p}_2 = \frac{217}{806} = 0.269231$$

$$[(\hat{p}_1 - \hat{p}_2) - Z_{\frac{\alpha}{2}}\hat{\sigma}_{\hat{p}_1 - \hat{p}_2}, (\hat{p}_1 - \hat{p}_2) + Z_{\frac{\alpha}{2}}\hat{\sigma}_{\hat{p}_1 - \hat{p}_2}]$$

where $\hat{\sigma}_{\hat{p}_1 - \hat{p}_2} = \sqrt{\dfrac{\hat{p}_1(1 - \hat{p}_1)}{n_1} + \dfrac{\hat{p}_2(1 - \hat{p}_2)}{n_2}}$

$$[(0.34647 - 0.269231) - Z_{\frac{0.05}{2}}\hat{\sigma}_{\hat{p}_1 - \hat{p}_2}, (0.34647 - 0.269231) - Z_{\frac{0.05}{2}}\hat{\sigma}_{\hat{p}_1 - \hat{p}_2}]$$

where $\hat{\sigma}_{\hat{p}_1 - \hat{p}_2} = \sqrt{\dfrac{0.34647(1 - 0.34647)}{609} + \dfrac{0.269231(1 - 0.269231)}{806}} = 0.024817$

The 95% confidence interval means that $\alpha = 0.05$ and $\alpha/2 = 0.025 \Rightarrow Z_{0.025} = 1.96$. Thus, $100(1 - 0.05)\%$ CI is: $(0.077239 - 1.96 \times 0.024817, 0.077239 + 1.96 \times 0.024817) = (0.028597, 0.125881)$.

Exercise 12.11

$n_1 = 1035, n_2 = 1106; x_1 = 149, x_2 = 103;$

$$\hat{p}_1 = \frac{102}{1035} = 0.143961; \hat{p}_2 = \frac{93}{1106} = 0.093128$$

$$[(\hat{p}_1 - \hat{p}_2) - Z_{\frac{\alpha}{2}}\hat{\sigma}_{\hat{p}_1 - \hat{p}_2}, (\hat{p}_1 - \hat{p}_2) + Z_{\frac{\alpha}{2}}\hat{\sigma}_{\hat{p}_1 - \hat{p}_2}]$$

where $\hat{\sigma}_{\hat{p}_1 - \hat{p}_2} = \sqrt{\dfrac{\hat{p}_1(1 - \hat{p}_1)}{n_1} + \dfrac{\hat{p}_2(1 - \hat{p}_2)}{n_2}}$

$$[(0.143961 - 0.093128) - Z_{\frac{0.05}{2}}\hat{\sigma}_{\hat{p}_1 - \hat{p}_2}, (0.143961 - 0.093128) - Z_{\frac{0.05}{2}}\hat{\sigma}_{\hat{p}_1 - \hat{p}_2}]$$

where $\hat{\sigma}_{\hat{p}_1 - \hat{p}_2} = \sqrt{\dfrac{0.143961(1 - 0.143961)}{1035} + \dfrac{0.093128(1 - 0.093128)}{1106}} = 0.01398$

The 95% confidence interval means that $\alpha = 0.05$ and $\alpha/2 = 0.025 \Rightarrow Z_{0.025} = 1.96$. Thus, $100(1 - 0.05)\%$ CI is: $(0.050833 - 1.96 \times 0.01398, 0.050833 + 1.96 \times 0.01398) = (0.023433, 0.078233)$.

Exercise 12.13

$$n_A = 1645, n_B = 1356; x_A = 844; x_B = 546;$$

$$\hat{p}_A = \frac{844}{1645} = 0.51307; \hat{p}_B = \frac{546}{1356} = 0.402655$$

$$[(\hat{p}_A - \hat{p}_B) - Z_{\frac{\alpha}{2}}\hat{\sigma}_{\hat{p}_A - \hat{p}_B}, (\hat{p}_A - \hat{p}_B) + Z_{\frac{\alpha}{2}}\hat{\sigma}_{\hat{p}_A - \hat{p}_B}]$$

where $\hat{\sigma}_{\hat{p}_A - \hat{p}_B} = \sqrt{\dfrac{\hat{p}_A(1 - \hat{p}_A)}{n_A} + \dfrac{\hat{p}_B(1 - \hat{p}_B)}{n_B}}$

$$[(0.51307 - 0.402655) - Z_{\frac{0.10}{2}}\hat{\sigma}_{\hat{p}_1 - \hat{p}_2}, (0.51307 - 0.402655) - Z_{\frac{0.10}{2}}\hat{\sigma}_{\hat{p}_1 - \hat{p}_2}]$$

where $\hat{\sigma}_{\hat{p}_A - \hat{p}_B} = \sqrt{\dfrac{0.51307(1 - 0.51307)}{1645} + \dfrac{0.402655(1 - 0.402655)}{1356}} = 0.01814523$

The 90% confidence interval means that $\alpha = 0.10$ and $\alpha/2 = 0.05 \Rightarrow Z_{0.05} = 1.645$. Thus, $100(1 - 0.10)\%$ CI is: $(0.110415 - 1.645 \times 0.01814523, 0.110415 + 1.645 \times 0.01814523) = (0.08056, 0.1402613)$.

Exercise 12.15

	Readmitted	Were Not Readmitted	Total
Early Discharged	167	1091	1258
Late Discharge	76	1304	1380

Let E = early discharge and L = late discharge.

$$n_e = 1258, n_l = 1380; x_e = 167; x_l = 76;$$

$$\hat{p}_e = \frac{167}{1258} = 0.13275; \hat{p}_l = \frac{76}{1380} = 0.055072$$

$$[(\hat{p}_e - \hat{p}_l) - Z_{\frac{\alpha}{2}}\hat{\sigma}_{\hat{p}_e - \hat{p}_l}, (\hat{p}_e - \hat{p}_l) + Z_{\frac{\alpha}{2}}\hat{\sigma}_{\hat{p}_e - \hat{p}_l}]$$

where $\hat{\sigma}_{\hat{p}_e - \hat{p}_l} = \sqrt{\dfrac{\hat{p}_e(1 - \hat{p}_e)}{n_e} + \dfrac{\hat{p}_l(1 - \hat{p}_l)}{n_l}}$

$$[(0.13275 - 0.055072) - Z_{\frac{0.01}{2}}\hat{\sigma}_{\hat{p}_e - \hat{p}_l}, (0.13275 - 0.055072) - Z_{\frac{0.01}{2}}\hat{\sigma}_{\hat{p}_e - \hat{p}_l}]$$

where $\hat{\sigma}_{\hat{p}_e - \hat{p}_l} = \sqrt{\dfrac{0.13275(1 - 0.13275)}{1258} + \dfrac{0.055072(1 - 0.055072)}{1380}} = 0.011368$

The 99% confidence interval means that $\alpha = 0.01$ and $\alpha/2 = 0.005 \Rightarrow Z_{0.005} = 2.575$. Thus, $100(1 - 0.01)\%$ CI is: $(0.077678 - 2.575 \times 0.011368, 0.077678 + 2.575 \times 0.011368) = (0.048406, 0.10695)$.

Exercise 12.17

$n_{2000} = 300$, $n_{2005} = 400$; $x_{2000} = 211$, $x_{2005} = 302$;

$$\hat{p}_{2000} = \frac{210}{300} = 0.703333; \hat{p}_{2005} = \frac{302}{400} = 0.755 \Rightarrow \hat{p}_{2005} - \hat{p}_{2000} = 0.05167$$

$$[(0.833333 - 0.685714) - Z_{\frac{0.02}{2}}\hat{\sigma}_{\hat{p}_m - \hat{p}_f}, (0.833333 - 0.685714) - Z_{\frac{0.02}{2}}\hat{\sigma}_{\hat{p}_m - \hat{p}_f}]$$

where $\hat{\sigma}_{\hat{p}_m - \hat{p}_f} = \sqrt{\dfrac{0.703333(1 - 0.703333)}{300} + \dfrac{0.755(1 - 0.755)}{400}} = 0.034029$

The 98% confidence interval means that $\alpha = 0.02$ and $\alpha/2 = 0.01 \Rightarrow Z_{0.01} = 2.33$. Thus, $100(1 - 0.02)\%$ CI is: $(0.05167 - 2.33 \times 0.034029, 0.05167 + 2.33 \times 0.034029) = (-0.02762, 0.13095)$.

Exercise 12.19

$n_{MA} = 1002$, $n_{MN} = 1010$; $x_{MA} = 466$, $x_{MN} = 312$;

$$\hat{p}_{MA} = \frac{466}{1002} = 0.46507; \hat{p}_{MN} = \frac{312}{1010} = 0.308911 \Rightarrow \hat{p}_{MA} - \hat{p}_{MN} = 0.156159$$

$$[(0.46507 - 0.308911) - Z_{\frac{0.04}{2}}\hat{\sigma}_{\hat{p}_{MA} - \hat{p}_{MN}}, (0.46507 - 0.308911) - Z_{\frac{0.04}{2}}\hat{\sigma}_{\hat{p}_{MA} - \hat{p}_{MN}}]$$

where $\hat{\sigma}_{\hat{p}_{MA} - \hat{p}_{MN}} = \sqrt{\dfrac{0.46507(1 - 0.46507)}{1002} + \dfrac{0.308911(1 - 0.308911)}{1010}} = 0.02144$

The 96% confidence interval means that $\alpha = 0.04$ and $\alpha/2 = 0.02 \Rightarrow Z_{0.02} = 2.05$. Thus, $100(1 - 0.04)\%$ CI is: $(0.156159 - 2.05 \times 0.02144, 0.156159 + 2.05 \times 0.02144) = (0.112208, 0.20011)$.

Exercise 12.21

Group 1: those who got credit cards after high school but before entering college
Group 2: those who received credit cards while in high school

$n_1 = 2306$, $n_2 = 1102$; $x_1 = 471$, $x_2 = 121$;

$$\hat{p}_1 = \frac{471}{2306} = 0.20425; \hat{p}_2 = \frac{121}{1102} = 0.1098;$$

$$[(\hat{p}_1 - \hat{p}_2) - Z_{\frac{\alpha}{2}}\hat{\sigma}_{\hat{p}_1 - \hat{p}_2}, (\hat{p}_1 - \hat{p}_2) + Z_{\frac{\alpha}{2}}\hat{\sigma}_{\hat{p}_1 - \hat{p}_2}]$$

where $\hat{\sigma}_{\hat{p}_1 - \hat{p}_2} = \sqrt{\dfrac{\hat{p}_1(1 - \hat{p}_1)}{n_1} + \dfrac{\hat{p}_2(1 - \hat{p}_2)}{n_2}}$

$$[(0.20425 - 0.1098) - Z_{\frac{0.02}{2}}\hat{\sigma}_{\hat{p}_1 - \hat{p}_2}, (0.20425 - 0.1098) - Z_{\frac{0.02}{2}}\hat{\sigma}_{\hat{p}_1 - \hat{p}_2}]$$

where $\hat{\sigma}_{\hat{p}_1 - \hat{p}_2} = \sqrt{\dfrac{0.20425(1 - 0.20425)}{2306} + \dfrac{0.1098(1 - 0.1098)}{1102}} = 0.012617;$

The 96% confidence interval means that $\alpha = 0.04$ and $\alpha/2 = 0.02 \Rightarrow Z_{0.02} = 2.05$. Thus, $100(1 - 0.04)\%$ CI is: $(0.094449 - 2.05 \times 0.012617, 0.094449 + 2.05 \times 0.012617) = (0.068585, 0.120314)$.

Exercise 12.23

$$\overline{X}_A = \frac{209}{15} = 13.93333; \overline{X}_B = \frac{122}{16} = 7.625;$$

$$s_A^2 = \frac{\sum_{i=1}^{n_1} X_A^2 - n(\overline{X}_A)^2}{n_A - 1} = \frac{3171 - 15(13.9333)^2}{15 - 1} = 18.49524$$

$$s_B^2 = \frac{\sum_{i=1}^{n_1} X_B^2 - n(\overline{X}_B)^2}{n_B - 1} = \frac{1222 - 16(7.625)^2}{16 - 1} = 19.45$$

$$S_P^2 = \frac{(n_A - 1)s_A^2 + (n_B - 1)s_B^2}{n_A + n_B - 2} = \frac{(15 - 1)(18.49524) + (16 - 1)(19.45)}{15 + 16 - 2}$$

$$= 18.98908 \Rightarrow s_p = \sqrt{18.98908} = 4.183943$$

98% CI $\Rightarrow 100(1 - 0.02)\%$ CI $\Rightarrow \alpha = 0.02; \alpha/2 = 0.01;$

$t_{(15+14-2, 0.025)} = 2.462$

$$100(1 - \alpha)\% \text{ CI for} \mu_1 = \overline{X}_A - \overline{X}_B \pm t_{(n_A+n_B-2, \alpha/2)} \cdot s_P^2 \sqrt{\frac{1}{n_A} + \frac{1}{n_B}}$$

$$98\% \text{ CI for } \mu_A - \mu_B = 13.93333 - 7.625 \pm 2.462 \times 4.183943 \times \sqrt{\frac{1}{15} + \frac{1}{16}}$$

$$= (2.6062, 10.0104)$$

Exercise 12.25

Solution:
Let A = males; B = females.

$$\overline{X}_A = \frac{825.4}{12} = 68.78333; \overline{X}_B = \frac{752.2}{12} = 62.68333$$

$$s_A^2 = \frac{\sum_{i=1}^{n_1} X_A^2 - n(\overline{X}_A)^2}{n_A - 1} = \frac{56872.2 - 12(68.7833)^2}{12 - 1} = 8.94879$$

$$s_B^2 = \frac{\sum_{i=1}^{n_1} X_B^2 - n(\overline{X}_B)^2}{n_B - 1} = \frac{47265.3 - 12(62.68333)^2}{12 - 1} = 10.44697$$

$$S_P^2 = \frac{(n_A - 1)s_A^2 + (n_B - 1)s_B^2}{n_A + n_B - 2} = \frac{(12 - 1)(8.94879) + (12 - 1)(10.44697)}{12 + 12 - 2}$$

$$= 9.697879 \Rightarrow s_p = \sqrt{9.697879} = 3.114142$$

99% CI $\Rightarrow 100(1 - 0.01)\%$ CI $\Rightarrow \alpha = 0.01; \alpha/2 = 0.005;$

$t_{(12+12-2, 0.005)} = 2.819$

$$100(1 - \alpha)\% \text{ CI for } \mu_1 = \overline{X}_A - \overline{X}_B \pm t_{(n_A+n_B-2, \alpha/2)} \cdot s_P^2 \sqrt{\frac{1}{n_A} + \frac{1}{n_B}}$$

$$99\% \text{ CI for } \mu_A - \mu_B = 68.78333 - 62.68333 \pm 2.819 \times 3.114142 \times \sqrt{\frac{1}{12} + \frac{1}{12}}$$

$$= (2.5161, 9.6839)$$

Exercise 12.27

$$\bar{y}_1 = \frac{8210}{16} = 513.125; \bar{y}_2 = \frac{7020}{13} = 540;$$

$$s_1^2 = \frac{4214492 - 16(513.125)^2}{16 - 1} = 115.7167; s_2^2 = \frac{3801858 - 13(540)^2}{13 - 1} = 921.50$$

$$\hat{\sigma}_{\bar{y}_1 - \bar{y}_2} = \sqrt{\left[\frac{115.7167}{16} + \frac{921.5}{13}\right]} = 8.83838;$$

$$\nu = \frac{\left[\frac{115.7167}{16} + \frac{921.5}{13}\right]^2}{\left(\frac{1}{16-1}\right)\left(\frac{115.7167}{16}\right)^2 + \left(\frac{1}{13-1}\right)\left(\frac{921.5}{13}\right)^2} = 14.45325 \approx 14;$$

$$95\% \text{ CI} = (\bar{Y}_1 - \bar{Y}_2) \pm t_{(14, 0.025)} \times \hat{\sigma}_{\bar{y}_1 - \bar{y}_2}$$

$$(540 - 513.125) \pm 2.1448 \times 8.83838 = (7.9184, 45.8316)$$

Exercise 12.29

$$\bar{y}_1 = \frac{88.5}{14} = 6.321429; \bar{y}_2 = \frac{73.5}{14} = 5.25;$$

$$s_1^2 = \frac{612.25 - 14(6.321429)^2}{14 - 1} = 4.061813;$$

$$s_2^2 = \frac{395.75 - 14(5.25)^2}{14 - 1} = 0.759615$$

$$\hat{\sigma}_{\bar{y}_1 - \bar{y}_2} = \sqrt{\left[\frac{4.061813}{14} + \frac{0.759615}{14}\right]} = 0.586846;$$

$$\nu = \frac{\left[\frac{4.061813}{14} + \frac{0.759615}{14}\right]^2}{\left(\frac{1}{14-1}\right)\left(\frac{4.061813}{14}\right)^2 + \left(\frac{1}{14-1}\right)\left(\frac{0.759615}{14}\right)^2} = 17.6981 \approx 17;$$

$$90\% \text{ CI} = (\bar{Y}_1 - \bar{Y}_2) \pm t_{(17, 0.05)} \times \hat{\sigma}_{\bar{y}_1 - \bar{y}_2}$$

$$(6.321429 - 5.25) \pm 1.7396 \times 0.586846 = (0.0528596, 2.089998)$$

Exercise 13.1

$$\bar{X}_A = \frac{303.7}{12} = 25.30833; \bar{X}_B = \frac{303.5}{14} = 21.67857;$$

$$s_A^2 = \frac{\sum_{i=1}^{n_A} X_{Ai}^2 - n_A(\bar{X}_A)^2}{n_1 - 1} = \frac{7837.49 - 12(25.30833)^2}{12 - 1} = 13.75902;$$

$$s_B^2 = \frac{\sum_{i=1}^{n_B} X_{Bi}^2 - n_B(\bar{X}_B)^2}{n_2 - 1} = \frac{6842.33 - 14(21.67857)^2}{14 - 1} = 20.22181$$

$$s_P^2 = \frac{(n_A - 1)s_A^2 + (n_B - 1)s_B^2}{n_A + n_B - 2} = \frac{(12 - 1)(13.75902) + (14 - 1)(20.22181)}{12 + 14 - 2}$$

$$= 17.2597 \Rightarrow s_p = \sqrt{17.2597} = 4.154479$$

The hypothesis to be tested is the type represented by Case 2: $H_0 : \mu_A = \mu_B$ vs. $H_1 : \mu_A > \mu_B$

The test statistic is:

$$t_c = \frac{\overline{X}_A - \overline{X}_B}{s_P\sqrt{\frac{1}{n_A} + \frac{1}{n_B}}} = \frac{25.30833 - 21.67857}{4.154479\sqrt{\frac{1}{12} + \frac{1}{14}}} = 2.2209$$

The test statistic is distributed as t with $n_A + n_B - 2$ degrees of freedom. Since we are testing at a 5% level of significance, then in the t-table (Appendix 3), we read $t_{(12+14-2,\, 0.05)} = t_{(24,\, 0.05)}$ whose value is 1.711.

Decision: Since $t_c > t_{(24,\, 0.05)}$ we do reject the null, and conclude that the mean height for palm seedlings of variety A is greater than the mean height of palm seedlings of variety 2 at the 5% level of significance.

Exercise 13.3

Manila	Sisal	Manila-2200	Sisal-2200	(Manila-2200)^2	(Sisal-2200)^2
2218	2224	18	24	324	576
2230	2236	30	36	900	1296
2209	2219	9	19	81	361
2206	2215	6	15	36	225
2227	2237	27	37	729	1369
2211	2230	11	30	121	900
2234	2250	34	50	1156	2500
2245	2251	45	51	2025	2601
2244	2253	44	53	1936	2809
2224	2221	24	21	576	441
2241	2247	41	47	1681	2209
2231	2237	31	37	961	1369
	2218		18		
	2219		19		
	2239		39		

$$\overline{X}_A = \frac{320}{12} = 26.6667; \quad \overline{X}_B = \frac{496}{15} = 33.06667;$$

$$s_A^2 = \frac{\sum_{i=1}^{n_A} X_{Ai}^2 - n_A(\overline{X}_A)^2}{n_1 - 1} = \frac{10526 - 12(26.6667)^2}{12 - 1} = 181.1515$$

$$s_B^2 = \frac{\sum_{i=1}^{n_B} X_{Bi}^2 - n_B(\overline{X}_B)^2}{n_2 - 1} = \frac{18862 - 15(33.06667)^2}{15 - 1} = 175.781$$

$$s_P^2 = \frac{(n_A - 1)s_A^2 + (n_B - 1)s_B^2}{n_A + n_B - 2} = \frac{(12 - 1)(181.1515) + (15 - 1)(175.781)}{12 + 15 - 2}$$

$$= 178.144 \Rightarrow s_p = \sqrt{178.144} = 13.34706$$

The hypothesis to be tested is the type represented by Case 2: $H_0{:}\mu_A = \mu_B$ vs. $H_1{:}\mu_A > \mu_B$.

The test statistic is:

$$t_c = \frac{\overline{X}_A - \overline{X}_B}{s_p\sqrt{\dfrac{1}{n_A} + \dfrac{1}{n_B}}} = \frac{33.06667 - 26.6667}{13.34706\sqrt{\dfrac{1}{12} + \dfrac{1}{15}}} = 1.23808.$$

The test statistic is distributed as t with $n_A + n_B - 2$ degrees of freedom. Since we are testing at a 5% level of significance, then in the t-table (Appendix 3), we read $t_{(12+15-2,\,0.025)} = t_{(25,\,0.025)}$ whose value is 1.708.

Decision: Since $t_c < t_{(25,\,0.025)}$ we do not reject the null, and conclude that the mean breaking strength of ropes made with sisal is not greater than the mean breaking strength of ropes made with at 5% level of significance. The claim of the company is rejected.

Exercise 13.5

A1C Before	A1C After	d_i	d_i^2
7.9	6.9	−1	1
7.8	7.1	−0.7	0.49
8.3	7.5	−0.8	0.64
7.4	7.3	−0.1	0.01
7.3	7.3	−0	0
7.4	7.5	0.1	0.01
7.8	6.8	−1	1
7.6	6.7	−0.9	0.81
7.3	6.8	−0.5	0.25
7.4	6.7	−0.7	0.49
7.3	7.1	−0.2	0.04
7.4	7	−0.4	0.16
7.2	7.4	0.2	0.04
TOTAL→		**−6**	**4.94**

$$H_0 : \mu_d = 0 \text{ vs. } H_1 : \mu_d < 0$$

$$\bar{d} = \frac{-6}{13} = -0.461538;$$

$$s_d^2 = \frac{\sum\limits_{i=1}^{n} d_i^2 - n\bar{d}^2}{n - 1} = \frac{4.94 - 13(-0.461538)^2}{13 - 1} = 0.180897$$

$$t_{cal} = \frac{-0.461538}{\sqrt{(0.180897/13)}} = -3.912581$$

From the tables, $t_{(12,\,0.05)} = 1.782$. $t_{cal} < -t_{(12,\,0.05)}$, so we reject the null hypothesis. The diet/exercise regimen has significantly lowered the A1C levels in six months at a 5% level of significance.

Exercise 13.7

Keyboard Skills Before	Keyboard Skills After	d_i	d_i^2
50	55	5	25
55	60	5	25
59	56	−3	9
60	61	1	1
45	49	4	16
48	50	2	4
45	46	1	1
55	61	6	36
58	67	9	81
60	65	5	25
62	61	−1	1
45	49	4	16
58	59	1	1
61	67	6	36
55	70	15	225
44	52	8	64
TOTAL→		68	566

$$H_0 : \mu_d = 0 \text{ vs. } H_1 : \mu_d > 0$$

$$\bar{d} = \frac{68}{16} = 4.25; \quad s_d^2 = \frac{\sum\limits_{i=1}^{n} d_i^2 - n\bar{d}^2}{n - 1} = \frac{566 - 16(4.25)^2}{16 - 1} = 18.46667$$

$$t_{cal} = \frac{4.25}{\sqrt{(18.4667/16)}} = 3.956$$

From the tables, $t_{(15, 0.01)} = 2.602$. $t_{cal} > t_{(15, 0.01)}$, so we reject the null hypothesis. The training has significantly increased the keyboard skills in two weeks at a 1% level of significance.

Exercise 13.9

$$H_0: \mu_1 = \mu_2 \text{ vs. } H_1: \mu_1 \neq \mu_2$$

$$\overline{Y}_1 = \frac{481}{15} = 32.06667; \overline{Y}_2 = \frac{696}{15} = 46.4;$$

$$s_1^2 = \frac{15791 - 15(32.06667)^2}{15 - 1} = 26.20952$$

$$s_2^2 = \frac{33832 - 15(46.4)^2}{15 - 1} = 109.8286$$

$$\nu = \frac{\left[\dfrac{26.20952}{15} + \dfrac{109.8286}{15}\right]^2}{\left(\dfrac{1}{15-1}\right)\left(\dfrac{26.20952}{15}\right)^2 + \left(\dfrac{1}{15-1}\right)\left(\dfrac{109.8286}{15}\right)^2} = 20.3219 \approx 20.$$

$$t_{cal} = \frac{(\overline{X} - \overline{Y})}{\sqrt{\left(\dfrac{s_1^2}{n_1} + \dfrac{s_2^2}{n_2}\right)}} = \frac{46.4 - 32.06667}{\sqrt{\dfrac{26.20952}{15} + \dfrac{109.8286}{15}}} = 4.7595$$

$$t_{cal} > t_{(20, 0.005)} = 2.845;$$

We reject H_0; the means are not equal at a 1% level of significance.

Exercise 13.11

$$\overline{X}_1 = \frac{7332.03}{35} = 209.487; \overline{X}_2 = \frac{6042.28}{35} = 172.637;$$

$$s_1^2 = \frac{\sum\limits_{i=1}^{n} X_{1i}^2 - n_1 \overline{X}_1^2}{n_1 - 1} = \frac{1549887 - 35 \times 209.487^2}{35 - 1} = 409.5639;$$

$$s_2^2 = \frac{\sum\limits_{i=1}^{n} X_{2i}^2 - n_2 \overline{X}_2^2}{n_2 - 1} = \frac{1059491 - 35 \times 4.217^2}{35 - 1} = 481.3917;$$

$$\hat{\sigma}_{\overline{x}_1 - \overline{x}_2} = \sqrt{\frac{409.5639}{35} + \frac{481.3917}{35}} = 5.054381$$

$$\nu = \frac{\left[\dfrac{409.5639}{35} + \dfrac{481.3917}{35}\right]^2}{\left(\dfrac{1}{35-1}\right)\left(\dfrac{409.5639}{35}\right)^2 + \left(\dfrac{1}{35-1}\right)\left(\dfrac{481.3917}{35}\right)^2} = 67.56089 \approx 67$$

$$H_0{:}\mu_A = \mu_B \text{ vs. } H_0{:}\mu_A > \mu_B$$

$$t_{cal} = \frac{(\overline{X}_A - \overline{X}_B)}{\hat{\sigma}_{\overline{x}_A - \overline{x}_B}} = \frac{209.487 - 172.637}{5.045381} = 7.303709$$

Since $\alpha = 1\%$, $t_{(\nu,\,\alpha)} = t_{(67,\,0.01)} = 2.383$, we reject H_0; conclude that the mean recovery time for Hospital A is greater than the mean recovery time for Hospital B at 1% level of significance.

Exercise 13.13

$$\overline{y}_1 = \frac{204}{14} = 14.57143; \; \overline{y}_2 = \frac{202}{16} = 12.625$$

$$s_1^2 = \frac{3128 - 14(14.57143)^2}{14 - 1} = 11.95604; \; s_2^2 = \frac{2660 - 16(12.625)^2}{16 - 1} = 7.31667$$

$$\hat{\sigma}_{\overline{y}_1 - \overline{y}_2} = \sqrt{\frac{11.95604}{14} + \frac{7.31667}{16}} = 1.45118$$

$$\nu = \frac{\left[\dfrac{11.95604}{14} + \dfrac{7.31667}{16}\right]^2}{\left(\dfrac{1}{14 - 1}\right)\left(\dfrac{11.95604}{14}\right)^2 + \left(\dfrac{1}{16 - 1}\right)\left(\dfrac{7.31667}{16}\right)^2} = 24.54923 \approx 24$$

$$H_0{:}\mu_{01} = \mu_{02} \text{ vs. } H_0{:}\mu_{01} > \mu_{02}$$

$$t_{cal} = \frac{(\overline{X}_{01} - \overline{X}_{02})}{\hat{\sigma}_{\overline{x}_{01} - \overline{x}_{02}}} = \frac{14.57143 - 12.625}{1.145118} = 1.699763$$

Since $\alpha = 5\%$, $t_{(\nu,\,\alpha)} = t_{(24,\,0.05)} = 1.711$, we accept and do not reject H_0; we conclude that the mean study time of STAT 319_01 students is not more than the mean study time of STAT 319_02 students at 5% level of significance.

Exercise 13.15

	Readmitted	Were Not Readmitted	Total
Men	167	1340	1500
Women	83	1055	1138

Let m = men and w = women.

$$n_m = 1500, \, n_w = 1138; \, x_m = 160; \, x_w = 83;$$

$$\hat{p}_m = \frac{160}{1500} = 0.106667; \, \hat{p}_w = \frac{83}{1138} = 0.072935;$$

$$\overline{p} = \frac{160 + 83}{1500 + 1138} = 0.092115;$$

$$\hat{\sigma}_{\hat{p}_m - \hat{p}_w} = \sqrt{\overline{p}(1 - \overline{p})\left(\frac{1}{n_m} + \frac{1}{n_w}\right)}$$

$$= \sqrt{0.092115(1 - 0.092115)\left(\frac{1}{1500} + \frac{1}{1138}\right)} = 0.011368$$

$$H_0 : p_m = p_w \text{ vs. } H_1 : p_m > p_w$$

$$Z_{cal} = \frac{0.106667 - 0.072935}{0.011368} = 2.9673$$

Classical: Since $Z_{cal} > Z_{0.01}$ we reject H_0 and conclude proportion of males readmitted is more than the proportion of females readmitted at 1% level of significance.

p-value: $P(Z > Z_{cal}) = P(Z > 2.9673) = 0.0015 \ll 0.01$, so we reject H_0 and conclude proportion of males readmitted is more than the proportion of females readmitted at 1% level of significance.

Exercise 13.17

$$n_{2000} = 300, n_{2005} = 400; x_{2000} = 211; x_{2005} = 302;$$

$$\hat{p}_{2000} = \frac{211}{300} = 0.703333; \hat{p}_{2005} = \frac{302}{400} = 0.755;$$

$$\bar{p} = \frac{211 + 302}{300 + 400} = 0.732857$$

$$\hat{\sigma}_{\hat{p}_{2000} - \hat{p}_{2005}} = \sqrt{\bar{p}(1 - \bar{p})\left(\frac{1}{n_{2000}} + \frac{1}{n_{2005}}\right)}$$

$$= \sqrt{0.732857(1 - 0.732857)\left(\frac{1}{300} + \frac{1}{400}\right)}$$

$$= 0.033794$$

$$H_0 : p_{2000} = p_{2005} \text{ vs. } H_1 : p_{2000} < p_{2005}$$

$$Z_{cal} = \frac{0.703333 - 0.755}{0.033794} = -1.52887$$

Classical: $Z_{0.05} = 1.645 \Rightarrow -Z_{0.05} = -1.645;$

Since $Z_{cal} > -Z_{0.05}$ we do not reject H_0 and conclude proportion of students retained did not increase between 2000 and 2005 at 5% level of significance.

p-value: $P(Z > Z_{cal}) = P(Z < -1.52) = 0.06426 > 0.05$, so we do not reject H_0 and conclude proportion of students retained by SCSU did not increase between 2000 and 2005 at 5% level of significance.

Exercise 13.19

$$n_{MA} = 1002; n_{MN} = 1010; x_{MA} = 466; x_{MN} = 313;$$

$$\hat{p}_{MA} = \frac{466}{1002} = 0.46507; \hat{p}_{MN} = \frac{313}{1010} = 0.309901;$$

$$\bar{p} = \frac{466 + 313}{1002 + 1010} = 0.387177$$

$$\hat{\sigma}_{\hat{p}_{MA} - \hat{p}_{MN}} = \sqrt{\bar{p}(1 - \bar{p})\left(\frac{1}{n_{MA}} + \frac{1}{n_{MN}}\right)}$$

$$= \sqrt{0.387177(1 - 0.387177)\left(\frac{1}{1002} + \frac{1}{1010}\right)} = 0.021719$$

$$H_0{:}p_{MA} = p_{MN} \text{ vs. } H_1{:}p_{MA} \neq p_{MN}$$

$$Z_{cal} = \frac{0.46507 - 0.309901}{0.021719} = 7.144352$$

Classical: $\alpha = 0.01$; $Z_{0.005} = 2.575$;

Since $Z_{cal} >>> Z_{0.005}$ we reject H_0 and conclude proportion of 8th graders from MA who take Algebra I is significantly greater than the proportion of MN 8th graders who take Algebra I at 1% level of significance.

p-value $= P(Z < -7.144) + P(Z > 7.144) \approx 0$, we reject H_0 and conclude proportion of 8th graders from MA who take Algebra I is significantly greater than the proportion of MN 8th graders who take Algebra I at 1% level of significance.

Exercise 14.1

H_0: The rotten apples follow the binomial distribution with probability of rottenness of 11.65%.

Vs.

H_1: The rotten apples do not follow the binomial distribution with probability of rottenness of 11.65%.

Using binomial with $P = 0.1165$, and $n = 12$, we find the proportions that should fall into 0, 1, 2, . . . , 7 rotten apples. We also multiply the proportions by 1,200 to obtain expected counts as follows:

Number of Rotten Apples	0	1	2	3	4	5	6	7
Frequency	334	400	300	118	43	2	2	1
Proportion Using Binomial (12,0.1165)	0.226193	0.357914	0.259574	0.114093116	0.03385	0.007142	0.001099	0.000124
Expected	271.4313	429.4974	311.4889	136.9117393	40.62025	8.570022	1.318403	0.149012

For the chi-square approximation to be valid, the expected frequency should be at least 5. The test is not valid for small samples, and also, if some of the counts are less than five, we need to combine some categories in the tails to make up at least an expected value of 5. This is clearly so when number of rotten apples are 6 and 7. We merge cases when rotten apples are 5, 6, and 7 into one as in the following table:

Number of Rotten Apples	0	1	2	3	4	5, 6, 7
Frequency	308	426	300	118	43	5
Proportion Using Binomial (12,0.1165)	0.226193	0.357914	0.259574	0.114093116	0.03385	0.008365
Expected	271.4313	429.4974	311.4889	136.9117393	40.62025	10.03744

$$\chi_{cal} = \frac{(308-271.4313)^2}{271.4313} + \frac{(426-429..44974)^2}{429.4974} + \frac{(300-311.9117)^2}{311.9117}$$
$$+ \frac{(118-136.9117)^2}{136.9117} + \frac{(43-40.62025)^2}{40.62025} + \frac{(5-10.0374)^2}{10.0374} = 10.6588$$

There are six categories, so we read chi-square at $6-1 = 5$ degrees of freedom.
From the tables $\chi^2_{(5,0.05)} = 11.070 = 16.91898$. Thus, $\chi^2_{cal} < \chi^2_{(5,0.05)}$, so we do not reject the null hypothesis: the rotten apples follow the binomial distribution with rotten proportion of rottenness of 11.65%.

Exercise 14.3

H_0: Marital status is independent of alcohol consumption.
H_1: Marital status is not independent of alcohol consumption.
Table 1: The observed values, row and column totals
To obtain the expected e_{ij} for each cell, we use

$$e_{ij} = \frac{R_i \times C_j}{N} \Rightarrow e_{11} = \frac{312 \times 277}{967} = 89.37332, \ e_{12} = \frac{312 \times 284}{967} = 91.63185 \dots$$
$$e_{34} = \frac{242 \times 111}{967} = 27.778697.$$

The result is the following table:
Table 2: Expected values, e_{ij}

Expected			
89.37332	91.63185	95.18097	35.81385729
64.16546	65.78697	68.33506	25.71251293
54.13961	55.50776	57.6577	21.69493278
69.32161	71.07342	73.82627	27.778697

Test statistic is: $\chi^2_{cal} = \sum\limits_{i=1}^{3}\sum\limits_{j=1}^{4} \frac{(o_{ij}-e_{ij})^2}{e_{ij}}$

$$\chi^2_{cal} = \frac{(65-89.37332)^2}{89.37332} + \frac{(113-91.63185)^2}{91.63185} + \dots + \frac{(15-27.77892)^2}{27.77892} = 75.6141$$

The degrees of freedom is $df = (r-1)(c-1) = (4-1)(4-1) = 9$
From the tables for level of significance $= 0.01$, we read $\chi^2_{(9,0.01)} = 21.666$. Since $\chi^2_{cal} > \chi^2_{(9,0.01)}$, we reject H_0. We conclude that the levels of drinking are independent of marital status.

Exercise 14.5

H_0: There is no association between age of the driver and the number of accidents he or she has had.
Vs.
H_1: There is an association between age of the driver and the number of accidents he or she has had.

	Number of Accidents			
Age of Driver	**1**	**2**	**≥3**	R_i
16–21	66	50	10	126
22–35	49	21	15	85
≥35	35	18	18	71
C_j	150	89	43	282

Based on the above figures in the table above and using the formula: $e_{ij} = \dfrac{R_i \times C_j}{N}$ we obtain the expected frequencies as shown in the following table:

Expected values:

67.02	39.8	19.21
45.21	26.8	12.96
37.77	22.4	10.83

Test statistic is: $\chi^2_{cal} = \sum\limits_{i=1}^{3} \sum\limits_{j=1}^{4} \dfrac{(o_{ij} - e_{ij})^2}{e_{ij}}$

$$\chi^2_{cal} = \frac{(66 - 67.02)^2}{67.02} + \frac{(50 - 39.8)^2}{39.8} + \ldots + \frac{(18 - 10.83)^2}{10.83} = 14.7936$$

The degrees of freedom is $df = (r-1)(c-1) = (3-1)(3-1) = 4$

From the tables for level of significance $= 0.01$, we read $\chi^2_{(4,0.01)} = 13.2767$. Since $\chi^2_{cal} > \chi^2_{(4,0.01)}$, we reject H_0. We conclude there is an association between age of driver and the number of accidents he or she has had. **The P-value = 0.005149, obtained by typing = CHIDIST(14.7936,4) in a cell in EXCEL. The conclusion reached by comparing the value of the test statistic with the critical value is confirmed by the P-value which is less than the level of significance, signaling that we should reject the null hypothesis.**

Exercise 14.7

For the data to be Poisson, then the parameter λ is estimated as the mean.

Number of Times (x_i)	0	1	2	3	Sum
Count (f_i)	40	30	18	15	103
$f_i x_i$	0	30	36	45	121

$$\bar{x} = \frac{\sum f_i x_i}{\sum f_i} = \frac{111}{103} = 1.07767$$

The hypothesis to be tested is:

H_0: The number of absences of the students for introductory STAT class is Poisson distributed with $\lambda = 1.07767$.

Vs.

H_1: The number of absences of the students for introductory STAT class is not Poisson distributed with $\lambda = 1.07767$.

Assume data is Poisson distributed with $\lambda = 1.07767$; then we calculate the probabilities for x based on it, and the expected value for each $x(f_e)$:

Number of Times of Absences (x_i)	0	1	2	≥ 3	Sum
Count (o_i)	40	30	18	15	**103**
P(x_i)	0.340388	0.366826	0.197658	0.095128	**1.0**
e_i	35.0599337	37.7830387	20.3588237	9.7982	**103**
χ^2	0.69607248	1.60325093	0.27329915	2.7616	**5.334223**

In the table, chi-square was obtained as:

$$\chi^2 = \sum \frac{(o_i - e_i)^2}{e_i}$$

From the tables, the critical value is $\chi^2_{[2,0.01]} = 9.21034$. This value χ^2 calculated, the sum across all values of x, (5.334223) is less than the critical value, so we do not reject the null hypothesis and conclude that the data is not significant: The number of absences of the students for introductory STAT class probably follows the Poisson distribution with $\lambda = 1.07767$. The P-value $= 0.069453$; the P-value can be obtained by typing in a cell in EXCEL = **CHIDIST(5.334,2)**. **The conclusion reached by comparing the value of the test statistic with the critical value is confirmed by the P-value which is greater than the level of significance, indicating that we should accept the null hypothesis.**

Exercise 14.9

H_0: The flavors of the yoghurt are equally favored.
Vs.
H_1: The flavors of the yoghurt are not equally favored.
We assume that 1 out of 6 of every consumer would prefer the flavors.

Flavors	1	2	3	4	5	6	Total
Preferences	40	46	38	40	24	22	**210**
Expected	35	35	35	35	35	35	**210**
χ^2	0.714286	3.457143	0.257143	0.714286	3.457143	4.828571	**13.42857**

From the tables, the critical value is $\chi^2_{[5,0.01]} = 15.1$. The value χ^2 calculated in the earlier table, that is, the sum across all flavors (13.42857) is less than the critical value, so we do not reject the null hypothesis and conclude that the data are not significant. The flavors are probably favored at 1% level of significance. The P-value $= 0.019667$. The P-value was obtained by typing in a cell in EXCEL = **CHIDIST(13.42857,5)**. **The conclusion reached by comparing the value of the test statistic with the critical value is confirmed by the P-value which is greater than the level of significance, signaling that we should accept the null hypothesis.**

Exercise 15.1

$H_0 : \alpha_1 = \alpha_2 = \alpha_3 = \alpha_4 = \alpha_5 = \alpha_6 = \alpha_7 = 0$ *vs* H_1 : at least one α_i is different; $i = 1, 2, ..., 7$.

Shifts	1	2	3	4	Totals
1	23	11	27	18	79
2	8	10	5	18	41
3	23	13	16	21	73
4	8	9	4	4	25
5	16	12	3	14	45
6	20	8	12	8	48
7	19	12	15	9	55

$$SS_{total} = 23^2 + 11^2 + ... + 15^2 + 9^2 - \frac{366^2}{7(4)} = 1091.857$$

$$SS_{shift} = \frac{79^2 + 41^2 + 73^2 + 25^2 + 45^2 + 48^2 + 55^2}{4} - \frac{366^2}{7(4)} = 523.3571$$

$$SS_{Error} = 1091.857 - 523.3571 = 568.5$$

Source	DF	Sum of Squares	Mean Square	F Ratio
Shift	6	523.357143	87.226190	3.22
Error	21	568.500000	27.071429	
Total	27	1091.857143		

From the tables, we see that $F(6, 21, 0.05) = 2.5727$. The F-ratio $> F(6, 21, 0.05)$, so we reject the null hypothesis; the means from the shifts are not equal at 5% level of significance.

Exercise 15.3

The hypothesis being tested in ANOVA is:

$H_0 : \alpha_1 = \alpha_2 = \alpha_3 = \alpha_4 = \alpha_5 = 0$ *vs* H_1 : at least one α_i is different; $i = 1, 2, 3, 4, 5$.

	Samples										Total
Factor	1	2	3	4	5	6	7	8	9	10	
1	10	12	10	11	12	9	11	13	10	9	107
2	14	14	14	12	16	10	14	15	10	15	134
3	17	13	15	12	16	10	14	15	10	15	137
4	19	15	19	14	23	12	17	17	15	19	170
5	20	19	23	16	20	18	21	22	18	20	197
Total											745

$$SS_{total} = 10^2 + 12^2 + 10^2 ... + 18^2 + 20^2 - \frac{745^2}{5(10)} = 722.5$$

$$SS_{factor} = \frac{107^2 + 134^2 + 137^2 + 170^2 + 197^2}{10} - \frac{745^2}{5(10)} = 487.8$$

$$SS_{Error} = 722.5 - 487.8 = 234.7$$

ANOVA

Source	SS	df	MS	F
factor	487.8	4	121.95	23.3820
error	234.7	45	5.2156	
total	722.5	49		

From the tables, $F(4,45,0.01) = 3.7674$. This F-ratio is greater than $F(4,45,0.01)$, therefore we reject the null hypothesis, H_0.

Exercise 15.5

$H_0 : \mu_1 = \mu_2 = \mu_3 = \mu_4 = \mu_5$ vs. H_1 : at one μ_i is different ($i = 1, 2, 3, 4, 5$)

Varieties of Melon	Sample					
	1	2	3	4	5	Total
I	6.2	6	5.7	5.9	6.3	30.1
II	6.7	6.9	7.2	6.3	7.3	34.4
III	4.6	5.2	4.5	5.6	4.9	24.8
IV	5.8	5	5.6	5.8	5	27.2
V	6	5.6	5.3	6.6	6	29.5
Total						146

$$SS_{total} = 6.2^2 + 6.0^2 + 5.7^2 ... + 6.6^2 + 6.0^2 - \frac{146^2}{5(5)} = 13.58$$

$$SS_{varieties} = \frac{30.1^2 + 34.4^2 + 24.8^2 + 27.2^2 + 29.5^2}{5} - \frac{146^2}{5(5)} = 10.26$$

$$SS_{Error} = 13.58 - 10.26 = 3.32$$

ANOVA

Source	Sum of Squares	df	Mean Square	F Value
Varieties	10.26	4	2.565	15.45
Error	3.32	20	0.166	
Total	13.58	24		

From the tables, $F(4,20,0.05) = 2.8661$. This F-ratio of 15.45 is greater than $F(4,20,0.05)$, therefore we reject the null hypothesis, H_0. The means for the pomegranate from five countries are not equal at 1% level of significance.

Exercise 15.7

Varieties of Melon	Sample							Total
	1	2	3	4	5	6	7	
Hales Best Jumbo	6.36	6.75	4.66	6.2	7.7	7.45	5.5	44.62
Hearts of Gold	3.1	4.25	5.25	3.35	5.3			21.35
Ambrosia	4.65	5.12	4.35	5.26	4.49			23.87
Athena	6.8	5.8	5.9	6.8	6.0	6.23		37.53
Honey Bun Hybrid	5.0	5.26	5.33	5.56	5.0			26.15
Sweet N Early Hybrid	5.71	5.8	5.75	6.0	5.77	5.92		34.95
Total								188.37

The hypotheses to be tested in this case are:

$$H_0 : \alpha_1 = \alpha_2 = \alpha_3 = \alpha_4 = \alpha_5 = \alpha_6 = 0 \quad vs \quad H_1 : \text{at least one } \alpha_i \neq 0; \quad i = 1,2,3,4,5,6.$$

$$SS_{\text{total}} = 6.36^2 + 6.75^2 + 4.66^2 ... + 6^2 + 5.77^2 + 5.92^2 - \frac{188.37^2}{(7+5+6+5+6+5)} = 33.0985$$

$$S_{\text{melon}} = \frac{44.62^2}{7} + \frac{21.35^2}{5} + \frac{23.87^2}{6} + \frac{37.53^2}{5} + \frac{26.15^2}{6} + \frac{34.95^2}{5} - \frac{188.37^2}{(7+5+6+5+6+5)}$$

$$= 20.16171$$

$$SS_{\text{Error}} = 33.0985 - 20.16171 = 12.93679$$

Source	SS	df	MS	F	P-Value
Melon	20.16171	5	4.032341126	8.727477146	4.41E-05
Error	12.93679	28	0.462028265		
Total	33.0985	33			

With $F(0.01,5,28) = 3.9539$ and P-value $= 0.0000441$, we reject the null hypothesis and decide that there is significant melon variety effect, that is, the means for the melons are not equal at 2.5% level of significance.

Exercise 15.9

The hypotheses to be tested in this case are:

$$H_0 : \alpha_1 = \alpha_2 = \alpha_3 = \alpha_4 = \alpha_5 = 0 \quad vs \quad H_1 : \text{at least one } \alpha_i \neq 0; \quad i = 1,2,3,4,5$$

Palliative	1	2	3	4	5	Total
I	7.4	8.3	9.4	7	9.6	**41.7**
II	10.8	12.1	10	10.9	9.3	**53.1**
III	8.9	8.8	7.8	6.5	7.2	**39.2**
IV	6.8	6.8	6	7	6.5	**33.1**
V	9.1	9.5	7.8	8.5	9.4	**44.3**
Total						**211.4**

$$SS_{\text{total}} = 7.4^2 + 8.3^2 + \ldots + 6^2 + 7^2 + 6.5^2 - \frac{211.4^2}{5(5)} = 59.7416$$

$$SS_{\text{palliative}} = \frac{41.7^2 + 53.1^2 + 39.2^2 + 33.1^2 + 44.3^2}{5} - \frac{211.4^2}{5(5)} = 43.0496$$

$$SS_{\text{Error}} = 59.7416 - 43.0496 = 16.692$$

Source	SS	df	MS	F	P-Value
Palliative	43.0496	4	10.7624	12.89528	0.000024
Error	16.692	20	0.8346		
Total	59.7416	24			

With $F(0.01, 4, 20) = 4.4307$ and P-value $= 0.000024$, we reject the null hypothesis and decide that there is significant palliative effect, that is, the means for the palliatives are not equal at 1% level of significance.

Exercise 15.11

The hypotheses to be tested in this case are:

$$H_0 : \alpha_1 = \alpha_2 = \alpha_3 = \alpha_4 = 0 \quad vs \quad H_1 : \text{at least one } \alpha_i \neq 0; \; i = 1, 2, 3, 4$$

Filling Process	Sample							Total
	1	2	3	4	5	6	7	
I	16	16	16	17				65
II	20	19	19	19	20	20	18	135
III	20	18	18	20	19			95
IV	18	18	16	17	16	17		102
Total								397

ANOVA

Source	SS	df	MS	F	P-Value
Process	34.77597	3	11.59199	17.13303119	0.000016
Error	12.17857	18	0.676587		
Total	46.95455	21			

From the above analysis, we see that $MS_{Error} = 0.676587$;
$s = \sqrt{0.676587} = 0.822549; t_{(0.025,18)} = 2.101$.

Now, $LSD = t\left[\dfrac{\alpha}{2}, \sum\limits_{i=1}^{k} n_i - k\right] s \sqrt{\left(\dfrac{1}{n_i} + \dfrac{1}{n_j}\right)}$

From this we see that for the combinations *of i,j:*

i,j	LSD
4,6	1.1155
4,5	1.1593
4,7	1.0559
6,5	1.0201
6,7	0.9615
5,7	0.8114

Thus,

Process 1	Process 2	Process 3	Process 4
16.25	19.2857	19	17
$\bar{y}_{1.}$	$\bar{y}_{2.}$	$\bar{y}_{3.}$	$\bar{y}_{4.}$

The means at different levels of filling process can be arranged in order as follows:
$\bar{y}_{1.} < \bar{y}_{4.} < \bar{y}_{3.} < \bar{y}_{4.}$
We use the LSD in the earlier table to carry out the comparison as follows:

$\bar{y}_{1.} - \bar{y}_{4.} = 16.25 - 17 = -0.75 > -1.1155$ (not significant)
$\bar{y}_{1.} - \bar{y}_{3.} = 16.25 - 19 = -2.75 < -1.1593$ (significant)
$\bar{y}_{1.} - \bar{y}_{2.} = 16.25 - 19.2857 = -3.0357 < -1.0559$ (significant)

$\bar{y}_{4.} - \bar{y}_{3.} = 17 - 19 = -2.0 < -1.0201$ (significant)
$\bar{y}_{4.} - \bar{y}_{2.} = 17 - 19.2857 = -2.2857 < -0.9615$ (significant)

$\bar{y}_{3.} - \bar{y}_{2.} = 19 - 19.2857 = -0.2857 > -0.8114$ (not significant)

By this analysis, we group the means for the levels of the factors as follows:
$(\mu_1, \mu_4), (\mu_2 \mu_3)$.

Exercise 16.1

$H_0 : \alpha_1 = \alpha_2 = \alpha_3 = \alpha_4 = 0$ vs H_1 : at least one α_i is different; $i = 1,2,3,4$.
$H_0 : \beta_1 = \beta_2 = \beta_3 = \beta_4 = \beta_5 = 0$ vs H_1 : at least one β_j is different; $j = 1,2,3,4,5$.

Treatment	Blocks					Total
	1	**2**	**3**	**4**	**5**	
I	35	39	36	42	34	**186**
II	21	26	19	32	28	**126**
III	30	37	33	40	40	**180**
IV	24	22	23	35	33	**137**
Total	**110**	**124**	**111**	**149**	**135**	**629**

$$SS_{total} = 35^2 + 39^2 + 36^2 \ldots + 35^2 + 33^2 - \frac{629^2}{4(5)} = 926.95$$

$$SS_{treatment} = \frac{186^2 + 126^2 + 180^2 + 137^2}{5} - \frac{629^2}{4(5)} = 546.15$$

$$SS_{blocks} = \frac{186^2 + 126^2 + 180^2 + 137^2}{4} - \frac{629^2}{4(5)} = 273.70$$

$$SS_{Error} = 926.95 - 546.15 - 273.70 = 107.10$$

Source	SS	df	MS	F-Ratio
Treatment	546.15	3	182.05	20.40
Block	273.70	4	68.43	7.67
Error	107.10	12	8.93	
Total	926.95	19		

From the tables $F(3,12,0.01) = 5.9525$ and $F(4,12,0.01) = 5.4119$. Both treatment and block effects are significant. The null hypotheses for both are rejected in favor of the alternatives.

Exercise 16.3

We enter the data in MINITAB in columns C1–C3 which we rename Formulation, Experts, and Y, respectively, as follows:

```
MTB > name c1 'Formulation' c2 'Experts' c3 'Y'
MTB > read c1-c3
DATA> 1  1  9.0
DATA> 2  1  7.5
DATA> 3  1  6.5
DATA> 4  1  7.0
DATA> 1  2  7.0
DATA> 2  2  6.0
DATA> 3  2  4.3
DATA> 4  2  5.0
DATA> 1  3  8.0
DATA> 2  3  7.5
DATA> 3  3  5.0
DATA> 4  3  6.0
DATA> 1  4  7.0
```

Continued.

```
DATA> 2  4  6.0
DATA> 3  4  5.5
DATA> 4  4  5.0
DATA> 1  5  8.0
DATA> 2  5  7.0
DATA> 3  5  6.0
DATA> 4  5  6.0
DATA> end
20 rows read.
MTB > twoway c3 c1 c2
```

Two-Way ANOVA: Y Versus FORMULATION, EXPERTS

```
Source        DF      SS        MS       F       P
FORMULATION    3   16.7335   5.57783   42.93   0.000
EXPERTS        4    9.2730   2.31825   17.84   0.000
Error         12    1.5590   0.12992
Total         19   27.5655

S = 0.3604   R-Sq = 94.34%   R-Sq(adj) = 91.05%
```

Exercise 16.5

Treatments	Blocks							Total
	I	II	III	IV	V	VI	VII	
I	134	136	161	142	134	151	157	1015
II	120	120	130	130	130	136	136	902
III	129	144	151	136	120	144	145	969
IV	139	146	149	140	140	175	166	1055
Total	522	546	591	548	524	606	604	3491

$$SS_{total} = 134^2 + 136^2 + 161^2 ... + 175^2 + 166^2 - \frac{3941^2}{4(7)} = 4793.25$$

$$SS_{treatment} = \frac{1015^2 + 902^2 + 969^2 + 1055^2}{7} - \frac{3941^2}{4(7)} = 1849.25$$

$$SS_{blocks} = \frac{522^2 + 546^2 + 291^2 + 548^2 + 524^2 + 606^2 + 604^2}{4} - \frac{3941^2}{4(7)} = 2007.5$$

$$SS_{Error} = 4793.25 - 2007.5 - 1849.25 = 936.5$$

Source	SS	DF	MS	F
Diets	1849.25	3	616.4167	11.84784
Chicks	2007.5	6	334.5833	6.43086
Error	936.5	18	52.02778	
Total	4793.25	27		

Exercise 17.1

Experts	Formulations of the Cookie			Total
	1	**2**	**3**	
1	20,23 (43)	19,21 (40)	27,24 (51)	134
2	15,18 (33)	16,15 (31)	20,25 (45)	109
3	20,23 (43)	15,20 (35)	24,26 (50)	128
4	21,18 (39)	19,23 (42)	24,25 (49)	130
5	20,20 (40)	22,25 (47)	22,23 (45)	132
Total	**198**	**195**	**240**	**633**

The hypothesis to be tested in ANOVA are:

$H_0 : \alpha_1 = \alpha_2 = \alpha_3 = \alpha_4 = \alpha_5 = 0$ vs H_1 : at least one α_i is different; $i = 1,2,3,4,5$.

$H_0 : \beta_1 = \beta_2 = \beta_3 = 0$ vs H_1 : at least one β_j is different; $j = 1,2,3$.

$H_0 : (\alpha\beta)_{ij} = 0$ vs H_1 : at least one $(\alpha\beta)_{ij} \neq 0; i = 1,2,3,4,5; j = 1,2,3$.

$$SS_{\text{total}} = 20^2 + 23^2 + \ldots + 22^2 + 23^2 - \frac{633^2}{5(4)(2)} = 318.70$$

$$SS_{\text{experts}} = \frac{134^2 + 109^2 + 128^2 + 130^2 + 132^2}{(3)(2)} - \frac{633^2}{5(3)(2)} = 67.87$$

$$SS_{\text{Formulation}} = \frac{198^2 + 195^2 + 240^2}{(5)(2)} - \frac{633^2}{5(3)(2)} = 126.6$$

$$SS_{\text{Formulation*experts}} = \frac{33^2 + 31^2 + 45^2 + \ldots + 40^2 + 47^2 + 45^2}{2} - 67.87 - 126.6 - \frac{633^2}{5(3)(2)}$$

$$= 58.73$$

$$SS_{\text{Error}} = 318.7 - 67.87 - 126.6 - 58.73 = 65.5$$

ANOVA

Source	SS	df	MS	F
Experts(E)	67.87	4	16.97	3.885
Formula(F)	126.6	2	63.3	14.5
E*F	58.73	8	7.342	1.681
Error	65.5	15	4.367	
Total	318.7	29		

Exercise 17.3

The hypothesis to be tested is:

$H_0 : \alpha_1 = \alpha_2 = \alpha_3 = \alpha_4 = 0$ vs H_1 : at least one α_i is different; $i = 1, 2, 3.$

$H_0 : \beta_1 = \beta_2 = \beta_3 = 0$ vs H_1 : at least one β_j is different; $j = 1, 2, 3.$

$H_0 : (\alpha\beta)_{ij} = 0$ vs H_1 : at least one $(\alpha\beta)_{ij} \neq 0; i = 1, 2, 3, 4; j = 1, 2, 3.$

Minitab Analysis

Here instead of invoking the TWOWAY command on MINITAB, we invoke the ANOVA command. The ANOVA command enables us to specify in a sing statement, all the main effects and interactions.

```
MTB > name c1 'fertilizer' c2 'Watering' c3 'y'
MTB > read c1-c3
DATA> 1   1   30
DATA> 1   1   25
DATA> 1   1   29
DATA> 1   2   32
DATA> 1   2   34
DATA> 1   2   29
DATA> 1   3   25
DATA> 1   3   22
DATA> 1   3   23
DATA> 2   1   31
DATA> 2   1   37
DATA> 2   1   33
DATA> 2   2   31
DATA> 2   2   32
DATA> 2   2   30
DATA> 2   3   28
DATA> 2   3   24
DATA> 2   3   25
DATA> 3   1   33
DATA> 3   1   29
DATA> 3   1   24
DATA> 3   2   35
DATA> 3   2   41
DATA> 3   2   31
DATA> 3   3   30
DATA> 3   3   29
DATA> 3   3   29
DATA> 4   1   31
DATA> 4   1   30
DATA> 4   1   32
DATA> 4   2   33
DATA> 4   2   30
DATA> 4   2   31
DATA> 4   3   26
DATA> 4   3   29
DATA> 4   3   25
DATA> end
36 rows read.
MTB > print c1-c3
```

Data Display

Row	fertilizer	Watering	y
1	1	1	30
2	1	1	25
3	1	1	29
4	1	2	32
5	1	2	34
6	1	2	29
7	1	3	25
8	1	3	22
9	1	3	23
10	2	1	31
11	2	1	37
12	2	1	33
13	2	2	31
14	2	2	32
15	2	2	30
16	2	3	28
17	2	3	24
18	2	3	25
19	3	1	33
20	3	1	29
21	3	1	24
22	3	2	35
23	3	2	41
24	3	2	31
25	3	3	30
26	3	3	29
27	3	3	29
28	4	1	31
29	4	1	30
30	4	1	32
31	4	2	33
32	4	2	30
33	4	2	31
34	4	3	26
35	4	3	29
36	4	3	25

MTB > anova c4=c2|c3;

SUBC> means c2 c3.

ANOVA: y versus fertilizer, Watering

Factor	Type	Levels	Values
fertilizer	fixed	4	1, 2, 3, 4
Watering	fixed	3	1, 2, 3

Analysis of Variance for y

Source	DF	SS	MS	F	P
fertilizer	3	59.556	19.852	2.84	0.059
Watering	2	236.167	118.083	16.87	0.000
fertilizer*Watering	6	98.278	16.380	2.34	0.064
Error	24	168.000	7.000		
Total	35	562.000			

$S = 2.64575$ R-Sq $= 70.11\%$ R-Sq(adj) $= 56.41\%$

Means

fertilizer	N	y
1	9	27.667
2	9	30.111
3	9	31.222
4	9	29.667

Watering	N	y
1	12	30.333
2	12	32.417
3	12	26.250

Exercise 17.5

The hypothesis to be tested in ANOVA is:

$H_0 : \alpha_1 = \alpha_2 = \alpha_3 = 0$ vs H_1 : at least one α_i is different; $i = 1, 2, 3$.

$H_0 : \beta_1 = \beta_2 = \beta_3 = \beta_4 = 0$ vs H_1 : at least one β_j is different; $j = 1, 2, 3, 4$.

$H_0 : (\alpha\beta)_{ij} = 0$ vs H_1 : at least one $(\alpha\beta)_{ij} \neq 0$; $i = 1, 2, 3; j = 1, 2, 3, 4$.

MINITAB Solution:

```
MTB > name c1 'material' c2 'temperature' c3 'y'
MTB > read c1-c3
DATA> 1   1   130
DATA> 1   1   155
DATA> 1   1   74
DATA> 1   1   180
DATA> 1   2   34
```

Continued.

```
DATA>  1   2   40
DATA>  1   2   80
DATA>  1   2   75
DATA>  1   3   20
DATA>  1   3   70
DATA>  1   3   82
DATA>  1   3   58
DATA>  1   4   120
DATA>  1   4   100
DATA>  1   4   101
DATA>  1   4   115
DATA>  2   1   150
DATA>  2   1   188
DATA>  2   1   159
DATA>  2   1   126
DATA>  2   2   136
DATA>  2   2   122
DATA>  2   2   106
DATA>  2   2   115
DATA>  2   3   50
DATA>  2   3   60
DATA>  2   3   87
DATA>  2   3   86
DATA>  2   4   114
DATA>  2   4   70
DATA>  2   4   54
DATA>  2   4   115
DATA>  3   1   138
DATA>  3   1   110
DATA>  3   1   168
DATA>  3   1   160
DATA>  3   2   74
DATA>  3   2   120
DATA>  3   2   150
DATA>  3   2   139
DATA>  3   3   120
DATA>  3   3   96
DATA>  3   3   104
DATA>  3   3   82
DATA>  3   4   108
DATA>  3   4   110
DATA>  3   4   109
DATA>  3   4   112
DATA>  end
48 rows read.
MTB > anova y=c1|c2
```

ANOVA: y versus material, temperature

Factor	Type	Levels	Values
material	fixed	3	1, 2, 3
temperature	fixed	4	1, 2, 3, 4

```
Analysis of Variance for y
Source                  DF        SS        MS        F       P
material                 2     6996.2    3498.1     5.51   0.008
temperature              3    29419.8    9806.6    15.45   0.000
material*temperature     6     9546.7    1591.1     2.51   0.039
Error                   36    22846.0     634.6
Total                   47    68808.7


    S = 25.1915   R-Sq = 66.80%   R-Sq(adj) = 56.65%
```

Exercise 17.7

Factor A	Factor B			Total
	I	**II**	**III**	
I	10,10,9	8.9,8.5,8.7	7.8,7,7.1	77
II	8.4,8.8,7.7	7,5.6,6.7	9,8,8	69.2
Total	53.9	45.4	46.9	146.2

The hypothesis to be tested in ANOVA is:

$H_0 : \alpha_1 = \alpha_2 = \alpha_3 = 0$ vs H_1 : at least one α_i is different; $i = 1, 2, 3$.

$H_0 : \beta_1 = \beta_2 = 0$ vs H_1 : at least one β_j is different; $j = 1, 2, 3$.

$H_0 : (\alpha\beta)_{ij} = 0$ vs H_1 : at least one $(\alpha\beta)_{ij} \neq 0; i = 1, 2, 3; j = 1, 2$.

$$SS_{\text{total}} = 10^2 + 10^2 + .. + 9^2 + 8^2 + 8^2 - \frac{146.2^2}{2(3)(3)} = 22.4711$$

$$SS_A = \frac{76^2 + 69.2^2}{(3(3))} - \frac{146.2^2}{2(3)(3)} = 3.38$$

$$SS_B = \frac{53.9^2 + 45.5^2 + 46.9^2}{(2)(3)} - \frac{146.2^2}{2(3)(3)} = 6.8611$$

$$SS_{A*B} = \frac{29^2 + 26.1^2 + 21.9^2 + ... + 24.9^2 + 19.3^2 + 25^2}{3} - 3.38 - 6.8611 - \frac{146.2^2}{2(3)(3)} = 8.73$$

$$SS_{\text{Error}} = 22.4711 - 3.38 - 6.8611 - 8.73 = 3.5$$

ANOVA

Source	SS	df	MS	F
A	3.38	1	3.38	11.59
B	11.728	2	3.431	11.76
A*B	8.73	2	4.365	14.97
Error	3.5	12	0.2917	
total	22.4711	17		

$F(1, 12, 0.01) = 9.33$; $F(2, 12, 0.01) = 6.927$; Both main effects and interaction are significant at 1% level of significance. In the face of a significant interaction, the significance of the main effects is no longer important as both effects are acting together. An analysis of the interaction is called for, which is beyond the remit of this text.

Exercise 18.1

H_0: Data on fuel consumption for fuel types A and B are from the same or equal population.
Vs.
H_1: Data on fuel consumption for fuel types A and B are not from the same or equal population.

We combine and rank the data for both fuels, breaking the ties as demonstrated in earlier example. The ranked data are shown in the following:

Observation	Rank	Fuel A	Rank	Fuel B
1	8	29	24	38
2	9.5	30	17	33
3	12	31	14.5	32
4	12	31	21.5	37
5	6.5	28	19	35
6	5	27	26	39
7	2	22	27.5	40
8	3	24	24	38
9	4	25	21.5	37
10	6.5	28	30	43
11	27.5	40	20	36
12	24	38	14.5	32
13	17	33	12	31
14	1	19	29	42
15	9.5	30	17	33
Total	147.5		317.5	

The method employed is the normal approximation to the Wilcoxon Ranked Sum Test:

$$\text{Test statistic } Z = \frac{W - \frac{n_1(n_1 + n_2 + 1)}{2}}{\sqrt{\frac{n_1 n_2 (n_1 + n_2 + 1)}{12}}} = \frac{147.5 - \frac{15(15 + 15 + 1)}{2}}{\sqrt{\frac{15(15)(15 + 15 + 1)}{12}}} = -3.53$$

At level of significance of 1%, and for a two-sided test, the critical values are ±2.575. The Z calculated above is less than the critical value of −2.575, therefore we reject the null hypothesis and accept that the two set of consumptions A and B are not from same population.

Exercise 18.3

H_0: Data for titration times for School A and B are from the same or equal population.
Vs.
H_1: Data for titration times for School A and B are not from the same or equal population.

Rank	A	Rank	B
1	2.3	27	6.7
27	6.7	29.5	7.3
3	3.8	7.5	4.4
17	5	32	8.3
15.5	4.9	25	6.2
24	6.1	6	4.3
7.5	4.4	20.5	5.5
18	5.2	2	3.2
4	3.9	33	8.4
13	4.8	29.5	7.3
9	4.6	20.5	5.5
22.5	5.7	13	4.8
19	5.3	15.5	4.9
10.5	4.7	27	6.7
5	4.2	31	7.5
22.5	5.7	**319**	
13	4.8		
10.5	4.7		
242			

$$\text{Test statistic } Z = \frac{W - \dfrac{n_1(n_1+n_2+1)}{2}}{\sqrt{\dfrac{n_1 n_2(n_1+n_2+1)}{12}}} = \frac{242 - \dfrac{18(18+15+1)}{2}}{\sqrt{\dfrac{18(15)(18+15+1)}{12}}} = -2.31$$

At the level of significance of 5%, and for a two-sided test, the critical values are ± 1.96. The Z calculated above is less than the critical value of -1.96, therefore we reject the null hypothesis and conclude that the two set of titration times do not belong to the same or equal population.

Exercise 18.5

H_0: Data for performance for week1 and week2 are from the same or equal population.
Vs.
H_1: Data for performance for week1 and week2 are not from the same or equal population.

x1	x2	d	rank
13.47	8.3	−5.17	20
16.6	11.61	−4.99	19
14.46	9.66	−4.8	18
14.3	10.4	−3.9	15
17.9	14.3	−3.6	13
13.2	10.92	−2.28	8
12.93	10.88	−2.05	7
12.57	10.6	−1.97	6
11.3	10.44	−0.86	2
7.49	6.78	−0.71	1
6.45	8.08	1.63	3
7.43	9.11	1.68	4
7.34	9.23	1.89	5
6.54	8.83	2.29	9
7.45	10.09	2.64	10
7.58	10.36	2.78	11
7.36	10.21	2.85	12
7.92	11.78	3.86	14
7.83	12.09	4.26	16
7.88	12.36	4.48	17
8.39	13.91	5.52	21
8.88	14.77	5.89	22
8.99	15.26	6.27	23
11.39	18	6.61	24
9.08	15.79	6.71	25
10.39	17.17	6.78	26
9.42	16.67	7.25	27

From the data, the sum of positive ranks, $T^+ = 269$, while the sum of negative ranks, $T^- = 109$.

$$\text{Test statistic } Z = \frac{W - \frac{n(n+1)}{4}}{\sqrt{\frac{n(n+1)(2n+1)}{24}}} = \frac{269 - \frac{27(27+1)}{4}}{\sqrt{\frac{27(27+1)(2(27)+1)}{24}}} = 1.922$$

At the level of significance of 5%, and for a two-sided test, the critical values are ±1.96. The Z calculated above is less than the critical value of +1.96, therefore we fail to reject the null hypothesis and conclude that the two sets of weekly performance belong to the same or equal populations.

Exercise 18.7

We insert the ranks in parentheses as in the following table:

Methods	1	2	3	4	5	Sum of Ranks
1	24.95(2)	26.44(4)	28.57(7)	23.17(1)	26.34(3)	17
2	30.41(14)	29.23(11)	28.87(9)	32.95(18)	35.87(20)	72
3	26.49(5)	28.5(6)	33.13(19)	32.36(17)	28.95(10)	57
4	30.66(16)	29.72(12)	30.14(13)	30.44(15)	28.73(8)	64

Here, $N = 4(5) = 20$; $n_1 = n_2 = n_3 = n_4 = 5$; $k = 4$;
The test statistic is:

$$H = \frac{12}{N(N+1)}\left(\frac{S_1^2}{n_1} + \frac{S_2^2}{n_2} + \ldots + \frac{S_k^2}{n_k}\right)$$

$$H = \frac{12}{20(20+1)}\left(\frac{17^2}{5} + \frac{72^2}{5} + \frac{57^2}{5} + \frac{64^2}{5}\right) = 73.24$$

Critical value is $\chi^2_{[3,0.05]} = 7.81$. The test statistic value, is 73.24, is greater than the critical value. So we reject the null hypothesis that the methods have equal effect, and conclude that their effects are different.

Exercise 18.9

H_0: Data on ratings of cookie1 and cookie2 belong to the same or equal population. Vs.
H_1: Data on ratings of cookie1 and cookie2 do not belong to the same or equal population.
We subtract the rating for cookie2 from rating for cookie1.

Cookie 1	Cookie2	Diff	Ranks
91	69	22	30
94	69	25	31
65	81	−16	21.5
70	62	8	8.5

Continued.

Cookie 1	Cookie2	Diff	Ranks
82	74	8	8.5
81	80	1	1
50	70	−20	27
62	79	−17	24
75	81	−6	6.5
77	82	−5	4.5
76	78	−2	2
82	73	9	11
78	74	4	3
91	86	5	4.5
80	74	6	6.5
74	64	10	13.5
77	66	11	15
81	69	12	16
92	83	9	11
70	52	18	26
83	66	17	24
77	56	21	28.5
70	91	−21	28.5
77	90	−13	17.5
66	56	10	13.5
92	79	13	17.5
83	69	14	19
90	75	15	20
91	75	16	21.5
93	76	17	24
85	76	9	11

From the data, the sum of positive ranks, $T^+ = 364.5$, while the sum of negative ranks, $T^- = 131.5$.

$$\text{Test statistic } Z = \frac{W - \dfrac{n(n+1)}{4}}{\sqrt{\dfrac{n(n+1)(2n+1)}{24}}} = \frac{364.5 - \dfrac{31(31+1)}{4}}{\sqrt{\dfrac{31(31+1)(2(31)+1)}{24}}} = 2.28$$

At the level of significance of 5%, and for a two-sided test, the critical values are ±1.96. The Z calculated above (2.28) is greater than the critical value of +1.96, therefore we reject the null hypothesis and conclude that the two sets of ratings for the cookies do not belong to the same or equal population.